Frontiers in Dengue Virus Research

Edited by

Kathryn A. Hanley

Department of Biology
New Mexico State University
Las Cruces, NM
USA

and

Scott C. Weaver

Center for Tropical Diseases and Department of Pathology
University of Texas Medical Branch
Galveston, TX
USA

 Caister Academic Press

Copyright © 2010

Caister Academic Press
Norfolk, UK

www.caister.com

British Library Cataloguing-in-Publication Data
A catalogue record for this book is available from the British Library

ISBN: 978-1-904455-50-9

Cover image adapted from Figure 11.1. This figure was generated utilizing the expert
graphic design abilities of Dr Shannan L. Rossi (Center for Vaccine Research, University of
Pittsburgh).

Printed and bound in Great Britain by Cromwell Press Group, Trowbridge, UK

Contents

Contributors

Scott J. Balsitis
Division of Infectious Diseases and Vaccinology
School of Public Health
University of California
Berkeley, CA
USA

sbalsitis@berkeley.edu

Ralf Bartenschlager
Department of Molecular Virology
University of Heidelberg
Heidelberg
Germany

ralf_bartenschlager@med.uni-heidelberg.de

Shannon N. Bennett
Asia-Pacific Institute of Tropical Medicine and
Infectious Diseases
Department of Tropical Medicine
Medical Microbiology and Pharmacology
John A. Burns School of Medicine
Honolulu, HI
USA

sbennett@hawaii.edu

Irene Bosch
Massachusetts Institute of Technology
Harvard-MIT Division of Health Sciences and
Technology
Cambridge, MA
USA

ibosch@mit.edu

Derek Cummings
Department of Epidemiology
Johns Hopkins Bloomberg School of Public Health
Baltimore, MD
USA

dcumming@jhsph.edu

Anna P. Durbin
Center for Immunization Research
Johns Hopkins Bloomberg School of Public Health
Johns Hopkins University
Baltimore, MD
USA

adurbin@jhsph.edu

Timothy P. Endy
State University of New York
Upstate Medical University
Syracuse, NY
USA

endyt@upstate.edu

Jeremy Farrar
Oxford University Clinical Research Unit
Hospital for Tropical Diseases
Ho Chi Minh City
Vietnam
and
Centre for Tropical Medicine
University of Oxford
Headington
Oxford, UK

Jeremy.Farrar@ndm.ox.ac.uk

Andrea Gamarnik
Fundación Instituto Leloir-CONICET
Buenos Aires
Argentina

agamarnik@leloir.org.ar

Ana P. Goncalvez
Molecular Viral Biology Section and Hepatitis Viruses Section
Laboratory of Infectious Diseases
National Institute of Allergy and Infectious Diseases
National Institutes of Health
Bethesda, MD
USA

agoncalvez@niaid.nih.gov

Duane J. Gubler
Program in Emerging Infectious Diseases
Duke-NUS Graduate Medical School
Singapore
and
Asia-Pacific Institute for Tropical Medicine and Infectious Diseases
University of Hawaii
John A. Burns School of Medicine
Honolulu, HI
USA

gubler@hawaii.edu

Maria G. Guzman
Institute of Tropical Medicine 'Pedro Kouri'
PAHO/WHO Collaborating Center for the Study of Dengue and its Vector
Autopista Novia del Mediodía
Habana
Cuba

lupe@ipk.sld.cu

Kathryn A. Hanley
Department of Biology
New Mexico State University
Las Cruces, NM
USA

khanley@nmsu.edu

Eva Harris
Division of Infectious Diseases and Vaccinology
School of Public Health
University of California
Berkeley, CA
USA

eharris@berkeley.edu

Gustavo Kouri
Institute of Tropical Medicine 'Pedro Kouri'
PAHO/WHO Collaborating Center for the Study of Dengue and its Vector
Autopista Novia del Mediodía
Habana
Cuba

gkouri@ipk.sld.cu

Richard J. Kuhn
Markey Center for Structural Biology
Department of Biological Sciences
Purdue University
West Lafayette, IN
USA
and
Bindley Bioscience Center
Purdue University
West Lafayette, IN
USA

kuhnr@purdue.edu

Elisa La Bauve
Markey Center for Structural Biology
Department of Biological Sciences
Purdue University
West Lafayette, IN
USA

elabauve@purdue.edu

Ching-Juh Lai
Molecular Viral Biology Section
Laboratory of Infectious Diseases
National Institute of Allergy and Infectious Diseases
National Institutes of Health
Bethesda, MD
USA

clai@niaid.nih.gov

Mayuri
Markey Center for Structural Biology
Department of Biological Sciences
Purdue University
West Lafayette, IN
USA

mmayuri@purdue.edu

Sven Miller
3-V Biosciences
Zurich
Switzerland

millersven@gmx.de

Jorge L. Muñoz-Jordán
Centers for Disease Control and Prevention
Division of Vector Borne Infectious Diseases
Dengue Branch
San Juan
Puerto Rico

ckq2@cdc.gov

Eng-Eong Ooi
Program in Emerging Infectious Diseases
Duke-NUS Graduate Medical School
Singapore

engeong.ooi@duke-nus.edu.sg

R. Padmanabhan
Department of Microbiology and Immunology
Georgetown University School of Medicine
Washington, DC
USA

rp55@georgetown.edu

Robert H. Purcell
Hepatitis Viruses Section
Laboratory of Infectious Diseases
National Institute of Allergy and Infectious Diseases
National Institutes of Health
Bethesda, MD
USA

rpurcell@niaid.nih.gov

Ines Romero-Brey
Department of Molecular Virology
University of Heidelberg
Heidelberg
Germany

Ines_Romero-Brey@med.uni-heidelberg.de

Beatriz Sierra
Institute of Tropical Medicine 'Pedro Kouri'
PAHO/WHO Collaborating Center for the Study of
Dengue and its Vector
Autopista Novia del Mediodía
Habana
Cuba

siebet@ipk.sld.cu

Cameron Simmons
Oxford University Clinical Research Unit
Hospital for Tropical Diseases
Ho Chi Minh City
Vietnam
and
Centre for Tropical Medicine
University of Oxford
Headington
Oxford, UK

csimmons@oucru.org

Alex Y. Strongin
Inflammatory and Infectious Disease Center/Cancer
Research Center
Burnham Institute for Medical Research
La Jolla, CA
USA

strongin@burnham.org

Nikos Vasilakis
Center for Vaccine Research and Department of
Microbiology and Molecular Genetics
School of Medicine
University of Pittsburgh
Pittsburgh, PA
USA

nivasila@utmb.edu

Scott C. Weaver
Center for Tropical Diseases and Department of
Pathology
University of Texas Medical Branch
Galveston, TX
USA

sweaver@utmb.edu

Stephen S. Whitehead
Laboratory of Infectious Diseases
National Institute of Allergy and Infectious Diseases
National Institutes of Health
Bethesda, MD
USA

swhitehead@niaid.nih.gov

Preface

We had several reasons for creating a book that focuses on the frontiers in dengue virus research.

The first was that dengue virus, long an 'orphan' in terms of research, has finally been adopted by a worldwide network of scientists, and now seemed an opportune time to have a family reunion. As with most such gatherings, many important figures in the field were prevented from participating by scheduling and space constraints. Nonetheless, we are well pleased by the breadth of the contributors, both in their range of expertise and in their global distribution.

Second, both of us are field scientists and thus have first-hand experience in the tropics with the growing burden imposed by dengue disease. We greatly hope that this book will point the way to the development of vaccines, therapeutics and other strategies to prevent and control dengue disease. When a vaccine is finally available, we'll be first in line to volunteer!

Finally we wanted to create a book that would highlight new approaches and open the door to new investigators from a wide variety of disciplines who might consider taking up the study of dengue virus. It is critical that new techniques, novel perspectives and fresh eyes be brought to bear on the pressing questions identified in these chapters.

We owe thanks to the many people who helped to make this book possible. Working with the authors of each of the chapters was delightful and brought us a new appreciation of our colleagues. Moreover, we wish to particularly acknowledge Paige Adams, Nicole Arrigo, Andrew Haddow, Joan Jenney, Jyotsna Pandya, Nikos Vasilakis, and Sara Volk for assistance with proofreading the manuscript and Dora Salinas for her help in preparing the index. We must also thank Annette Griffin, Acquisitions Editor at Horizon Press, and Rhiannon Miller, Production Typesetter at Prepress Projects Ltd.

We dedicate this book to our spouses, Drs. Timothy Wright and Donna Weaver: 'Greater love hath no man (or woman!) than this, that a man lay down his own writing time for his spouse's.'

Kathryn A. Hanley
Scott C. Weaver

Part I

Introduction

Dengue Virus: Past, Present and Future

Timothy P. Endy, Scott C. Weaver and Kathryn A. Hanley

Abstract

A

Dengue

Dengue Virus: Past, Present and Future

Timothy P. Endy, Scott C. Weaver and Kathryn A. Hanley

Dengue Virus: Past, Present and Future

Timothy P. Endy, Scott C. Weaver and Kathryn A. Hanley

Dengue Virus: Past, Present and Future

1

Timothy P. Endy, Scott C. Weaver and Kathryn A. Hanley

Dengue Virus: Past, Present and Future

1

Timothy P. Endy, Scott C. Weaver and Kathryn A. Hanley

Dengue Virus: Past, Present and Future

1

Timothy P. Endy, Scott C. Weaver and Kathryn A. Hanley

Abstract

Dengue is an old disease caused by the mosquito-borne dengue viruses (DENV-1–4). In the last century, dengue has escalated in geographic distribution and disease severity to become now the most common arboviral infection of humans in the subtropical and subtropical regions of the world. In this chapter the authors discuss the historical aspects of dengue virus transmission and factors contributing to its evolution as one of the most important public health problems of this century.

Dengue fever – an old disease spreading with a new vengeance

Across much of the world, people bitten by an *Aedes aegypti* mosquito may soon find themselves prostrated by the high fever and severe joint pain that are the classic symptoms of dengue fever., and to some extent this is nothing new; reports of symptoms consistent with dengue date back over two millennia (Gubler, 2006). Why then is a book about frontiers in dengue virus needed at this time? The reason: although dengue is an old disease, recent decades have seen an unprecedented increase in the geographic range, incidence, and severity of dengue infection (Gubler, 2006; Kyle and Harris, 2008).

Dengue disease is caused by the four serotypes of mosquito-borne dengue virus (DENV-1–4), positive-sense RNA viruses belonging to the genus *Flavivirus*. Escalation of the dengue pandemic can largely be attributed to three factors: (i) increased urbanization and consequent urban detritus and population density leading to enhanced vector breeding and increased contact between humans and vectors, (ii) global invasion of the major mosquito vectors, *Aedes aegypti* and *Aedes albopictus,* leading to geographic spread and geographic overlap of all four dengue virus serotypes and (iii) interaction and evolution of the four serotypes themselves, resulting in greater disease severity (Gubler, 2006; Kyle and Harris, 2008). As a result of these changes, DENV is now the most common arboviral infection of humans in the subtropical and subtropical regions of the world. The World Health Organization (WHO) estimates that 2.5 billion people are risk from dengue with 50 million dengue infections worldwide every year (Anonymous, 2008). In 2007, there were more than 890,000 reported cases of dengue in the Americas, approximately 26,000 of which were the most severe form, dengue haemorrhagic fever (DHF). The WHO reports that dengue disease is endemic in more than 100 countries in Africa, the Americas, the Eastern Mediterranean, South-East Asia and the Western Pacific, with South-East Asia and the Western Pacific the most seriously affected. Approximately 500,000 people with DHF require hospitalization each year, of whom 2.5% die.

Scope and direction of frontiers in dengue virus research

It is our hope that this book will provide a foundation for the response to the public health emergency posed by dengue virus. We have made an effort not just to review the rapidly expanding dengue research literature, but also to identify the most pressing questions that remain to be answered about dengue biology and control. The remainder of this chapter provides an overview of the evolutionary history and epidemiology of dengue virus. The chapters in section two cover translation and processing of the dengue virus polyprotein, viral replication, and the role of the viral untranslated regions in regulation of genome synthesis and translation. Section three presents current knowledge on the pathogenesis of and host immune response to dengue illness, focusing on the role of host and virus determinants of susceptibility and dengue disease severity, changes in protein expression in infected hosts, virus modulation of the host immune response, and development of animal models in which to study dengue virus pathogenesis. Section four discusses the crucial topic of the epidemiology and evolutionary dynamics of DENV, with chapters on DENV–mosquito interactions, evolutionary dynamics of dengue virus, temporal and spatial dynamics of dengue virus transmission, and emergence of DENV from its ancestral, sylvatic cycle. Finally, section five addresses various approaches that are currently being developed in the control of dengue disease, including vaccines, novel drugs, and passive immunotherapy.

Evolution of the flaviviruses

It is not known with certainty when and where the progenitor of the approximately 80 species in the genus *Flavivirus* first arose, although geographic evidence suggests that this ancestral flavivirus may have first appeared in Africa. Over the course of speciation, the flaviviruses have shown substantial ecological diversification. Most notably, different lineages of flaviviruses adapted to different modes of transmission. A current phylogenetic tree of the genus *Flavivirus* (Fig. 1.1) shows that the basal-most lineages are viruses that have only been isolated from mosquitoes and are not known to infect vertebrates at all. This suggests that the ancestor of the genus may have been a 'mosquito-only' virus that later acquired the ability to infect vertebrates. The remaining flaviviruses are divided into vector-borne viruses of vertebrates, with major groups using ticks and mosquitoes for horizontal transmission, and another group that infects vertebrates without

the use of arthropod vectors. This topology does not suggest whether vector-borne or non-vector-borne transmission was ancestral, but the basal position of the 'mosquito-only' viruses suggests that mosquito-borne transmission among vertebrates may have preceded the loss of vector transmission. Tick-borne transmission may have evolved from a mosquito-borne lineage after the lineage that infects only vertebrates arose.

All of the flaviviruses known to be human pathogens are transmitted by vectors, and, with the exception of dengue virus, all are zoonoses (Karabatsos, 1985). The Japanese encephalitis virus (JEV) group, consisting of Japanese encephalitis virus, West Nile virus (WNV) and St Louis encephalitis (SLE), among others, is maintained in a cycle of transmission between passerine birds and *Culex* mosquitoes. Mammalian hosts, including humans, can be infected but are an evolutionary dead-end since the viraemia achieved is too low for subsequent transmission (Endy and Nisalak, 2002). The tick-borne encephalitis group is transmitted among rodents by a tick vector; as with the JEV group humans are a dead-end host (Endy and Nisalak, 2002). Yellow fever virus (YFV) is primarily maintained in a sylvatic cycle involving non-human primates and *Aedes* mosquitoes, but it has shown the capacity to adapt to transmission in urban areas using *Aedes aegypti* mosquitoes and humans as its primary reservoir. Such adaptation results in urban epidemics of yellow fever (Monath, 1988). Though each of the species described above shows intraspecific genetic variation, evolution has not led to the divergence of multiple serotypes, consequently humans that survive infection retain life-long protection against re-infection (Rice, 1996). Despite ecological variation among flavivirus species, the organization of the flavivirus genome is conserved throughout the genus. Each positive-sense, single-stranded RNA genome is approximately 11 kb in length and encodes a single polyprotein that is co- and post-translationally processed into three structural and seven nonstructural (NS) proteins (Fig. 1.2) (Rice, 1996). This processing event is described in detail in Chapter 2. The structural proteins consist of capsid (C), membrane (the mature form of the pre-membrane (prM) protein) and envelope (E). As discussed in Chapter 13, the E protein contains

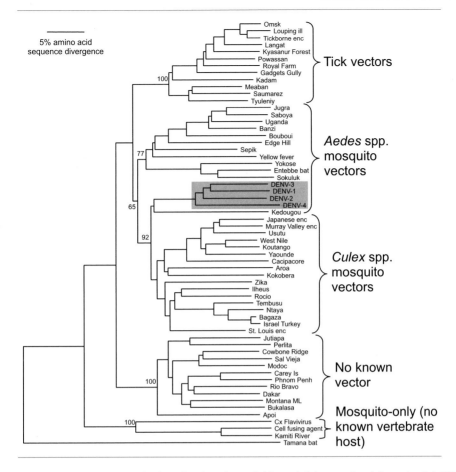

Figure 1.1 Phylogenetic tree is displayed using the neighbour joining method from partial NS5 amino acid sequences available in the GenBank library. Numbers indicate bootstrap values obtained from 1000 replicates for groups to the right. The four serotypes of dengue virus are shaded in grey.

the binding site for the as-yet unidentified cellular receptor; within the endosome E shifts from a homodimer to a homotrimer, enabling fusion with the cell membrane. The functions of some of the flavivirus non-structural proteins have been extensively studied. For example NS5 acts as the RNA-dependent RNA polymerase and possesses a nuclear localization sequence and methyltransferase activity and, as described in Chapter 2, NS2b and NS3 together act as the viral protease. However the function of most non-structural proteins is not well known. Current understanding of the role of each of these NS proteins in the flavivirus replication complex is described in Chapter 3. The genome is flanked at the 5′ and 3′ termini by untranslated regions whose binding facilitates genome synthesis and whose structure and function are discussed in Chapter 4. The

length of the UTRs varies considerably among different species. The structure of the flavivirus virion, a smooth sphere approximately ~500 Å in diameter, was first determined by Kuhn *et al.* (2002) using DENV and is described in further detail in Chapter 13.

Dengue virus evolution

The evolutionary path of dengue virus (discussed in detail in Chapter 9) differs in several important aspects from its flavivirus cousins, though dengue retains many of the same clinical characteristics such as production of severe fever, myalgias, headache, hepatitis, encephalitis and haemorrhage. The phylogeny of the flaviviruses sheds little light on the origin of DENV because the closest relatives include mosquito-borne viruses that occur in several continents (Fig. 1.1). However,

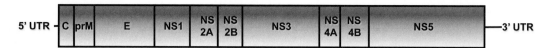

Figure 1.2 The dengue virus genome is displayed demonstrating the coding region flanked at the 5' and 3' termini by untranslated regions (UTRs). The coding region contains three structural genes [capsid (C), pre-membrane (prM), and envelope (E)] as well as seven non-structural (NS) genes. Subdivision of each region represents its relative size; the total genome is approximately 10.6 kb in length.

as described in Chapter 11, more detailed phylogenetic studies of DENV suggest an Asian origin, where sylvatic cycles between non-human primates and *Aedes* mosquitoes arose. Unlike the other flaviviruses however, DENV evolved into four antigenically and phylogenetically distinct serotypes: DENV-1, DENV-2, DENV-3 and DENV-4. Subsequently, each of these four serotypes emerged independently into an endemic cycle of transmission between humans and *Aedes albopictus* (Holmes and Twiddy, 2003). This endemic cycle is now both ecologically and evolutionarily separate from the sylvatic cycle. Thus, unlike other flaviviral pathogens, urban cycles of DENV can no longer be considered zoonotic.

It has been demonstrated that DENV evolves according to a molecular clock at a serotype- and genotype-specific rate (Twiddy *et al.*, 2003a), and that the transfer of DENV from a sylvatic cycle to sustained human transmission may have occurred on the order of 100 to 1500 years ago years ago (Wang *et al.*, 2000), suggesting that the current global pandemic of all four serotypes of DENV appeared during the past century (Twiddy *et al.*, 2003a). The contemporary genetic diversity seen in all four dengue serotypes is related to population growth, urbanization, and mass transport of both virus and its mosquito vector. Using an analytical technique based on coalescent theory, Holmes and Twiddy demonstrated that DENV-2 and DENV-3 experienced two phases of exponential growth (Twiddy *et al.*, 2003b). In the first phase and for most of their history, the dengue viruses experienced a low rate of exponential growth. Thirty years ago, the rate of growth of DENV-2 and DENV-3 suddenly increased by a factor of between 15 and 20.

Dengue virus epidemiology

The spread of *Aedes aegypti* mosquitoes through the slave trade and later through the movement of ships and goods during the Second World War facilitated the global expansion of dengue virus (see Chapter 8). The first descriptions of dengue fever characterized the eighteenth-century pandemic of dengue infection as described in 1780 by Benjamin Rush during a large outbreak of dengue fever in Philadelphia, Pennsylvania, in the USA (Rush, 1789). Dengue was thought to have been introduced in the USA as a consequence of the rum and slave trade between Africa and Caribbean ports. Dengue outbreaks occurred throughout the USA, the Caribbean and South America during the nineteenth and early twentieth centuries (Halstead, 1992). The second dengue pandemic was centred in the mining towns of northern Queensland, Australia, where boom towns and resulting *Aedes aegypti* population growth resulted in continuous dengue transmission from the 1870s until the First World War (Halstead, 1992). Dengue outbreaks were also occurring in the Eastern Mediterranean and resulted in a large epidemic in Greece during 1928. During the Second World War, dengue strains were carried by ships and soldiers from South-East Asia to Japan, the Pacific Islands, Philippines and Hawaii (Halstead, 1992). A new manifestation of severe dengue illness resulted, dengue haemorrhagic fever, first reported in the Philippines then later in Thailand during the 1950s (Halstead, 1992).

The discovery of the role of *Aedes aegypti* in the transmission and spread of yellow fever and the subsequent isolation of the virus and creation of an effective yellow fever vaccine introduced the concept of mosquito control as an effective measure to disrupt yellow fever transmission. Subsequently the International Health Board and the Rockefeller Foundation instituted mosquito control strategies including the use of a larvicidal, Paris Green, throughout the USA and Central and South America (Stapleton, 2004). These techniques were soon applied to malaria control and during the years from 1924 to 1925, funding for malaria prevention through the strategy of mosquito control doubled (Stapleton, 2004). The success of this programme in Italy during the 1920s set the stage for the global use of mosquito control in the prevention of malaria. The Second World War prompted the creation of the Rockefeller Foundation Health Commission in 1942 to support national defence and in particular malaria control for U.S. forces. The need for lousicides to combat typhus ushered in a new insecticide developed by the Swiss firm, Geigy, called dichlorodiphenyl-trichloroethane (DDT) (Stapleton, 2004). Led by Fred Soper, the Rockefeller team demonstrated the effectiveness of DDT as a lousicide and in disrupting typhus epidemics. DDT was soon used in aerial and ground spraying for Allied Forces during a malaria outbreak in Italy and was found to be a highly effective larvicide with a long environmental persistence. DDT subsequently became a key component of the World Health Organization's global malaria eradication campaign in 1955 (Stapleton, 2004). This campaign resulted in the elimination of both the malaria mosquito vector and *Aedes aegypti* throughout South America and the virtual elimination of malaria, yellow fever and dengue throughout the Americas. A reassessment of this global strategy by the WHO and the growing concerns of the environmental effects of DDT led to the end of the use of DDT as a mosquito control larvicide in 1969 (Nájera, 2001). The cessation of DDT-based mosquito control programmes in the Americas and the social disruption that resulted from the Second World War allowed the spread of DENV in Asia, the reintroduction and resurgence of *Aedes aegypti* throughout the Americas, and, consequently, resurgence of DENV, particularly South-East Asian strains, in the Americas.

The first two dengue pandemics were characterized by epidemics that produced severe outbreaks of fever, headache and myalgias, a clinical syndrome termed dengue fever. As waves of DENV-1 to -4 spread throughout the human population, especially in Asia, DENV adapted to be able to reach virus levels during a course of infection that allowed mosquitoes to become infected, thereby ensuring continued transmission of the virus. Chapter 8 discusses variation among vector species in their susceptibility to dengue and the potential selective effects of such

variation on viral replication; however, high levels of co-circulation among serotypes also posed a challenge for the persistence of each serotype. Consider a DENV-2 strain entering a population that had a high degree of pre-existing antibody to an established DENV, such as DENV-1. Pre-existing DENV-1 antibody, though not neutralizing, would under ordinary circumstances have provided significant heterotypic neutralization of DENV-2, potentially reducing viral levels in infected humans and thereby interrupting mosquito transmission. Thus, the presence of high levels of infection by multiple serotypes imposed significant selection for viruses that, via mutations in the E protein coat and changes in specific epitopes, were able to either fully escape the effects of heterotypic neutralization, or as is currently thought to be the case, to utilize these subneutralizing antibodies to enhance infection. This phenomenon of viral replicative enhancement due to subneutralizing heterotypic antibody is known as antibody-dependant enhancement (ADE) (Halstead, 1992). Since ADE results in higher viral loads, viruses with a particularly high tendency towards enhancement should have a selective advantage (Cummings *et al.*, 2005; Ferguson *et al.*, 1999a,b).

The ability of all DENV serotypes to utilize pre-existing heterotypic flavivirus antibody to enhance infection is a unique feature of DENV that is particularly common among South-East Asian strains. The tendency to be enhanced by heteroserotypic antibody distinguishes DENV from all other flaviviruses, and is the primary basis of DENV pathogenesis in severe dengue illness. During the third pandemic, this tendency of DENV to be enhanced in secondary dengue infection resulted in the clinical manifestation of a previously unrecognized sequelae of DENV infection – severe haemorrhagic disease and plasma leakage (Halstead *et al.*, 1963). First described as Philippine and Bangkok haemorrhagic fever during the 1950s, it is now recognized as dengue haemorrhagic fever (DHF).

Studies of dengue in Thailand

During the 1950s, the South-East Asian Treaty Organization (SEATO), in response to a cholera outbreak occurring throughout Asia, created a number of laboratories comprised of host-country and US scientists in Thailand, Malaysia, Bangladesh and Pakistan. The Thailand laboratory named the SEATO General Medical Research Project located in Bangkok, later re-named the Armed Forces Research Institute of Medical Sciences (AFRIMS) in 1977, formed a still ongoing 50-year relationship of Thai–US collaborators in the study of tropical infectious diseases. The discovery that Bangkok was experiencing an outbreak of a new clinical manifestation of dengue infection, dengue haemorrhagic fever by both Thai and US scientists, allowed the ongoing study of dengue in Thailand that spanned over a half of a century producing many of the seminal concepts of dengue virus transmission and disease severity. Early studies on DHF in Thailand established this as a unique clinical syndrome of DENV infection. Careful clinical studies of hospitalized children in Bangkok, Thailand demonstrated the clinical severity of DHF in producing thrombocytopenia, leucopenia, coagulopathy and plasma leakage (Nimmannitya *et al.*, 1969). Studies on DHF pathogenesis in the 1960s revealed its unique features in being largely a phenomenon of secondary DENV infections or in primary infection of infants, a function of declining maternal DENV antibody (Halstead, 1988). Classic studies in Thai children first established the role of enhancing antibody in the peripheral blood mononuclear cells of children in producing severe dengue illness and DHF (Kliks *et al.*, 1988). Prospective studies in hospitalized Thai children and in long-term cohort studies demonstrated the importance of dengue viral load and the T-cell response in determining dengue severity, the diversity of all four dengue serotypes circulating spatially and temporally in a well-defined geographic area and the role of subclinical dengue infection and its contribution to the overall burden of dengue illness (Anderson *et al.*, 2007; Endy *et al.*, 2002a,b, 2004).

The next step…

Despite our increased understanding of both virological and host factors of DENV and human infection, many questions remain about the virus–host interactions that result in severe dengue illness. At present, viral virulence factors and the mechanisms by which the South-East Asian strains produce DHF are not well characterized.

The role of the non-structural proteins in viral replication and escape from the host's innate immunity are discussed in detail in Chapters 3 and 7, respectively. Host genetic factors that predispose to DHF (discussed in Chapter 5), the role of heterotypic antibody in protection and enhancement of infection (discussed in Chapter 14) and a functional assay to detect this antibody are important scientific questions that are being explored. Many challenges remain in the quest to control DENV, including prediction of dengue transmission dynamics (Chapter 10), development of tractable, informative animal models (discussed in Chapter 6), development of novel therapies (Chapters 13 and 14), and the ultimate challenge (Chapter 12), development of a DENV vaccine that provides durable long-lasting protective immunity against infection.

Dengue has emerged in the last 60 years as a global health problem producing severe morbidity and mortality across the subtropical and tropical regions of the world. Considering that much of the emergence of dengue epidemics is due to population growth, continued spread of the vector *Aedes aegypti* and urbanization of the developing world, dengue will continue to grow as this century's most important public health problem.

References

Anderson, K.B., Chunsuttiwat, S., Nisalak, A., Mammen, M.P., Libraty, D. H., Rothman, A.L., Green, S., Vaughn, D.W., Ennis, F. A., and Endy, T.P. (2007). Burden of symptomatic dengue infection in children at primary school in Thailand: a prospective study. Lancet *369*, 1452–1459.

Anonymous (2008). Dengue and dengue hemorrhagic fever (Geneva: World Health Organization).

Cummings, D.A., Schwartz, I.B., Billings, L., Shaw, L.B., and Burke, D.S. (2005). Dynamic effects of antibody-dependent enhancement on the fitness of viruses. Proc. Natl. Acad. Sci. U.S.A. *102*, 15259–15264.

Endy, T.P., and Nisalak, A. (2002). Japanese encephalitis virus: ecology and epidemiology. Curr. Top. Microbiol. Immunol. *267*, 11–48.

Endy, T.P., Chunsuttiwat, S., Nisalak, A., Libraty, D.H., Green, S., Rothman, A.L., Vaughn, D. W., and Ennis, F.A. (2002a). Epidemiology of inapparent and symptomatic acute dengue virus infection: a prospective study of primary school children in Kamphaeng Phet, Thailand. Am. J. Epidemiol. *156*, 40–51.

Endy, T.P., Nisalak, A., Chunsuttiwat, S., Libraty, D.H., Green, S., Rothman, A.L., Vaughn, D.W., and Ennis, F.A. (2002b). Spatial and temporal circulation of dengue virus serotypes: a prospective study of primary school children in Kamphaeng Phet, Thailand. Am. J. Epidemiol. *156*, 52–59.

Endy, T.P., Nisalak, A., Chunsuttitwat, S., Vaughn, D.W., Green, S., Ennis, F.A., Rothman, A.L., and Libraty, D.H. (2004). Relationship of preexisting dengue virus (DV) neutralizing antibody levels to viremia and severity of disease in a prospective cohort study of DV infection in Thailand. J. Infect. Dis. *189*, 990–1000.

Ferguson, N., Anderson, R., and Gupta, S. (1999a). The effect of antibody-dependent enhancement on the transmission dynamics and persistence of multiple-strain pathogens. Proc. Natl. Acad. Sci. U.S.A. *96*, 790–794.

Ferguson, N.M., Donnelly, C.A., and Anderson, R.M. (1999b). Transmission dynamics and epidemiology of dengue: insights from age-stratified sero-prevalence surveys. Phil. Trans. R. Soc. Lond. B Biol. Sci. *354*, 757–768.

Gubler, D.J. (2006). Dengue/dengue haemorrhagic fever: history and current status. Novartis Found. Symp. *277*, 3–16; discussion 16–22, 71–13, 251–253.

Halstead, S. B. (1988). Pathogenesis of dengue: challenges to molecular biology. Science *239*, 476–481.

Halstead, S. B. (1992). The XXth century dengue pandemic: need for surveillance and research. World Health Stat. Q. *45*, 292–298.

Halstead, S.B., Yamarat, C., and Scanlon, J.E. (1963). The Thai hemorrhagic fever epidemic of 1962 (A preliminary report). J .Med. Assoc. Thai *46*, 449–462.

Holmes, E.C., and Twiddy, S.S. (2003). The origin, emergence and evolutionary genetics of dengue virus. Infect. Genet. Evol. 3, 19–28.

Karabatsos, N. ed. (1985). International catalogue of arthropod-borne viruses, 3rd edn (San Antonio, TX: American Society for Tropical Medicine and Hygiene).

Kliks, S.C., Nimmannitya, S., Nisalak, A., and Burke, D.S. (1988). Evidence that maternal dengue antibodies are important in the development of dengue hemorrhagic fever in infants. Am. J. Trop. Med. Hyg. *38*, 411–419.

Kuhn, R.J., Zhang, W., Rossmann, M.G., Pletnev, S.V., Corver, J., Lenches, E., Jones, C.T., Mukhopadhyay, S., Chipman, P.R., Strauss, E.G., Baker, T.S., and Strauss, J.H. (2002). Structure of dengue virus. Implications for flavivirus organization, maturation, and fusion. Cell *108*, 717–725.

Kyle, J.L., and Harris, E. (2008). Global spread and persistence of dengue. Annu. Rev. Microbiol. *62*, 71–92.

Monath, T.P. (1988). Yellow fever, In The Arboviruses: Epidemiology and Ecology, T.P. Monath, ed. (Boca Raton, FL: CRC Press), pp. 139–233.

Nájera, J.A. (2001). Malaria control: achievements, problems and strategies. Parassitologia *43*, 1–89.

Nimmannitya, S., Halstead, S.B., Cohen, S., and Margiotta, M.R. (1969). Dengue and chikungunya virus infection in man in Thailand, 1962–1964. I. Observations on hospitalized patients with hemorrhagic fever. Am. J. Trop. Med. Hyg. *18*, 954–971.

Rice, C.M., ed. (1996). Flaviviridae: The viruses and their replication (Philadelphia: Lippincott-Raven Publishers).

Rush, B. (1789). An account of the bilious remitting fever, as it appeared in Philadelphia, in the summer and autumn of the year 1780. In Medical inquires and observations (Philadelphia: Prichard and Hall), pp. 89–100.

Stapleton, D.H. (2004). Lessons of History? Anti-malaria strategies of the International Health Board and the Rockefeller Foundation from the 1920s to the era of DDT Public Health Reports *119*, 206–215.

Twiddy, S.S., Holmes, E. C., and Rambaut, A. (2003a). Inferring the rate and time-scale of dengue virus evolution. Mol. Biol. Evol. *20*, 122–129.

Twiddy, S.S., Pybus, O.G., and Holmes, E.C. (2003b). Comparative population dynamics of mosquito-borne flaviviruses. Infect. Genet. Evol. 3, 87–95.

Wang, E., Ni, H., Xu, R., Barrett, A.D., Watowich, S. J., Gubler, D.J., and Weaver, S.C. (2000). evolutionary relationships of endemic/epidemic and sylvatic dengue viruses. J. Virol. *74*, 3227–3234.

Part II

Dengue Virus Replication

Translation and Processing of the Dengue Virus Polyprotein

R. Padmanabhan and Alex Y. Strongin

Abstract

Positive-strand RNA viruses, including flaviviruses, generally utilize the translational machinery of the host to synthesize viral proteins either in a cap-dependent or cap-independent manner to produce polyprotein precursors which are then processed into mature proteins. Polyprotein processing is accomplished by the concerted action of host and viral proteases. While some viruses, such as the hepatitis C virus code for more than one protease to perform distinct functions, flaviviruses code for a novel two-component serine protease which participates in early and late stages of the viral life cycle. This chapter summarizes the state of knowledge of polyprotein processing by the viral protease since its discovery approximately 20 years ago.

Introduction

Positive-strand RNA viruses, subsequent to infection of their respective hosts, undergo translation in the cytoplasm either by cap-dependent or cap-independent mechanisms with the viral RNA genome serving as mRNA template for the host's translation machinery. Flaviviruses contain a type I cap at the 5′-end and no poly(A) tail at the 3′-end (Cleaves and Dubin, 1979; Wengler et al., 1978); (reviewed in (Chambers et al., 1990a); also see references therein) and the viral RNA is translated by a cap-dependent initiation and scanning the 5′-UTR. The initiation codon AUG is in a poor Kozak's consensus context. However, an inefficient translation initiation at this start codon is compensated by efficient start site selection mechanism mediated by a cis-acting RNA element such as cHP in the capsid coding region which is conserved among mosquito-borne and tick-borne flaviviruses (Clyde et al., 2008; Clyde and Harris, 2006). Moreover, both the 5′- and 3′-UTRs are required for translation when the host cell cap-dependent translation is blocked (Edgil and Harris, 2006; Edgil et al., 2006).

The genomes of tick-borne and mosquito-borne flaviviruses are translated to give rise to polyproteins which then undergo processing both co-translationally and post-translationally by the host and viral proteases into mature proteins (Chambers et al., 1990a). Crawford et al. provided the early experimental evidence that flavivirus polyproteins are synthesized as a large precursor polyprotein which is subsequently processed into mature polypeptides (Crawford and Wright, 1987). They detected high molecular weight forms of dengue virus type 2 (DENV-2)-specific polypeptides (130–250 kDa) in DENV-2-infected BHK21 cells (Cleaves, 1985; Ozden and Poirier, 1985) and Kunjin virus (KUNV)-specific polypeptides in KUNV-infected Vero cells labelled with hydroxyleucine (Crawford and Wright, 1987). During pulse-chase experiments, these high molecular weight polypeptides were converted into smaller and stable polypeptides, presumably the mature forms in the 10–98-kDa size range. However, addition of Zn^{2+} and the serine protease inhibitor, N-tosyl-l-phenylalanyl chloromethyl ketone (TPCK) during the pulse-labelling inhibited the conversion of high molecular weight polypeptides into smaller polypeptides. The complete sequence analysis of the yellow fever virus (YFV) genome provided the genetic evidence in support of this observation; the sequence information indicated that flaviviral RNA is ~11 kb in length containing 5′- and 3′-untranslated regions of approximately 95–135 and 114–650 nucleotides, respectively, and a single long open reading frame with a coding potential for a 3411 amino acids (Rice *et al.*, 1985). Reports of sequence analysis of several flavivirus RNAs followed, including DENV-4 (Mackow *et al.*, 1987; Zhao *et al.*, 1986), DENV-2 (Deubel *et al.*, 1988; Hahn *et al.*, 1988; Irie *et al.*, 1989; Yaegashi *et al.*, 1986), KUN (Coia *et al.*, 1988), and West Nile virus (WNV) (Castle *et al.*, 1985, 1986) which firmly established that flavivirus genomes share similar genomic organization. The 5′ and 3′ ends of the genome contain conserved dinucleotides AG and CU, respectively (reviewed in Chambers *et al.*, 1990a, and Chapter 4).

Processing of polyprotein by cellular protease(s)

Processing at the structural region of the polyprotein

The information gathered from the amino terminal sequence analysis of the mature structural proteins of YFV, St. Louis encephalitis virus (SLEV), DENV-2 (Bell *et al.*, 1985; Biedrzycka *et al.*, 1987), tick-borne encephalitis virus (TBEV) (Svitkin *et al.*, 1984), and non-structural proteins of YFV and SLEV (Rice *et al.*, 1986) as well as that of KUNV (Monckton and Westaway, 1982; Speight *et al.*, 1988; Speight and Westaway, 1989a,b) and WNV (Wengler *et al.*, 1978; Wengler *et al.*, 1990) allowed the construction of the gene order for the flavivirus polyprotein as C (capsid)-prM (precursor membrane)-E (envelope)-NS1–NS2A–NS2B–NS3–NS4A–NS4B–NS5 (reviewed in Chambers *et al.*, 1990a, and Chapter 4). The N-terminal quarter of the genome codes for the three structural proteins (C, prM, and E) and the remainder of the genome encodes the seven non-structural proteins, NS1 to NS5. The C-terminal regions of C, prM, and E contain hydrophobic amino acids, each of which functions as a signal sequence for insertion of the protein that follows, prM, E, and NS1, respectively, into the endoplasmic reticulum (ER) membrane. The

information gained from the N-terminal sequence analysis of prM, E and NS1 and the known consensus sequence for the signal peptidase-mediated cleavages suggested the involvement of an ER signal peptidase for cleavages at the C-prM, prM-E, and E-NS1 sites. Cleavage of C-prM junction by the ER signal peptidase results in membrane-bound intracellular form of the capsid protein ('C-anchored') (Lobigs, 1993; Yamshchikov and Compans, 1993). The viral protease is required for generation of the mature form of C protein (described in the section detailing the role of the viral protease in assembly and secretion of mature virions). For example, the N-terminal sequence of NS1 conforms to the '(−3, −1) rule' (Val-Gln-Ala/Asp) (Perlman and Halvorson, 1983; von Heijne, 1984). The *in vitro* translation of the genomic RNA or the coding regions of the structural protein precursors of DENV-4 and TBEV in the presence of microsomal membranes (Markoff, 1989; Nowak *et al.*, 1989; Ruiz-Linares *et al.*, 1989; Svitkin *et al.*, 1984; Wengler *et al.*, 1990) support the involvement of an ER signal peptidase. Using polyclonal antisera against YFV C, prM, E, and NS1 and metabolic labelling of YFV-infected mammalian cells, Chambers *et al.* determined that the structural proteins are produced by a rapid co-translational processing of the precursor polyprotein. An analysis of infected cells subjected to tunicamycin revealed that prM and NS1 each have two N-linked oligosaccharides (Chambers *et al.*, 1990b). The evidence, taken together, supported the role of the cellular signal peptidase in co-translational processing of the polyprotein in the ER although the role of the viral protease involvement in efficient cleavage of C-prM junction and regulation of nucleocapsid incorporation into mature virions were revealed only from the results of experiments with cloned infectious viruses or long polyprotein expression clones that included viral protease coding sequences (see below).

Cleavages at the E-NS1 and NS1–NS2A sites

Falgout *et al.* (1989) have characterized the sequence requirements for the cleavage of the NS1–NS2A site. A 4-kb cDNA was cloned into a recombinant vaccinia virus expression vector to encode the DENV-4 precursor polypeptide, C-prM-E-NS1–NS2A, NS1–NS2A and several subfragments of this precursor polypeptide containing N- and C-terminal deletion mutants. CV1 cells were infected with these recombinant vaccinia viruses and metabolically labelled with [^{35}S] methionine. Cell lysates were then immunoprecipitated using an NS1-specific antibody. Their results showed that the 24-residue hydrophobic region preceding the N-terminus of NS1 was necessary and sufficient for the translocation of NS1 into the ER and for the synthesis of a glycosylated form of NS1. This hydrophobic region serves as a signal sequence for the translocation of NS1 into the ER and is cleaved by the signal peptidase. The cleavage at the NS1–NS2A boundary also required the presence of the full-length NS2A because the incomplete NS1–NS2A (15%) and NS1–NS2A (49%) precursor constructs were not cleaved at the NS1–NS2A junction. Providing the intact NS1–NS2A *in trans* did not cause cleavage at the NS1–NS2A junction indicating that NS2A is required *in cis*. The requirement for the signal sequence suggests that the NS1–NS2A cleavage occurs at an intracellular site within the ER (Pethel *et al.*, 1992). To test the minimum requirements for NS1 cleavage, an NS1–NS2A precursor and a panel of mutants that contained the hydrophobic signal sequence for translocation of NS1 into the ER and a series of C-terminal deletions of most of the NS1, except the last 3–20 amino acid residue segment, were expressed using a recombinant vaccinia virus system (Hori and Lai, 1990). The cleavage of the NS1–NS2A site required the presence of the eight C-terminal amino acid residues of NS1. A comparison of this eight-residue sequence of DENV-4 NS1 with those of 12 other flaviviruses indicated the consensus sequence at the P8- P1 positions [according to the nomenclature of Schechter and Berger (Schechter and Berger, 1967)] to be Leu/Met(P8)-Val(P7)-X aa(P6)-Ser(P5)-X aa(P4)-Val(P3)-X aa(P2)-Ala(P1)↓ (where ↓ indicates the cleavage and X aa is any amino acid residue) (Hori and Lai, 1990). Further studies confirmed the requirement of the C-terminal eight amino acid residues for the NS1–NS2A cleavage and showed that a substitution at the P5 of Pro or Ala for Ser was tolerated. Some non-conserved

amino acids could be substituted without any effect on the cleavage efficiency whereas some other substitutions (for example, Gly or Glu for P4-Gln) reduced the cleavage efficiency (Pethel *et al.*, 1992). In addition to the eight C-terminal amino acids of NS1, the N-terminal amino acid residues of NS2A are also important for the NS1–NS2A cleavage. The N-terminus of NS2A follows a consensus cleavage site for the signal peptidase (von Heijne, 1984) but lacks the upstream hydrophobic sequence at the C-terminus of NS1. Since brefeldin A, an inhibitor of trafficking from the ER to the Golgi, did not block the NS1–NS2A cleavage but blocked the secretion of mature NS1, suggesting that cleavage occurs before NS1 passed through the Golgi and is secreted. *In vitro* transcription-rabbit reticulocyte-mediated translation system in the presence of dog pancreatic microsomal membranes produced the authentic NS1 and NS2A, and NS1 was glycosylated in the membrane pellet fraction (Falgout and Markoff, 1995). The membrane fraction retained the residual uncleaved precursor and cleaved NS2A upon treatment with sodium carbonate at pH 11.5, but the mature NS1 became soluble, indicating that NS1 is localized in the luminal side of the ER and NS2A is an integral ER membrane protein (Falgout and Markoff, 1995). By analysing the NS2A mutants for their effects on NS1–NS2A cleavage, Falgout *et al.* (Falgout and Markoff, 1995) concluded that NS2A is not a protease. Taken together, an ER-resident host protease, probably the signal peptidase, is involved in the NS1–NS2A cleavage (Falgout and Markoff, 1995).

Cleavage at the NS4A–NS4B site

The flaviviral NS4A and NS4B are hydrophobic proteins and their hydrophobicity pattern is conserved rather than their primary sequences. In DENV-2 infected cells NS4A–NS4B precursor appears in low amounts in mammalian cells but not mosquito cells (Preugschat and Strauss, 1991). The N-terminus of NS4A is generated by the viral serine protease (as described in detail in the next section). A signal sequence for the translocation of NS4B into the ER precedes the amino terminus of NS4B suggesting that the NS4A–NS4B cleavage occurs in the ER by a signal peptidase. NS4B is a 30-kDa protein which appears to be post-translationally modified to generate a ~28-kDa protein during *in vitro* translation and in DENV-2-infected as well as in recombinant vaccinia virus-infected BHK cells (Preugschat and Strauss, 1991). Using a recombinant vaccinia virus vector for the transient expression of a DENV-2 or DENV-4 polyprotein precursor NS3–NS4A–NS4B–NS5 (NS3→NS5) in mammalian cells, it was found that NS4A–NS4B cleavage did not occur unless NS2B, the cofactor for the NS3 protease catalytic domain (NS3pro; described in the following section), was also co-expressed (Cahour *et al.*, 1992; Zhang *et al.*, 1992). However, when the recombinant vaccinia virus polyprotein precursor, NS4A–NS4B–NS5 was expressed and the NS5-specific antibody was used to monitor the cleavage, NS4B–NS5 precursor was produced together with uncleaved polyprotein in the absence of the viral protease indicating that cleavage at the NS4A–NS4B site occurred (Cahour *et al.*, 1992). The NS4B–NS5 precursor protein could undergo processing only if NS2B–NS3pro was co-expressed (Cahour *et al.*, 1992). From the differences in the cleavage efficiency at the NS4A–NS4B site in the NS3→NS5 and NS4A→NS5 polyprotein precursors, the authors concluded that the signalase (ER resident signal peptidase) cleavage at the NS4A–NS4B site was inhibited in the presence of the full-length NS3 in the polyprotein (Cahour *et al.*, 1992) likely due to the influence of polyprotein conformation on the cleavage efficiency at the NS3–NS4A site as demonstrated (Zhang and Padmanabhan, 1993). In another study, cleavage at the NS4A–NS4B site was observed in a cell-free translation system mediated by rabbit reticulocyte lysate with a construct containing NS4B and the C-terminal 17 amino acid residues of NS4A that constitute the putative signal sequence for translocation of NS4B into the ER, (Lin *et al.*, 1993a). However, cleavage of NS4A–NS4B precursor was not observed in this cell-free translation system or in transient recombinant vaccinia virus expression system unless the viral protease was co-expressed (Lin *et al.*, 1993a). It was concluded that a prior cleavage at an internal site in NS4A (NS4A↓2K site) was required for the subsequent cleavage of the NS4A–NS4B site by the signal peptidase (Lin *et al.*, 1993a).

Processing of the polyprotein precursor by the viral serine protease

Identification of NS3 as a component of serine protease

A comparison of the nucleotide sequences of flavivirus genomes with the N-terminal amino acid sequences of the non-structural proteins indicates that the cleavage sites at NS2A–NS2B, NS2B–NS3, NS3–NS4A, NS4B–NS5 are conserved (Table 2.1). The amino acids preceding the cleavage sites at the P1 and the P2 positions are two basic residues, predominantly R-R, R-K, or occasionally a Q at the P2 position at the NS2B–NS3 site of the DENV-2 and DENV-4 polyproteins. The C-terminal side of the cleavage site (P1′ position) is occupied by a short chain amino acid residue such as G, A or S. A flavivirus-encoded protease was hypothesized to be involved in the cleavages of these sites (Krausslich and Wimmer, 1988; Rice *et al.*, 1985). However, the identification of the flavivirus protease and its biochemical characterization were not initiated until two groups (Bazan and Fletterick, 1989; Gorbalenya *et al.*, 1989) reported their findings that the amino-terminal region of flavivirus NS3 consisting of ~180 amino acid residues contains a highly conserved catalytic triad of the trypsin family of serine proteases (Fig. 2.1). Several groups experimentally verified these findings by *in vitro* transcription and translation as well as by *in vivo* assays using polyprotein precursors of non-structural proteins that included NS3 expressed in mammalian cells (Cahour *et al.*, 1992; Chambers *et al.*, 1990c, 1991; Falgout *et al.*, 1991; Preugschat *et al.*, 1990; Wengler *et al.*, 1991; Zhang *et al.*, 1992).

Chambers *et al.* (Chambers *et al.*, 1990c) used *in vitro* transcription and translation of the YFV polyprotein precursors containing NS2A*NS2B–NS3$_{181\,aa}$ or NS2B–NS3$_{181\,aa}$, the former containing 115 residues of NS2A (* indicates truncated NS2A) linked to the NS2B–NS3$_{181\,aa}$ precursor. The transcripts synthesized *in vitro* by T7 RNA polymerase were capped and translated in a rabbit reticulocyte lysate-mediated *in vitro* translation system. The expression of the NS2A*-NS2B–NS3$_{181\,aa}$ precursor yielded the cleavage products, NS2A*–NS2B, NS2B (15 kDa; N-terminal

Table 2.1 Amino acid sequences at the flaviviral polyprotein cleavage sites

	Capsid C	NS2A/NS2B	NS2B–NS3	NS3/NS4A	NS4B/NS5
WNV	$Q^{101}KKR{\downarrow}\mathbf{G}GTA^{108}$	$N^{1367}RKR{\downarrow}\mathbf{G}WPA^{1374}$	$Y^{1498}TKR{\downarrow}\mathbf{GG}VL^{1505}$	$S^{2117}GKR{\downarrow}SQIG^{2124}$	$G^{2522}LKR{\downarrow}\mathbf{GG}AK^{2529}$
JEV	$Q^{102}NKR{\downarrow}\mathbf{GG}NE^{109}$	$N^{1370}KKR{\downarrow}\mathbf{G}WPA^{1377}$	$T^{1501}TKR{\downarrow}\mathbf{GG}VF^{1508}$	$A^{2120}GKR{\downarrow}SAVS^{2127}$	$S^{2564}LKR{\downarrow}\mathbf{G}RPG^{2571}$
YFV	$R^{98}KRR{\downarrow}SHDV^{105}$	$F^{1351}GRR{\downarrow}SIPV^{1358}$	$G^{1481}ARR{\downarrow}SGDV^{1488}$	$E^{2104}GRR{\downarrow}\mathbf{G}AAE^{2111}$	$T^{2503}GRR{\downarrow}\mathbf{G}SAN^{2510}$
DV1	$R^{97}RKR{\downarrow}SVTM^{104}$	$W^{1341}GRK{\downarrow}SWPL^{1348}$	$K^{1471}KQR{\downarrow}SGVL^{1478}$	$A^{2090}GRR{\downarrow}SVSG^{2097}$	$G^{2489}GRR{\downarrow}\mathbf{G}TGA^{2496}$
DV2	$R^{97}RRR{\downarrow}TAGV^{104}$	$S^{1342}KKR{\downarrow}SWPL^{1349}$	$K^{1472}KQR{\downarrow}AGVL^{1479}$	$A^{2090}GRK{\downarrow}SLTL^{2097}$	$N^{2488}TRR{\downarrow}\mathbf{G}TGN^{2495}$
DV3	$K^{97}RKK{\downarrow}TSLC^{104}$	$L^{1340}KRR{\downarrow}SWPL^{1347}$	$Q^{1470}TQR{\downarrow}SGVL^{1477}$	$A^{2089}GRK{\downarrow}SIAL^{2096}$	$T^{2487}GKR{\downarrow}\mathbf{G}TGS^{2494}$
DV4	$G^{96}RKR{\downarrow}STIT^{103}$	$A^{1341}SRR{\downarrow}SWPL^{1348}$	$K^{1471}TQR{\downarrow}SGAL^{1478}$	$S^{2089}GRK{\downarrow}SITL^{2096}$	$T^{2484}PRR{\downarrow}\mathbf{G}TGT^{2491}$

The sequence of the natural cleavage sites of the NS3 protease in the capsid protein C and at the NS2A/NS2B, NS2B/NS3, NS3/NS4A, NS4A/NS4B and NS4B/NS5 boundaries of the polyprotein precursor. WNV, West Nile virus; JEV, Japanese encephalitis virus; YFV, Yellow fever virus; DV1–4, Dengue virus serotypes 1–4, respectively. The P1'/P2' Gly is shown in bold.

sequencing confirmed its identity), and NS3$_{181\,aa}$, in addition to a small amount of the unprocessed precursor (Chambers *et al.*, 1990c). The NS2B–NS3$_{181\,aa}$ species was not observed suggesting that the cleavage at the NS2B–NS3$_{181\,aa}$ site occurs prior to the NS2A–NS2B cleavage. These results provided evidence that there is protease activity resident in the NS2B–NS3$_{181\,aa}$ precursor that mediates the cleavage at the NS2A*–NS2B and NS2B–NS3 sites. Mutagenesis of the YFV NS3 serine protease catalytic triad (His53-Asp77-Ser138) residue, Ser138Cys, significantly reduced the yield of the NS3$_{181\,aa}$ and NS2B products and mutation of other residues (His53 and Asp77 or Ser138 to Ala) blocked the NS2B–NS3 cleavage. Moreover, the cleavages at the NS2A–NS2B and NS2B–NS3 sites were dilution insensitive (concentration independent), suggesting that these cleavages are intramolecular and occur in '*cis*'. The importance of the serine protease function in the viral life cycle was shown by the mutation of Ser138 to Cys in the context of RNA genome which was shown to be not viable and no revertants were observed (Chambers *et al.*, 1990c).

Similarly Preugschat *et al.* examined the cleavages at the NS2A–NS2B and NS2B–NS3 sites of the DENV-2 precursor polypeptide (Preugschat *et al.*, 1990). In this study, the construct that produced a T7 RNA polymerase-catalysed transcript contained the coding region for the first 37 amino acid residues of the capsid protein with or without the upstream 5' untranslated region (UTR), the C-terminal 118 aa of NS2A, all of NS2B and the first 610 aa of NS3. *In vitro* translation of the precursor RNA mediated by rabbit reticulocyte lysates showed that the insertion of 5'-UTR of DENV-2 upstream of the coding region of the precursor protein enhanced expression fivefold *in vitro* and by an additional twofold if the first 37 amino acids of the capsid coding sequence was fused in-frame with the upstream 5'-UTR, suggesting that the 5'-UTR and the authentic translation initiation codon of DENV-2 polyprotein improved the expression of the polyprotein precursor (Preugschat *et al.*, 1990). Similar stimulatory effects on translational efficiency of YFV precursor protein *in vitro* were also reported earlier (Ruiz-Linares *et al.*, 1989). The stimulatory effects of 5'-UTR and the N-terminal capsid coding sequence on

```
                 1                                                50
Den NS3pro (1)   -AGVLWDVPSPPPVGKAE-LEDGAYRIKQKGILGYSQIGAGVYKEGTFHT
JEV NS3pro (1)   -GGVFWDTPSPKPCSKGD-TTTGVYRIMARGILGTYQAGVGVMYENVFHT
WNV NS3pro (1)   -GGVLWDTPSPKEYKKGD-TTTGVYRIMTRGLLGSYQAGAGVMVEGVFHT
YFV NS3pro (1)   SGDVLWDIPTPKIIEECEHLEDGIYGIFQSTFLGASQRGVGVAQGGVFHT
Consensus  (1)   GGVLWDTPSPKPI KGD TTTGVYRIMQRGILGSYQAGVGVM EGVFHT

                 51                                               100
Den NS3pro (1)   MWHVTRGAVLMHKGKRIEPSWADVKKDLISYGGGWKLEGEWKEGEEVQVL
JEV NS3pro (1)   LWHTTRGAAIMSGEGKLTPYWGSVKEDRIAYGGPWRFDRKWNGTDDVQVI
WNV NS3pro (1)   LWHTTKGAALMSGEGRLDPYWGSVKEDRLCYGGPWKLQHKWNGHDEVQMI
YFV NS3pro (1)   MWHVTRGAFLVRNGKKLIPSWASVKEDLVAYGGSWKLEGRWDGEEEVQLI
Consensus  (1)   LWHVTRGAALMSGGKKLDPYWASVKEDRIAYGGPWKLEGKWNG DEVQVI

                 101                                              150
Den NS3pro (1)   ALEPGKNPRAVQTKPGLFRT-NTGTIGAVSLDFSPGTSGSPIVDKKGKVV
JEV NS3pro (1)   VVEPGKAAVNIQTKPGVFRT-PFGEVGAVSLDYPRGTSGSPILDSNGDII
WNV NS3pro (1)   VVEPGKNVKNVQTKPGVFKT-PEGEIGAVTLDYPTGTSGSPIVDKNGDVI
YFV NS3pro (1)   AAVPGKNVVNVQTKPSLFKVRNGGEIGAVALDYPSGTSGSPIVNRNGEVI
Consensus  (1)   VVEPGKNVVNVQTKPGLFKT P GEIGAVSLDYPSGTSGSPIVDKNGDVI

                 151            170          185    192
Den NS3pro (1)   GLYGNGVVTRSGAYVSAIAQTEKSIEDNPEIEDDIFRKRRL-
JEV NS3pro (1)   GLYGNGVELGDGSYVSAIVQGDRQEEPVPEAYTPNMLRKRQM
WNV NS3pro (1)   GLYGNGVIMPNGSYISAIVQGERMEEPAPAGFEPEMLRKKQI
YFV NS3pro (1)   GLYGNGILVGDNSFVSAISQTEVKEEGKEELQEIPTMLKKG-
Consensus  (1)   GLYGNGVILGDGSYVSAIVQTER EEP PEIFEP MLRKKQI
```

Figure 2.1 Alignment of NS3-pro sequences of representative subfamily members of mosquito-borne flaviviruses. Gaps are introduced for alignment of conserved residues in DENV, JEV, WNV and YFV. Consensus sequence is shown at the bottom. The residue numbering corresponds to YFV and its sequence is uninterrupted. The Genbank accession numbers are: DENV-2, ACB87136.1; JEV, AAD16275.1; WNV, NP_041724.2; YFV, NP776004.1.

efficient translation is likely related at least in part to the role of cHP element identified in this region in a more recent study (Clyde and Harris, 2006).

In these studies, contrary to the processing of the YFV precursor reported by Chambers *et al.* (1990c) (see above), the cleavage at NS2A–NS2B occurred first as the NS2B–NS3 processing intermediate could be readily detected (Preugschat *et al.*, 1990). However, there were similarities to the pattern reported by Chambers *et al.* (1990c) as well, specifically cleavages at NS2A–NS2B and NS2B–NS3 occurred *in cis* as demonstrated by the dilution insensitive processing of the precursor and by failure of the catalytic triad mutant precursor to undergo processing *in trans* by co-expression of the active form of the precursor. The identity of the products of processing, NS2B and NS3, were verified by N-terminal sequencing (Preugschat *et al.*, 1990).

Requirement of NS2B as a cofactor for the NS3 serine protease catalytic domain

Using a transient recombinant vaccinia virus expression system, a YFV polyprotein in which the signal sequence from the E protein was fused to the coding sequence of NS2A–NS2B–NS3–NS4A–NS5$_{356aa}$ and expressed in BHK cells, was shown to undergo processing (Chambers *et al.*, 1991). A polyprotein precursor containing NS2B at the N-terminus was also efficiently processed; however, the precursor containing the Ser138 mutation in the catalytic triad of NS3 (Fig. 2.1), or lacking the NS2B sequence, failed to undergo processing. In the latter case, in the presence of NS2B supplied *in trans*, polyprotein processing occurred (Chambers *et al.*, 1991). In another study, a DENV-2 polyprotein with NS3 at the N-terminus and full-length NS5 at the C-terminus (NS3→NS5) was expressed using the recombinant vaccinia virus vector under the control of the T7 promoter and encephalomyocarditis virus (EMCV) leader region for cap-independent translation in monkey kidney (CV1) cells. This polyprotein did not undergo processing unless NS2B was provided *in trans*. Mutation of the catalytic triad residue, H51A in DENV-2 NS3 blocked processing even if NS2B was co-expressed (Zhang *et al.*, 1992). In a follow-up study, polyprotein precursors containing NS2B, NS2A–NS2B, and NS1–NS2A–NS2B fused to the NS3→NS5 at the N-terminus were constructed in the recombinant vaccinia virus vector (Zhang and Padmanabhan, 1993). The DENV-2 polyprotein precursors that could undergo self-processing in mammalian cells were examined. The longest polyprotein precursor containing the coding sequences of all of the NS proteins in the order NS1–NS2A–NS2B–NS3–NS4A–NS4B–NS5 (NS1→NS5) could undergo efficient processing in mammalian cells without any detectable processing intermediates. However, when the precursors, NS2A→NS5, NS2B→NS5, or NS2B and NS3→NS5 were co-expressed, although the *cis* cleavage at NS2B/NS3 was efficient, the NS3/NS4A cleavage was incomplete giving rise to NS3–NS4A intermediate. Truncation of NS1 in the N-terminus of the NS1→NS5 precursor or NS5 at the C-terminus also affected the cleavage at NS3/NS4A site, suggesting that the polyprotein conformation in the ER membrane, where the polyprotein processing is thought to occur, plays a role in efficient processing at NS3–NS4A site in mammalian cells (Zhang and Padmanabhan, 1993). A similar conclusion was reached in other studies; for example, a YFV polyprotein or Murray Valley encephalitis virus polyprotein (sig2A→NS5$_{356}$) in which the C-terminal 23 aa of E protein and the first 11 aa of NS1 fused to 356 aa of NS5 was expressed as a vaccinia virus recombinant in mammalian cells. Cleavage at NS3/NS4A was inefficient producing an uncleaved NS3–NS4A intermediate (Lin *et al.*, 1993b; Lobigs, 1992) which is in agreement with the role of polyprotein conformation for efficient cleavage at NS3–NS4A site as described above (Zhang and Padmanabhan, 1993). It was further shown that mutagenesis by amino acid substitutions at P2, P1, and P1′ positions of YFV NS3/NS4A and NS4B/NS5 sites influenced the cleavages at these sites but not at the NS2A/

NS2B and NS2B/NS3 sites (Lin *et al.*, 1993b). In that study, the results showed that G at the P1′ of NS4B/NS5 site could be replaced only by S or A, or the substitution of R at the P1 to K, Q, N, or H, but the substitution of E eliminated the cleavage. Substitution of the P2 R of NS4B/NS5 site by a polar or hydrophobic residue is tolerated to various extents of partial cleavage and the P1 R of NS3/NS4A site with E or P abolished the cleavage.

Subsequent studies also demonstrated the requirement for NS2B as a cofactor for the catalytic activity of the NS3 serine protease domain in dilution-insensitive *cis* cleavages at NS2A/NS2B and NS2B/NS3 sites, and for *trans* cleavages at these sites as well as at NS4A/NS2K, NS4B/NS5, and anchored C (Amberg *et al.*, 1994; Lin *et al.*, 1993a,b). A short polyprotein consisting of DENV-4 NS2A, NS2B, and the N-terminal 184 amino acid residues of NS3 was also able to undergo processing during its expression using a recombinant vaccinia virus system in CV1 cells. A deletion in NS2B blocked this processing, but processing was restored by providing NS2B *in trans* (Falgout *et al.*, 1991). Moreover, a WNV polyprotein encoding NS2A fused at the N-terminus with the influenza haemagglutinin signal sequence, NS2B, and the N-terminal region of NS3 that included an intact catalytic triad expressed by *in vitro* transcription and translation in the presence of microsomal membranes was processed by cleavages at the NS2A/NS2B and NS2B/NS3 sites (Wengler *et al.*, 1991).

NS2B interacts functionally and physically with the NS3 protease domain *in vitro* and *in vivo*

To identify the region of NS2B that is required for a cofactor function, deletions within the DENV-4 NS2B gene of NS2B–NS3(30%) precursor were cloned into a recombinant vaccinia virus expression vector under the control of the T7 promoter and the EMCV leader. Expression of these mutants and analysis of efficiency of cleavage at NS2B/NS3 site showed that a conserved 40-residue hydrophilic region of NS2B flanked by hydrophobic regions (Figs. 2.2 and 2.3) is necessary for the NS3 protease activity (Falgout *et al.*, 1993). A similar conclusion was reached in a study in which the YFV NS2B–NS3$_{181aa}$

```
DEN-1  SWPLNEGIMAVGIVSILLSSLLKNDVP-LAGPLIAGGMLIACYVISGSSADLSLEKAAEV
DEN-3  SWPLNEGVMAVGLVSILASSLLRNDVP-MAGPLVAGGLLIACYVITGTSADLTVEKAADV
DEN-4  SWPLNEGIMAVGLVSLLGSALLKNDVP-LAGPMVAGGLLLAAYVMSGSSADLSLEKAANV
DEN-2  SWPLNEAIMAVGMVSILASSLLKNDIP-MTGPLVAGGLLTVCYVLTGRSADLELERAADV
JEV    GWPATEFLSAVGLMFAIVGGLAELDIESMSIPFMLAGLMAVSYVVSGKATDMWLERAADI
MVEV   GWPATEVLTAVGLMFAIVGGLAELDIDSMSVPFTIAGLMLVSYVISGKATDMWLERAADV
KUN    GWPATEVMTAVGLMFAIVGGLAELDIDSMAIPMTIAGLMFAAFVISGKSTDMWIERTADI
WNV    GWPATEVMTAVGLMFAIVGGLAELDIDSMAIPMTIAGLMFAAFVISGKSTDMWIERTADI
YFV    SIPVNEALAAAGLVGVLAG-LAFQEMENFLGPIAVGGLLMMLVSVAGRVDGLELKKLGEV

DEN-1  SWEEEAEHSGASHNILVEVQDDGTMKIKDEERDDTLTILLKATLLAVSGVYPLSIPATLF
DEN-3  TWEEEAEQTGVSHNLMITVDDDGTMRIKDDETENILTVLLKTALLIVSGIFPYSIPATML
DEN-4  QWDEMADITGSSPIIEVKQDEDGSFSIRDVEETNMITLLVKLALITVSGLYPLAIPVTMT
DEN-2  KWEDQAEISGSSPILSITISEDGSMSIKNEEEEQTLTILIRTGLLVISGLFPVSIPITAA
JEV    SWEMDAAITGSSRRLDVKLDDDGDFHLIDDPGVPWKVWVLRMSCIGLAALTPWAIVPAAF
MVEV   SWEAGAAITGTSERLDVQLDDDGDFHLLNDPGVPWKIWVLRMTCLSVAAITPWAILPSAF
KUN    SWEGDAEITGSSERVDVRLDDDGNFQLMNDPGAPWKIWMLRMACLAISAYTPWAILPSVV
WNV    TWESDAEITGSSERVDVRLDDDGNFQLMNDPGAPWKIWMLRMACLAISAYTPWAILPSVI
YFV    SWEEEAEISGSSARYDVALSEQGEFKLLSEEKVPWDQVVMTSLALVGAALHPFALLLVLA

DEN-1  VWYFWQKK--KQR
DEN-3  VWHTWQKQ--TQR
DEN-4  LWYMWQVK--TQR
DEN-2  AWYLWEVK--KQR
JEV    GYWLTLKT--TKR
MVEV   GYWLTLKY--TKR
KUN    GFWITLQY--TKR
WNV    GFWITLQY--TKR
YFV    GWLFHVRG—ARR
```

Figure 2.2 Alignment of NS2B sequences from mosquito-borne flaviviruses. Gaps are introduced for alignment of conserved residues in DENV-1–4, JEV, MVEV, KUNV, WNV and YFV. The Genbank accession numbers are: DENV-1, ACJ04274.1; DENV-3, AAW51411.2; DENV-4, AAW51421.1; DENV-2, ACB87136.1; JEV, AAD16275.1; MVEV, CAA27184.1; KUN, AAP78941.1; WNV, NP_041724.2; YFV, NP776004.1

precursor containing mutations in the NS2B gene was analysed for efficiency of *cis* cleavage at the NS2B/NS3 site using cell-free translation. The region that is required for NS2B/NS3 cleavage was the same as the conserved domain of 40 aa identified by Falgout *et al.* (1993). In addition, the effects of these deletion mutations of NS2B on cleavages at the 2A/2B, 3/4A, 4A/4B, and 4B/5 sites were also analysed using the YFV sigNS2A–NS5$_{356}$ precursor (Chambers *et al.*, 1993). Mutations within the conserved domain of 40 aa abolished cleavages at these sites, and only the unprocessed precursor was produced. In that study, the interaction between the wild type (WT) NS2B and NS3$_{181}$ was also demonstrated by using co-immunoprecipitation. Under non-denaturing conditions, the anti-NS2B could precipitate WT NS2B but not NS3$_{181\,aa}$ when these two proteins were co-expressed. However, the anti-NS3$_{181\,aa}$ could co-precipitate both NS3$_{181\,aa}$

and NS2B but not the NS2B containing the mutant 40 aa, suggesting that these two proteins interact and the conserved hydrophilic domain of NS2B is important for this interaction. Mutations in NS2B that interfered with this interaction also abolished *trans* cleavages mediated by the protease at 4A/2K and NS4B/5 sites. It was further shown that genomes carrying deletion mutants in each of the three hydrophobic regions as well as in the conserved hydrophilic domain were defective in the production of infectious virus (Chambers *et al.*, 1993). Thus, the active flavivirus protease is a heterodimer of NS2B and NS3 serine protease domain.

The interaction between NS2B and NS3pro was demonstrated independently *in vitro* as well as *in vivo* in mammalian cells for DENV-2 (Arias *et al.*, 1993) and Japanese encephalitis virus (JEV) (Jan *et al.*, 1995). In the latter study, an internal cleavage within NS2A of the precursor, NS2A–

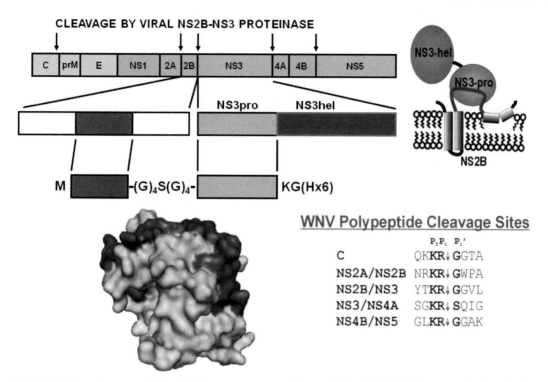

CLEAVAGE BY VIRAL NS2B-NS3 PROTEINASE

WNV Polypeptide Cleavage Sites

	$P_2 P_1$ P_1'
C	QKKR↓GGTA
NS2A/NS2B	NRKR↓GWPA
NS2B/NS3	YTKR↓GGVL
NS3/NS4A	SGKR↓SQIG
NS4B/NS5	GLKR↓GGAK

Figure 2.3 The structure of the WNV precursor polyprotein and the NS2B–NS3 (protease-helicase). The cleavage sites are shown by the arrows. The viral NS2B–NS3 protease-helicase is a membrane-bound protein attached to the membrane through the NS2B cofactor. The recombinant NS2B–NS3 protease construct, which was used for crystallization and atomic resolution structural analysis, included the central, hydrophilic part of NS2B linked to the NS3 protease sequence via a flexible GGGGSGGGG linker. The construct was C-terminally tagged with a Hisx6 tag. The crystal structure has demonstrated that NS2B (dark grey) tightly interacts with the NS3 protease domain (grey).

NS2B–NS3$_{322\,aa}$ was reported which was dependent on NS2B and NS3 protease components (Jan et al., 1995). The precursor, JEV NS2A–dNS2B–NS3$_{322\,aa}$ containing a deletion of NS2B can be cleaved by supplying NS2B or NS2A–NS2B from JEV or DENV-4 NS2B in trans whereas the defective DENV-4 NS2A–dNS2B–NS3$_{322\,aa}$ precursor was not cleaved by JEV NS2B. However, cleavage of the defective DENV-4 precursor occurred if the C-terminal 80- aa of JEV NS2B were substituted with the corresponding region from DENV-4, suggesting that the C-terminal part of NS2B is required for activation of the NS3 protease domain. However, it is not clear why DENV-4 NS2B was able to cleave the JEV NS2A–dNS2B–NS3$_{322\,aa}$ precursor but the JEV NS2B failed to cleave the defective DENV-4 NS2A–dNS2B–NS3$_{322\,aa}$ precursor. Further work is necessary to dissect the mechanism of activation of the NS3 protease domain by NS2B cofactor.

Role of the viral protease in assembly and secretion of mature virions

In the processing of the C-prM junction, there is a consensus viral protease cleavage site upstream from the hydrophobic signal sequence for translocation of prM. The cleavage of the C-prM junction by signalase in the ER is inefficient and delayed until the NS2B/NS3pro cleaves the C-terminal region of the capsid protein. The net positive charge in the protease cleavage site inhibits signalase cleavage and the release of prM (Lobigs, 1993; Stocks and Lobigs, 1998; Yamshchikov and Compans, 1995). In YFV, the viral protease cleavage site at the COOH-terminal region of the C protein contains RKRR↓. Substitution mutations of the basic residues to Ala were made and analysed for the trans cleavage of this site by the NS2B–NS3pro in vitro by transient expression in BHK21 cells using the C-prM precursor, and by infectivity using a full-length clone.

Mutants that were defective in the cleavage at this site by the viral protease were also incapable of producing infectious particles which suggested that the efficient production of prM was dependent on the active viral protease cleavage at this site (Amberg and Rice, 1999). A mutation in the C-terminal region of the YFV prM signal sequence, VPQAQA uncoupled the requirement of the viral protease-mediated cleavage of C protein for efficient cleavage of the C-prM by signalase (Lee *et al.*, 2000). The enhanced signalase cleavage of the VPQAQA mutant in the context of the full-length infectious YFV clone was lethal for infectivity of the virus. Revertants emerged containing single amino acid substitutions in the YFV prM signal sequence. Although the growth properties of the VPQAQA variants remained essentially similar, their neurovirulence was attenuated compared to the WT (Lee *et al.*, 2000). Using MVEV as a model, the importance of the viral protease in the cleavage of the C-prM junction by signalase and the mechanism of nucleocapsid incorporation into budding flavivirus particles were further investigated (Lobigs and Lee, 2004). The VPQAQA mutation in MVEV produced an increased level of subviral particles that failed to incorporate nucleocapsid because of a partial inhibition of viral protease cleavage at the C-terminus of the capsid protein. Growth of the PQAQA mutant in C6/36 cells produced revertants, PQAQV and LQAQA that showed reduced efficiency of signalase cleavage of C-prM junction albeit not to the same extent as in WT virus. The efficient cleavage of the capsid was restored in these mutants. The authors suggested that the inefficient cleavage of C-prM and dependence on the viral protease to release the capsid protein may be regulated in the flavivirus life cycle to prevent the loss of prM from the virion assembly site and the generation of VLPs instead of mature particles (Lobigs and Lee, 2004). In the TBEV capsid protein, there are two viral protease consensus sites. A mutagenesis study using an infectious clone showed that the downstream site is sufficient for nucleocapsid assembly and secretion (Schrauf *et al.*, 2008).

Functional and biochemical characterization of the viral protease

The NS2B cofactor domain of several flaviviruses displays common hydrophobic regions flanking a conserved hydrophilic region which plays an essential role in the interaction of NS2B with the NS3pro domain and the enzymatic activity of the protease. To determine the amino acid residues involved in the *cis* and the *trans* cleavage activity of the protease, single amino acid substitutions in the DENV-2 NS2B–NS3pro were introduced and their effects on the cleavage of the NS2B/NS3 site in the precursor were analysed using transient expression in mammalian cells (Valle and Falgout, 1998). Substitutions of highly conserved residues to A abolished the NS3 protease activity. Two of the five substrate binding residues that are conserved among flavivirus NS3 proteins, specifically Y150 and G153, could not be replaced by A. However, Y150 and the conserved residue, N152, could be replaced by conservative substitutions. Based on a comparison with trypsin, D129 was proposed to be located at the bottom of the substrate binding pocket interacting with the basic amino acid at the substrate cleavage site. D129 and F130 could be readily substituted. A significant level of cleavage in D129E/S/A and a low but detectable level of cleavage in D129K/R/L mutants still occurred. Based on these results the authors concluded that D129 is not a major substrate binding determinant. A number of NS3 mutations and their effects on protease activity are summarized in Table 2.2 (Valle and Falgout, 1998). The C-terminal amino acids, [89]EEEE, in the DENV-2 NS2B cofactor peptide are also important because a sequential deletion of these residues resulted in gradual loss of cleavage at the NS2B–NS3 junction site (Wu *et al.*, 2003).

Mutational analysis of the YFV NS2B and NS3pro domains revealed that the conserved residues [52]ELKK[55] in NS2B (Figs. 2.1 and 2.2) and [21]ED[22] and H47 in NS3 were important for the protease-dependent cleavage. [21]ED[22] mutants of the NS3pro were also important for interaction with NS2B (Droll *et al.*, 2000). [52]ELKK[55] and [21]ED[22] mutations affected NS2A/NS2B and NS2B/NS3 cleavages more than any other protease-sensitive sites in the polyprotein. Deletion of 2, 6, or 11 residues from the N-terminus of NS3 did not affect the *trans* cleavages in the presence of WT NS2B as determined by the production of NS4B. Deletion of 14 and 17 residues had only a minimal effect; deletion of 20 residues of NS3

Table 2.2 Mutational effects of amino acid residues within the two-component flaviviral proteases

Virus	NS2B	NS3pro	Mutational effects	Assay	Reference
DENV-2		D129K/R/L F130A/S G133A T134D S135A/C G136A G148A L149A/R Y150A/V/H G151A N152A/Q G153A/V	Mutations which result in very low or undetectable (underlined) 2B/3 cleavage	Transient expression	Valle and Falgout (1998)
DENV-2		$E^{17}E^{19}D^{20}$ $G^{32}Y^{33}I^{36}$ $K^{63}R^{64}E^{66}$ $E^{179}D^{180}D^{181}$	Residues mutated to A; Protease activity was detected	Transient expression	Matusan et al. (2001)
		$V^{95}Q^{96}$	No cleavage at 2B/3 site		
DENV-2		$G^{32}Y^{33}I^{36}$ $E^{91}E^{93}E^{94}$	Mutations of hydrophobic and charged residues to A were put into infectious clone. The viruses were temperature sensitive (33°C) with low titres	Infectivity assays	Matusan et al. (2001)
YFV	^{45}AGRVD→AGAVA		Revertants in NS2B with intermediate (^{int}) or parental(p) plaque morphology; $A^{47}E^{(int)}$, $A^{45}S^{(p)}$,$A^{49}S^{(int)}$	Infectivity assays	Chambers et al. (2005)
	^{52}ELKK→ALAK →ELAK		Revertant in NS2B-$A^{54}V^{(int)}$ Revertant in NS3: $R^{62}W^{(p)}$		
	^{122}LFHVR→LFAVA		Revertant in NS2B: $L^{122}R^{(p)}$		
		^{20}LEDG→LADG →LEAG	Revertant in NS3:$A^{21}E^{(p)}$ Revertant in NS3:$A^{22}D^{(p)}$		

(including ^{18}EH^{19}), however, caused a moderate decrease while a 22 residue deletion (including ^{21}ED^{22}) abolished the *trans* cleavage activity. Thus, the clusters of amino acid residues ^{18}EH^{19} and ^{21}ED^{22} are important for the NS2B interaction with NS3 and protease activity (Droll *et al.*, 2000). In a later study, the same group analysed 46 charged amino acid to alanine mutations in YFV NS2B and NS3pro by infectivity assays. Eight mutants exhibited a small plaque or a no-plaque phenotype. Revertants arose in six mutants in which same site or second site mutations occurred in NS2B except one that was observed in the NS3pro domain. These results reveal the

importance of these charged amino acids in the NS2B/NS3pro functional interactions.

The importance of the charged amino acid clusters (E17A, E19A, D20A, K63A, R64A, E66A, E91A, E93A, E94A, E169A, K170A, E173A, E179A, D180A, D181A) as well as two hydrophobic regions (G32A, Y33A and V95A, Q96A) of DENV-2 NS2B and NS3pro was also analysed by alanine mutagenesis and transient expression of the mutants in COS cells followed by metabolic labelling and immunoprecipitation (Matusan *et al.*, 2001). Five mutants were also analysed in the context of full-length genome. V95A Q96A resulted in no cleavage at the NS2B/

NS3 site. Viruses containing mutations of residues 32–36 and 91–94 were temperature-sensitive and were recovered (at 33°C) in low titres (Matusan *et al.*, 2001) (Table 2.2).

Establishment of *in vitro* protease assays using purified NS2B and NS3pro components

Several studies involving flavivirus proteases included the N-terminal portion of NS3 protein containing ~180 amino acids (Chambers *et al.*, 1991; Preugschat *et al.*, 1990; Wengler *et al.*, 1991). The region C- terminal to the protease domain has a conserved domain found in the DEXH family of RNA helicases. To define the boundaries of these two distinct functional domains of the DENV-2 NS3 protein, the NS2B and the NS2B/NS3 cleavage site, linked to the amino-terminal portion of the NS3 protein containing 183-, 176-, and 170- to 164- aa each differing by a single amino acid residue, were expressed *in vitro* using the coupled transcription/translation system in the presence of [^{35}S]-labelled methionine (Li *et al.*, 1999). This study established that the minimal NS3pro domain required for NS2B/NS3 cleavage consists of the first 167- aa of the NS3 protease domain although NS3pro domain of 183 aa has optimal protease activity (Li *et al.*, 1999).

With the aim of establishing an *in vitro* protease assay for DENV-2 protease, NS2B and NS3pro components were co-expressed using a recombinant vaccinia virus system in BHK-21 cells. The protease activity involved in the cleavage of NS4B–NS5$_{trunc}$ precursor protein was localized in the microsomal membrane fraction of BHK cells (Clum *et al.*, 1997). To study the role of the membranes in the NS2B/NS3 *cis* cleavage, the expression of the NS2B–NS3pro precursor was examined using the coupled *in vitro* transcription/translation system followed by SDS-PAGE and autoradiography in the presence and absence of canine pancreatic microsomal membranes. The results indicated that a co-translational insertion of NS2B into microsomal membranes augmented *cis* cleavage of the NS2B/NS3 site in the NS2B–NS3pro as well as the trans cleavage of a deletion mutant of the NS2B essential domain, ndNS2BH-NS3pro. The products, NS2B and NS3pro, were membrane-associated and NS2B

was shown to be an integral membrane protein in the ER (Clum *et al.*, 1997). This membrane requirement was eliminated by the deletion of the hydrophobic regions (see Fig. 2.3) flanking the highly conserved ~45-amino acid hydrophilic domain of NS2B (Clum *et al.*, 1997). This region of NS2B has previously been shown to be required for the interaction with the NS3pro domain and for the protease activity (Chambers *et al.*, 1993; Falgout *et al.*, 1993). In contrast to the co-translational membrane requirement for the efficient processing of the DENV-2 NS2B/NS3 site, the processing of the YFV NS2B/NS3 site was not enhanced by the addition of the membranes to the coupled transcription/translation system (Chambers *et al.*, 2005). Although the hydrophobic regions are dispensable for protease activity *in vitro*, they are essential for infectivity in the context of full-length genome (Chambers *et al.*, 2005) suggesting that the hydrophobic region plays an essential role in localizing the viral protease in a cytoplasmic membrane compartment.

A removal of the hydrophobic regions from the NS2B–NS3pro precursor allowed the expression of the precursor (containing 13 amino acids from the C- terminus of NS2B as a spacer between the NS2B hydrophilic domain and the NS3pro domain) in *E. coli*. The precursor, NS2BH-NS3pro with a His-tag at the N-terminus of NS2B was expressed in *E. coli* in the form of insoluble inclusion bodies. Purification by metal affinity column chromatography, denaturation and refolding yielded a protease in which the NS2B/NS3 site was cleaved and the NS2B and NS3pro components were non-covalently associated in a complex that was active in cleaving the NS4B–NS5$_{trunc}$ substrate *in trans* (Yusof *et al.*, 2000) whereas the NS3pro domain alone with a N-terminal His-tag expressed and purified from *E. coli* in a similar manner was inactive. Using the *E. coli*-expressed and purified DENV-2 protease, an *in vitro* protease assay with a fluorogenic peptide substrate containing two basic amino acid residues at P1 and P2 and the 4-methylcoumaryl-7-amide (MCA) at the P1′ positions was established. Among the four substrates, the optimum substrate, Boc-Gly-Arg-Arg-MCA had a k_{cat}/K_m (M^{-1} s^{-1}) of 172 with the NS2BH/NS3pro complex which was ~3300-fold more active than the NS3pro domain alone. For the expression of the WNV protease in *E. coli*, a similar strategy was used except that the length of the spacer was 5 residues from the C-terminus of NS2B. With this change, the precursor underwent *cis* cleavage even before purification and the protease components were purified as a non-covalently associated active NS2BH/NS3pro complex (Mueller *et al.*, 2007).

Leung *et al.* (Leung *et al.*, 2001) reported expression and purification of DENV-2 NS2B cofactor domain linked covalently to the NS3pro domain with elimination of the NS2B/NS3 site (CF40.NS3pro) or insertion of a –G4–S–G4– linker (CF40glyNS3pro). The authors used the chromogenic hexapeptide substrates spanning P6-P1 of the DENV-2 polyprotein cleavage sites (NS2A/NS2B, NS2B/NS3, NS3/NS4A and NS4B/NS5) with the *p*-nitroanilide moiety at the P1′ position for *in vitro* protease assays. Although the Km values of the two non-cleavable forms of the proteases did not change significantly, the k_{cat}/K_m values for the CF40glyNS3pro were better by 2-fold (NS3/NS4A), ~3-fold (NS2B/NS3 and NS4B/NS5) and 4.7-fold (NS2A/NS2B) compared with the CF40.NS3pro (Leung *et al.*, 2001). This strategy for expression of flavivirus protease was followed in most of the studies done subsequent to Leung *et al.* (Leung *et al.*, 2001) including the elucidation of the crystal structures of the DENV-2 and WNV proteases (Erbel *et al.*, 2006). The classical serine protease inhibitor, bovine pancreatic trypsin inhibitor (BPTI or aprotinin) was found to be a potent inhibitor of DENV-2 (Aleshin *et al.*, 2007; Leung *et al.*, 2001; Mueller *et al.*, 2007).

The NS2BH–NS3pro of all four DENV serotypes were expressed in *E. coli* as the non-cleavable forms (CF40glyNS3pro$_{185 aa}$) and the differences in the substrate specificity at the P4-P1 positions were determined using positional scanning of tetrapeptide libraries with two fixed positions containing the fluorogenic group (7-amino-4-carbamoylmethylcoumarin; ACMC) at the P1′ position. The k_{cat}/K_m values of the DENV-1 protease for the BZ-nKRR-ACMC (n=norleucine) versus the tripeptide substrate (Boc-Gly-Arg-Arg-MCA) used in an earlier study (Yusof *et al.*, 2000) were 51,800 and 290 $M^{-1} s^{-1}$, respectively (a 150-fold

improvement in the catalytic efficiency). The substrate specificity of the protease revealed a strong preference for P1: R/K, P2: R>T>G>N>K, P3: K>R>N, and P4: n>L>K>X (n= norleucine; X=any amino acid). The prime site substrate specificity was for a small and polar residue at the P1′ and the P3′ sites whereas the P2′ and the P4′ sites showed no preferences. The intramolecular cleavage at NS2A/NS2B preceded that at the NS2B/NS3 site. These results reveal that the natural DENV polyprotein cleavages have suboptimal sequences, especially at the NS2B/NS3 site, with a Q at the P2 position. Moreover, the cleavage efficiency at the four sites differs so that sufficient amounts of the processing intermediates with a potentially distinct function in the viral replication and assembly could accumulate. It is highly probable that natural cleavage sites are not necessarily occupied by optimal residues and that this parameter has a high physiological significance. The differential rates of cleavage at the four major cleavage sites may guide an ordered processing of DENV polypeptide, yield sufficient intermediates with desired function, and synchronize viral processing, replication and assembly. Little information is available regarding identification and the role of processing intermediates in flavivirus life cycle and future studies in this direction are needed.

WNV CF40glyNS3pro was also expressed in a similar manner (Leung *et al.*, 2001) and the biochemical properties of the protease were studied (Nall *et al.*, 2004). The WNV protease was 2-fold more active against substrates with K rather than R at the P2 position; in contrast, the DENV protease prefers R over K at the P2 position (Chappell *et al.*, 2006; Mueller *et al.*, 2007; Shiryaev *et al.*, 2007b). N-84 is within H-bonding distance from the P2-K. This residue is either polar or acidic amino acid (D or E) and it interacts with the P2-R or K residue or Q in the NS2B/NS3 site of DENV of the substrate (T or S in NS2B of DENV-1–4, E in YFV, D in JEV/MVE or N in WNV/SLEV). The S3 pocket is not well-defined in the molecular model and is largely solvent-exposed. A range of residues at the P3 position (R, G, T, L) occupy the shallow groove of the S3 pocket extending towards the S1; P4–S4 interactions are predominantly hydrophobic;

V154 and M156 of NS3 and L87 of NS2B participate in this interaction. Chappell et al. (Chappell et al., 2006) analysed 12 mutants each of the NS2B and NS3pro domains (NS2B: V75A/F, N84A/D/E/L/S, Q86A/E/L, L87A/F; NS3pro: T111F/L, D129A/E, V154F/L, I155F, M156A, I162F, A164S/V and V166L). Q86 in WNV is either polar or basic residue in other viruses (E, S, K, R, and H) and makes an H-bond with the O atom of the P4 backbone carbonyl. NS2B V75 is conserved and interacts with the P5 or P6 residue of the substrate. NS3 T111 is within H-bonding distance from NS2B T69 and contributes to the cofactor binding. I162 is part of the hydrophobic S4 pocket. A164 and V166 are proposed to contribute to the folding of the protease and hence A164S/V and V166L had only minor effects.

The recombinant NS2BglyNS3pro protein was sensitive to autoproteolysis (Chappell et al., 2007; Shiryaev et al., 2007a). In fact, the crystal structure of WNV NS2BglyNS3pro shows that the N-terminus of NS3pro is short by 18 amino acids (Erbel et al., 2006) providing evidence for autoproteolysis that occurred during crystallization. Bera et al. (2007) expressed WNV $NS2B_{40}G_4SG_4HM.NS3_{FL}$ (full length) in E. coli and identified two autoproteolytic cleavages catalysed by the protease. Cleavage at the NS2B/NS3 site occurred rapidly at $G4SG4HM\downarrow GG ... (NS3_{FL})$ before the purification of the protease and more slowly at the second site at $R^{459}\downarrow G^{460}$ (Bera et al., 2007). In that study, to distinguish an intramolecular (cis) from trans cleavages, a dodecapeptide spanning the NS2B/NS3 cleavage site fused to CFP and YFP at the N- and C-termini was used as a substrate for the protease (CFP-LQYTKR↓GGVLWD-YFP). Both autoproteolytic cleavages were determined to be intramolecular and dilution insensitive (Bera et al., 2007). Recombinant proteases of the type $G4SG4HM\downarrow GG ... (NS3_{FL})$ from other flaviviruses, including YFV, DENV-2, DENV-4 and JEV, all underwent autoproteolysis at NS2B/NS3 as well as at an internal site within NS3 (Bera et al., 2007). The fragments of proteolysis were associated as shown by metal affinity and gel filtration chromatography. In another study, the autocleavage of the WNV recombinant protease occurred at the $G^{49}\downarrow G^{50}$ site in which the NS2B was fused through its C-terminal sequence of PWK[48]-G4SG4 linker to the NS3pro domain (Shiryaev et al., 2007a). K48A mutation produced a stable protease in a non-cleavable form (Shiryaev et al., 2007a). Chappell et al. (Chappell et al., 2007) also identified autoproteolytic sites with their recombinant WNV protease containing a G4SG4 linker. An improved and stable expression was achieved in which four non-essential amino acids from the vicinity of the G4SG4 linker were deleted ($NS2B_{44}G_4SG_3NS3pro$). A full-length $NS2B_{44}G_4SG_3NS3$ was also constructed and optimized for the expression of a stable protein (Chappell et al., 2007). Future refinement of stable expression of full-length NS2B–NS3 in a suitable expression vector should prove useful for structural and functional studies.

In contrast to autoproteolysis of the recombinant NS2BglyNS3pro protein, Melino et al. (2006) used a limited proteolysis approach using Asp-N endoproteinase to identify the minimal domain necessary for protease activity. The biochemical and biophysical characterization of the minimal essential domain of DENV-2 protease (CF40glyNS3pro) included fluorescence resonance energy transfer (FRET), CD and NMR spectroscopy. The FRET assay utilized a depsipeptide, 2-ABZ-Gly-Arg-Arg-φ[COO] Ala-Tyr (3-NO2)-Asp-OH which alone was inert to Asp-N endoproteinase. However, the Asp N-endoproteinase digested the CF40glyNS3pro to a limit fragment of 20 kDa that was stable up to 24 h. The residual protease activity remained as 45% after an overnight incubation at 25°C. The N-terminal sequence of the stable proteolysis product having 60% activity was NH_2-DVPSPPPMG (CFNS3d). This fragment resolved into two peaks in RP-HPLC (Pa and Pb) which corresponded to NH_2-GS(H)$_6$GSADL ... E80 of NS2B and D6-E179(NS3pro), respectively. The CD and NMR data revealed the presence of a flexible region of the NS2B. The D50-E80 fragment of NS2B remained bound to the D6-E179 fragment of NS3pro during gel-filtration chromatography. NS3pro domain (D6-E179) has a stable conformation as the only cleavage occurred at E180. D50, D58, and D63 of NS2B are protected from cleavage suggesting less solvent exposure and more tight binding to the NS3pro domain whereas D81-L96 does not interact tightly with the NS3pro (Melino et al., 2006).

Another study elucidated the role of φX3φ motif in NS2B in cofactor binding to the NS3pro domain (Niyomrattanakit et al., 2004). Mutation of the conserved W residue of NS2B resulted in an inactive protease (Niyomrattanakit et al., 2004) and this residue is in the part of NS2B protected from the Asp-N endoproteinase digestion.

The substrate specificity of the DENV-2 protease was also examined using internally quenched fluorescent peptides containing Abz (o-aminobenzoic acid) and 3-nitrotyrosine (nY). The peptide, Abz-RRRRSAGnY-amide (P4-P3') representing the viral protease cleavage site within the capsid protein was efficiently cleaved by the viral protease with kcat/km of $11.087\,mM^{-1}s^{-1}$. The substrate peptide Abz-KKQRAGVLnY-amide representing the DENV-2 NS2B/NS3 site also exhibited a high catalytic efficiency (Niyomrattanakit et al., 2006). Recently, Gouvea et al. (2007) characterized the NGC strain of DENV-2 NS2BglyNS3pro peptide substrates and compared the properties to that of a DENV-2 isolate from a Brazilian dengue fever case and to the human furin enzyme using (FRET). The FRET peptides had Abz and EDDnp [N-(2,4-dinitrophenyl)-ethylene diamine] at the N- and C-terminal ends containing the polyprotein cleavage site sequences including the viral protease-sensitive internal cleavage sites of capsid, NS3 and NS4A proteins (C_{int}, $NS3_{int}$ and $NS4A_{int}$). The cleavage occurred after the last C-terminal basic residue when a consecutive run of basic amino acids was present in the cleavage site (i.e. RRRR in the C_{int} site). This result also suggests that basic residues at the P1' position are not tolerated. The catalytic efficiency of the protease from the NGC strain was 2-fold higher than that of the Brazilian isolate. The best substrate with the highest k_{cat}/K_m value was Abz-AKRRSQ-EDDnp ($31.8\,mM^{-1}s^{-1}$). The FRET peptide substrates were also sensitive to human furin and the k_{cat}/K_m values were higher than those of DENV-2 protease: RTSKKR↓SWPLNEQ (2A/2B);~6-fold), LNRRRR↓TAGMIIQ (C_{int}; 68-fold), SAAQRR↓GRIGRNQ ($NS3_{int}$; 6-fold), and HRREKR↓SVALQ (prM/M; 3000-fold). Thus, rapid and quantitative methods have been developed for in vitro protease assays to determine the substrate specificity and identify inhibitors using high throughput methods.

Structural confirmation of the essential role of the NS2B cofactor

In general, the structure of the two-component NS2B–NS3 protease from West Nile and dengue viruses, which have been recently solved, is an excellent model for flaviviral protease activation, with important implications for a structure-based inhibitor design for this entire class of flaviviruses (Aleshin et al., 2007; Erbel et al., 2006).

The WNV and DENV NS2B–NS3 proteases has a chymotrypsin-like fold with the active site located at the interface of the N- and C-terminal lobes. The C-terminal part of the NS2B fragment (residues 64–96 in WNV) forms a belt that wraps around NS3pro, ending in a β-hairpin (β2–β3) that augments the upper β-barrel and inserts its tip directly into the protease active site. The importance of this interaction is clearly shown by the deleterious effects of the mutation of NS2B residues L-75 and I-79 (conserved hydrophobic residues in flaviviruses) in DENV-2 (Niyomrattanakit et al., 2004), which lie on the inner surface of the invading cofactor β-hairpin and anchor it to a hydrophobic cleft on NS3. It is clear from the structural data that the presence of the NS2B cofactor is essential for the functional activity of the flaviviral NS3 protease catalytic domain.

The organization of the flaviviral NS3 protease active site

The active site of chymotrypsin-like proteases includes three conserved elements: (1) a classic H–D–S catalytic triad (H51, D75, and S135 in WNV), whose precise arrangement in space is required to enhance the nucleophilicity of the serine hydroxyl group; (2) the 'oxyanion hole', which stabilizes the developing negative charge on the scissile peptide carbonyl oxygen in the transition state; and (3) the substrate binding β-strands E2 and B1, which help to position the substrate in the active site. In the WNV NS2B–NS3 protease-aprotinin complex, the catalytic triad adopts a productive geometry that is virtually identical to that observed in the trypsin-aprotinin complex (PDB 2FTL) (Pasternak et al., 1999). Thus, the S135 OH–H51 Nε2 distance (2.8 Å) and the H51 Nδ1–D75 CO^{2-} distance (2.7 Å) are indicative of strong H-bonds, as is the distance (2.8 Å) between the S135 hydroxyl oxygen and the (nominally) scissile peptide carbonyl of aprotinin. The main chain of aprotinin residues 13–19 (PCKARII) forms antiparallel β-sheet interactions with strands E2B and B1 of NS2B–NS3, while the side chains occupy the presumed subsites S3–S1 and S19–S49, exactly as observed in the trypsin–aprotinin complex. The main-chain conformations of residues occupying the S3 and S2 sites (P3 and P2) are closely superimposable with those of the peptidic inhibitor-bound DENV protease.

Substrate-induced fit in the oxyanion hole of NS2B–NS3pro

Intriguingly, aprotinin binding induces a catalytically competent conformation of the 'oxyanion hole' in the WNV NS3 protease. This hole, lined by main-chain nitrogens (from G133 to S135 in WNV) is improperly formed in the substrate-free and peptidic inhibitor-bound structures; in those structures the peptide bond between T132 and G133 is flipped. The flipped bond creates a helical (3_{10}) conformation for residues 131–135 that is stabilized by two hydrogen bonds, which are absent in the productive conformation. It is possible that the non-productive conformation of the oxyanion hole is energetically favoured in the absence of a substrate, acquiring the productive conformation only in the presence of a substrate with an appropriate P1' residue. These structural data suggest that an 'induced fit' mechanism (Koshland, 1958) contributes to substrate specificity or enzyme turnover or both in the NS2B–NS3 protease.

The structural rationale for the substrate specificity of the two-component NS2B–NS3 protease

The aprotinin-bound WNV structure (Aleshin et al., 2007) mimics a classic Michaelis–Menten complex, allowing for an authentic view of the enzyme–substrate pre-cleavage complex. WNV and related flaviviral proteases have a requirement for basic residues at both the P1 and P2 positions of their substrates and G, S, or T at the P1' position. In the aprotinin-bound NS2B–NS3 protease structure, the P2 side chain (R/K) interacts directly with N84 of NS2B and indirectly with a negatively charged surface created by the invading hairpin of NS2B, including D80 and D82,

as previously proposed (Otlewski et al., 2001). NS2B binding also helps to define the S1 pocket in the WNV protease by changing the conformation of NS3 residues 116–132, such that D129 is introduced into the base of the pocket, where it can salt-bridge with the P1 R/K. Accordingly, in WNV or DENV proteases, mutation of D-129 to E or A abrogates catalysis (Chappell et al., 2005). In trypsin, D189 plays an analogous role (Hedstrom, 2002; Perona and Craik, 1995), consistent with its similar P1 specificity. The in vitro protease activity assays, in which E. coli-derived D129E/S/A mutants were used, do not completely agree with the results obtained in cell-based assays (Valle and Falgout, 1998). The pocket is well formed but only large enough to accommodate small P1' side chains such as the consensus residues G, S, or T. S and T have the potential to H-bond to the main-chain C=O of the adjacent A36, rationalizing the P1' preference for these residues, while glycine leaves enough space for a water molecule. The inability of the A side chain to interact with the H-bond could explain its rare appearance in flavivirus polyprotein cleavage sequences. In trypsin, no such pocket is formed because a disulphide bridge occupies this site and, as a result, trypsin has no specificity at the P1' position. It appears that the additional H-bond in the WNV protease as compared to the DENV protease places stringent restraints on the substrate dihedral angles, allowing for limited 'wiggle room' for the P1' side chain, which is positioned close to the catalytic His; this fact could explain the preference for glycine.; see also Functional and biochemical characterization of the viral protease). These differences appear to be related to the incomplete folding of the protease mutants in E. coli (Valle and Falgout, 1998). The S1' pocket in NS2B–NS3pro comprises a cavity between strand B1 and the helical turn (residues 50–53) following strand C1 and is lined on one side by the catalytic H residue and on the other by an invariant G residue (G37 in WNV

Furthermore, T132 in WNV NS3pro forms a hydrogen bond with the peptide bond involving the P1' residue. This bond is unique for WNV because in the DENV NS3 polypeptide chain the P132 residue which occupies this position is incapable of making a similar hydrogen bond. Accordingly, the hydrogen bond involving T132 stabilizes the backbone conformations of the P1'/P2' residues, allowing their side chains to make tight contacts with H51 and T132 of WNV NS3pro. These parameters limit the mobility of the P2' residues, thus leading to the preferred G at the P2' position of WNV protease.

Alternating conformations of NS2B–NS3 protease

Both the inhibitor (aprotinin)-bound structure of the catalytically potent NS2B–NS3 protease and the inhibitor-free structure of the inert NS2B–NS3 protease (bearing a mutation of the essential H residue of the catalytic triad) have been solved recently (Aleshin et al., 2007). The major difference between the inhibitor-bound and inhibitor-free WNV NS2A–NS3 protease structures is in the conformation of the NS2B cofactor. In the inhibitor-free structure, the β1 strand of NS2B is formed as in the aprotinin complex, augmenting the β-barrel of the N-terminal lobe of the NS3 protease moiety. Beyond strand β1, however, the last NS2B residue that adopts the same conformation as in the aprotinin complex is W62, which is buried in a pocket on the C-terminal lobe of NS3. WNV W62 (or DENV W61) appears to act as an 'anchor' for the NS2B cofactor: accordingly, the mutation of either W61 (Niyomrattanakit et al., 2004) or of the NS3 residue, Q96 (Matusan et al., 2001), which lines the base of the W62 acceptor pocket, does not permit catalytic activity. Following W62, the NS2B chain adopts a new conformation in an inhibitor-free structure: a new turn-and-a-half of helix ('α1') is followed by an abrupt reversal of chain direction at G70 (conserved as G or A in flaviviruses) followed by a four-residue β-strand (called β1' in the original publication (Aleshin et al., 2007) that augments the central β-sheet of the NS3 C-terminal domain by making main-chain interactions with β-strand B2A. The chain then continues in the reverse direction, towards the N-terminus of the NS2B sequence, i.e. opposite from that in the inhibitor-bound structure. Despite these major changes in the interaction with NS2B, the NS3 coordinates are very close in all of the solved flavivirus protease structures. It is now clear that both the WNV and DENV proteases can adopt two distinct cofactor-bound conformations, thus raising the strong possibility that the two conformations are a conserved feature of all flaviviruses.

Future perspectives

Currently, there are multiple peptide-based approaches aiming to elucidate the mechanisms involved in flaviviral polyprotein processing in a wide spectrum of host cells. The structural studies of the NS2B–NS3pro have laid a foundation for structure-based drug design. This knowledge will be essential for the discovery and optimization of flaviviral inhibitors targeted to the initial steps of the viral multiplication cycle. Further studies are required to establish whether selective or wide-range inhibitors of viral proteases or both are required for anti-flaviviral therapy of the individual infection types. The interactions of the NS2B cofactor with the NS3pro domain represent a unique target for drug design because interfering with these interactions is likely to lead to the novel allosteric inhibitors of viral proteases which, as opposed to the active site-targeting inhibitors, will not cross-react with the host serine proteases.

Conclusions

Since the discovery that the flavivirus NS3 protein includes a trypsin-like serine protease catalytic triad, excellent progress has been made in understanding the mode of action in cis and in trans cleavages, substrate specificity, and the mechanism of activation of the protease domain by the viral cofactor protein. Moreover, solving the crystal structures of several flavivirus proteases could lead to the successful development of antiviral therapeutics in the near future. Still more information remains to be uncovered regarding the regulation and kinetics of the viral protease in the infected host, host cell responses to protease expression, and compartmentalization of the polyprotein processing and viral replication.

Acknowledgements

The work in authors' laboratories was supported by grants from National Institutes of Health AI 070791 and AI 577045 to R.P., and AI061139, AI055789, RR020843 and MH077601 to A.Y.S. The authors sincerely apologize for the omission of a number of other important citations in this chapter because of strict space constraints.

References

Aleshin, A.E., Shiryaev, S.A., Strongin, A.Y., and Liddington, R.C. (2007). Structural evidence for regulation and specificity of flaviviral proteases and evolution of the Flaviviridae fold. Protein Sci. *16*, 795–806.

Amberg, S.M., Nestorowicz, A., McCourt, D.W., and Rice, C.M. (1994). NS2B-3 proteinase-mediated processing in the yellow fever virus structural region: *in vitro* and *in vivo* studies. J. Virol. *68*, 3794–3802.

Amberg, S.M., and Rice, C.M. (1999). Mutagenesis of the NS2B–NS3-mediated cleavage site in the flavivirus capsid protein demonstrates a requirement for coordinated processing. J. Virol. *73*, 8083–8094.

Arias, C.F., Preugschat, F., and Strauss, J.H. (1993). Dengue 2 virus NS2B and NS3 form a stable complex that can cleave NS3 within the helicase domain. Virology *193*, 888–899.

Bazan, J.F., and Fletterick, R.J. (1989). Detection of a trypsin-like serine protease domain in flaviviruses and pestiviruses. Virology *171*, 637–639.

Bell, J.R., Kinney, R.M., Trent, D.W., Lenches, E.M., Dalgarno, L., and Strauss, J.H. (1985). Amino-terminal amino acid sequences of structural proteins of three flaviviruses. Virology *143*, 224–229.

Bera, A.K., Kuhn, R.J., and Smith, J.L. (2007). Functional characterization of cis and trans activity of the Flavivirus NS2B–NS3 protease. J. Biol. Chem. *282*, 12883–12892.

Biedrzycka, A., Cauchi, M.R., Bartholomeusz, A., Gorman, J.J., and Wright, P.J. (1987). Characterization of protease cleavage sites involved in the formation of the envelope glycoprotein and three non-structural proteins of dengue virus type 2, New Guinea C strain. J. Gen. Virol. *68*, 1317–1326.

Cahour, A., Falgout, B., and Lai, C.J. (1992). Cleavage of the dengue virus polyprotein at the NS3/NS4A and NS4B/NS5 junctions is mediated by viral protease NS2B–NS3, whereas NS4A/NS4B may be processed by a cellular protease. J. Virol. *66*, 1535–1542.

Castle, E., Leidner, U., Nowak, T., and Wengler, G. (1986). Primary structure of the West Nile flavivirus genome region coding for all nonstructural proteins. Virology *149*, 10–26.

Castle, E., Nowak, T., Leidner, U., and Wengler, G. (1985). Sequence analysis of the viral core protein and the membrane-associated proteins V1 and NV2 of the flavivirus West Nile virus and of the genome sequence for these proteins. Virology *145*, 227–236.

Chambers, T.J., Droll, D.A., Tang, Y., Liang, Y., Ganesh, V.K., Murthy, K.H., and Nickells, M. (2005). Yellow fever virus NS2B–NS3 protease: characterization of charged-to-alanine mutant and revertant viruses and analysis of polyprotein-cleavage activities. J. Gen. Virol. *86*, 1403–1413.

Chambers, T.J., Grakoui, A., and Rice, C.M. (1991). Processing of the yellow fever virus nonstructural polyprotein: a catalytically active NS3 proteinase domain and NS2B are required for cleavages at dibasic sites. J. Virol. *65*, 6042–6050.

Chambers, T.J., Hahn, C.S., Galler, R., and Rice, C.M. (1990a). Flavivirus genome organization, expression, and replication. Annu. Rev. Microbiol. *44*, 649–688.

Chambers, T.J., McCourt, D.W., and Rice, C.M. (1990b). Production of yellow fever virus proteins in infected cells: identification of discrete polyprotein species and analysis of cleavage kinetics using region-specific polyclonal antisera. Virology *177*, 159–174

Chambers, T.J., Nestorowicz, A., Amberg, S.M., and Rice, C.M. (1993). Mutagenesis of the yellow fever virus NS2B protein: effects on proteolytic processing, NS2B–NS3 complex formation, and viral replication. J. Virol. *67*, 6797–6807.

Chambers, T.J., Weir, R.C., Grakoui, A., McCourt, D.W., Bazan, J.F., Fletterick, R.J., and Rice, C.M. (1990c). Evidence that the N-terminal domain of nonstructural protein NS3 from yellow fever virus is a serine protease responsible for site-specific cleavages in the viral polyprotein. Proc. Natl. Acad. Sci. U.S.A. *87*, 8898–8902.

Chappell, K.J., Nall, T.A., Stoermer, M.J., Fang, N.X., Tyndall, J.D., Fairlie, D.P., and Young, P.R. (2005). Site-directed mutagenesis and kinetic studies of the West Nile Virus NS3 protease identify key enzyme–substrate interactions. J. Biol. Chem. *280*, 2896–2903.

Chappell, K.J., Stoermer, M.J., Fairlie, D.P., and Young, P.R. (2006). Insights to substrate binding and processing by West Nile Virus NS3 protease through combined modeling, protease mutagenesis, and kinetic studies. J. Biol. Chem. *281*, 38448–38458.

Chappell, K.J., Stoermer, M.J., Fairlie, D.P., and Young, P.R. (2007). Generation and characterization of proteolytically active and highly stable truncated and full-length recombinant West Nile virus NS3. Protein Expr. Purif. *53*, 87–96.

Cleaves, G.R. (1985). Identification of dengue type 2 virus-specific high molecular weight proteins in virus-infected BHK cells. J. Gen. Virol. *66*, 2767–2771.

Cleaves, G.R., and Dubin, D.T. (1979). Methylation status of intracellular dengue type 2 40 S RNA Virology *96*, 159–165.

Clum, S., Ebner, K.E., and Padmanabhan, R. (1997). Cotranslational membrane insertion of the serine proteinase precursor NS2B–NS3(Pro) of dengue virus type 2 is required for efficient *in vitro* processing and is mediated through the hydrophobic regions of NS2B. J. Biol. Chem. *272*, 30715–30723.

Clyde, K., Barrera, J., and Harris, E. (2008). The capsid-coding region hairpin element (cHP) is a critical determinant of dengue virus and West Nile virus RNA synthesis. Virology *379*, 314–323.

Clyde, K., and Harris, E. (2006). RNA secondary structure in the coding region of dengue virus type 2 directs translation start codon selection and is required for viral replication. J. Virol. *80*, 2170–2182.

Coia, G., Parker, M.D., Speight, G., Byrne, M.E., and Westaway, E.G. (1988). Nucleotide and complete amino acid sequences of Kunjin virus: definitive gene order and characteristics of the virus-specified proteins. J. Gen. Virol. *69*, 1–21.

Crawford, G.R., and Wright, P.J. (1987). Characterization of novel viral polyproteins detected in cells infected by the flavivirus Kunjin and radiolabelled in the presence of the leucine analogue hydroxyleucine. J. Gen. Virol. *68*, 365–376.

Deubel, V., Kinney, R.M., and Trent, D.W. (1988). Nucleotide sequence and deduced amino acid sequence of the nonstructural proteins of dengue type 2 virus, Jamaica genotype: comparative analysis of the full-length genome. Virology 165, 234–244.

Droll, D.A., Krishna Murthy, H.M., and Chambers, T.J. (2000). Yellow fever virus NS2B–NS3 protease: charged-to-alanine mutagenesis and deletion analysis define regions important for protease complex formation and function. Virology 275, 335–347.

Edgil, D., and Harris, E. (2006). End-to-end communication in the modulation of translation by mammalian RNA viruses. Virus Res. 119, 43–51.

Edgil, D., Polacek, C., and Harris, E. (2006). Dengue virus utilizes a novel strategy for translation initiation when cap-dependent translation is inhibited. J. Virol. 80, 2976–2986.

Erbel, P., Schiering, N., D'Arcy, A., Renatus, M., Kroemer, M., Lim, S.P., Yin, Z., Keller, T.H., Vasudevan, S.G., and Hommel, U. (2006). Structural basis for the activation of flaviviral NS3 proteases from dengue and West Nile virus. Nat. Struct. Mol. Biol. 13, 372–373.

Falgout, B., Chanock, R., and Lai, C.J. (1989). Proper processing of dengue virus nonstructural glycoprotein NS1 requires the N-terminal hydrophobic signal sequence and the downstream nonstructural protein NS2a. J. Virol. 63, 1852–1860

Falgout, B., and Markoff, L. (1995). Evidence that flavivirus NS1–NS2A cleavage is mediated by a membrane-bound host protease in the endoplasmic reticulum. J. Virol. 69, 7232–7243.

Falgout, B., Miller, R.H., and Lai, C.-J. (1993). Deletion analysis of dengue virus type 4 nonstructural protein NS2B: Identification of a domain required for NS2B–NS3 protease activity. J. Virol. 67, 2034–2042.

Falgout, B., Pethel, M., Zhang, Y.M., and Lai, C.J. (1991). Both nonstructural proteins NS2B and NS3 are required for the proteolytic processing of dengue virus nonstructural proteins. J. Virol. 65, 2467–2475.

Gorbalenya, A.E., Donchenko, A.P., Koonin, E.V., and Blinov, V.M. (1989). N-terminal domains of putative helicases of flavi- and pestiviruses may be serine proteases. Nucleic Acids Res. 17, 3889–3897.

Gouvea, I.E., Izidoro, M.A., Judice, W.A., Cezari, M.H., Caliendo, G., Santagada, V., dos Santos, C.N., Queiroz, M.H., Juliano, M.A., Young, P.R. Fairlie, D.P., and Juliano, L. (2007). Substrate specificity of recombinant dengue 2 virus NS2B–NS3 protease: influence of natural and unnatural basic amino acids on hydrolysis of synthetic fluorescent substrates. Arch. Biochem. Biophys. 457, 187–196.

Hahn, Y.S., Galler, R., Hunkapiller, T., Dalrymple, J.M., Strauss, J.H., and Strauss, E.G. (1988). Nucleotide sequence of dengue 2 RNA and comparison of the encoded proteins with those of other flaviviruses. Virology 162, 167–180.

Hedstrom, L. (2002). Serine protease mechanism and specificity. Chem. Rev. 102, 4501–4524.

Hori, H., and Lai, C.J. (1990). Cleavage of dengue virus NS1–NS2A requires an octapeptide sequence at the C terminus of NS1. J. Virol. 64, 4573–4577.

Irie, K., Mohan, P.M., Sasaguri, Y., Putnak, R., and Padmanabhan, R. (1989). Sequence analysis of cloned dengue virus type 2 genome (New Guinea-C strain). Gene 75, 197–211.

Jan, L.R., Yang, C.S., Trent, D.W., Falgout, B., and Lai, C.J. (1995). Processing of Japanese encephalitis virus non-structural proteins: NS2B–NS3 complex and heterologous proteases. J. Gen. Virol. 76, 573–580.

Koshland, D.E. (1958). Application of a theory of enzyme specificity to protein synthesis. Proc. Natl. Acad. Sci. U.S.A. 44, 98–104.

Krausslich, H.G., and Wimmer, E. (1988). Viral proteinases. Annu. Rev. Biochem. 57, 701–754.

Lee, E., Stocks, C.E., Amberg, S.M., Rice, C.M., and Lobigs, M. (2000). Mutagenesis of the signal sequence of yellow fever virus prM protein: enhancement of signalase cleavage In vitro is lethal for virus production. J. Virol. 74, 24–32.

Leung, D., Schroder, K., White, H., Fang, N.X., Stoermer, M.J., Abbenante, G., Martin, J.L., Young, P.R., and Fairlie, D.P. (2001). Activity of recombinant dengue 2 virus NS3 protease in the presence of a truncated NS2B co-factor, small peptide substrates, and inhibitors. J. Biol. Chem. 276, 45762–45771.

Li, H., Clum, S., You, S., Ebner, K.E., and Padmanabhan, R. (1999). The serine protease and RNA-stimulated nucleoside triphosphatase and RNA helicase functional domains of dengue virus type 2 NS3 converge within a region of 20 amino acids. J. Virol. 73, 3108–3116.

Lin, C., Amberg, S.M., Chambers, T.J., and Rice, C.M. (1993a). Cleavage at a novel site in the NS4A region by the yellow fever virus NS2B-3 proteinase is a prerequisite for processing at the downstream 4A/4B signalase site. J. Virol. 67, 2327–2335

Lin, C., Chambers, T.J., and Rice, C.M. (1993b). Mutagenesis of conserved residues at the yellow fever virus 3/4A and 4B/5 dibasic cleavage sites: Effects on cleavage efficiency and polyprotein processing. Virology 192, 596–604.

Lobigs, M. (1992). Proteolytic processing of a Murray Valley encephalitis virus non-structural polyprotein segment containing the viral proteinase: accumulation of a NS3–4A precursor which requires mature NS3 for efficient processing. J. Gen. Virol. 73, 2305–2312.

Lobigs, M. (1993). Flavivirus premembrane protein cleavage and spike heterodimer secretion require the function of the viral proteinase NS3. Proc. Natl. Acad. Sci. U.S.A. 90, 6218–6222.

Lobigs, M., and Lee, E. (2004). Inefficient signalase cleavage promotes efficient nucleocapsid incorporation into budding flavivirus membranes. J. Virol. 78, 178–186.

Mackow, E., Makino, Y., Zhao, B.T., Zhang, Y.M., Markoff, L., Buckler-White, A., Guiler, M., Chanock, R., and Lai, C.J. (1987). The nucleotide sequence of dengue type 4 virus: analysis of genes coding for nonstructural proteins. Virology 159, 217–228.

Markoff, L. (1989). In vitro processing of dengue virus structural proteins: cleavage of the pre-membrane protein. J. Virol. 63, 3345–3352.

Matusan, A.E., Kelley, P.G., Pryor, M.J., Whisstock, J.C., Davidson, A.D., and Wright, P.J. (2001). Mutagenesis of the dengue virus type 2 NS3 proteinase and the production of growth-restricted virus. J. Gen. Virol. 82, 1647–1656.

Melino, S., Fucito, S., Campagna, A., Wrubl, F., Gamarnik, A., Cicero, D.O., and Paci, M. (2006). The active essential CFNS3d protein complex. FEBS J. 273, 3650–3662.

Monckton, R.P., and Westaway, E.G. (1982). Restricted translation of the genome of the flavivirus Kunjin in vitro. J. Gen. Virol. 63 (Pt 1), 227–232.

Mueller, N.H., Yon, C., Ganesh, V.K., and Padmanabhan, R. (2007). Characterization of the West Nile virus protease substrate specificity and inhibitors. Int. J. Biochem. Cell Biol. 39, 606–614.

Nall, T.A., Chappell, K.J., Stoermer, M.J., Fang, N.X., Tyndall, J.D., Young, P.R., and Fairlie, D.P. (2004). Enzymatic characterization and homology model of a catalytically active recombinant West Nile virus NS3 protease. J. Biol. Chem. 279, 48535–48542.

Niyomrattanakit, P., Winoyanuwattikun, P., Chanprapaph, S., Angsuthanasombat, C., Panyim, S., and Katzenmeier, G. (2004). Identification of residues in the dengue virus type 2 NS2B cofactor that are critical for NS3 protease activation. J. Virol. 78, 13708–13716.

Niyomrattanakit, P., Yahorava, S., Mutule, I., Mutulis, F., Petrovska, R., Prusis, P., Katzenmeier, G., and Wikberg, J.E. (2006). Probing the substrate specificity of the dengue virus type 2 NS3 serine protease by using internally quenched fluorescent peptides. Biochem. J. 397, 203–211.

Nowak, T., Farber, P.M., and Wengler, G. (1989). Analyses of the terminal sequences of West Nile virus structural proteins and of the in vitro translation of these proteins allow the proposal of a complete scheme of the proteolytic cleavages involved in their synthesis. Virology 169, 365–376.

Otlewski, J., Jaskolski, M., Buczek, O., Cierpicki, T., Czapinska, H., Krowarsch, D., Smalas, A.O., Stachowiak, D., Szpineta, A., and Dadlez, M. (2001). Structure–function relationship of serine protease-protein inhibitor interaction. Acta Biochim. Pol. 48, 419–428.

Ozden, S., and Poirier, B. (1985). Dengue virus induced polypeptide synthesis. Brief report. Arch. Virol. 85, 129–137.

Pasternak, A., Ringe, D., and Hedstrom, L. (1999). Comparison of anionic and cationic trypsinogens: the anionic activation domain is more flexible in solution and differs in its mode of BPTI binding in the crystal structure. Protein Sci. 8, 253–258.

Perlman, D., and Halvorson, H.O. (1983). A putative signal peptidase recognition site and sequence in eukaryotic and prokaryotic signal peptides. J Mol. Biol. 167, 391–409.

Perona, J.J., and Craik, C.S. (1995). Structural basis of substrate specificity in the serine proteases. Protein Sci. 4, 337–360.

Pethel, M., Falgout, B., and Lai, C.J. (1992). Mutational analysis of the octapeptide sequence motif at the NS1–NS2A cleavage junction of dengue type 4 virus. J. Virol. 66, 7225–7231

Preugschat, F., and Strauss, J.H. (1991). Processing of nonstructural proteins NS4A and NS4B of dengue 2 virus in vitro and in vivo. Virology 185, 689–697.

Preugschat, F., Yao, C.W., and Strauss, J.H. (1990). In vitro processing of dengue virus type 2 nonstructural proteins NS2A., NS2B., and NS3. J. Virol. 64, 4364–4374.

Rice, C.M., Aebersold, R., Teplow, D.B., Pata, J., Bell, J.R., Vorndam, A.V., Trent, D.W., Brandriss, M.W., Schlesinger, J.J., and Strauss, J.H. (1986). Partial N-terminal amino acid sequences of three nonstructural proteins of two flaviviruses. Virology 151, 1–9.

Rice, C.M., Lenches, E.M., Eddy, S.R., Shin, S.J., Sheets, R.L., and Strauss, J.H. (1985). Nucleotide sequence of yellow fever virus: implications for flavivirus gene expression and evolution. Science 229, 726–733.

Ruiz-Linares, A., Cahour, A., Despres, P., Girard, M., and Bouloy, M. (1989). Processing of yellow fever virus polyprotein: role of cellular proteases in maturation of the structural proteins. J. Virol. 63, 4199–4209.

Schechter, I., and Berger, A. (1967). On the size of the active site in proteases. I. Papain. Biochem. Biophys. Res. Commun. 27, 157–162.

Schrauf, S., Schlick, P., Skern, T., and Mandl, C.W. (2008). Functional analysis of potential carboxy-terminal cleavage sites of tick-borne encephalitis virus capsid protein. J. Virol. 82, 2218–2229.

Shiryaev, S.A., Aleshin, A.E., Ratnikov, B.I., Smith, J.W., Liddington, R.C., and Strongin, A.Y. (2007a). Expression and purification of a two-component flaviviral proteinase resistant to autocleavage at the NS2B–NS3 junction region. Protein Expr. Purif. 52, 334–339.

Shiryaev, S.A., Kozlov, I.A., Ratnikov, B.I., Smith, J.W., Lebl, M., and Strongin, A.Y. (2007b). Cleavage preference distinguishes the two-component NS2B–NS3 serine proteinases of Dengue and West Nile viruses. Biochem. J. 401, 743–752.

Speight, G., Coia, G., Parker, M.D., and Westaway, E.G. (1988). Gene mapping and positive identification of the non-structural proteins NS2A., NS2B., NS3, NS4B and NS5 of the flavivirus Kunjin and their cleavage sites. J. Gen. Virol. 69, 23–34.

Speight, G., and Westaway, E.G. (1989a). Carboxy-terminal analysis of nine proteins specified by the flavivirus Kunjin: evidence that only the intracellular core protein is truncated. J. Gen. Virol. 70, 2209–2214.

Speight, G., and Westaway, E.G. (1989b). Positive identification of NS4A., the last of the hypothetical nonstructural proteins of flaviviruses. Virology 170, 299–301

Stocks, C.E., and Lobigs, M. (1998). Signal peptidase cleavage at the flavivirus C-prM junction: dependence on the viral NS2B-3 protease for efficient processing requires determinants in C., the signal peptide, and prM. J. Virol. 72, 2141–2149.

Svitkin, Y.V., Lyapustin, V.N., Lashkevich, V.A., and Agol, V.I. (1984). Differences between translation products of tick-borne encephalitis virus RNA in cell-free systems from Krebs-2 cells and rabbit reticulocytes: involvement of membranes in the processing of nascent precursors of flavivirus structural proteins. Virology 135, 536–541.

Valle, R.P., and Falgout, B. (1998). Mutagenesis of the NS3 protease of dengue virus type 2. J. Virol. *72*, 624–632.

von Heijne, G. (1984). How signal sequences maintain cleavage specificity. J. Mol. Biol. *173*, 243–251.

Wengler, G., Czaya, G., Farber, P.M., and Hegemann, J.H. (1991). *In vitro* synthesis of West Nile virus proteins indicates that the amino-terminal segment of the NS3 protein contains the active centre of the protease which cleaves the viral polyprotein after multiple basic amino acids. J. Gen. Virol. *72*, 851–858.

Wengler, G., Wengler, G., and Gross, H.J. (1978). Studies on virus-specific nucleic acids synthesized in vertebrate and mosquito cells infected with flaviviruses. Virology *89*, 423–437.

Wengler, G., Wengler, G., Nowak, T., and Castle, E. (1990). Description of a procedure which allows isolation of viral nonstructural proteins from BHK vertebrate cells infected with the West Nile flavivirus in a state which allows their direct chemical characterization. Virology *177*, 795–801

Wu, C.F., Wang, S.H., Sun, C.M., Hu, S.T., and Syu, W.J. (2003). Activation of dengue protease autocleavage at the NS2B–NS3 junction by recombinant NS3 and GST-NS2B fusion proteins. J. Virol. Methods *114*, 45–54.

Yaegashi, T., Vakharia, V.N., Page, K., Sasaguri, Y., Feighny, R., and Padmanabhan, R. (1986). Partial sequence analysis of cloned dengue virus type 2 genome. Gene *46*, 257–267.

Yamshchikov, V.F., and Compans, R.W. (1993). Regulation of the late events in flavivirus protein processing and maturation. Virology *192*, 38–51.

Yamshchikov, V.F., and Compans, R.W. (1995). Formation of the flavivirus envelope: role of the viral NS2B–NS3 protease. J. Virol. *69*, 1995–2003.

Yusof, R., Clum, S., Wetzel, M., Murthy, H.M., and Padmanabhan, R. (2000). Purified NS2B/NS3 serine protease of dengue virus type 2 exhibits cofactor NS2B dependence for cleavage of substrates with dibasic amino acids *in vitro*. J. Biol. Chem. *275*, 9963–9969.

Zhang, L., Mohan, P.M., and Padmanabhan, R. (1992). Processing and localization of Dengue virus type 2 polyprotein precursor NS3–NS4A–NS4B–NS5. J. Virol. *66*, 7549–7554.

Zhang, L., and Padmanabhan, R. (1993). Role of protein conformation in the processing of dengue virus type 2 nonstructural polyprotein precursor. Gene *129*, 197–205.

Zhao, B., Mackow, E., Buckler White, A., Markoff, L., Chanock, R.M., Lai, C.J., and Makino, Y. (1986). Cloning full-length dengue type 4 viral DNA sequences: analysis of genes coding for structural proteins. Virology *155*, 77–88.

The Dengue Virus Replication Complex   3

Sven Miller, Ines Romero-Brey and Ralf Bartenschlager

Abstract

Replication of all positive-stranded RNA viruses investigated so far occurs in close association with virus-induced intracellular membrane structures. Dengue virus (DENV), as a member of the family *Flaviviridae,* also induces such extensive rearrangements of intracellular membranes, called replication complex (RC). These RCs seem to contain viral proteins, viral RNA and host cell factors. However, the biogenesis of the RC and the three-dimensional organization is to the most part unclear. In this chapter we summerize the current state of research as regards (1) the individual steps involved in DENV replication, (2) the viral and host cell proteins involved in DENV replication, (3) the intracellular sites of DENV replication, (4) the architecture of the DENV RCs and (5) the host cell nucleus and DENV replication. Finally, we present a hypothetical model of the biogenesis of the DENV replication complex.

Introduction

Replication of the mosquito-transmitted human pathogen dengue virus (DENV) takes place in the cytoplasm of infected host cells at virus-induced membrane structures. These rearranged membranes, which contain specific DENV proteins, viral RNA, and putative host cell factors involved in the production of progeny RNA, are called a replication complex (RC). Although replication of DENV and other members of the *Flaviviridae* family has been extensively studied, the precise architecture of the DENV RC as well as its biogenesis is still not known. In this chapter we discuss the role of the RC during DENV replication and summarize recent knowledge about its putative composition and biogenesis. Furthermore, we highlight what remains to be done to unravel the process of DENV replication in detail.

Individual steps involved in dengue virus RNA synthesis

As described above (see Chapter 1) DENV contains a single-stranded genomic RNA of positive polarity ([+]RNA). This genome mimics a cellular mRNA molecule even though it lacks a polyadenylated 3′-end (poly-A tail). For this reason, the viral [+]RNA is infectious. A prerequisite for viral genome replication is the expression and processing of the viral polyprotein (see Chapter 2), since the host cell does not encode for all factors required for DENV replication. After initial translation, viral and putative host cell factors assemble into the membrane-bound viral RC (Fig. 3.1A). During virus replication viral [+]RNA serves as a template for the production of a genome length minus-strand RNA ([−]RNA), which is used as a template for the generation of progeny [+]RNA genomes. By using metabolic-labelling studies Cleaves and co-authors found that – in addition to [+]RNA – two further species of viral RNA exist in DENV-infected cells (Cleaves *et al.,* 1981). These are the so called replicative form (RF) comprising double-stranded RNase-resistant RNA (dsRNA) as well as a partially RNase-resistant RNA or replicative intermediate (RI), which are both precursors of the genome [+] RNA. In the current model of DENV replication it is assumed that during replication the newly

A

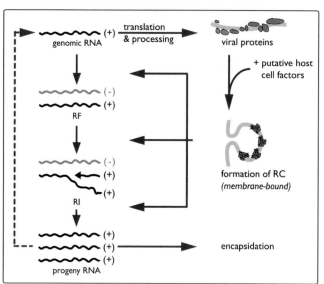

B

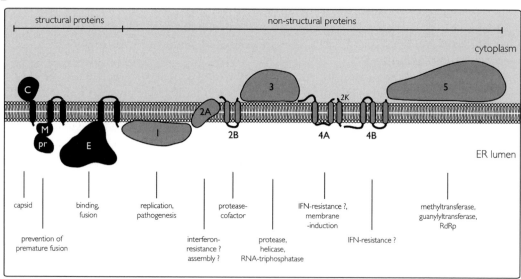

Figure 3.1 Dengue virus RNA replication strategy and topology and function of Dengue virus proteins. **A** After initial translation of the incoming viral [+]RNA viral replicase proteins, putative host cell factors, and the [+]RNA-template assemble into the membrane bound replication complex (RC). During replication the viral [+]RNA serves as a template for the production of a genome length [–]RNA that remains base-paired with its [+]RNA template thus forming the double-stranded replicative form (RF). The RF functions as a template for the generation of multiple [+]RNA copies via the replicative intermediate (RI), which most likely contains duplex regions and newly synthesized [+]RNA displaced by nascent strands undergoing elongation. The newly produced [+]RNA is released from the RI and either is packed into progeny virus particles or attaches to ribosomes to initiate a new translation cycle (dashed line). **B** Putative membrane topology of DENV structural (dark grey) and non-structural (light grey) proteins. Functions of the individual proteins are indicated; assumed functions are labelled with a question mark (adapted from Bartenschlager and Miller, 2008).

synthesized [−]RNA remains base-paired with its [+]RNA template forming the double-stranded RF (Fig. 3.1A). The RF functions as a template for the generation of multiple [+]RNA strands via the RI, which most likely contains duplex regions as well as single stranded regions assumed to arise from strand displacement by the viral replicase. Viral RNA synthesis is semi-conservative and asymmetric, with plus strands accumulating in about 10-fold excess over minus strands (Cleaves *et al.*, 1981; Lindenbach, 2001). Molecular details of how the viral RNA-dependent RNA polymerase (RdRp) recognizes and specifically amplifies the DENV genome remain unclear. However, it has been shown that genome cyclization seems to be a prerequisite for this process (Alvarez *et al.*, 2005a,b) (see Chapter 4 for details). Following amplification, the newly produced [+]RNA is released from the RI and either attaches to ribosomes to initiate a new translation cycle or is packaged into progeny virus particles.

Viral proteins involved in dengue virus replication

Since eukaryotic cells do not express enzymes that are able to replicate DENV [+]RNA, the virus has to encode its own replicase factors. Using the incoming viral [+]RNA as a template, the viral genome is translated by the host cell machinery giving rise to a polyprotein immediately after infection. This polyprotein is co- and post-translationally processed into three structural proteins (C, prM, and E) and seven non-structural (NS) proteins (NS1, NS2A, NS2B, NS3, NS4A, NS4B, and NS5; see Chapter 2 for more details) (Fig. 3.1B). Whereas the structural proteins are components of extracellular mature virus particles, the NS proteins are expressed only in the infected host cell and are not packaged to detectable levels into mature particles. Several studies indicate that most if not all NS proteins are involved in the replication of the DENV genome, whereas the structural proteins are not required for this process. This observation is supported by the finding that subgenomic DENV replicons are capable of autonomous replication in transfected cells (Ng *et al.*, 2001; Pang *et al.*, 2001a, b; Alvarez *et al.*, 2006; Sven Miller and Ralf Bartenschlager; unpublished results). Replicons are subgenomic viral RNAs, in which the genes encoding for some or all of the

structural proteins are deleted. Upon transfection of eukaryotic cells with *in vitro* transcribed DENV replicon RNA the encoded NS proteins are translated by the host cell translation machinery and assemble into the viral RC. Thereupon, new subgenomic RNA is synthesized within the RC by the semi-conservative and asymmetric process described above. Thus, only NS proteins are required for RNA replication. For this reason, replicons have been shown to be useful tools to study viral replication in the absence of virion assembly and maturation (Lohmann *et al.*, 1999; Khromykh *et al.*, 2000). However, although the role of some NS proteins in replication has been studied in detail, others are still poorly characterized (see Fig. 3.1B).

The best-characterized DENV NS proteins required for viral replication are NS5 and NS3. NS5 – the largest of the DENV NS proteins ($\approx$ 103 kDa) carries the viral RdRp activity in its C-terminal domain. In addition, NS5 contains a N-terminal domain that is involved in RNA capping, possessing both (guanine-N7)-methyltransferase and nucleoside-2′-O methyltransferase activities required for sequential methylation of the cap structure present at the 5′ end of the flavivirus RNA (Koonin, 1993; Tan *et al.*, 1996; Ackermann and Padmanabhan, 2001; Egloff *et al.*, 2002, 2007; Nomaguchi *et al.*, 2003; Kroschewski *et al.*, 2008). In a recent study it has been shown that mutations that severely reduced (> 90%) or totally abolished both methyltransferase activities *in vitro* abolished viral replication, demonstrating the requirement of at least one of these two activities for viral replication (Kroschewski *et al.*, 2008). The polymerase activity of DENV NS5 RdRp could be confirmed biochemically with recombinant protein (Tan *et al.*, 1996; Ackermann and Padmanabhan, 2001). Under these *in vitro* conditions, the major product was self-primed copy-back RNA. However, RNA generated by *de novo* initiation was also observed, resembling the *in vivo* situation.

In addition to NS5, also DENV NS3 ($\approx$ 70 kDa) was found to exert multiple functions during DENV replication. On the one hand, NS3 acts as the viral serine protease needed for polyprotein-processing through its N-terminal domain (see Chapter 2 for more details), requiring a segment of 40 residues of the small hydrophobic protein NS2B as a cofactor (Bazan and Fletterick, 1989; Gorbalenya et al., 1989a; Chambers et al., 1990; Murthy et al., 1999, 2000). Specifically, the amino-terminal 180 aa of this multifunctional protein complexed with NS2B are responsible for cleavages on the cytoplasmic side of the ER membrane at the junctions between NS2A–NS2B, NS2B–NS3, NS3–NS4A and NS4B–NS5 as well as at internal sites within C, NS2A, NS3 and NS4A. On the other hand, NS3 contains a 5′ RNA-triphosphatase (RTP), a nucleoside triphosphatase (NTPase) and a helicase activity in the C terminal region of the molecule (Gorbalenya et al., 1989a; Gorbalenya et al., 1989b; Li et al., 1999; Wengler, 1993; Bartelma and Padmanabhan, 2002). These nucleic acid-modifying activities rely on a common active centre in the C-terminal part of NS3. The helicase is assumed to be required for unwinding of RNA during replication, whereas the RTPase was proposed to dephosphorylate the 5′-end of genomic RNA before addition of the 5′-cap (Wengler, 1993; Bartelma and Padmanabhan, 2002; Benarroch et al., 2004). The essential role of the helicase for DENV replication has been confirmed by mutagenesis of the corresponding domain (Matusan et al., 2001). It could be shown that NS3 forms a complex with NS5, and that this interaction stimulates the RTPase as well as the NTPase activity of NS3 (Kapoor et al., 1995; Cui et al., 1998; Johansson et al., 2001; Yon et al., 2005).

Apart from NS5 and NS3 most if not all other DENV NS proteins seem to be involved in replication in vivo. The first indication that the viral glycoprotein NS1 (≈ 46 kDa), which is secreted from infected cells in oligomeric form, could be involved in RNA replication came from the observation that NS1 localizes to the intracellular sites of RNA synthesis (see below) (Mackenzie et al., 1996). Even if its role in this process is yet unclear, trans-complementation experiments revealed that NS1 functions at a very early stage in RNA replication (Lindenbach and Rice, 1997). Furthermore, it was found that an interaction between NS1 and NS4A appears to be required for replicase function (Lindenbach and Rice, 1999). Beside its role in RNA synthesis NS1 has been recognized as an important target for immune recognition and contributor to pathogenesis.

In addition to the comparatively large NS proteins NS1, NS3, and NS5, DENV encodes for four small and very hydrophobic membrane-associated NS proteins NS2A (≈ 22 kDa), NS2B (≈ 14 kDa), NS4A (≈ 16 kDa), and NS4B (≈ 27 kDa). As mentioned above, NS2B forms a stable complex with NS3 and acts as a cofactor for the NS2B-3 serine protease (Falgout et al., 1991). Little is known about the function of NS2A, NS4A, and NS4B. It has been shown that these proteins contribute to the inhibition of the interferon alpha/beta response of the infected host cell (Munoz-Jordan et al., 2003; Munoz-Jordan et al., 2005). Even if it is not clear yet which role NS2A plays during DENV replication, it is thought that NS2A of another flavivirus – namely Kunjin virus (KUNV) – localizes to the sites of RNA synthesis (Mackenzie et al., 1998) and may coordinate the shift between RNA packaging and RNA replication (Khromykh et al., 2001; Leung et al., 2008). Whether NS2A of DENV carries out the same functions remains to be clarified.

Reverse genetic studies with DENV replicons (Sven Miller and Ralf Bartenschlager, unpublished results) as well as localization studies (Miller et al., 2006; Umareddy et al., 2006; Miller et al., 2007) suggest that NS4A and NS4B are both required for DENV replication. Evidence has been obtained that NS4B modulates DENV replication via an interaction with the helicase domain of NS3 (Umareddy et al., 2006), whereas NS4A appears to play a key role in the induction of membrane alterations, which may serve as a scaffold to anchor the viral replication complex (Miller et al., 2007) (see below).

Intracellular sites of DENV replication

Microscopic studies

Replication of flaviviruses is thought to occur in association with cytoplasmic membranes (Boulton and Westaway, 1976), and numerous viral proteins involved in RNA replication of different flaviviruses have been found to localize to these intracellular membrane structures (Table 3.1). By using antibodies raised against poly(I)-poly(C) – a synthetic polymer of inosine that resembles the dsRNA-intermediate produced during flavivirus infection – Ng et al. (1983)

Table 3.1 Localization of flavivirus non-structural proteins within infected cells. The table summarizes the results of immunofluorescence microscopic and electron microscopic studies of flavivirus-infected cells

		ER	Dot-like structures (IF)	Nucleus	VP (EM)	CM (EM)	Cell surface
NS1	DENV		X [1]		X [1]		X [2]
	KUNV		X [3]		X [3]		
	Others						
NS2A	DENV						
	KUNV		X [4]		X [4]		
	Others						
NS2B	DENV	X [15]	X [15]				
	KUNV		X [3]			X [3]	
	Others						
NS3	DENV		X [5, 6]				
	KUNV		X [3]		X [3]	X [3]	
	Others			X [7]			
NS4A	DENV	X [8]	X [8]				
	KUNV		X [4]		X [4]	X [4]	
	Others						
NS4B	DENV	X [5]	X [5, 6]				
	KUNV	X [9]		X [9]			
	Others						
NS5	DENV		X [5]	X [5, 10–12]			
	KUNV		X [13]		X [13]	X [13]	
	Others			X [7, 14]			

Numbers refer to original references. 1, Mackenzie *et al.* (1996); 2, Winkler *et al.* (1989); 3, Westaway *et al.* (1997b); 4, Mackenzie *et al.* (1998); 5, Miller *et al.* (2006); 6, Umareddy *et al.* (2006); 7, Uchil *et al.* (2006); 8, Miller *et al.* (2007); 9, Westaway *et al.* (1997a); 10, Brooks *et al.* (2002); 11, Kapoor *et al.* (1995); 12, Pryor *et al.* (2007); 13, Mackenzie *et al.* (2007); 14, Buckley *et al.* (1992); 15, S Miller and R Bartenschalger, unpublished results.

have identified the likely cellular location of RNA replication in flavivirus-infected cells via indirect immunofluorescence. They could show that the sites of RNA-replication in case of flavivirus KUNV extended from the perinuclear region into a reticular membrane network, which included numerous small cytoplasmic foci. The observed staining pattern in part corresponds to grossly rearranged membranes of the rough endoplasmic reticulum (ER). A comparable dot-like staining pattern of dsRNA could also be found in DENV-infected cells (Mackenzie *et al.*, 1996; Miller *et al.*, 2006) (Fig. 3.2). By using immunoelectron microscopy of cryo sections of DENV-infected Vero cells, Mackenzie and co-workers showed that the dot-like structures identified by immunofluorescence microscopy correspond to accumulations of virus-induced membrane alterations at the ultrastructural level. These modified membranes, which stained positive

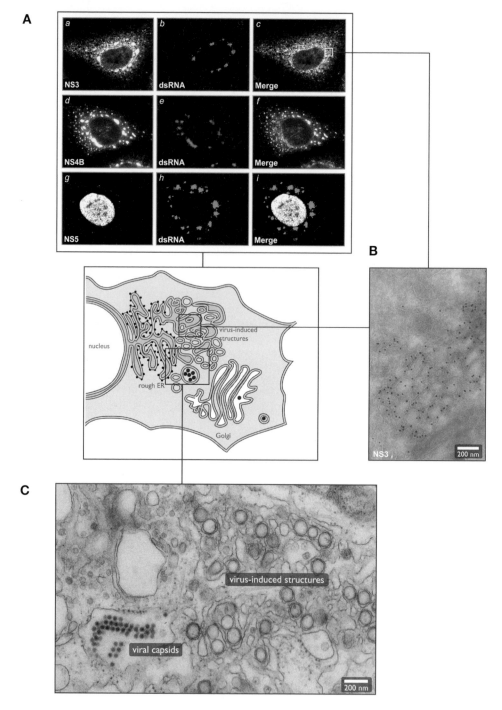

Figure 3.2 Microscopic analyses of Dengue virus-infected cells. **A** Colocalization studies of DENV NS proteins with dsRNA, an intermediate of viral RNA replication. At 24 h p.i. infected cells were fixed and processed for immunolabelling. Used antibodies are given in the lower left of each picture. Merged pictures are shown on the right. The dot-like structures in the cytoplasm of infected cells represent the sites of viral replication (adapted from Miller *et al.*, 2006). **B** Accumulations of virus-induced membrane alterations as shown by the ultrathin cryo-sections of cells 24 h p.i. NS3 was detected by immunogold labelling. Strong labelling of vesicle-accumulations is visible. **C** Overview of DENV-induced membrane alterations in the cytoplasm of infected cells. Resin-embedded sections of cells 24 h p.i. reveal double membrane vesicles and viral particles inside membranous structures (adapted from Bartenschlager and Miller, 2008). A colour version of this figure is located in the plate section at the back of the book.

for dsRNA, consist of accumulations of double membrane vesicles (each vesicle approximately 50–100 nm in diameter) that have already been observed previously in DENV-infected mosquito cells (Barth, 1992). According to the degree of ordered structure, the observed membrane alterations have been designated vesicle packets (VPs) or smooth membrane structures (SMS). By using a combination of non-isotopic *in situ* hybridization and electron microscopy of DENV-infected cells it could be confirmed that dsRNA, which is produced during DENV-replication, localizes to the virus-induced VP/SMS but also to the rough ER (Grief *et al.*, 1997). In addition to dsRNA, also DENV NS1 was found to localize to the VP/SMS, suggesting that they may represent the intracellular sites of DENV replication (Mackenzie *et al.*, 1996). In support of this conclusion DENV NS2B, NS3, NS4A, and NS4B were found by using immunofluorescence microscopy to localize together with dsRNA to cytoplasmic, virus-induced dot-like structures (Kapoor *et al.*, 1995; Miller *et al.*, 2006; Umareddy *et al.*, 2006; Miller *et al.*, 2007; Sven Miller and Ralf Bartenschlager; unpublished results), indicating that all of these DENV proteins localize to the viral RC (see Fig. 3.2). However, detailed studies to unambiguously define the localization of all of these proteins at the ultra-structural level (e.g. by using immunoelectron microscopy of thawed cryo sections) are still lacking. Therefore, we do not know whether all DENV NS proteins localize to exactly the same or different substructures. More detailed information is available for the related flavivirus KUNV, which is the best-characterized flavivirus with respect to the induction of intracellular membrane rearrangements and the composition of the flavivirus RC (for review see (Mackenzie, 2005)). In KUNV-infected cells – in addition to VP/SMS – so-called convoluted membranes (CM) and paracrystalline structures (PC) have been observed, which reside adjacent to the VP/SMS. Although the VP/SMS and CM/PC represent distinct cellular compartments, they appear to be linked via interconnections with the rough ER. KUNV proteins NS1, NS2A, NS3, NS4A, NS5 and also dsRNA localize to VP/SMS (Westaway *et al.*, 1997b; Mackenzie *et al.*, 1998). In addition, NS2B, NS3, NS4A and NS5 localize to CM/PC, whereas NS4B was found at prolifer-

ated ER and also within the nucleus (Westaway *et al.*, 1997a). Since the viral serine-protease NS2B–NS3 colocalizes with the CM/PC, it was proposed that these structures represent the sites of protein translation and polyprotein cleavage, whereas colocalization of the viral replicase components as well as dsRNA within the VP/SMS suggest that viral RNA replication occurs at this site. The membrane reorganization induced by KUNV might therefore give rise to adjacent, but distinct, subcellular structures where different steps of the viral replication cycle are carried out (Westaway *et al.*, 2003). Although detailed studies are lacking, the association of DENV RNA as well as several DENV replicase proteins with VP/SMS in infected cells suggests that they represent the replication sites also in case of DENV (Mackenzie *et al.*, 1996; Miller *et al.*, 2006; Umareddy *et al.*, 2006; Miller *et al.*, 2007). However, differences in the localization of some KUNV and DENV NS proteins have been observed. In case of KUNV, NS4B seems not to be associated with the active RC (Chu and Westaway, 1992) but was rather found in the nucleus, in perinuclear membranes, in ER membranes and often adjacent to the CM or PC structures (Westaway *et al.*, 1997a). In contrast, DENV NS4B could not be detected in the nucleus, but colocalizes with other NS proteins and viral dsRNA to virus-induced membrane structures in the cytoplasm of infected cells, which seem to be the sites of RNA replication. Within these structures NS4B is thought to modulate DENV replication via an interaction with the helicase domain of NS3 (Umareddy *et al.*, 2006). We can therefore assume that DENV NS4B is actively involved in viral RNA replication. In addition to NS4B, the subcellular localization of DENV and KUNV NS5 differs in infected cells. Whereas KUNV NS5 localizes exclusively to virus-induced membrane structures in the cytoplasm (including the sites of RNA replication) (Mackenzie *et al.*, 2007), the majority of DENV NS5 localizes to the nucleus (see below) (Kapoor *et al.*, 1995; Miller *et al.*, 2006; Pryor *et al.*, 2006, 2007). The reasons for these differences are not clear, but may reflect adaptations to particular host requirements in spite of the overall comparable replication strategies. Thus, it is not yet clear if the model that was developed for the architecture and function of the KUNV RC is also valid for DENV. Further morphological as well as biochemical studies are required to unravel the complex architecture and biogenesis of the DENV RC and to highlight differences with other flaviviruses.

Biochemical studies

Immunoelectron microscopy studies of cryo sections have shown that DENV RF RNA is often associated with virus-induced VPs. However, the resolution of this technique is too low to exactly decipher the orientation and architecture of the viral RC. Therefore, biochemical analyses have been performed to substantiate these observations. The results support the assumption that replication occurs in double-layered membranous compartments. The individual compartments can be distinguished based on differential sensitivity to proteases and nucleases after detergent treatment (Uchil and Satchidanandam, 2003). The bounding membrane enclosing the VPs was readily solubilized with non-ionic detergents, rendering enzymatically active protein component(s) of the RC partially sensitive to trypsin, but allowing limited access for nucleases only to the genomic [+]RNA and single stranded tails of the RI. The RF co-sedimented in high-speed centrifugation experiments from non-ionic detergent extracts of virus-induced heavy membrane fractions along with the released individual inner membrane vesicles also containing NS3. The RF remained nuclease-resistant even after solubilization of the inner VP membrane with ionic detergents. All of the viral RNA species became nuclease-sensitive following membrane disruption and trypsin treatment, suggesting that proteins coat the viral genomic [+]RNA as well as RF within these membranous sites. Thus, it might be possible that newly formed VP-associated [+]RNA is oriented outwards, while the RF is located inside the virus-induced vesicles. However, topology and quantity of the individual DENV replicase proteins responsible for RdRp activity is still not known, since trypsin treatment, even in the absence of detergents, degraded most of the replicase proteins, without concomitant loss of replicase activity. This result suggests that only a small proportion of replicase proteins is required for RNA replication, a finding that is consistent with recent data for hepatitis C virus (Miyanari

et al., 2003; Quinkert *et al.*, 2005). Another problem in determining the exact localization of the protein components of the RC is the insufficient resolution of immunoelectron microscopy of cryo sections. For this reason, more detailed studies using high-resolution techniques are required to completely resolve the structure of the DENV RC.

Composition of virus-induced membrane structures

Even if the nature of the DENV-induced membrane structures is not known, the differential susceptibility to solubilization by detergents of the outer and inner membranes of these structures harbouring the viral RC suggests that they are derived from different host cell organelles. However, alterations in membrane properties can also be brought about by differential incorporation of viral and/or cellular proteins into these membranes or by specific post-translational modifications of membrane proteins or lipids.

More detailed analyses of KUNV have shown that each of the virus-induced membrane structures contains a different subset of host cell proteins (Mackenzie *et al.*, 1999). Thus, the trans-Golgi marker β-1,4-galactosyltransferase (GalT; and also *trans*-Golgi network markers p230 and TGN46) localized to the VP and is especially concentrated in membranes of the vesicle within the packet and not the bounding membrane itself (Mackenzie, 2005). In contrast, ERGIC-53, a marker for the intermediate compartment (IC), did not localize to the VP, but accumulated within the CM/PC. Protein disulphide isomerase (PDI; an rER resident protein) was found to label rER continuous with the CM/PC, but was not significantly found within these structures (Mackenzie *et al.*, 1999). These results let assume that a subset of *trans*-Golgi network/trans-Golgi proteins associates intimately with the ER/intermediate compartment-derived CM/PC and that the CM/PC are physically linked with the ER. Thus far, comparable co-localization studies have not been performed with DENV infected cells, but recent biochemical studies indicated that the membranes that are involved in DENV replication possess lipid raft properties (Lee *et al.*, 2008). However, in contrast to KUNV, in immunofluorescence studies no co-localization of DENV-induced structures with trans-Golgi or trans-Golgi network marker proteins could be found, whereas individual replicase-components colocalized with markers of the ER (Miller *et al.*, 2006; Miller *et al.*, 2007). Further studies will be required to determine the origin of the DENV-induced structures and to find out whether differences between DENV and other flaviviruses exist.

Formation of Dengue virus-induced membrane structures

Although DENV and other flaviviruses have long been known to induce membrane rearrangements during infection, the mechanisms that are responsible for the formation of the induced structures are still unknown. Recent studies indicate that the small and hydrophobic *trans*-membrane protein NS4A might be involved in this process. It is known that the NH_2-terminus of NS4A is generated in the cytoplasm by the viral two-component protease NS2B-3, whereas the COOH-terminal 23 amino acid residues of NS4A seem to act as a signal sequence for the translocation of NS4B into the lumen of the endoplasmic reticulum (Fig. 3.3). This signal sequence (designated the 2K-fragment) is removed from the NH_2-terminus of NS4B by the host signalase in the ER lumen. Signalase cleavage at the 2K-4B site requires a prior NS2B-3 protease-mediated cleavage at the so-called 4A/2K site just NH_2-terminal of the 2K-fragment (Lin *et al.*, 1993). Recently it was found that the expression of DENV NS4A lacking the 2K-fragment and thus representing the fully processed NS4A, induced cytoplasmic membrane rearrangements and led to a redistribution of the ER as observed in infected cells (Miller *et al.*, 2007). These results provide strong evidence that DENV processing at the 4A-2K site is required for the induction of membrane alterations. A comparable role for NS4A-2K in intracellular membrane-rearrangements was also described for KUNV (Roosend aal *et al.*, 2006). Even if these studies are promising, it is still not clear how NS4A can promote the formation of the remarkable membrane alterations observed in infected cells. One possibility is that this viral protein is able to induce the observed structures on its own. Since NS4A contains a hydrophobic stretch of amino acids that seems to be closely associated with intracellular membranes but does

not span these membranes completely (Miller *et al.*, 2007), it is possible that the corresponding amino acids insert only into the luminal leaflet of the ER membrane and act like a wedge, resulting in membrane curvature. This effect may be amplified by NS4A homo-oligomerization (Mackenzie *et al.*, 1998; Miller *et al.*, 2007) similar to what has been described for the Hepatitis C virus NS4B protein (Yu *et al.*, 2006) and several other membrane curvature inducing proteins (Ford *et al.*, 2002; Zimmerberg and Kozlov, 2006). Alternatively, NS4A alone or (in conjunction with) other components of the viral RC could interact with cellular proteins and thus modify cellular pathways involved in intracellular transport and membrane biogenesis. Recently, one such interaction between the viral RC and host cell factors potentially involved in intracellular trafficking was described. Chua *et al* found that DENV NS3 interacts specifically with nuclear receptor binding protein (NRBP), a host cell protein influencing trafficking between the ER and the Golgi apparatus (Chua *et al.*, 2004). This interaction leads to a redistribution of NRBP from the cytoplasm to the perinuclear region. Thus, it might well be that DENV exploits host cell pathways involved in membrane trafficking via an interaction of NS3 with NRBP. However, in this study only NS3 was expressed which in the absence of NS2B is not membrane associated, arguing that the colocalization should be repeated with the NS2B/3 complex. Nevertheless, membrane rearrangements have been observed by immunofluorescence microscopy in cells individually expressing DENV NS3 (Chua *et al.*, 2004). Although comparable structures have not been observed in NS3-expressing cells by other groups (Sven Miller and Ralf Bartenschlager; unpublished results), it might well be that NS4A as well as NS3 are both involved in the induction of membrane structures serving as a scaffold for anchoring the viral RC. In a highly speculative scenario membrane-bound NS4A could induce curvature by inserting like a wedge into intracellular membranes and oligomerization. In addition, the interaction of NS3 with NRBP could further help to induce vesiculation by an interaction with host cell trafficking pathways. Thus, the collaboration of different viral and host cell factors would ensure a highly efficient process.

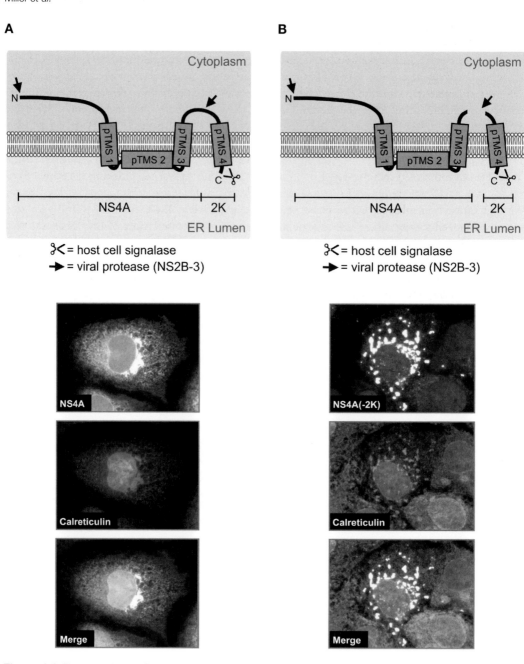

Figure 3.3 Dengue virus NS4A: a putative inducer of membrane alterations. **A** DENV NS4A-2K is a polytopic trans-membrane protein. Its N-terminal one-third is localized in the cytoplasm. The predicted trans-membrane segments (pTMS) 1 and 4 span the membrane from the cytoplasmic to the luminal site, whereas pTMS3 seems to span the lipid bilayer from the luminal to the cytoplasmic site. pTMS2 most probably does not span the membrane but is closely associated with the luminal site of the lipid bilayer. NS4A is cleaved off from the preceding NS3 protein during polyprotein processing by the viral NS2B-3 protease. It also cleaves at the NS4A-2K site within NS4A and removes pTMS4 (the '2K-fragment'). This cleavage is a prerequisite for processing by the host cell signalase at the 2K-4B site. Immunofluorescence studies have shown that expression of NS4A-2K in eukaryotic cells leads to a reticular staining pattern at ER membranes (lower panel). **B** Expression of the mature form of NS4A (lacking the 2K fragment) induces ER derived dot-like structures in the cytoplasm (lower panel). Since these structures resemble the membrane-alterations observed in DENV-infected cells, NS4A might be responsible for the induction of membrane structures that harbour the DENV RC (adapted from Miller *et al.*, 2007). A colour version of this figure is located in the plate section at the back of the book.

Architecture of the dengue virus replication complex

Biochemical analysis and computer predictions suggest that four of the seven DENV NS proteins contain membrane-spanning hydrophobic stretches of amino acids that are thought to mediate the integration of these proteins into cellular membranes (Fig. 3.1B) (Lindenbach et al., 2007). For this reason it is assumed that the integral-membrane proteins NS2A, NS2B, NS4A, and NS4B are required to ensure the direct or indirect membrane-association of the viral RC. Indeed, many interactions between DENV or other flavivirus NS proteins, as well as interactions between NS proteins and viral RNA, have been described (Fig. 3.4). For instance, it was found that the multifunctional protein NS1

interacts with the small membrane-anchored protein NS4A and that this interaction appears to be critical for replicase function (Lindenbach and Rice, 1999). Besides interacting with NS1, NS4A was also found to homo-oligomerize in KUNV (Mackenzie et al., 1998) and we can assume that DENV NS4A is able to interact with itself as well (Miller et al., 2007). Recently, an interaction between the integral membrane protein DENV NS4B and the helicase domain of NS3 has been described (Umareddy et al., 2006). The authors suggested that NS4B modulates DENV replication via this interaction. DENV NS3 was also found to interact with the RdRp NS5 (Kapoor et al., 1995; Johansson et al., 2001), the protease cofactor NS2B (Falgout et al., 1991; Arias et al., 1993; Clum et al., 1997), and viral RNA (Arias et

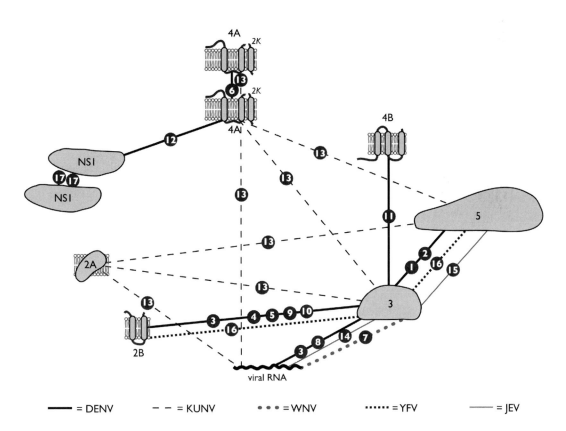

Figure 3.4 Protein–protein and protein–RNA interactions of flavivirus replicase components. Interactions between different flavivirus NS proteins and NS proteins with viral RNA are indicated. Line patterns specified in the bottom are used to discriminate between the interactions found for the different flaviviruses. Numbers refer to the following original references: 1, Johansson et al. (2001); 2, Kapoor et al. (1995); 3, Arias et al. (1993); 4, Falgout et al. (1991); 5, Clum et al. (1997); 6, Miller et al. (2007); 7, Wengler et al. (1991); 8, Cui et al. (1998); 9, Matusan et al. (2001); 10, Xu et al. (2005); 11, Umareddy et al. (2006); 12, Lindenbach and Rice (1999); 13, Mackenzie et al. (1998); 14, Chen et al. (1997); 15, Jan et al. (1995); 16, Chambers et al. (1993); 17, Winkler et al. (1988).

al., 1993). It could be shown that the C-terminal region of NS3 (aa 303–618) is important to mediate the interaction with the N-terminal S-adenosyl-methionine region of NS5 (residues 320–368). However, DENV NS5 exists in a hypophosphorylated, cytoplasmic form and is in addition found in a hyperphosphorylated form in the nucleus of infected cells (see below). Since only the cytoplasmic form of NS5 appears to interact with NS3, differential phosphorylation may control NS5–NS3 interaction and thus activity of the viral RNA replicase (Kapoor *et al.*, 1995), presumably by modulating NTPase activity of NS3 (Cui *et al.*, 1998).

For the interaction of the viral protease NS3 with its cofactor NS2B the N-terminal 184 amino acid residues of NS3 are sufficient (Arias *et al.*, 1993). Apart from activation of the protease domain, NS2B tethers the protease complex to intracellular membranes via the trans-membrane segment in NS2B (Clum *et al.*, 1997). DENV NS3 also interacts with viral RNA. By using band shift assays, Cui *et al.* demonstrated that DENV NS3 forms a complex with the stem–loop structure in the 3′-non-coding region, although sites outside the stem–loop may also participate in binding (Cui *et al.*, 1998). While these studies provide first insights into the architecture and activity of DENV RC, we can assume that more interactions take place. Based on the high similarity to other flaviviruses, studies performed with these viruses can be used as a guide (Fig. 3.4), even if it is not yet clear how general these results are. For instance, Mackenzie *et al.* have found that KUNV NS2A strongly binds to the 3′-end of the RNA genome and also to the replicase components NS3 and NS5 (Mackenzie *et al.*, 1998). Also KUNV NS4A has the ability to bind to the 3′-end of KUNV RNA, even though this interaction is weaker than in case of NS2A. NS4A also has the ability to interact with itself and with most of the other KUNV NS proteins, including NS3 and NS5. In this respect NS4A may serve as a central 'organizer' of the RC of flaviviruses.

The data discussed above show clearly that protein–protein and protein–RNA interactions between the individual components of the viral RC are important to mediate the assembly and functionality of the RC. The strategy of expressing all viral proteins as a polyprotein evidently facilitates targeting and assembly of all factors to the same location. The strictly regulated processing events involved in generating the individual viral components (see Chapter 2) may further contribute to the proper assembly of the membrane-associated RC. The importance of tightly regulated RC formation is exemplified by the fact that it is difficult to *trans*-complement individual components of the RC of DENV (Sven Miller and Ralf Bartenschlager, unpublished results) and other [+]RNA viruses (Khromykh *et al.*, 1999a,b, 2000; Appel *et al.*, 2005). These findings suggest that the RC is a rather tight complex, with limited exchange of viral and eventually also cellular components. However, a more detailed fine molecular mapping of interactions between DENV proteins, host cell factors, and RNA is important both as a means to characterize them to reveal their significance in the replication mechanism and also to investigate the suitability of these interaction sites as specific targets for new antiviral compounds. Modern microscopy techniques such as live-cell microscopy using fluorescently labelled replicase components and host cell factors may help to achieve this aim.

Host cell proteins involved in dengue virus replication

Because of their limited genetic content, viruses rely on the host cell for productive replication. However, there persists a dearth of information pertaining to host cell proteins involved in the biogenesis and function of the DENV RC. Up to now only very few host cell factors involved in DENV replication have been identified and in most cases their exact role for DENV replication is not known. It was found that human as well as mosquito La protein binds to the 3′-end of DENV [+]RNA and [−]RNA and relocates to the cytoplasm of infected cells (De Nova-Ocampo *et al.*, 2002; Yocupicio-Monroy *et al.*, 2003; Garcia-Montalvo *et al.*, 2004; Yocupicio-Monroy *et al.*, 2007). The human La protein also interacts with two components of the viral RC, namely NS3 and NS5 (Garcia-Montalvo *et al.*, 2004). This interaction that was found in human and mosquito cells and the fact that the presence of La protein in an *in vitro* replication system inhibited RNA synthesis in a dose-dependent manner suggests that the La protein plays a crucial role during DENV replication. The use of UV-induced cross-linking has enabled identification of additional proteins interacting with DENV RNA (De Nova-Ocampo et al., 2002). Among them are the translation elongation factor-1alpha (EF-1alpha) that binds to the complete 3'-untranslated region and to the CS1-L-SL region, and the polypyrimidine tract binding protein (PTBP). Binding of the latter to the DENV RNA genome is considered to be essential for viral RNA replication, suggesting that PTBP acts as a RNA chaperone maintaining RNA structure in a conformation that favours viral replication, while EF-1alpha may function as an RNA helicase. Additional host cell factors that bind with high specificity and high affinity to sequences within the untranslated regions of the DENV genome have been described recently (Paranjape and Harris, 2007). Among them are the Y box-binding protein-1 (YB-1) and the heterogeneous nuclear ribonucleoproteins (hn-RNPs), hnRNP A1, hnRNP A2/B1, and hnRNP Q, which all bind to the DENV 3'-untranslated region. Analysis of the impact of YB-1 on DENV replication indicated that this protein interferes with DENV RNA translation, suggesting that YB-1 is a restriction factor.

Another host cell factor interacting with a component of the DENV RC is nuclear receptor binding protein (NRBP), a host cell protein previously identified as an adaptor protein containing a kinase homology domain (Chua et al., 2004). As described above, NRBP appears to interact with NS3 and may assist in the induction of alterations of the ER membrane harbouring the viral RC.

Taking advantage of high-throughput screening methods, the c-Src protein kinase was recently identified as an important host cell factor involved in the DENV assembly. This example illustrates the power of modern high-content screening approaches and we can expect that further cellular factors and pathways required for DENV replication will be discovered with these methods.

The host cell nucleus and dengue virus replication

Although flavivirus replication occurs in cytoplasmic RCs, some NS proteins have been detected in the nucleus of infected cells (for review see (Bartenschlager and Miller, 2008)). For example, the majority of DENV NS5 is found in the nucleus already at very early time points after infection (Kapoor et al., 1995). Even if it is not completely understood which role the flavivirus NS proteins play within the nucleus, it should be kept in mind that only a minor amount of viral protein produced in infected cells appear to be involved in RNA replication (Miyanari et al., 2003; Uchil and Satchidanandam, 2003; Quinkert et al., 2005). Thus, it is possible that 'surplus' proteins carry out other functions within the nucleus. For example, it is assumed that nuclear NS5 suppresses IL-8 production, thus protecting DENV against the antiviral activity of this cytokine (Pryor et al., 2007). The transport of NS5 to the nucleus was described to be an active process that requires the recognition of a bipartite nuclear localization signal (NLS) within NS5 by the importin alpha/beta and importin beta-1 nuclear transport proteins of the host cell (Brooks et al., 2002). In addition to its ability to be transported into the nucleus, NS5 can also be actively exported from the nucleus, for which the nuclear export receptor exportin 1 (CRM1) is required (Pryor et al., 2006). Beside NS5, a recent study suggests that also NS3 and presumably up to 20% of the active viral RC resides in the nucleus of DENV-infected cells (Uchil et al., 2006). However, a nuclear localization of NS3 could not be confirmed by other groups (Westaway et al., 2003; Miller et al., 2006; Sven Miller and Ralf Bartenschlager, unpublished results). Beside NS5 and NS3 also the DENV C protein could be detected in the nuclei and nucleoli of many infected cells lines (Tadano et al., 1989; Bulich and Aaskov, 1992; Wang et al., 2002). Although the mechanism of its nuclear translocation is completely unknown, it has been recently reported that several regions of this structural protein contribute to its nuclear localization (Sangiambut et al., 2008). Since C is found in the nucleus already at very early time points after infection it is possible that it is involved in regulating the DENV replication cycle and apoptotic cell death (Marianneau et al., 1997; Avirutnan et al., 1998; Andersen et al., 2005; Boisvert et al., 2007).

A better understanding of nuclear-localized flavivirus proteins could lead to the development of new recombinant vaccines based on mutants that are deficient in nuclear trafficking of DENV proteins, thus leading to an attenuation of the virus. Furthermore, nuclear localized viral proteins could be an ideal tool to unravel the biological details of nuclear processes and functions.

Biogenesis of the dengue virus replication complex: a hypothetical model

Despite the possibility that a minor fraction of DENV replication could take place in the nucleus of infected host cells, the majority of the viral RCs are associated with virus-induced membrane alterations in the cytoplasm. However, it is not yet clear whether, in case of DENV, the VP/SMS are the intracellular sites of viral replication and whether polyprotein-processing occurs in association with CM, as was described for KUNV. Nonetheless, based on analogy to other [+] RNA viruses, especially KUNV, a hypothetical model for the biogenesis of the DENV RC can be developed (Fig. 3.5): incoming viral [+]RNA is translated at the rough ER, and the polyprotein is co- and post-translationally processed into the individual components. After completion of

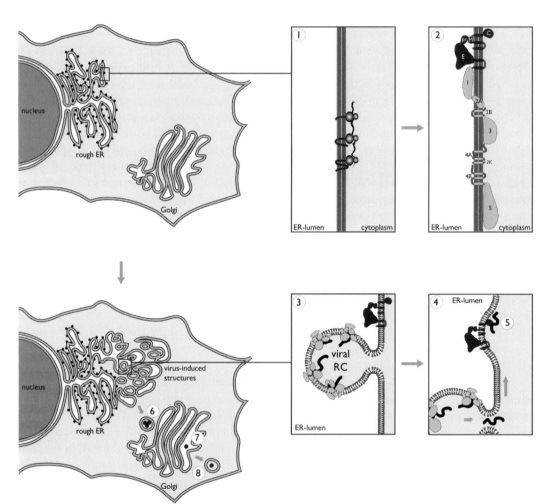

Figure 3.5 Hypothetical model of the biogenesis of the Dengue virus replication complex. Incoming viral [+]RNA is translated by ribosomes at the rough ER (1), and the polyprotein is co- and post-translationally processed into the individual components (2). Viral proteins as well as putative host cell factors may start to assemble into the viral RC, and the formation of membranous vesicles induced by NS4A and virus-induced alterations of host cell trafficking pathways may take place. RNA replication is assumed to occur within the virus-induced vesicular structures (3). Newly generated viral RNA is released from the RC by an unknown mechanism (4) and may be either used as mRNA for translation or associate with the viral core protein for assembly of progeny virus particles (5). Virion assembly proceeds via budding into the ER lumen (6). During release the particles cross the Golgi (7), where maturation occurs. Subsequently, the particles leave the cell at the plasma membrane via exocytosis (8).

initial translation viral proteins – eventually in concert with host cell factors – may assemble into the viral RC. By analogy to KUNV one can assume that NS5 binds to the 3'-end of the [+]RNA molecule together with or prior to binding of NS3 and NS2A. This RNA-protein complex resides on the cytoplasmic face of membranes through an interaction with viral integral membrane proteins such as NS4A and NS2A (Mackenzie et al., 1998; Mackenzie, 2005). In parallel to RC-formation the induction of membrane alterations by viral protein NS4A and further RC-components altering host cell trafficking pathways, might take place. The RC-bound RNA might become wrapped in the VP/SMS and contained within induced invaginations where efficient amplification of the viral RNA could be carried out (Mackenzie, 2005). The encasement of the DENV RF by membranous vesicles may reflect the need for the virus to prevent or reduce the induction of dsRNA-mediated host defence mechanisms such as interferon-induced pathways and RNA interference. In addition, the placement of the RF inside the VP/SMS would allow the reuse of the RF and improve replication efficiency due to the increase of the local concentration of replicase components. Newly generated viral RNA might be either transported to the CM/PC for translation and processing of new viral proteins or could associate with core protein to trigger assembly of progeny virus particles. For KUNV it is known that translation and RNA replication are coupled and replication of a nascent RNA molecule is required for packaging (Khromykh et al., 2001). This coupling may assure that defective (non-replicating) genomes are counterselected. Virion assembly may proceed via budding into the ER lumen. During release the particles cross the Golgi, where maturation occurs. Subsequently, they leave the cell at the plasma membrane via exocytosis. It should be noted that this model has recently been supported by a high-resolution electron tomography-based study of the DENV-induced replication sites (Welsch et al., 2009).

Concluding remarks

Extensive research during the last decades has contributed to a better understanding of the individual steps of the DENV replication cycle. However, much work remains to be done to unravel further details of this complex process. For instance, the biogenesis of DENV-induced membrane structures and the formation of the RC is not known. Furthermore, we do not known the 3D-structure of the virus-induced mini-organelles, how the membranous structures are connected with each other and the cellular endomembrane system, how newly produced viral [+]RNA is transported within the induced structures and how it is packed into progeny virus particles. In addition, a detailed understanding of how the individual factors of the viral RC interact with each other and how the assembly of the RC is regulated is lacking. A combination of classical biochemical and microscopic methods in combination with advanced techniques such as genome-wide RNA interference studies, as well as modern microscopy techniques (e.g. tomography and live cell imaging), will provide further insights into all these aspects of DENV replication. In addition, more detailed studies are required to characterize the role of nuclear-localized DENV proteins for viral replication. In the end, an improved knowledge of the molecular mechanisms of DENV replication may hopefully help to develop an effective vaccine as well as therapeutic strategies to treat infections with this insidious pathogen.

References

Ackermann, M., and Padmanabhan, R. (2001). De novo synthesis of RNA by the dengue virus RNA-dependent RNA polymerase exhibits temperature dependence at the initiation but not elongation phase. J. Biol. Chem. 276, 39926–39937.

Alvarez, D.E., De Lella Ezcurra, A.L., Fucito, S., and Gamarnik, A.V. (2005a). Role of RNA structures present at the 3'-UTR of dengue virus on translation, RNA synthesis, and viral replication. Virology 339, 200–212.

Alvarez, D.E., Lodeiro, M.F., Luduena, S.J., Pietrasanta, L.I., and Gamarnik, A.V. (2005b). Long-range RNA–RNA interactions circularize the dengue virus genome. J. Virol. 79, 6631–6643.

Alvarez, M., Rodriguez-Roche, R., Bernardo, L., Vazquez, S., Morier, L., Gonzalez, D., Castro, O., Kouri, G., Halstead, S.B., and Guzman, M.G. (2006). Dengue hemorrhagic Fever caused by sequential dengue 1–3 virus infections over a long time interval: Havana epidemic, 2001–2002. Am. J. Trop. Med. Hyg. 75, 1113–1117.

Appel, N., Herian, U., and Bartenschlager, R. (2005). Efficient rescue of hepatitis C virus RNA replication by trans-complementation with nonstructural protein 5A. J. Virol. 79, 896–909.

Arias, C.F., Preugschat, F., and Strauss, J.H. (1993). Dengue 2 virus NS2B and NS3 form a stable complex that can cleave NS3 within the helicase domain. Virology *193*, 888–899.

Bartelma, G., and Padmanabhan, R. (2002). Expression, purification, and characterization of the RNA 5′-triphosphatase activity of dengue virus type 2 nonstructural protein 3. Virology *299*, 122–132.

Bartenschlager, R., and Miller, S. (2008). Molecular aspects of Dengue virus replication. Future Microbiol. *3*, 155–165.

Barth, O.M. (1992). Replication of dengue viruses in mosquito cell cultures – a model from ultrastructural observations. Memorias do Instituto Oswaldo Cruz *87*, 565–574.

Bazan, J.F., and Fletterick, R.J. (1989). Detection of a trypsin-like serine protease domain in flaviviruses and pestiviruses. Virology *171*, 637–639.

Benarroch, D., Selisko, B., Locatelli, G.A., Maga, G., Romette, J.L., and Canard, B. (2004). The RNA helicase, nucleotide 5′-triphosphatase, and RNA 5′-triphosphatase activities of Dengue virus protein NS3 are Mg^{2+}-dependent and require a functional Walker B motif in the helicase catalytic core. Virology *328*, 208–218.

Boulton, R.W., and Westaway, E.G. (1976). Replication of the flavivirus Kunjin: proteins, glycoproteins, and maturation associated with cell membranes. Virology *69*, 416–430.

Brooks, A.J., Johansson, M., John, A.V., Xu, Y., Jans, D.A., and Vasudevan, S.G. (2002). The interdomain region of dengue NS5 protein that binds to the viral helicase NS3 contains independently functional importin beta 1 and importin alpha/beta-recognized nuclear localization signals. J. Biol. Chem. *277*, 36399–36407.

Buckley, A., Gaidamovich, S., Turchinskaya, A., and Gould, E.A. (1992). Monoclonal antibodies identify the NS5 yellow fever virus non-structural protein in the nuclei of infected cells. J. Gen. Virol. *73*, 1125–1130.

Bulich, R., and Aaskov, J.G. (1992). Nuclear localization of dengue 2 virus core protein detected with monoclonal antibodies. J. Gen. Virol. *73*, 2999–3003.

Chambers, T.J., Hahn, C.S., Galler, R., and Rice, C.M. (1990). Flavivirus genome organization, expression, and replication. Annu. Rev. Microbiol. *44*, 649–688.

Chambers, T.J., Nestorowicz, A., Amberg, S.M., and Rice, C.M. (1993). Mutagenesis of the yellow fever virus NS2B protein: effects on proteolytic processing, NS2B–NS3 complex formation, and viral replication. J. Virol. *67*, 6797–6807.

Chen, C.J., Kuo, M.D., Chien, L.J., Hsu, S.L., Wang, Y.M., and Lin, J.H. (1997). RNA–protein interactions: involvement of NS3, NS5, and 3′ noncoding regions of Japanese encephalitis virus genomic RNA. J. Virol. *71*, 3466–3473.

Chu, P.W., and Westaway, E.G. (1992). Molecular and ultrastructural analysis of heavy membrane fractions associated with the replication of Kunjin virus RNA Arch. Virol. *125*, 177–191.

Chua, J.J., Ng, M.M., and Chow, V.T. (2004). The nonstructural 3 (NS3) protein of dengue virus type 2 interacts with human nuclear receptor binding protein and is associated with alterations in membrane structure. Virus Res. *102*, 151–163.

Cleaves, G.R., Ryan, T.E., and Schlesinger, R.W. (1981). Identification and characterization of type 2 dengue virus replicative intermediate and replicative form RNAs. Virology *111*, 73–83.

Clum, S., Ebner, K.E., and Padmanabhan, R. (1997). Cotranslational membrane insertion of the serine proteinase precursor NS2B–NS3(Pro) of dengue virus type 2 is required for efficient *in vitro* processing and is mediated through the hydrophobic regions of NS2B. J. Biol. Chem. *272*, 30715–30723.

Cui, T., Sugrue, R.J., Xu, Q., Lee, A.K., Chan, Y.C., and Fu, J. (1998). Recombinant dengue virus type 1 NS3 protein exhibits specific viral RNA binding and NTPase activity regulated by the NS5 protein. Virology *246*, 409–417.

De Nova-Ocampo, M., Villegas-Sepulveda, N., and del Angel, R.M. (2002). Translation elongation factor-1alpha, La, and PTB interact with the 3′ untranslated region of dengue 4 virus RNA Virology *295*, 337–347.

Egloff, M.P., Benarroch, D., Selisko, B., Romette, J.L., and Canard, B. (2002). An RNA cap (nucleoside-2′-O-)-methyltransferase in the flavivirus RNA polymerase NS5: crystal structure and functional characterization. Embo J. *21*, 2757–2768.

Egloff, M.P., Decroly, E., Malet, H., Selisko, B., Benarroch, D., Ferron, F., and Canard, B. (2007). Structural and functional analysis of methylation and 5′-RNA sequence requirements of short capped RNAs by the methyltransferase domain of dengue virus NS5. J. Mol. Biol. *372*, 723–736.

Falgout, B., Pethel, M., Zhang, Y.M., and Lai, C.J. (1991). Both nonstructural proteins NS2B and NS3 are required for the proteolytic processing of dengue virus nonstructural proteins. J. Virol. *65*, 2467–2475.

Ford, M.G., Mills, I.G., Peter, B.J., Vallis, Y., Praefcke, G.J., Evans, P.R., and McMahon, H.T. (2002). Curvature of clathrin-coated pits driven by epsin. Nature *419*, 361–366.

Garcia-Montalvo, B.M., Medina, F., and del Angel, R.M. (2004). La protein binds to NS5 and NS3 and to the 5′ and 3′ ends of Dengue 4 virus RNA Virus Res. *102*, 141–150.

Gorbalenya, A.E., Donchenko, A.P., Koonin, E.V., and Blinov, V.M. (1989a). N-terminal domains of putative helicases of flavi- and pestiviruses may be serine proteases. Nucleic Acids Res. *17*, 3889–3897.

Gorbalenya, A.E., Koonin, E.V., Donchenko, A.P., and Blinov, V.M. (1989b). Two related superfamilies of putative helicases involved in replication, recombination, repair and expression of DNA and RNA genomes. Nucleic Acids Res. *17*, 4713–4730.

Grief, C., Galler, R., Cortes, L.M., and Barth, O.M. (1997). Intracellular localisation of dengue-2 RNA in mosquito cell culture using electron microscopic in situ hybridisation. Arch. Virol. *142*, 2347–2357.

Jan, L.R., Yang, C.S., Trent, D.W., Falgout, B., and Lai, C.J. (1995). Processing of Japanese encephalitis virus non-structural proteins: NS2B–NS3 complex and heterologous proteases. J. Gen. Virol. *76*, 573–580.

Johansson, M., Brooks, A.J., Jans, D.A., and Vasudevan, S.G. (2001). A small region of the dengue virus-encoded RNA-dependent RNA polymerase, NS5, confers interaction with both the nuclear transport receptor importin-beta and the viral helicase, NS3. J. Gen. Virol. *82*, 735–745.

Kapoor, M., Zhang, L., Ramachandra, M., Kusukawa, J., Ebner, K.E., and Padmanabhan, R. (1995). Association between NS3 and NS5 proteins of dengue virus type 2 in the putative RNA replicase is linked to differential phosphorylation of NS5. J. Biol. Chem. *270*, 19100–19106.

Khromykh, A.A., Sedlak, P.L., Guyatt, K.J., Hall, R.A., and Westaway, E.G. (1999a). Efficient trans-complementation of the flavivirus kunjin NS5 protein but not of the NS1 protein requires its coexpression with other components of the viral replicase. J. Virol. *73*, 10272–10280.

Khromykh, A.A., Sedlak, P.L., and Westaway, E.G. (1999b). trans-Complementation analysis of the flavivirus Kunjin ns5 gene reveals an essential role for translation of its N-terminal half in RNA replication. J. Virol. *73*, 9247–9255.

Khromykh, A.A., Sedlak, P.L., and Westaway, E.G. (2000). cis- and trans-acting elements in flavivirus RNA replication. J. Virol. *74*, 3253–3263.

Khromykh, A.A., Varnavski, A.N., Sedlak, P.L., and Westaway, E.G. (2001). Coupling between replication and packaging of flavivirus RNA: evidence derived from the use of DNA-based full-length cDNA clones of Kunjin virus. J. Virol. *75*, 4633–4640.

Koonin, E.V. (1993). Computer-assisted identification of a putative methyltransferase domain in NS5 protein of flaviviruses and lambda 2 protein of reovirus. J. Gen. Virol. *74 (Pt 4)*, 733–740.

Kroschewski, H., Lim, S.P., Butcher, R.E., Yap, T.L., Lescar, J., Wright, P.J., Vasudevan, S.G., and Davidson, A.D. (2008). Mutagenesis of the dengue virus type 2 NS5 methyltransferase domain. J. Biol. Chem. *283*, 19410–19421.

Lee, C.J., Lin, H.R., Liao, C.L., and Lin, Y.L. (2008). Cholesterol effectively blocks entry of flavivirus. J. Virol. *82*, 6470–6480.

Leung, J.Y., Pijlman, G.P., Kondratieva, N., Hyde, J., Mackenzie, J.M., and Khromykh, A.A. (2008). The role of nonstructural protein Ns2a in flavivirus assembly. J. Virol. *82*, 4731–4741.

Li, H., Clum, S., You, S., Ebner, K.E., and Padmanabhan, R. (1999). The serine protease and RNA-stimulated nucleoside triphosphatase and RNA helicase functional domains of dengue virus type 2 NS3 converge within a region of 20 amino acids. J. Virol. *73*, 3108–3116.

Lin, C., Amberg, S.M., Chambers, T.J., and Rice, C.M. (1993). Cleavage at a novel site in the NS4A region by the yellow fever virus NS2B-3 proteinase is a prerequisite for processing at the downstream 4A/4B signalase site. J. Virol. *67*, 2327–2335.

Lindenbach, B.D., and Rice, C.M. (1997). Trans-Complementation of yellow fever virus NS1 reveals a role in early RNA replication. J. Virol. *71*, 9608–9617.

Lindenbach, B.D., and Rice, C.M. (1999). Genetic interaction of flavivirus nonstructural proteins NS1 and NS4A as a determinant of replicase function. J. Virol. *73*, 4611–4621.

Lindenbach, B.D., and Rice, C.M. (2001). Fields virology, 4th edn. (Philadelphia: Lippincott-Raven).

Lohmann, V., Korner, F., Koch, J., Herian, U., Theilmann, L., and Bartenschlager, R. (1999). Replication of subgenomic hepatitis C virus RNAs in a hepatoma cell line. Science *285*, 110–113.

Mackenzie, J. (2005). Wrapping things up about virus RNA replication. Traffic *6*, 967–977.

Mackenzie, J.M., Jones, M.K., and Westaway, E.G. (1999). Markers for trans-Golgi membranes and the intermediate compartment localize to induced membranes with distinct replication functions in flavivirus-infected cells. J. Virol. *73*, 9555–9567.

Mackenzie, J.M., Jones, M.K., and Young, P.R. (1996). Immunolocalization of the dengue virus nonstructural glycoprotein NS1 suggests a role in viral RNA replication. Virology *220*, 232–240.

Mackenzie, J.M., Kenney, M.T., and Westaway, E.G. (2007). West Nile virus strain Kunjin NS5 polymerase is a phosphoprotein localized at the cytoplasmic site of viral RNA synthesis. J. Gen. Virol. *88*, 1163–1168.

Mackenzie, J.M., Khromykh, A.A., Jones, M.K., and Westaway, E.G. (1998). Subcellular localization and some biochemical properties of the flavivirus Kunjin nonstructural proteins NS2A and NS4A. Virology *245*, 203–215.

Matusan, A.E., Pryor, M.J., Davidson, A.D., and Wright, P.J. (2001). Mutagenesis of the Dengue virus type 2 NS3 protein within and outside helicase motifs: effects on enzyme activity and virus replication. J. Virol. *75*, 9633–9643.

Miller, S., Kastner, S., Krijnse-Locker, J., Buhler, S., and Bartenschlager, R. (2007). The non-structural protein 4A of dengue virus is an integral membrane protein inducing membrane alterations in a 2K-regulated manner. J. Biol. Chem. *282*, 8873–8882.

Miller, S., Sparacio, S., and Bartenschlager, R. (2006). Subcellular localization and membrane topology of the dengue virus type 2 non-structural protein 4B. J. Biol. Chem. *281*, 8854–8863.

Miyanari, Y., Hijikata, M., Yamaji, M., Hosaka, M., Takahashi, H., and Shimotohno, K. (2003). Hepatitis C virus non-structural proteins in the probable membranous compartment function in viral genome replication. J. Biol. Chem. *278*, 50301–50308.

Munoz-Jordan, J.L., Laurent-Rolle, M., Ashour, J., Martinez-Sobrido, L., Ashok, M., Lipkin, W.I., and Garcia-Sastre, A. (2005). Inhibition of alpha/beta interferon signaling by the NS4B protein of flaviviruses. J. Virol. *79*, 8004–8013.

Munoz-Jordan, J.L., Sanchez-Burgos, G.G., Laurent-Rolle, M., and Garcia-Sastre, A. (2003). Inhibition of interferon signaling by dengue virus. Proc. Natl. Acad. Sci. U.S.A. *100*, 14333–14338.

Murthy, H.M., Clum, S., and Padmanabhan, R. (1999). Dengue virus NS3 serine protease. Crystal structure and insights into interaction of the active site with substrates by molecular modeling and structural analysis of mutational effects. J. Biol. Chem. *274*, 5573–5580.

Murthy, H.M., Judge, K., DeLucas, L., and Padmanabhan, R. (2000). Crystal structure of Dengue virus NS3 protease in complex with a Bowman-Birk inhibitor:

implications for flaviviral polyprotein processing and drug design. J. Mol. Biol. *301*, 759–767.

Ng, M.L., Pedersen, J.S., Toh, B.H., and Westaway, E.G. (1983). Immunofluorescent sites in vero cells infected with the flavivirus Kunjin. Arch. Virol. *78*, 177–190.

Ng, M.L., Tan, S.H., and Chu, J.J. (2001). Transport and budding at two distinct sites of visible nucleocapsids of West Nile (Sarafend) virus. J. Med. Virol. *65*, 758–764.

Nomaguchi, M., Ackermann, M., Yon, C., You, S., and Padmanabhan, R. (2003). *De novo* synthesis of negative-strand RNA by Dengue virus RNA-dependent RNA polymerase *in vitro*: nucleotide, primer, and template parameters. J. Virol. *77*, 8831–8842.

Pang, X., Zhang, M., and Dayton, A.I. (2001a). Development of dengue virus replicons expressing HIV-1 gp120 and other heterologous genes: a potential future tool for dual vaccination against dengue virus and HIV. BMC Microbiol. *1*, 28.

Pang, X., Zhang, M., and Dayton, A.I. (2001b). Development of Dengue virus type 2 replicons capable of prolonged expression in host cells. BMC Microbiol. *1*, 18.

Paranjape, S.M., and Harris, E. (2007). Y box-binding protein-1 binds to the dengue virus 3'-untranslated region and mediates antiviral effects. J. Biol. Chem. *282*, 30497–30508.

Pryor, M.J., Rawlinson, S.M., Butcher, R.E., Barton, C.L., Waterhouse, T.A., Vasudevan, S.G., Bardin, P.G., Wright, P.J., Jans, D.A., and Davidson, A.D. (2007). Nuclear localization of dengue virus nonstructural protein 5 through its importin alpha/beta-recognized nuclear localization sequences is integral to viral infection. Traffic *8*, 795–807.

Pryor, M.J., Rawlinson, S.M., Wright, P.J., and Jans, D.A. (2006). CRM1-dependent nuclear export of dengue virus type 2 NS5. Novartis Foundation symposium *277*, 149–161; discussion 161–143, 251–143.

Quinkert, D., Bartenschlager, R., and Lohmann, V. (2005). Quantitative analysis of the hepatitis C virus replication complex. J. Virol. *79*, 13594–13605.

Roosend aal, J., Westaway, E.G., Khromykh, A., and Mackenzie, J.M. (2006). Regulated Cleavages at the West Nile Virus NS4A-2K-NS4B Junctions Play a Major Role in Rearranging Cytoplasmic Membranes and Golgi Trafficking of the NS4A Protein. J. Virol. *80*, 4623–4632.

Sangiambut, S., Keelapang, P., Aaskov, J., Puttikhunt, C., Kasinrerk, W., Malasit, P., and Sittisombut, N. (2008). Multiple regions in dengue virus capsid protein contribute to nuclear localization during virus infection. J. Gen. Virol. *89*, 1254–1264.

Tadano, M., Makino, Y., Fukunaga, T., Okuno, Y., and Fukai, K. (1989). Detection of dengue 4 virus core protein in the nucleus. I. A monoclonal antibody to dengue 4 virus reacts with the antigen in the nucleus and cytoplasm. J. Gen. Virol. *70*, 1409–1415.

Tan, B.H., Fu, J., Sugrue, R.J., Yap, E.H., Chan, Y.C., and Tan, Y.H. (1996). Recombinant dengue type 1 virus NS5 protein expressed in Escherichia coli exhibits RNA-dependent RNA polymerase activity. Virology *216*, 317–325.

Uchil, P.D., Kumar, A.V., and Satchidanandam, V. (2006). Nuclear localization of flavivirus RNA synthesis in infected cells. J. Virol. *80*, 5451–5464.

Uchil, P.D., and Satchidanandam, V. (2003). Architecture of the flaviviral replication complex. Protease, nuclease, and detergents reveal encasement within double-layered membrane compartments. J. Biol. Chem. *278*, 24388–24398.

Umareddy, I., Chao, A., Sampath, A., Gu, F., and Vasudevan, S.G. (2006). Dengue virus NS4B interacts with NS3 and dissociates it from single-stranded RNA. J. Gen. Virol. *87*, 2605–2614.

Wang, S.H., Syu, W.J., Huang, K.J., Lei, H.Y., Yao, C.W., King, C.C., and Hu, S.T. (2002). Intracellular localization and determination of a nuclear localization signal of the core protein of dengue virus. J. Gen. Virol. *83*, 3093–3102.

Welsch, S., Miller, S., Romero-Brey, I., Merz, A., Bleck, C.K.E., Walther, P., Fuller, S.D., Antony, C., Krijnse-Locker, J., and Bartenschlager, R. (2009) Composition and three-dimensional architecture of the Dengue virus replication and assembly sites. Cell Host Microbe *4*, 365–375.

Wengler, G. (1993). The NS3 nonstructural protein of flaviviruses contains an RNA triphosphatase activity. Virology *197*, 265–273.

Wengler, G., Czaya, G., Farber, P.M., and Hegemann, J.H. (1991). *In vitro* synthesis of West Nile virus proteins indicates that the amino-terminal segment of the NS3 protein contains the active centre of the protease which cleaves the viral polyprotein after multiple basic amino acids. J. Gen. Virol. *72 (Pt 4)*, 851–858.

Westaway, E.G., Khromykh, A.A., Kenney, M.T., Mackenzie, J.M., and Jones, M.K. (1997a). Proteins C and NS4B of the flavivirus Kunjin translocate independently into the nucleus. Virology *234*, 31–41.

Westaway, E.G., Mackenzie, J.M., Kenney, M.T., Jones, M.K., and Khromykh, A.A. (1997b). Ultrastructure of Kunjin virus-infected cells: colocalization of NS1 and NS3 with double-stranded RNA, and of NS2B with NS3, in virus-induced membrane structures. J. Virol. *71*, 6650–6661.

Westaway, E.G., Mackenzie, J.M., and Khromykh, A.A. (2003). Kunjin RNA replication and applications of Kunjin replicons. Adv. Virus Res. *59*, 99–140.

Winkler, G., Maxwell, S.E., Ruemmler, C., and Stollar, V. (1989). Newly synthesized dengue-2 virus nonstructural protein NS1 is a soluble protein but becomes partially hydrophobic and membrane-associated after dimerization. Virology *171*, 302–305.

Winkler, G., Randolph, V.B., Cleaves, G.R., Ryan, T.E., and Stollar, V. (1988). Evidence that the mature form of the flavivirus nonstructural protein NS1 is a dimer. Virology *162*, 187–196.

Xu, T., Sampath, A., Chao, A., Wen, D., Nanao, M., Chene, P., Vasudevan, S.G., and Lescar, J. (2005). Structure of the Dengue virus helicase/nucleoside triphosphatase catalytic domain at a resolution of 2.4 A. J. Virol. *79*, 10278–10288.

Yocupicio-Monroy, M., Padmanabhan, R., Medina, F., and del Angel, R.M. (2007). Mosquito La protein binds to the 3' untranslated region of the positive and negative polarity dengue virus RNAs and relocates to the cytoplasm of infected cells. Virology *357*, 29–40.

Yocupicio-Monroy, R.M., Medina, F., Reyes-del Valle, J., and del Angel, R.M. (2003). Cellular proteins from human monocytes bind to dengue 4 virus minus-strand 3′ untranslated region RNA. J. Virol. *77*, 3067–3076.

Yon, C., Teramoto, T., Mueller, N., Phelan, J., Ganesh, V.K., Murthy, K.H., and Padmanabhan, R. (2005). Modulation of the nucleoside triphosphatase/RNA helicase and 5′-RNA triphosphatase activities of Dengue virus type 2 nonstructural protein 3 (NS3) by interaction with NS5, the RNA-dependent RNA polymerase. J. Biol. Chem. *280*, 27412–27419.

Yu, G.Y., Lee, K.J., Gao, L., and Lai, M.M. (2006). Palmitoylation and polymerization of hepatitis C virus NS4B protein. J. Virol. *80*, 6013–6023.

Zimmerberg, J., and Kozlov, M.M. (2006). How proteins produce cellular membrane curvature. Nat. Rev. Mol. Cell Biol. *7*, 9–19.

Role of the Dengue Virus 5′ and 3′ Untranslated Regions in Viral Replication

4

Andrea Gamarnik

Abstract

As with all plus-stranded RNA viruses, dengue virus (DENV) genomic RNA is infectious. Transfection of full-length DENV RNA genome into a susceptible cell triggers a complete cycle of viral replication. Construction of cDNA clones together with reverse genetics has proven to be a valuable tool to uncover genetic determinants of viral replication and to understand the function of the viral untranslated regions (UTRs). Translation initiation and initiation of RNA synthesis occur at the 5′ and 3′ terminal regions of the genome, respectively, and rely on complex RNA–RNA and RNA- protein interactions. The DENV 5′-UTR contains two defined RNA structures, Stem–loop A and Stem–loop B, which have distinct functions during the process of viral RNA synthesis. The viral 3′-UTR contains three domains with conserved sequences and structures. In these domains, there are RNA elements that are essential for the replication process and other elements that act as enhancers of the process. The 5′ and 3′ terminal regions of the viral RNA also carry inverted complementary sequences that mediate long-range RNA–RNA interactions and genome cyclization. It has been demonstrated for dengue and other flaviviruses that the circular conformation of the genome is a crucial determinant for viral replication. In the last few years, a great deal has been learned about the mechanisms by which the viral UTRs function during DENV replication. In this chapter, I discuss the current understanding of the function of different RNA structures of the DENV UTRs and provide some
working models of how genome cyclization enables DENV RNA synthesis.

The 5′ untranslated region

Sequence and structures of DENV 5′-UTR

The 5′-UTR sequences of the four dengue virus serotypes (DENV-1 to DENV-4) are between 95 to 101 nucleotides long and contain a type I cap structure at the 5′ end (m7GpppAm). Alignment of the 5′ terminal regions from different DENV serotypes shows high sequence conservation (Fig. 4.1A). RNA folding predictions of these sequences indicate the formation of a large 5′ terminal stem–loop (SLA) followed by a smaller stem–loop structure (SLB), which ends in the translation initiation codon (Fig. 4.1B). These two stem–loop structures are separated by an oligo(U) sequence. The SLA is predicted to have a Y shaped structure, which was recently confirmed by enzymatic and chemical probing (Lodeiro et al., 2009; Polacek et al., 2009). These studies indicate the presence of double stranded regions interrupted by bulges, and highly reactive single stranded regions in agreement with the presence of a side stem–loop and a top loop.

Using DENV-2 as a reference, the predicted SLA shows three helical regions (S1, S2, and S3), a side stem–loop (SSL), and a top loop (TL) (Fig. 4.1B). Co-variations within the three helical regions and in the stem of the SSL that conserve the structural elements were found when the predicted SLA structures of the four serotypes

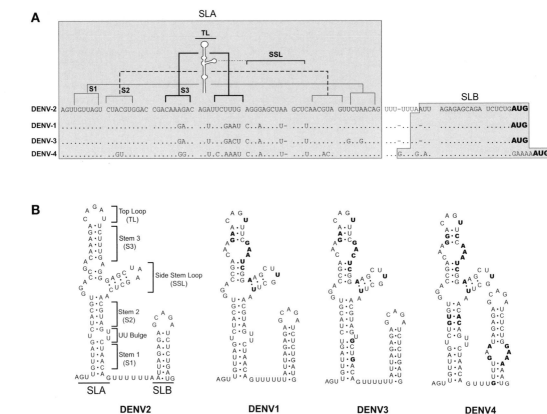

Figure 4.1 Comparison of the nucleotide sequence and secondary structure of the 5′-UTR from different DENV serotypes. **A** Alignment of the consensus sequence of the 5′-UTR of DENV-1, DENV-2, DENV-3, and DENV-4. Sequences corresponding to the SLA and SLB are indicated with grey boxes. In the SLA region, the sequence corresponding to the predicted secondary structures stem 1 (S1), stem 2 (S2), stem 3 (S3), top loop (TL), and side stem–loop (SSL) are indicated on the top. The translation start codon AUG is in bold. **B** Comparative RNA secondary structure analysis predicted for the 5′-UTR of the four DENV serotypes. Nucleotide variations between DENV-2 and the other types are indicated in bold.

were analysed, suggesting an important role of these elements in viral replication. The bottom part of the SLA, including S1 and S2, displays the highest conservation. Between these two helical regions there is a conserved UU bulge in DENV-1, 2, and 4, while there is a single U bulge in DENV-3 (Fig. 4.1B). The sequence and predicted structure of the SSL and S3 show the highest variation. Regarding the SSL structure, DENV-2 has a predicted stem with one base pair more than that predicted for the other 3 serotypes. In addition, the sequence of the predicted loop of SSL in DENV-2 is UAA, while in the other serotypes there is a tetraloop CUUG. The helical region S3 in DENV-2 also differs from that in the other serotypes. While in DENV-2 S3 contains six contiguous base parings, in the other DENV

types the helical region is interrupted by three mismatches (Fig. 4.1B).

Interestingly, the conserved structural elements described within the SLA of DENV are also found at the 5′ end of other members of the flavivirus genus. In a pioneering study by Brinton and Dispoto, the 5′-UTR sequences of different mosquito-borne flaviviruses were compared. This study indicated that conserved secondary structures were present at the 5′ end of West Nile virus (WNV), Saint Louis encephalitis virus (SLEV), DENV, yellow fever virus (YFV), and Murray Valley encephalitis virus (MVEV) (Brinton and Dispoto, 1988). More recently, the predicted structures at the 5′ end of the genomes of tick-borne flaviviruses and flaviviruses with no known vector were found to be similar to that observed in the mosquito-borne flavivirus (Gritsun and Gould, 2007; Gritsun *et al.*, 1997; Leyssen *et al.*, 2002; Mandl *et al.*, 1993). The high conservation of the structure present at the 5′ end of flavivirus genomes provided the first evidence of a functional role of this RNA element in viral replication.

Structural elements of the SLA that are crucial for DENV replication

Cahour and co-workers reported the first functional analysis of the role of 5′-UTR sequences in flavivirus replication (Cahour *et al.*, 1995). In this study, deletions from 5 to 25 nucleotides were incorporated throughout the 5′-UTR by mutagenesis of an infectious DNA copy of DENV-4. The dominant effect of the deletions appeared to be at the level of RNA synthesis, and many of the mutations were found to be lethal. More recently, using a DENV-2 infectious clone, each of the conserved structural elements within the SLA was modified and the resulting RNAs were tested for their ability to replicate in transfected cells (Lodeiro *et al.*, 2009). Mutations at each side of stem S1 or S2, disrupting base pairings, impaired viral replication. Revertant viruses carrying spontaneous mutations were found to partially reconstitute the helical regions. In addition, mutations at both sides of the stems that maintained the structure but changed the nucleotide sequence replicated efficiently. These observations, together with the sequence conservation, and the co-variations observed in different sequences within S1 and S2 (Fig. 4.1B), indicate a requirement of the bottom part of the SLA for its function during viral replication. Deletion or substitution of the UU bulge between the helical regions S1 and S2, indicated that this element is essential for replication, and that at least one U residue in the bulge is required. Substitution of the UU bulge for an AA bulge gave rise to revertant viruses with a UA bulge, confirming the requirement of at least one U in that position. Deletion of the SSL abolished viral replication, however, changing the sequence of the stem, the sequence of the loop, or modifying the length of the stem resulted in RNAs that were able to replicate (Lodeiro *et al.*, 2009). Regarding the helical region S3, substitutions disrupting three base pairs present in the middle of S3, yielded viral RNAs that replicated as efficiently

as the parental RNA. However, disruption of the closing base pair at the top of S3 was lethal, suggesting that although S3 tolerates large variations, structural features of this region are important for viral replication. Finally, substitution of four nucleotides of the TL resulted in spontaneous insertions that partially restored the sequence as well as the structure of the loop, and rescued viral replication.

The infectivity of DENV-2 RNAs with mutations in the SLA indicated that certain elements in this structure are highly constrained while others are more permissive to change. There is an evident correlation between the significant variability of S3 and SSL in different DENV isolates and the high tolerance observed for replication of viral RNAs carrying mutations within these structures. In addition, the high conservation within S1 and S2 also correlates with the limited tolerance to mutations observed in these helical regions (Lodeiro *et al.*, 2009).

The 3′ untranslated region

Sequence and predicted structures of DENV 3′-UTRs

The 3′-UTRs of DENV genomes are highly conserved in sequence and structure (Shurtleff *et al.*, 2001). The length of these regions ranges from 470 nucleotides in the case of DENV-1 to about 385 nucleotides in DENV-4. DENV-2 and DENV-3 3′-UTRs are about 450 and 430 nucleotides, respectively. In the 3′-UTR, three domains can be described according to the predicted secondary structures (Olsthoorn and Bol, 2001; Proutski *et al.*, 1997; Rauscher *et al.*, 1997). Domain I is located immediately after the stop codon of NS5, and is considered the most variable region within the viral 3′-UTRs (VR, Fig. 4.2A). It exhibits extensive size variation between serotypes; it can be from more than 120 nucleotides in the case of DENV-1, with imperfect repeats absent in the other serotypes, to less than 50 nucleotides in DENV-4 (Shurtleff *et al.*, 2001; Silva *et al.*, 2008; Zhou *et al.*, 2006). Deletions and nucleotide variations were also observed in the variable regions within the same serotype (Aquino *et al.*, 2006; Roche *et al.*, 2007; Vasilakis *et al.*, 2008). Domain II is of moderate conservation, comprising several hairpin motifs,

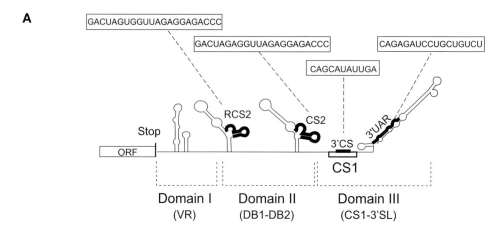

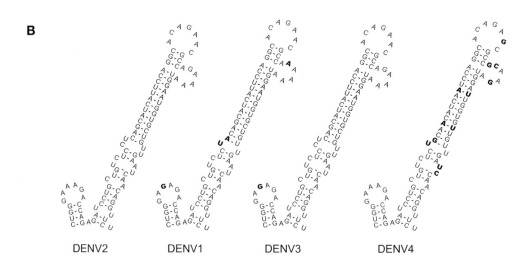

Figure 4.2 Conserved RNA sequences and structures in DENV 3'-UTRs. **A** Schematic representation of RNA elements found at the 3'-UTR of DENV genome. The predicted secondary structures of the three defined domains are indicated: domain I (variable region, VR), domain II (dumbbell structures, DB), and domain III (conserved sequence CS1 and 3' SL). In addition, the location and the sequence of each of the conserved elements corresponding to RCS2, CS2, 3' CS, and 3' UAR are shown. **B** Comparative RNA secondary structure analysis predicted for the 3' SL of the four DENV serotypes. Nucleotide variations between DENV-2 and the other types are indicated in bold.

including a characteristic dumbbell (DB) structure in DENV-1, DENV-2, and DENV-3, and a turret structure proposed in DENV-4 (Shurtleff *et al.*, 2001; Romero *et al.*, 2006). The DB element contains a conserved sequence named CS2, motif of 22 nucleotides, present in all mosquito borne flaviviruses (Gritsun and Gould, 2006; Hahn *et al.*, 1987; Olsthoorn and Bol, 2001; Romero *et al.*, 2006). This DB structure is duplicated in tandem, containing a repeated CS2 (RCS2) located about 65 nucleotides upstream of CS2 (Fig. 4.2A). Although the RNA structures within domains I and II are considered dispensable for flavivirus replication, these elements are believed to serve as replication enhancers. Genomes bearing deletions of domains I and II exhibit decreased viral RNA synthesis and attenuation (Alvarez *et al.*, 2005a; Bredenbeek *et al.*, 2003; Lo *et al.*, 2003; Mandl *et al.*, 1998; Men *et al.*, 1996). Domain III is the most conserved region of the 3'-UTR, bearing a conserved sequence CS1 followed by a terminal stem–loop structure (3' SL). In CS1 there is a cyclization sequence involved in long range RNA–RNA interactions between the ends of the viral genome (Hahn *et al.*, 1987). The 3' terminal structure contains a short stem–loop of 14 nucleotides followed by a large stem–loop of 79 nucleotides. The two adjacent structures involve 93 nucleotides and are referred to as 3' SL (Fig. 4.2B). The large stem–loop is interrupted by bulges and mismatches in conserved locations. In addition, the sequence of the bottom part of the 3' SL contains a sequence known as 3' UAR, which is complementary to a sequence located at the 5'-UTR, also involved in genome cyclization (Alvarez *et al.*, 2008; Alvarez *et al.*, 2005b). The presence of the 3' SL has been supported by secondary structure predictions, co-variation analysis, and biochemical probing in dengue and other members of the Flaviviridae family (Blight and Rice, 1997; Brinton *et al.*, 1986; Deng and Brock, 1993; Grange *et al.*, 1985; Hahn *et al.*, 1987; Proutski *et al.*, 1997; Rice *et al.*, 1985; Takegami *et al.*, 1986).

The role of different domains of the 3'-UTR in DENV replication

Deletions of different sequences of the DENV 3'-UTRs have demonstrated that elements of domain I and domain II are not essential for viral replication, while CS1 and the 3' SL are necessary for viral viability. Using an infectious cDNA of DENV-4, Men and co-workers introduced a series of deletions from 30 to 262 nucleotides of the viral 3'-UTR (Men *et al.*, 1996). These mutations were generated using a unique ApaI restriction site located between RCS2 and CS2. Complete or partial deletions of CS2, RCS2, or the VR were introduced. Eight different mutated RNAs were transfected in simian LLC-MK2 and C6/36 mosquito cells. The maximum 3' boundary of any deletion that gave rise to virus progeny was 113 nucleotides from the 3' end of the genome, preserving CS1 and the 3' SL. A mutant including all RCS2 and the VR or a mutant with a deletion of the complete CS2 produced viral progeny but the replication was delayed compared to the parental virus. In addition, a deletion of CS2 and CS1 failed to produce plaques and viral progeny was not detected, confirming the essential role of CS1. In this study, it was demonstrated that DENV-4 tolerated deletions of more than half of the 3'-UTR, provided that the deleted sequences were upstream of the CS1 and the 3' SL (Men *et al.*, 1996).

The requirement for different structures in DENV replication was also investigated using a DENV-2 replicon system carrying a luciferase gene replacing the structural proteins (Alvarez *et al.*, 2005a). Using this construct, defined RNA domains of the 3'-UTR were deleted. Each of the dumbbell structures containing CS2 and RCS2 were deleted separately or together. In addition, the variable region or the 450 nucleotides of the complete 3'-UTR were removed. Translation of the transfected RNA and replicon amplification were determined in BHK and C6/36 cells by measuring luciferase activity as a function of time. Translation of the transfected RNAs carrying different deletions of the 3'-UTR was efficient, suggesting that these structures were not necessary for the first rounds of translation. In contrast, important defects in amplification of the replicons were observed. As expected, the deletion of the complete 3'-UTR yielded RNAs that were unable to replicate. Deletion of the VR decreased RNA replication 10-fold in BHK cells while it did not affect replicon amplification in C6/36 cells. Replicons carrying deletions of the structures that contain CS2 or RCS2 replicated 100 fold less efficiently than the parental RNA in both mosquito and hamster cells. In addition, deletion of both structures including CS2 and RCS2 yielded replicons with undetectable RNA amplification.

In the same study, the deletions of the 3'-UTR were performed in the context of the full-length cDNA clone of DENV-2. Replication of the viral RNAs was evaluated by immunofluorescence in transfected cells. In these experiments, replication of viruses with deletions of the VR, CS2, or RCS2 were delayed but viral progeny was obtained 3 days after transfection. In contrast to the results obtained with the replicon system, replication of the mutant carrying a deletion of both DB structures including CS2 and RCS2 showed immunofluorescence-positive cells 10 days after transfection, and although the titres were low, viral progeny was recovered. The sublethal phenotype observed with this mutant explains the undetectable levels of RNA synthesis at 72 h post-transfection of the replicon carrying the same deletion. Because the two DB structures contain similar conserved sequences, it is possible that they perform the same function and that the loss of one can be compensated by the other. However, deletion of both structures results in a profound negative effect on DENV replication (Alvarez et al., 2005a).

More recently, the role of different sequences of the VR was examined using a DENV-1 infectious clone. The VR was divided into two subregions: a 5′ terminal hypervariable region (HVR) and a 3′ terminal semivariable region (SVR). Recombinant DENV-1 deletion of either or both subregions exhibited reduced growth properties in mammalian cells compared with the parental virus (Tajima et al., 2007). Substitution of the HVR with unrelated sequences resulted in viruses that replicated efficiently; however, replacement of the SVR with unrelated sequences did not recover the phenotype of the parental virus, suggesting different roles for HVR and SVR.

The results obtained from different laboratories, employing DENV-4, DENV-2, and DENV-1 infectious clones are in agreement and demonstrated that sequences and or structures within domain I and domain II enhance viral replication, while domain III is essential. In addition, the replicon system helped to define a role of these 3′-UTR elements during amplification of the viral RNA.

The essential role of the 3′ SL

A similar 3′ stem–loop structure can be predicted at the 3′ termini of all flavivirus genomes. Biochemical evidence for the existence of the 3′ SL structure in solution was provided using WNV RNA sequences (Brinton et al., 1986). Numerous studies have demonstrated an essential role of this domain in flavivirus replication (for review see (Markoff, 2003)). In DENVs the 3′ SL contains a total of 93 nucleotides. The long stem is interrupted by mismatches in conserved locations, a six nucleotide bulge is predicted near the top, and the loop contains seven nucleotides (Fig. 4.2B). Between serotypes only few nucleotide variations can be observed. Most of the nucleotide variations can be found in positions predicted to be single stranded; however, in DENV-4 two

co-variations in the long stem that maintain the structure can be observed. An invariable region of 16 nucleotides known as 3′ UAR is present in the small stem–loop at the 5′ of the 3′ SL that extends into the bottom of the large stem–loop. This sequence is complementary to the 5′ UAR element located at the viral 5′-UTR, which mediates long range RNA–RNA interactions between the ends of the genome (Alvarez et al., 2008; Alvarez et al., 2005b).

To investigate the role of different structural elements and nucleotide sequences of the 3′ SL in viral replication, mutagenesis of an infectious clone of DENV-2 was used (Zeng et al., 1998). Replacement of the 93 terminal nucleotides of DENV-2 genome corresponding to the complete 3′ SL by the analogous sequence of WNV was sublethal for viral replication. While DENV-2 wild type produced 10^6 pfu/ml in monkey cells, the chimeric virus never achieved titres greater than 100 pfu/ml, indicating that a specific nucleotide sequence, not merely the secondary structure of the 3′ SL, was a determinant for replication efficiency. In order to define which nucleotides of the 3′ SL were necessary for efficient replication, the DENV-2 3′ SL was partially restored. Construction of chimeric viruses containing segments of the 3′ SL of WNV in the context of DENV-2 genome indicated that the wild type bottom half of the long stem was a requirement for viral viability. In contrast, exchanging the top half of the 3′ SL or the adjacent small stem–loop of 14 nucleotides was well tolerated (Zeng et al., 1998).

All vector-borne flavivirus genomes contain a pentanucleotide sequence CACAG at the top of the large stem–loop as well as a dinucleotide CU sequence at the 3′ end. Mutation of the pentanucleotide in different flavivirus genomes indicated that nucleotides at specific locations are necessary for viral replication, while others can be substituted without losing viral viability (Elghonemy et al., 2005; Khromykh et al., 2003; Silva et al., 2007; Tilgner et al., 2005; Yu and Markoff, 2005). The top loop of the large stem–loop in DENVs has a highly conserved ACAGAAC sequence. In DENV-4 a single nucleotide variation A to G can be found (ACAGA<u>G</u>C). The conservation of three G–C base pairs closing the top loop suggests that a stable stem is necessary for correct presentation of the loop sequence. Using Kunjin virus (KUNV) replicons and WNV infectious clones, a requirement for the 3′ terminal CU nucleotides for viral replication has been demonstrated (Khromykh *et al.*, 2003; Tilgner and Shi, 2004). In a more recent study employing DENV-2 infectious RNAs, a series of 3′ end deletions were evaluated for viral replication (Teramoto *et al.*, 2008). Transfection of the mutated RNAs resulted in viruses with different 3′ end nucleotides. Consensus sequences were selected to maintain CU or UU terminus, suggesting a preference of U residue at the 3′ end of the viral genome.

Little is known about the molecular details of how the 3′ SL participates in the mechanism of viral RNA synthesis. It is possible that sequences and structures of the 3′ SL provide the recognition site for assembly of the RNA replication complex, or it could be the binding site for host proteins that facilitate this process. For example, *in vitro* studies have demonstrated the binding of cellular and viral proteins to the 3′ SL RNA. Binding of EF-1α, PTB, YB-1, and human La autoantigen to the 3′ SL has been demonstrated (Blackwell and Brinton, 1997; De Nova-Ocampo *et al.*, 2002; Garcia-Montalvo *et al.*, 2004; Paranjape and Harris, 2007; Yocupicio-Monroy *et al.*, 2003, 2007). However, it remains unclear how these cellular proteins participate in viral replication. In addition, specific nucleotides of the 3′ SL could interact with other RNA elements of the viral genome to provide the correct conformation for initiation of RNA synthesis.

DENV 3′-UTRs and host-dependent enhancement of viral replication

Like most arthropod borne viruses, DENV can cause significant damage when they infect vertebrate cells, yet in most cases mosquito cells sustain persistent DENV infection. It is plausible that both host and viral factors are responsible for differential growth of DENV in mosquito and mammalian cells. It was reported that RNA sequences within the VR of the 3′-UTR enhances DENV-2 replicon RNA synthesis in BHK cells but not in mosquito cells (Alvarez *et al.*, 2005a). Furthermore, recombinant full-length genomes carrying a 155 nucleotide deletion corresponding to the VR (DENV-2ΔVR) showed delayed replication in BHK cells. Comparison of growth

curves of DENV-2 and DENV-2ΔVR showed differences of about 100-fold in viral titres at 24 h post infection. In contrast, replication of DENV-2ΔVR in C6/36 cells was as efficient as the parental virus (Alvarez *et al.*, 2005a). In agreement with these observations, a large deletion in the 3′-UTR of DENV-4 including the VR was reported to cause differential growth in mosquito and LLC-MK2 cells (Men *et al.*, 1996). In a different report, using a DENV-1 infectious clone, it was demonstrated that deletion of subregions of the VR delayed viral replication in Vero cells without altering replication in mosquito cells (Tajima *et al.*, 2007).

Mutations in other regions of the 3′-UTR, outside of the VR, were also shown to have a host dependent regulation of viral replication. A mutation in the 3′ SL in DENV-2 was shown to have a host range-restricted phenotype in mosquito cells (Zeng *et al.*, 1998). Substitution of the bottom half of the DENV-2 3′ SL for the analogous sequences of the WNV resulted in viruses with delayed replication in monkey cells. However, IF positive cells were detected at 3 days post transfection and a large part of the monolayer was positive by day 10. In contrast, the same mutant grew very poorly in mosquito cells. The titres obtained in C6/36 cells were about 1000 to 10000 fold lower than that obtained in monkey cells (Zeng *et al.*, 1998). A similar host restricted phenotype was observed when the mutations in the 3′ SL were introduced in infectious clones of other DENV serotypes (Markoff *et al.*, 2002). In addition, deletion of specific nucleotides in the central region of the 3′-UTR of DENV-3 showed a strong host range restriction phenotype, with complete loss of replication in C6/36 mosquito cells despite robust replication in Vero cells (DENV-3Δ30/31, Blaney *et al.*, 2008). Furthermore, in a different report, chimeric viruses were constructed to investigate a possible correlation between pathogenesis and viral replication in culture. In these studies, viruses carrying the 3′-UTR of an American DENV-2 in the context of an Asian DENV-2 isolate yielded viruses with impaired replication in Vero cells but efficient replication in mosquito cells (Cologna and Rico-Hesse, 2003).

The finding that RNA elements present at the 3′-UTR differentially modulate viral replication in insect and mammalian cells, suggests that specific host proteins/RNAs bind to sequences and/or structures within 3'-UTR to enhance viral replication. Identification of these host factors will help to understand the requirements for DENV replication in different cell types.

DENV untranslated regions and viral translation

Similar to cellular mRNAs, DENV genomes contain a methylated cap structure at the 5' end. However, the viral RNAs are not polyadenylated. Because the poly(A) tail plays an important role in RNA stability and enhancement of translation initiation, it is likely that DENV has alternative mechanisms that facilitate those processes. The contribution of 5' and 3' terminal regions of DENV on translation was investigated by several groups using either mRNAs encoding a luciferase gene flanked by the viral 5' and 3'-UTRs or DENV replicons (Alvarez et al., 2005a, 2008; Chiu et al., 2005; Holden and Harris, 2004; Lodeiro et al., 2009). Transfection of mRNAs encoding luciferase into Vero and BHK cells indicated that the viral 3'-UTR elements, particularly the 3' SL structure, enhance translation (Chiu et al., 2005; Holden and Harris, 2004). In contrast, in DENV replicons competent for translation and RNA replication, there was no significant effect on translation when the 5' and 3'-UTRs were mutated or specific domains were deleted (Alvarez et al., 2005a). Mutations within SLA or SLB in the 5'-UTR or deletions of each of the 3'-UTR domains (the VR, each of the dumbbell structures, or the 3' SL) did not compromise viral translation. The different results obtained using the two reporter systems may be attributable to a role of viral 3'-UTR elements on elongation or termination of translation, processes that have not yet been analysed.

In the infected cell, viral RNA is translated in different conditions throughout the viral life cycle. Upon infection, translation of the input RNA depends on cis-acting elements of the RNA and the host translation machinery. However, after several rounds of translation and RNA synthesis, new viral RNA molecules have to be translated in the presence of viral proteins, in a host cell modified by viral products. In this latter scenario, a complex interaction between viral and host factors influence viral RNA translation.

Interestingly, DENV translation has been shown to occur under circumstances that inhibit cellular cap-dependent translation, suggesting that in those conditions the virus could initiate translation independently of the cap at the 5' end (Edgil et al., 2006). Treatment of cells with inhibitors of cellular cap-dependent translation (LY294002, wortmannin, or eIF4E-specific siRNAs) did not affect the level of DENV protein synthesis or the yield of infectious DENV progeny (Edgil et al., 2006). It was proposed that in cells competent for cap-dependent translation, DENV translates via its 5' m7G-cap structure. However, under conditions in which the cap-binding protein or other cellular translation factors are limiting, viral translation initiates by an alternative mechanism.

The start codon for translation of DENV-1, 2, and 3, was found to be in a poor Kozak initiation context and therefore it is expected to be utilized inefficiently (Clyde and Harris, 2006). It has been demonstrated that a predicted RNA hairpin structure downstream of the DENV-2 start codon ('cHP' for C-coding region hairpin) enhances translation from the first AUG in a position-dependent manner (Clyde and Harris, 2006). Disruption of the cHP resulted in undetectable virus titres, whereas restoration of base pairing by compensatory mutations rescued viral replication. More recently, it was demonstrated that this RNA structure is also necessary for RNA replication (Clyde et al., 2008).

Interaction between the 5'- and 3'-UTRs

Inverted complementary sequences in the flavivirus genomes

Flavivirus genomes possess inverted complementary sequences at the ends of the RNA, similar to those observed in the negative strand RNA bunya-, arena- and orthomyxoviruses (Flick and Hobom, 1999; Kohl et al., 2004; Mir et al., 2006; Mir and Panganiban, 2005; Perez and de la Torre, 2003). These inverted complementary sequences have been suggested to allow the ends of the RNA to associate through base pairing, leading to circular conformations of the viral genomes.

The presence of 5' and 3' complementary sequences (CS) in the genome of flaviviruses was first noticed by Hahn et al., who defined an original 8 nucleotide core sequence conserved among mosquito borne flaviviruses (Hahn *et al.*, 1987). The 5′ CS element was found within the viral ORF, coding for the amino-terminal part of protein C. The complementary 3′ CS was identified within the 3′-UTR, just upstream of the conserved terminal 3′ SL. The requirement for complementarity between 5′ and 3′ CS elements for viral replication was demonstrated using infectious clones and replicon systems of KUNV, DENV, and WNV. Mutations at the 5′ or 3′ CS regions impaired RNA synthesis without altering translation of the input RNA, providing the first evidence for a role of genome cyclization during RNA replication. In these experiments it was demonstrated that reconstitution of base pairings between 5′–3′ CS with foreign nucleotide sequences was sufficient to rescue replicon and viral RNA replication, confirming that complementarity rather than the nucleotide sequence *per se* is essential for RNA synthesis. Using functional assays, the originally recognized 8 nt core CS was extended to 10, 11 and 18 nts for WNV, DENV, and YFV respectively (Alvarez *et al.*, 2005a; Corver *et al.*, 2003; Khromykh *et al.*, 2001; Lo *et al.*, 2003).

Additional complementary sequences between the ends of mosquito-borne flavivirus genomes, outside of CS, were noticed using folding prediction algorithms (Hahn *et al.*, 1987; Khromykh *et al.*, 2001; Thurner *et al.*, 2004). A sequence located just upstream of the translation initiator AUG at the 5′-UTR was found to be complementary to a region present within the stem of the 3′ SL. This pair of complementary sequence was named cyclization sequence 5′–3′ UAR, (the name stands for upstream AUG region) (Alvarez *et al.*, 2005b).

In the case of tick-borne flaviviruses, two pairs of complementary sequences, 5′–3′ CSA and 5′–3′ CSB, were observed by prediction analysis (Khromykh *et al.*, 2001; Mandl *et al.*, 1993). The 5′ CSA is located upstream of the initiator AUG and is complementary to the 3′ CSA located within the stem of the 3′ SL, which is reminiscent of the location of 5′–3′ UAR in the MBFV. The 5′–3′ CSB sequences are in similar locations as the MBFV 5′–3′ CS. Recent studies using TBEV RNAs demonstrated that 5′–3′ CSA hybridization is essential for viral RNA synthesis, while

no crucial function was connected with the CSB elements (Kofler *et al.*, 2006). Sequence complementarity between the ends of the genome of flaviviruses with no known vectors was also reported. For Modoc virus (MODV), predicted cyclizations elements were observed upstream of the translation initiation codon and in the coding sequence of the capsid protein at the 5′ terminal region, and upstream and within the 3′ SL at the 3′ end of the genome (Leyssen *et al.*, 2002).

Two pairs of complementary sequences mediate RNA–RNA interaction between the ends of the DENV genome

Direct contact between the terminal regions of the DENV-2 genome was first observed using psoralen/UV cross-linking assays (You *et al.*, 2001). In this report, the first 230 nucleotides of the viral genome were radiolabelled and mixed with the unlabelled 454 nucleotide long RNA corresponding to the viral 3′-UTR. Psoralen treatment and UV cross-linking showed the formation of a complex between the two RNA molecules. More recently, an electrophoretic mobility shift assay was employed to study the interaction between the ends of DENV genome (Alvarez *et al.*, 2005b). In this report, an RNA molecule containing the first 160 nucleotides of the viral genome was incubated with a radiolabelled RNA molecule corresponding to the last 106 nucleotides of the viral RNA. Formation of a defined RNA–RNA complex with slower mobility than the probe was observed only in the presence of Mg^{2+}. RNA titration experiments indicated high affinity between the two RNA molecules with an apparent dissociation constant (K_d) of 8 nM.

Based on the predicted secondary structure of a molecule including both ends of the DENV-2 genome, two pairs of complementary sequences were suggested to be involved in the RNA–RNA interaction (5′–3′ CS and 5′–3′ UAR, Fig. 4.3). Because 5′ and 3′ UAR nucleotides are in predicted secondary structures, hybridization of these regions include substantial changes of the viral 5′ and 3′-UTRs. 5′ UAR nucleotides form part of the SLB, and 3′ UAR includes 3 nucleotides of the stem of the short hairpin at the 5′ end of the 3′ SL and the bottom half of the large stem of the 3′ SL. Thus, upon base pairings between 5′ and 3′ UAR,

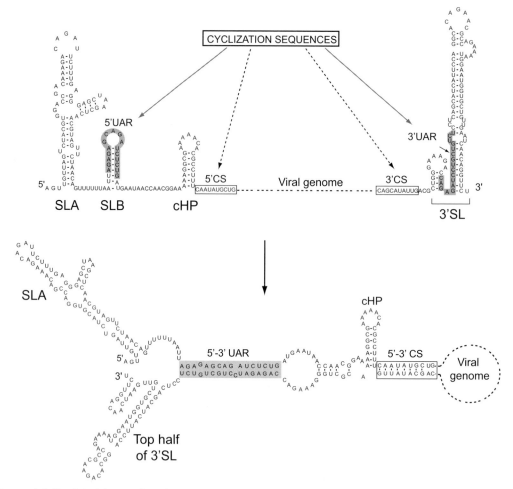

Figure 4.3 Predicted interactions between the ends of the DENV-2 genome. Sequence and predicted secondary structure of 5' and 3' terminal sequences of DENV are shown. On the top, the structural elements located at the 5' end: stem–loop A (SLA), stem–loop B (SLB), and capsid region hairpin (cHP) are shown linked by a dash line, representing the viral genome, to the 3' stem–loop (3' SL) structure. The sequence corresponding to the 5' UAR and 3' UAR are marked in grey, and the sequences corresponding to 5' CS and 3' CS are indicated with boxes. The bottom structure shows the predicted structural changes upon hybridization of 5'–3' UAR and 5'–3' CS sequences. Note that the predicted SLA and cHP are maintained, while the SLB opens. In the 3' SL, the short hairpin at the 5' end and the bottom half of the large stem–loop disappear, while the structure of the top half of the 3' SL is maintained as marked in the structure.

the predicted SLB as well as the short hairpin at the 5' end of the 3' SL disappear (Fig. 4.3). In addition, the bottom half of the large stem–loop of the 3' SL opens, releasing the last 17 nucleotides of the viral genome, which forms a less stable stem–loop structure. The top half of the large stem–loop is maintained, conserving the topology of the conserved pentanucleotide. Structural changes in the viral RNA upon 5' 3' hybridization were recently confirmed using RNA probing analysis (Polacek et al., 2009).

The requirement of 5'–3' CS and 5'–3' UAR sequences in the formation of an RNA–RNA complex in vitro was analysed by mutagenesis (Alvarez et al., 2005b). To investigate the requirement of 5'–3' CS, substitutions generating 2, 4, 5, or 7 mismatches within the CS region were incorporated in the RNA molecule corresponding to the 5' terminal region of the genome and binding of this molecule to the radiolabelled 3' RNA was evaluated. The dissociation constants increased from 8 to 100, 200, or more than 500 nM when the mutated RNAs were used, indicating an essential role of CS complementarity in RNA–RNA interaction. In contrast to the 5′–3′ CS region, the second pair of complementary sequences (5′–3′ UAR) has the potential to form 15 base pairings that are interrupted by a C bulge and a G–G mismatch (Fig. 4.3). Substitutions of single nucleotides interfering with 5′–3′ UAR complementarity also decreased the affinity between the two RNA molecules. Accumulation of three mismatches within 5′ UAR was sufficient to abolish RNA–RNA complex formation (Alvarez *et al.*, 2005b). These studies demonstrated that at least two pairs of complementary sequences, 5′–3′ CS and 5′–3′ UAR, are necessary for the interaction between the ends of the DENV RNA.

DENV genome cyclization

Long range RNA–RNA interactions between the ends of an RNA molecule carrying DENV sequences was confirmed by visualization of individual molecules using AFM (Alvarez *et al.*, 2005b). This technology allows imaging of single molecules with nanometre resolution. The microscope has a force sensor tip that is put in contact with the sample molecule. The force experienced between the tip and the sample causes a deflection that can be used to provide a topographical map of the molecule surface. Model RNA molecules of 2 kb carrying the 5′ and 3′ terminal sequences of DENV-2 were obtained by *in vitro* transcription and mounted on a mica surface for visualization by tapping mode AFM in air. These molecules acquired secondary and tertiary compact structures that precluded visualization of long-range RNA–RNA contacts (image I in Fig. 4.4). To overcome this problem, the RNA molecules were hybridized with shorter antisense molecules to generate elongated double-stranded RNA segments. This strategy allowed visualizing RNA molecules with single-stranded 5′ and 3′ ends separated by double-stranded, elongated central regions. Molecules in linear and circular conformation were observed (images II and III in Fig. 4.4). Statistical analysis indicated that 45% of the intact molecules were found in circular conformation and 13% corresponded to linear molecules forming head-to-tail dimers. Control molecules carrying deletions of 3′ CS and 3′ UAR were only found in linear forms. These experiments confirm

that DENV 5′ and 3′ terminal sequences are able to mediate long-range interactions, circularizing the RNA molecule. In the same report, cyclization of full-length DENV-2 RNA was also demonstrated. In this case, the 10.7 kb RNA was *in vitro* transcribed, treated with DNases, purified, and used for AFM. In order to visualize intramolecular interactions, an antisense RNA segment corresponding to the coding sequence of NS4B and NS5 was hybridized. AFM analysis of these molecules revealed both linear and circular conformations of the RNA, similar to the results obtained with the model molecule. These studies were performed in the absence of proteins, however, viral or cellular proteins could modulate these RNA–RNA interactions. As noted above, several viral and cellular proteins have been identified to bind the 5′ and/or the 3′ ends of flavivirus genome (Blackwell and Brinton, 1997; De Nova-Ocampo *et al.*, 2002; Garcia-Montalvo *et al.*, 2004; Ta and Vrati, 2000). Thus, it remains to be defined whether these proteins participate in the regulation of genome cyclization.

Requirement of 5′–3′ UAR interaction for DENV RNA synthesis

The function of 5′–3′ UAR complementarity in DENV replication was originally reported using an infectious DENV-2 clone (Alvarez *et al.*, 2005b). More recently, the role of 5′–3′ UAR interaction in RNA synthesis was investigated using a DENV-2 replicon system (Alvarez *et al.*, 2008; Alvarez *et al.*, 2006). In these studies, substitutions of one or two nucleotides of 5′ UAR and/or 3′ UAR were designed to disrupt or reconstitute the potential for 5′–3′ hybridization. The mutations at the 5′ end were introduced at the loop or at both sides of the stem of SLB, avoiding alteration of the predicted stem–loop structure. In addition, the mutations at the 3′ UAR were designed to restore complementarity with the respective mutated 5′ UAR sequence. In this report 21 mutated replicons were designed. Translation and RNA replication were analysed after RNA transfection in BHK cells. Translation of the mutated replicons was not significantly different from the control RNA. In contrast, a strong correlation between the ability of the RNA to maintain 5′–3′ UAR interaction and the efficiency of RNA synthesis was observed. The replicons predicted to have a

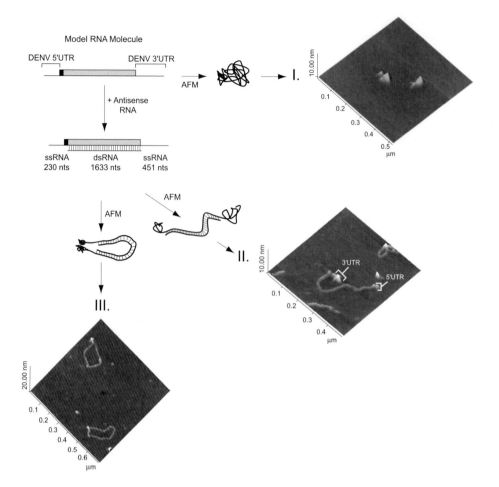

Figure 4.4 Single molecule analysis using atomic force microscopy indicates cyclization of model RNA molecule carrying the 5′ and 3′ terminal sequences of the DENV genome. At the top, a schematic representation of an RNA molecule of 2.3 kb carrying the viral 5′ and 3′ terminal sequences is included. The representative image I shows two single molecules that acquire compact conformations due to intramolecular interactions. Hybridization of this RNA with an antisense RNA of 1.6 kb results in molecules that bear a double stranded region flanked by the single stranded sequences corresponding to the viral elements. The presence of the double stranded RNA region allows visualization of 5′–3′ end contacts and cyclization of the model molecules. The representative image II shows a single RNA molecule in linear conformation in which the 5′-UTR and the 3′-UTR are indicated. These two regions can be differentiated due to their apparent volumes. The representative image III shows two RNA molecules in circular conformation.

disrupted 5′–3′ UAR interaction amplified RNA between 200 and 3000 fold less than the WT replicon. In addition, of 13 replicon, in which UAR complementarity was reconstituted, 12 RNAs replicated efficiently (Alvarez *et al.*, 2008). There was one exception (replicon Mut 7). This mutant maintained the predicted 5′–3′ UAR complementarity but replication was impaired. *In vitro* RNA binding assays confirmed that Mut 7 retained the ability to form RNA–RNA complexes between the 5′ and 3′ terminal sequences of the viral genome. To understand the cause of the impaired replication of Mut 7, the same substitutions present in this replicon were introduced in an infectious cDNA clone of DENV-2. Transfection of this RNA resulted in viable viruses containing one spontaneous mutation at the viral 3′-UTR. Surprisingly, the mutation was located outside of 3′ UAR, while the sequences within 5′ and 3′ UAR were identical to the transfected RNA. The mutation found was able to reconstitute the stem of the predicted short hairpin formed at the 5′ end of the 3′ SL. As described above, upon 5′–3′ UAR hybridization the bottom half of the large stem of the 3′ SL as well as the stem of the short hairpin open (Fig. 4.3). Because 5′–3′ UAR hybridization and the formation of the small hairpin are mutually exclusive structures, and both appear to be necessary for viral replication, the information reported supports the idea that linear and circular forms of the viral RNA co-exist, and that dynamic conformations of the viral genome might participate in RNA synthesis. In the same report, different mutations within 5′ and 3′ UAR were introduced in the context of an infectious DENV-2 clone to disrupt complementarity. Transfection of these RNAs into BHK cells resulted in viruses with spontaneous mutations reconstituting 5′–3′ UAR complementarity. The increased replication capacity of viruses with restored UAR interactions confirmed the role of these tertiary RNA contacts in DENV replication. Furthermore, most of the single nucleotide substitutions within UAR were tolerated for DENV-2 replication as long as complementarity was maintained. However, alignment of sequences of DENV isolates from different serotypes indicates absolute sequence conservation of UAR nucleotides, without co-variations, indicating that those nucleotides are under strong selective pressure; suggesting an additional role of these sequences in DENV replication in the natural host.

DENV RNA polymerase activity and the SLA promoter

DENV RNA synthesis requires cis-acting RNA elements present at the 5′ and 3′-UTRs. Deletions or mutations of SLA or SLB impaired viral RNA synthesis. In addition, while the 3′ SL and the 3′ CS1 region are essential for RNA synthesis, other elements of the 3′-UTR act as enhancers of the process. Understanding how these RNA structures regulate and promote RNA synthesis is necessary to elucidate the molecular mechanism of viral replication.

Studies performed by Padmanabhan and collaborators provided the first indication that the DENV RNA dependent RNA polymerase (RdRp) requires both ends of the viral genome for polymerase activity (You and Padmanabhan, 1999). Using cytoplasmic extracts of DENV-infected mosquito cells, the template activity of

exogenous RNAs carrying the 5′ and 3′ terminal sequences of the viral genome was analysed. In these reactions, RNA products corresponding to 1X and 2X the size of the template were obtained. The 1X product was a result of *de novo* initiation, while the 2X RNA was the product of snapback initiation, which yielded a double stranded RNA with a single stranded hairpin link. In these studies, it was demonstrated that both the 5′ and the 3′ ends of the RNA were necessary for polymerase activity because the 3′-UTR alone was inactive as template. In a different report, a similar requirement of the ends of the DENV RNA was demonstrated using a recombinant NS5 protein (Ackermann and Padmanabhan, 2001). In this study, conditions that favoured *de novo* RNA synthesis were investigated. It was found that the relative amounts of the *de novo* and the snap back products were modulated by the temperature of the reaction (Ackermann and Padmanabhan, 2001). In addition, the polarity of the RNA products obtained in the DENV polymerase assays was determined by RNase H mapping (You and Padmanabhan, 1999). DNA oligonucleotides of either plus or minus polarity were hybridized in specific locations with the radiolabelled RNA product and digested with RNase H. Digestion was only observed when an oligonucleotide was hybridized with the RNA product. Analysis of the appearance and the length of the radiolabelled RNA fragments indicated that the polymerase synthesized an RNA of minus polarity, complementary to the template, ruling out a terminal transferase activity in the experimental conditions used (You and Padmanabhan, 1999).

More recently, the activity of the recombinant DENV RdRp was investigated using viral and non-viral heteropolymeric RNAs. The RdRp was able to copy specifically RNA templates containing viral sequences, while it did not recognize cellular mRNAs. An RNA molecule carrying the 5′ and 3′ terminal sequences of the viral genome flanking the coding sequence of the structural proteins was a very active template (Filomatori et al., 2006). To define whether there were RNA structures in this template that specifically promoted polymerase activity, different RNA molecules were designed and tested in the in vitro polymerase assay. It was demonstrated that an RNA containing the first 160 nucleotides of the viral genome was sufficient to promote high polymerase activity. This RNA molecule contained the sequences corresponding to SLA, SLB, cHP, and the 5' CS. Deletions or mutations of each of these elements indicated that the SLA was the only viral element necessary to promote RNA synthesis *in vitro*. Although the polymerase must initiate RNA synthesis at the 3' end of the viral genome, the promoter element for polymerase activity was found at the 5' end of the genome.

Direct binding of the DENV RdRp to the 5' terminal sequences, including SLA, was demonstrated by mobility shift and filter binding assays (Filomatori *et al.*, 2006). Stable RNA-polymerase complexes were obtained with a dissociation constant of ~15 nM. The RdRp also interacted with an RNA–RNA complex formed between the 5' and the 3' terminal sequences of the viral genome (5' RNA-3' RNA). In native polyacrylamide gels, three different complexes were observed: 5' RNA-3' RNA, 5' RNA-3' RNA-RdRp, and 5' RNA-RdRp. The binary and ternary complexes including the viral polymerase showed similar dissociation constants, suggesting that incorporation of the 3' RNA molecule to the complex did not alter the binary interaction between the 5' RNA and the RdRp. In addition, interaction of the DENV RdRp to circular and linear form of RNAs carrying viral sequences was also demonstrated by tapping mode AFM. In these studies, visualization of single molecules of a highly purified polymerase preparation revealed binding of this protein to model RNA molecules (Filomatori *et al.*, 2006).

Evidence of a functional interaction between the SLA and the viral RdRp in infected cells was obtained using full-length DENV-2 RNAs with mutations throughout the SLA (Filomatori *et al.*, 2006). In most of the cases, mutations that alter the structure or the sequence of SLA abolished viral RNA synthesis and impaired *in vitro* polymerase activity. When mutants in the top loop of SLA that were inactive in the *in vitro* polymerase assay were introduced in the full-length genome and transfected into BHK cells, revertant viruses were obtained with spontaneous mutations in the SLA that restored both viral replication in cell culture and *in vitro* polymerase activity.

DENV RNA cyclization and polymerase activity

The *in vitro* activity of the DENV RdRp was shown to be highly dependent on: i) the presence of an intact SLA and ii) the length of the template used (Filomatori *et al.*, 2006). RNAs carrying the SLA at the 5' end of up to 500 nucleotides in length were very active templates; however, longer RNAs were inactive, unless the two pairs of complementary sequences (5'–3' CS and 5'–3' UAR) were included at the ends of the template. Therefore, in the absence of the cyclization sequences the polymerase activity was dependent on the length of the template, but in the presence of the cyclization sequences this activity became independent of the length of the RNA. Based on these observations, it was proposed that RNA circularization was necessary for the SLA-RdRp complex to initiate RNA synthesis at the 3' end of long RNA templates. Because the viral genome is about 11 kb, the ability of the SLA to promote RNA synthesis at the 3' end of the genome would be absolutely dependent on long range RNA–RNA interactions.

An experimental *in vitro* system that recapitulates the initiation of RNA synthesis at the 3' end of the genome was based on a trans-initiation assay, which is dependent on both SLA and the RNA–RNA interactions (Fig. 4.5A). As described above, the DENV polymerase is able to use the 5' terminal region but not the 3' end of the viral genome as template for *in vitro* RNA synthesis (Filomatori *et al.*, 2006; You and Padmanabhan, 1999). However, when both molecules are present, they form an RNA–RNA complex in which the 5' RNA molecule provides the SLA promoter in trans to copy the 3' RNA molecule (Alvarez *et al.*, 2008; Filomatori *et al.*, 2006; Lodeiro *et al.*, 2009; You and Padmanabhan, 1999). In this reaction, the two RNA molecules are copied, resulting in two different products (Fig. 4.5A). It has been demonstrated that this trans-initiation activity is absolutely dependent on both 5'–3' CS and 5'–3' UAR complementarity (Alvarez *et al.*, 2008; You and Padmanabhan, 1999). Therefore, the initiation of RNA synthesis at the 3' end of the 3' RNA molecule occurs in an RNA–RNA complex in which rearrangements around the SLB and the 3' SL are similar to that predicted in the viral genome (Fig. 4.3). This

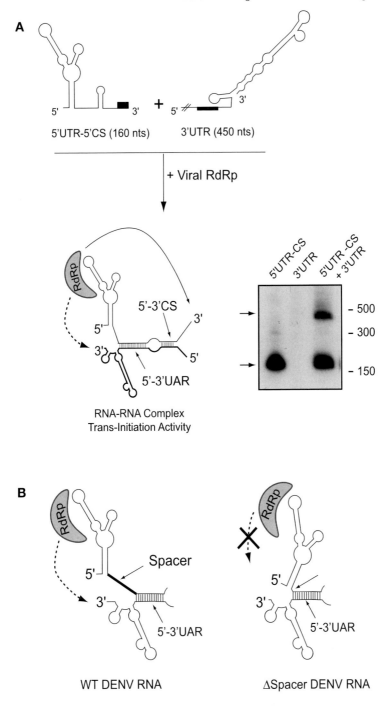

Figure 4.5 Representation of the *in vitro* trans-initiation activity of the viral polymerase. **A** Formation of an RNA–RNA complex between the first 160 nucleotides of the viral genome (5'-UTR-5' CS) and the viral 3'-UTR allows the RNA-dependent RNA polymerase (RdRp) to initiate RNA synthesis at the 3' end of both molecules, as indicated schematically in the figure. On the right, a representative denatured polyacrylamide gel showing the radiolabelled RNA products obtained after incubation of the viral RdRp with the templates described at the top. **B** Schematic representation of the role of a spacer sequence for RNA synthesis by the recombinant RdRp. The oligo(U) track spacer located between the SLA and the hybridized 5'–3' UAR sequences allows accommodation of the two ends of the genome for the RdRp to initiate RNA synthesis at the 3'-UTR. In contrast, an RNA molecule carrying an intact SLA but a deletion of the oligo(U) spacer is unable to promote RNA synthesis in trans by the RdRp.

trans-initiation assay provides a useful tool to investigate the complex interactions between the viral RdRp and the two ends of the genome. It will be important to develop a more complex assay including other viral proteins as well as cellular factors of the replication complex to reconstitute the process of initiation of minus strand RNA synthesis that resembles the process in the infected cell.

Regulation of genome integrity and synthesis

Maintenance of the 3′ end nucleotides of the DENV genome

The ability of the DENV RNA synthesis machinery to reconstitute 3′ end sequences of mutated viral RNAs has recently been reported (Teramoto et al., 2008). In this study, a series of 3′ end deletions in the DENV genome were evaluated for viral replication. Sequence analysis of the recovered viruses showed that, at early times, RNA molecules displayed heterogeneous lengths, however, this diversity decreased with time until a consensus sequence emerged. The preferred 3′ terminal sequences selected in culture were UU, CU, UCU or UUCU. These findings suggested that a repair of 3′ terminal nucleotides occurred and viral RNAs competent for replication were selected. A possible mechanism proposed for this process involves a non-templated nucleotide addition at the 3′ end of the genome by a terminal transferase activity of the viral RNA polymerase or a host polymerase, which will result in a population of RNAs with different 3′ end sequences. The RNA sequences that support viral replication more efficiently would be enriched and selected. Because the viral RNA can be susceptible to the loss of terminal nucleotides by the activity of 3′-> 5′ exoribonucleases in the cytoplasm of the infected cell, a mechanism of 3′ end repair could be necessary to protect the integrity of the viral genome.

A spacer sequence downstream of the SLA enhances DENV RNA synthesis

Between the SLA structure and the 5′ UAR sequence at the viral 5′-UTR there is an oligo(U) track conserved in DENV and other flavivirus genomes (Fig. 4.1). Deletion of six U residues downstream of the SLA in the infectious clone of DENV-2 resulted in viral attenuation (Lodeiro et al., 2009). Transfection of the mutated RNA into BHK cells required 6 days to infect 30% of the monolayer as determined by immunofluorescence, while the parental RNA was 100% DENV antigen positive by day 3. RT-PCR and sequencing analysis of the recovered viruses indicated that the 6U deletion was maintained. Infection of fresh cells and sequential passages of this virus confirmed that the deletion was retained as well as the attenuated phenotype. It was demonstrated that the nucleotide sequence in this region was not an important determinant for replication because replacement of the 6Us for 6As yielded viruses with phenotypes that were indistinguishable from that of the parental virus. The 6U deletion appeared to have a direct effect on viral RNA synthesis. Incorporation of the 6U deletion into a DENV-2 replicon system indicated that translation of the input RNA was unaffected while RNA synthesis was about 40 fold lower than the parental replicon (Lodeiro et al., 2009).

The mechanism by which the oligo(U) track enhanced DENV RNA synthesis in transfected cells was investigated testing the ability of the 5′-UTR with or without the U track to a) interfere with in vitro RNA–RNA interactions between the ends of the genome, and b) alter the viral polymerase activity using 5′ terminal RNA templates in cis or 3′-UTR templates in trans (Lodeiro et al., 2009). These studies indicated that the long-range RNA–RNA interactions and the SLA promoter activity in cis were unaffected in the mutated RNA. However, the trans-initiation activity was seriously compromised when the oligo(U) spacer was shorter than 3 nucleotides. Because this spacer is located between the SLA and the hybridized 5′-3′ UAR in the complex formed between the 5′ and the 3′ terminal regions of the viral genome (Fig. 4.5B), it is likely that the polymerase requires specific RNA structures downstream of the promoter element to initiate RNA synthesis in trans using the 3′-UTR molecule as template.

Model for DENV minus strand RNA synthesis

A current model for DENV negative strand RNA synthesis includes the SLA promoter at the 5'-UTR, two pairs of cyclization sequences (5'–3' CS and 5'–3' UAR), a spacer element between the 5' UAR and the SLA, and the 3' SL structure. During RNA replication, the viral NS5 protein binds the promoter SLA at the 5' end of the genome, 11 kb away from the 3' initiation site. Cyclization of the viral genome mediated by long-range RNA–RNA interactions brings the 3' end near the SLA-NS5 complex (Fig. 4.6). In this circular conformation of the RNA, the active RdRp reaches the 3' end of the genome to initiate minus strand RNA synthesis. The 3' SL is an essential element for viral replication, however, the molecular details by which this structure participates during minus strand RNA synthesis is still unclear. Hybridization of 5'–3' UAR mediates 5'–3' interaction but also promotes a rearrangement of the structure around the 3' SL (Fig. 4.3). The predicted changes include destabilization of the small hairpin at the 5' end of the 3' SL and disruption of the bottom stem of the large stem of the 3' SL. Interestingly, recent reports have provided evidence that the structure at the 3' SL predicted in the linear form of the RNA is also essential for

viral replication (Alvarez *et al.*, 2008; Teramoto *et al.*, 2008). DENV RNAs carrying mutations that would disrupt the bottom of the large stem of the 3' SL or mutants that were predicted to destabilize the stem of the small hairpin resulted in revertant viruses that restored those structures. These observations provide strong evidence that 3' SL structures predicted in the circular and linear conformations of the viral genome co-exist, and both appeared to be necessary for DENV RNA synthesis.

It is still intriguing why DENV genome evolved a promoter for RNA replication at the 5' end of the RNA. The requirement of genome cyclization may provide advantages for viral replication such as control mechanisms to amplify only full-length templates or coordination of translation, RNA synthesis, and RNA packaging by overlapping signals involved in these processes. Further analysis of the RNA conformations required in each viral process will help to clarify the molecular details by which DENVs replicate their genomes.

Manipulation of the UTRs as a strategy for control of viral replication

DENV UTR mutations and viral attenuation

Manipulation of the DENV 5' and 3'-UTRs have been shown to be a valuable strategy to generate attenuated vaccine candidates (see Chapter 12). Whitehead and collaborators have produced a recombinant DENV that harbours a 30 nucleotide deletion including the CS2 conserved sequence located in domain II of the 3'-UTR (Fig. 4.2). Viruses carrying this deletion resulted in reduced replication in rhesus monkeys (Blaney *et al.*, 2006). Comparison of DENV-4 and DENV-4Δ30 virus replication in mosquito indicated that the Δ30 mutation restricted dissemination from the midgut to the head of infected mosquitoes, which is important for viral transmission (Troyer *et al.*, 2001). The Δ30 mutation also conferred a reduction in replication of DENV-4 and DENV-1 in SCID-HuH7 mice (Blaney *et al.*, 2002). DENV-4Δ30 and DENV-1Δ30 are vaccine candidates that are currently being tested in clinical trials (for review see (Whitehead *et al.*, 2007)). When the

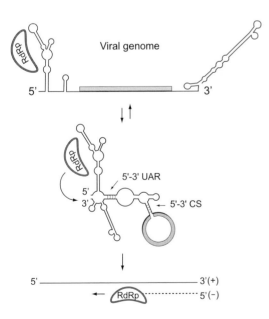

Figure 4.6 Model proposed for DENV minus strand RNA synthesis. The viral RNA-dependent RNA polymerase (RdRp) binds to the promoter SLA at the 5' end of the genome. Long-range RNA–RNA mediated by 5'–3' UAR and 5'–3' CS interactions mediate genome cyclization that facilitates polymerase initiation at the 3' end of the RNA.

Δ30 mutation was introduced into cDNA clones of DENV-2 and 3, the attenuation was not as pronounced as that observed with DENV-1 and 4 (Blaney *et al.*, 2004a; Blaney *et al.*, 2004b). Based on these observations, chimeric viruses with sequences from different serotypes carrying the Δ30 mutation have been designed and they are being investigated to obtain tetravalent formulations (Whitehead *et al.*, 2003).

Mutations within the DENV 5′-UTR have been also designed to generate attenuated viruses (Butrapet *et al.*, 2000; Cahour *et al.*, 1995; Lodeiro *et al.*, 2009; Sirigulpanit *et al.*, 2007). The DENV-2 PDK-53 vaccine candidate contains three mutations that are responsible for attenuation. One of these mutations is located within the helical region S2 of the SLA structure. The single C57U mutation resulted in small plaque phenotype, decreased replication in mosquito cells, and attenuation of neurovirulence in newborn mice (Butrapet *et al.*, 2000). A more recent study found that a double mutant C57U–G58C replicated more efficiently than the C57U mutant (Sirigulpanit *et al.*, 2007). Because both mutations generate mismatches in the S2 region of SLA, the mechanism by which the combination of these two substitutions increase viral replication is still unclear.

A DENV vaccine must be able to protect against all four circulating serotypes, thus dissecting the requirement of conserved RNA elements in different DENVs could aid the process of designing rationally attenuated viruses. It has been demonstrated that the SLA structure present at the 5′-UTR can be exchangeable between DENVs resulting in fully functional viruses. In addition, the SLB sequence is absolutely conserved among the four DENV types (Villordo and Gamarnik, 2008). Thus, it will be interesting to design a universal mutated 5′-UTR that causes attenuation and which could be introduced into each of the four serotypes to generate vaccine candidates, similar to the strategy that has been recently explored for the 3′-UTR (Blaney *et al.* 2008).

Viral inhibition by antisense oligonucleotides directed to the DENV UTRs

Antisense oligonucleotides targeted to different locations of the DENV genome have been used

to successfully suppress viral gene expression and replication (for review see (Stein and Shi, 2008)). The effect of modified oligonucleotides targeting the viral genome was first tested by microinjection into monkey cells and then infection with DENV-2. More than 50% reduction in virus levels at 24 h post infection was achieved with oligomers directed to the AUG translation start region or the sequence of the 3′ SL (Raviprakash *et al.*, 1995). In contrast, compounds targeting sequences in NS5 or the 3′ CS region were found to be ineffective.

More recent studies have used phosphorodiamidate morpholino oligomers (PMOs) to inhibit DENV replication. The PMOs do not cross the plasma membranes of cultured cells and therefore some type of assisted delivery is required in order to use them in cell culture experiments. However, conjugated PMOs with an arginine-rich peptide achieve uptake into mammalian cells in culture by simple incubation at 37°C. A peptide-PMO (PPMO) directed against the cyclization sequence 3′ CS demonstrated efficient inhibition of the four DENV serotypes (Kinney *et al.*, 2005). In addition, a PPMO directed to the 5′ SLA was highly effective against DENV-1, DENV-2, and DENV-3. Lower efficacy was observed with DENV-4, likely due to a two-nucleotide difference with respect to the other serotypes in the sequence targeted by this PPMO (nucleotides 1–20, Fig. 4.1). The effect of other three PPMOs directed to the translation start region, the 5′ CS, and the 3′ SL were also tested, but all of them showed lower efficacy in inhibiting viral replication when compared with the ones directed to the 5′ SLA and the 3′ CS sequences.

A DENV replicon system was used to study the step of viral replication affected by the PPMOs (Holden *et al.*, 2006). The compounds directed to the 3′ CS had little effect on translation but greatly diminished RNA synthesis. In contrast, the 5′ SLA PPMO reduced more than 90% the levels of viral translation. Another PPMO designed to target the top loop of the 3′ SL was very effective in decreasing the viral titres of DENV types 1, 2, and 3, while it was inactive against DENV-4, probably due to several mismatches between the PPMO and the target region. These findings indicate that antisense approaches targeting conserved 5′ and/

or 3′-UTR sequences may be useful as an antiviral strategy.

Perspectives

Although a great deal has been learned in the last few years about the role of the RNA elements of the 5′ and 3′-UTRs in DENV replication, many questions still remain. Does the virus use different mechanisms for minus and plus strand RNA synthesis? Are there specific RNA elements at the terminal regions of both strands involved in RNA synthesis of the complementary strand? In this regard, tools to discriminate between these two processes in infected cells are still missing; therefore, an important challenge is to develop *in vivo* assays to be able to determine minus strand RNA synthesis in conditions in which plus strand RNA is not produced.

The molecular mechanism by which the 3′ SL participates in viral replication is still unclear. It is of great interest to define the interplay between the promoter SLA, the 3′ SL, the viral NS5 polymerase, and host proteins during RNA synthesis. Although it has been demonstrated that genome cyclization through long-range RNA–RNA interactions is essential for viral RNA synthesis, new observations support the idea that linear and circular conformations of the viral RNA co-exist and that both are necessary for DENV replication. It is likely that the viral genome is a dynamic molecule that acquires different RNA conformations through the viral life cycle. It is important to continue unravelling which secondary and tertiary structures of the viral RNA participate in each step of viral replication. In particular, it is important to explore whether circular or linear forms of the RNA participate in the regulation of translation, genome replication, and/or RNA encapsidation.

References

Ackermann, M., and Padmanabhan, R. (2001). *De novo* synthesis of RNA by the dengue virus RNA-dependent RNA polymerase exhibits temperature dependence at the initiation but not elongation phase. J. Biol. Chem. *276*, 39926–39937.

Alvarez, D.E., De Lella Ezcurra, A.L., Fucito, S., and Gamarnik, A.V. (2005a). Role of RNA structures present at the 3′-UTR of dengue virus on translation, RNA synthesis, and viral replication. Virology *339*, 200–212.

Alvarez, D.E., Filomatori, C.V., and Gamarnik, A.V. (2008). Functional analysis of dengue virus cyclization sequences located at the 5′ and 3′-UTRs. Virology *375*, 223–235.

Alvarez, D.E., Lodeiro, M.F., Filomatori, C.V., Fucito, S., Mondotte, J.A., and Gamarnik, A.V. (2006). Structural and functional analysis of dengue virus RNA. Novartis Found Symp *277*, 120–132; discussion 132–135, 251–253.

Alvarez, D.E., Lodeiro, M.F., Luduena, S.J., Pietrasanta, L.I., and Gamarnik, A.V. (2005b). Long-range RNA–RNA interactions circularize the dengue virus genome. J. Virol. *79*, 6631–6643.

Aquino, V.H., Anatriello, E., Goncalves, P.F., E.V., D.A. S., Vasconcelos, P.F., Vieira, D.S., Batista, W.C., Bobadilla, M.L., Vazquez, C., Moran, M., and Figueiredo, L.T. (2006). Molecular epidemiology of dengue type 3 virus in Brazil and Paraguay, 2002–2004. Am. J. Trop. Med. Hyg. *75*, 710–715.

Blackwell, J.L., and Brinton, M.A. (1997). Translation elongation factor-1 alpha interacts with the 3′ stem–loop region of West Nile virus genomic RNA. J. Virol. *71*, 6433–6444.

Blaney, J.E., Jr., Hanson, C.T., Firestone, C.Y., Hanley, K.A., Murphy, B.R., and Whitehead, S.S. (2004a). Genetically modified, live attenuated dengue virus type 3 vaccine candidates. Am. J. Trop. Med. Hyg. *71*, 811–821.

Blaney, J.E., Jr., Hanson, C.T., Hanley, K.A., Murphy, B.R., and Whitehead, S. S. (2004b). Vaccine candidates derived from a novel infectious cDNA clone of an American genotype dengue virus type 2. BMC Infect Dis *4*, 39.

Blaney, J.E., Jr., Johnson, D.H., Manipon, G.G., Firestone, C.Y., Hanson, C.T., Murphy, B.R., and Whitehead, S.S. (2002). Genetic basis of attenuation of dengue virus type 4 small plaque mutants with restricted replication in suckling mice and in SCID mice transplanted with human liver cells. Virology *300*, 125–139.

Blaney, J.E., Jr., Durbin, A.P., Murphy, B.R., and Whitehead, S.S. (2006). Development of a live attenuated dengue virus vaccine using reverse genetics. Viral Immunol *19*, 10–32.

Blaney, J.E., Jr., Sathe, N.S., Goddard, L., Hanson, C.T., Romero, T.A., Hanley, K.A., Murphy, B.R., and Whitehead, S.S. (2008). Dengue virus type 3 vaccine candidates generated by introduction of deletions in the 3′ untranslated region (3′-UTR) or by exchange of the DENV-3 3′-UTR with that of DENV-4. Vaccine *26*, 817–828.

Blight, K.J., and Rice, C.M. (1997). Secondary structure determination of the conserved 98-base sequence at the 3′ terminus of hepatitis C virus genome RNA. J. Virol. *71*, 7345–7352.

Bredenbeek, P.J., Kooi, E.A., Lindenbach, B., Huijkman, N., Rice, C.M., and Sp aan, W.J. (2003). A stable full-length yellow fever virus cDNA clone and the role of conserved RNA elements in flavivirus replication. J. Gen. Virol. *84*, 1261–1268.

Brinton, M.A., and Dispoto, J.H. (1988). Sequence and secondary structure analysis of the 5′-terminal region of flavivirus genome RNA Virology *162*, 290–299.

Brinton, M.A., Fernandez, A.V., and Dispoto, J.H. (1986). The 3'-nucleotides of flavivirus genomic RNA form a conserved secondary structure. Virology *153*, 113–121.

Butrapet, S., Huang, C.Y., Pierro, D.J., Bhamarapravati, N., Gubler, D.J., and Kinney, R.M. (2000). Attenuation markers of a candidate dengue type 2 vaccine virus, strain 16681 (PDK-53), are defined by mutations in the 5' noncoding region and nonstructural proteins 1 and 3. J. Virol. *74*, 3011–3019.

Cahour, A., Pletnev, A., Vazielle-Falcoz, M., Rosen, L., and Lai, C.J. (1995). Growth-restricted dengue virus mutants containing deletions in the 5' noncoding region of the RNA genome. Virology *207*, 68–76.

Chiu, W.W., Kinney, R.M., and Dreher, T.W. (2005). Control of translation by the 5'- and 3'-terminal regions of the dengue virus genome. J. Virol. *79*, 8303–8315.

Clyde, K., Barrera, J., and Harris, E. (2008). The capsid-coding region hairpin element (cHP) is a critical determinant of dengue virus and West Nile virus RNA synthesis. Virology *379*, 314–323.

Clyde, K., and Harris, E. (2006). RNA secondary structure in the coding region of dengue virus type 2 directs translation start codon selection and is required for viral replication. J. Virol. *80*, 2170–2182.

Cologna, R., and Rico-Hesse, R. (2003). American genotype structures decrease dengue virus output from human monocytes and dendritic cells. J. Virol. *77*, 3929–3938.

Corver, J., Lenches, E., Smith, K., Robison, R.A., Sando, T., Strauss, E.G., and Strauss, J.H. (2003). Fine mapping of a cis-acting sequence element in yellow fever virus RNA that is required for RNA replication and cyclization. J. Virol. *77*, 2265–2270.

De Nova-Ocampo, M., Villegas-Sepulveda, N., and del Angel, R.M. (2002). Translation elongation factor-1alpha, La, and PTB interact with the 3' untranslated region of dengue 4 virus RNA Virology *295*, 337–347.

Deng, R., and Brock, K.V. (1993). 5' and 3' untranslated regions of pestivirus genome: primary and secondary structure analyses. Nucleic Acids Res *21*, 1949–1957.

Edgil, D., Polacek, C., and Harris, E. (2006). Dengue virus utilizes a novel strategy for translation initiation when cap-dependent translation is inhibited. J. Virol. *80*, 2976–2986.

Elghonemy, S., Davis, W.G., and Brinton, M.A. (2005). The majority of the nucleotides in the top loop of the genomic 3' terminal stem–loop structure are cis-acting in a West Nile virus infectious clone. Virology *331*, 238–246.

Filomatori, C.V., Lodeiro, M.F., Alvarez, D.E., Samsa, M.M., Pietrasanta, L., and Gamarnik, A.V. (2006). A 5' RNA element promotes dengue virus RNA synthesis on a circular genome. Genes Dev *20*, 2238–2249.

Flick, R., and Hobom, G. (1999). Interaction of influenza virus polymerase with viral RNA in the 'corkscrew' conformation. J. Gen. Virol. *80 (Pt 10)*, 2565–2572.

Garcia-Montalvo, B.M., Medina, F., and del Angel, R.M. (2004). La protein binds to NS5 and NS3 and to the 5' and 3' ends of Dengue 4 virus RNA Virus Res. *102*, 141–150.

Grange, T., Bouloy, M., and Girard, M. (1985). Stable secondary structures at the 3'-end of the genome of yellow fever virus (17 D vaccine strain). FEBS Lett *188*, 159–163.

Gritsun, T.S., and Gould, E.A. (2006). Direct repeats in the 3' untranslated regions of mosquito-borne flaviviruses: possible implications for virus transmission. J. Gen. Virol. *87*, 3297–3305.

Gritsun, T.S., and Gould, E.A. (2007). Origin and evolution of flavivirus 5'-UTRs and panhandles: trans-terminal duplications? Virology *366*, 8–15.

Gritsun, T.S., Venugopal, K., Zanotto, P. M., Mikhailov, M.V., Sall, A.A., Holmes, E.C., Polkinghorne, I., Frolova, T.V., Pogodina, V.V., Lashkevich, V.A., and Gould, E. A. (1997). Complete sequence of two tick-borne flaviviruses isolated from Siberia and the UK: analysis and significance of the 5' and 3'-UTRs. Virus Res. *49*, 27–39.

Hahn, C.S., Hahn, Y.S., Rice, C.M., Lee, E., Dalgarno, L., Strauss, E.G., and Strauss, J. H. (1987). Conserved elements in the 3' untranslated region of flavivirus RNAs and potential cyclization sequences. J Mol Biol *198*, 33–41.

Hanley, K.A., Manlucu, L.R., Manipon, G.G., Hanson, C.T., Whitehead, S.S., Murphy, B.R., and Blaney, J.E., Jr. (2004). Introduction of mutations into the nonstructural genes or 3' untranslated region of an attenuated dengue virus type 4 vaccine candidate further decreases replication in rhesus monkeys while retaining protective immunity. Vaccine *22*, 3440–3448.

Holden, K.L., and Harris, E. (2004). Enhancement of dengue virus translation: role of the 3' untranslated region and the terminal 3' stem–loop domain. Virology *329*, 119–133.

Holden, K.L., Stein, D.A., Pierson, T.C., Ahmed, A.A., Clyde, K., Iversen, P.L., and Harris, E. (2006). Inhibition of dengue virus translation and RNA synthesis by a morpholino oligomer targeted to the top of the terminal 3' stem–loop structure. Virology *344*, 439–452.

Khromykh, A.A., Kondratieva, N., Sgro, J.Y., Palmenberg, A., and Westaway, E.G. (2003). Significance in replication of the terminal nucleotides of the flavivirus genome. J. Virol. *77*, 10623–10629.

Khromykh, A.A., Meka, H., Guyatt, K.J., and Westaway, E.G. (2001). Essential role of cyclization sequences in flavivirus RNA replication. J. Virol. *75*, 6719–6728.

Kinney, R.M., Huang, C.Y., Rose, B.C., Kroeker, A.D., Dreher, T.W., Iversen, P.L., and Stein, D.A. (2005). Inhibition of dengue virus serotypes 1 to 4 in vero cell cultures with morpholino oligomers. J. Virol. *79*, 5116–5128.

Kofler, R.M., Hoenninger, V.M., Thurner, C., and Mandl, C.W. (2006). Functional analysis of the tick-borne encephalitis virus cyclization elements indicates major differences between mosquito-borne and tick-borne flaviviruses. J. Virol. *80*, 4099–4113.

Kohl, A., Dunn, E.F., Lowen, A.C., and Elliott, R.M. (2004). Complementarity, sequence and structural elements within the 3' and 5' non-coding regions of the Bunyamwera orthobunyavirus S segment determine promoter strength. J. Gen. Virol. *85*, 3269–3278.

Leyssen, P., Charlier, N., Lemey, P., Billoir, F., Vandamme, A.M., De Clercq, E., de Lamballerie, X., and Neyts, J. (2002). Complete genome sequence, taxonomic assignment, and comparative analysis of the untranslated regions of the Modoc virus, a flavivirus with no known vector. Virology 293, 125–140.

Lo, M.K., Tilgner, M., Bernard, K.A., and Shi, P.Y. (2003). Functional analysis of mosquito-borne flavivirus conserved sequence elements within 3′ untranslated region of West Nile virus by use of a reporting replicon that differentiates between viral translation and RNA replication. J. Virol. 77, 10004–10014.

Lodeiro, M.F., Filomatori, C.V., and Gamarnik, A.V. (2009). Structural and functional studies of the promoter element for dengue virus RNA replication. J. Virol. 83, 993–1008.

Mandl, C.W., Holzmann, H., Kunz, C., and Heinz, F.X. (1993). Complete genomic sequence of Powassan virus: evaluation of genetic elements in tick-borne versus mosquito-borne flaviviruses. Virology 194, 173–184.

Mandl, C.W., Holzmann, H., Meixner, T., Rauscher, S., Stadler, P.F., Allison, S.L., and Heinz, F. X. (1998). Spontaneous and engineered deletions in the 3′ noncoding region of tick-borne encephalitis virus: construction of highly attenuated mutants of a flavivirus. J. Virol. 72, 2132–2140.

Markoff, L. (2003). 5′ and 3′ NCRs in Flavivirus RNA, Vol 60, Elsevier Academic Press).

Markoff, L., Pang, X., Houng Hs, H.S., Falgout, B., Olsen, R., Jones, E., and Polo, S. (2002). Derivation and characterization of a dengue type 1 host range-restricted mutant virus that is attenuated and highly immunogenic in monkeys. J. Virol. 76, 3318–3328.

Men, R., Bray, M., Clark, D., Chanock, R.M., and Lai, C.J. (1996). Dengue type 4 virus mutants containing deletions in the 3′ noncoding region of the RNA genome: analysis of growth restriction in cell culture and altered viremia pattern and immunogenicity in rhesus monkeys. J. Virol. 70, 3930–3937.

Mir, M.A., Brown, B., Hjelle, B., Duran, W.A., and Panganiban, A.T. (2006). Hantavirus N protein exhibits genus-specific recognition of the viral RNA panhandle. J. Virol. 80, 11283–11292.

Mir, M.A., and Panganiban, A.T. (2005). The hantavirus nucleocapsid protein recognizes specific features of the viral RNA panhandle and is altered in conformation upon RNA binding. J. Virol. 79, 1824–1835.

Olsthoorn, R.C., and Bol, J.F. (2001). Sequence comparison and secondary structure analysis of the 3′ noncoding region of flavivirus genomes reveals multiple pseudoknots. RNA 7, 1370–1377.

Paranjape, S.M., and Harris, E. (2007). Y box-binding protein-1 binds to the dengue virus 3′-untranslated region and mediates antiviral effects. J. Biol. Chem. 282, 30497–30508.

Perez, M., and de la Torre, J.C. (2003). Characterization of the genomic promoter of the prototypic arenavirus lymphocytic choriomeningitis virus. J. Virol. 77, 1184–1194.

Polacek, C., Foley, J. E., and Harris, E. (2009). Conformational changes in the solution structure of the dengue virus 5′ end in the presence and absence of the 3′ untranslated region. J. Virol. 83, 1161–1166.

Proutski, V., Gould, E.A., and Holmes, E.C. (1997). Secondary structure of the 3′ untranslated region of flaviviruses: similarities and differences. Nucleic Acids Res 25, 1194–1202.

Rauscher, S., Flamm, C., Mandl, C.W., Heinz, F.X., and Stadler, P.F. (1997). Secondary structure of the 3′-noncoding region of flavivirus genomes: comparative analysis of base pairing probabilities. RNA 3, 779–791.

Raviprakash, K., Liu, K., Matteucci, M., Wagner, R., Riffenburgh, R., and Carl, M. (1995). Inhibition of dengue virus by novel, modified antisense oligonucleotides. J. Virol. 69, 69–74.

Rice, C.M., Lenches, E.M., Eddy, S. R., Shin, S.J., Sheets, R.L., and Strauss, J.H. (1985). Nucleotide sequence of yellow fever virus: implications for flavivirus gene expression and evolution. Science 229, 726–733.

Roche, C., Cassar, O., Laille, M., and Murgue, B. (2007). Dengue-3 virus genomic differences that correlate with in vitro phenotype on a human cell line but not with disease severity. Microbes Infect 9, 63–69.

Romero, T.A., Tumban, E., Jun, J., Lott, W.B., and Hanley, K.A. (2006). Secondary structure of dengue virus type 4 3′ untranslated region: impact of deletion and substitution mutations. J. Gen. Virol. 87, 3291–3296.

Shurtleff, A.C., Beasley, D.W., Chen, J.J., Ni, H., Suderman, M.T., Wang, H., Xu, R., Wang, E., Weaver, S.C., Watts, D.M., et al. (2001). Genetic variation in the 3′ noncoding region of dengue viruses. Virology 281, 75–87.

Silva, P.A., Molenkamp, R., Dalebout, T.J., Charlier, N., Neyts, J. H., Sp aan, W.J., and Bredenbeek, P.J. (2007). Conservation of the pentanucleotide motif at the top of the yellow fever virus 17D 3′ stem–loop structure is not required for replication. J. Gen. Virol. 88, 1738–1747.

Silva, R.L., de Silva, A.M., Harris, E., and MacDonald, G.H. (2008). Genetic analysis of Dengue 3 virus subtype III 5′ and 3′ non-coding regions. Virus Res. 135, 320–325.

Sirigulpanit, W., Kinney, R.M., and Leardkamolkarn, V. (2007). Substitution or deletion mutations between nt 54 and 70 in the 5′ non-coding region of dengue type 2 virus produce variable effects on virus viability. J. Gen. Virol. 88, 1748–1752.

Stein, D.A., and Shi, P.Y. (2008). Nucleic acid-based inhibition of flavivirus infections. Front Biosci 13, 1385–1395.

Ta, M., and Vrati, S. (2000). Mov34 protein from mouse brain interacts with the 3′ noncoding region of Japanese encephalitis virus. J. Virol. 74, 5108–5115.

Tajima, S., Nukui, Y., Takasaki, T., and Kurane, I. (2007). Characterization of the variable region in the 3′ nontranslated region of dengue type 1 virus. J. Gen. Virol. 88, 2214–2222.

Takegami, T., Washizu, M., and Yasui, K. (1986). Nucleotide sequence at the 3′ end of Japanese encephalitis virus genomic RNA Virology 152, 483–486.

Teramoto, T., Kohno, Y., Mattoo, P., Markoff, L., Falgout, B., and Padmanabhan, R. (2008). Genome 3′-end repair in dengue virus type 2. 14, 2645–2656.

Thurner, C., Witwer, C., Hofacker, I.L., and Stadler, P.F. (2004). Conserved RNA secondary structures in Flaviviridae genomes. J. Gen. Virol. *85*, 1113–1124.

Tilgner, M., Deas, T.S., and Shi, P.Y. (2005). The flavivirus-conserved penta-nucleotide in the 3′ stem–loop of the West Nile virus genome requires a specific sequence and structure for RNA synthesis, but not for viral translation. Virology *331*, 375–386.

Tilgner, M., and Shi, P.Y. (2004). Structure and function of the 3′ terminal six nucleotides of the west nile virus genome in viral replication. J. Virol. *78*, 8159–8171.

Troyer, J.M., Hanley, K.A., Whitehead, S.S., Strickman, D., Karron, R.A., Durbin, A. P., and Murphy, B. R. (2001). A live attenuated recombinant dengue-4 virus vaccine candidate with restricted capacity for dissemination in mosquitoes and lack of transmission from vaccinees to mosquitoes. Am. J. Trop. Med. Hyg. *65*, 414–419.

Vasilakis, N., Fokam, E.B., Hanson, C.T., Weinberg, E., Sall, A.A., Whitehead, S.S., Hanley, K.A., and Weaver, S.C. (2008). Genetic and phenotypic characterization of sylvatic dengue virus type 2 strains. Virology *377*, 296–307.

Villordo, S.M., and Gamarnik, A.V. (2008). Genome cyclization as strategy for flavivirus RNA replication. Virus Res. *139*, 230–239.

Whitehead, S.S., Blaney, J.E., Durbin, A.P., and Murphy, B.R. (2007). Prospects for a dengue virus vaccine. Nat Rev Microbiol *5*, 518–528.

Whitehead, S.S., Hanley, K.A., Blaney, J.E., Jr., Gilmore, L.E., Elkins, W.R., and Murphy, B.R. (2003). Substitution of the structural genes of dengue virus type 4 with those of type 2 results in chimeric vaccine candidates which are attenuated for mosquitoes, mice, and rhesus monkeys. Vaccine *21*, 4307–4316.

Yocupicio-Monroy, M., Padmanabhan, R., Medina, F., and del Angel, R.M. (2007). Mosquito La protein binds to the 3′ untranslated region of the positive and negative polarity dengue virus RNAs and relocates to the cytoplasm of infected cells. Virology *357*, 29–40.

Yocupicio-Monroy, R.M., Medina, F., Reyes-del Valle, J., and del Angel, R.M. (2003). Cellular proteins from human monocytes bind to dengue 4 virus minus-strand 3′ untranslated region RNA. J. Virol. *77*, 3067–3076.

You, S., Falgout, B., Markoff, L., and Padmanabhan, R. (2001). *In vitro* RNA synthesis from exogenous dengue viral RNA templates requires long range interactions between 5′- and 3′-terminal regions that influence RNA structure. J. Biol. Chem. *276*, 15581–15591.

You, S., and Padmanabhan, R. (1999). A novel *in vitro* replication system for Dengue virus. Initiation of RNA synthesis at the 3′-end of exogenous viral RNA templates requires 5′- and 3′-terminal complementary sequence motifs of the viral RNA. J. Biol. Chem. *274*, 33714–33722.

Yu, L., and Markoff, L. (2005). The topology of bulges in the long stem of the flavivirus 3′ stem–loop is a major determinant of RNA replication competence. J. Virol. *79*, 2309–2324.

Zeng, L., Falgout, B., and Markoff, L. (1998). Identification of specific nucleotide sequences within the conserved 3′-SL in the dengue type 2 virus genome required for replication. J. Virol. *72*, 7510–7522.

Zhou, Y., Mammen, M.P., Jr., Klungthong, C., Chinnawirotpisan, P., Vaughn, D.W., Nimmannitya, S., Kalayanarooj, S., Holmes, E.C., and Zhang, C. (2006). Comparative analysis reveals no consistent association between the secondary structure of the 3′-untranslated region of dengue viruses and disease syndrome. J. Gen. Virol. *87*, 2595–2603.

Part III

Pathogenesis and Host Immunity

Host and Virus Determinants of Susceptibility and Dengue Disease Severity

5

Maria G. Guzman, Beatriz Sierra, Gustavo Kouri, Jeremy Farrar and Cameron Simmons

Abstract
The pathogenesis of dengue infection is complicated and, despite intensive research, not well understood. Here, we review the burden of disease and the host (age, ethnicity, co-morbidities, immune and genetic factors) and viral (epidemiological and genetic factors) determinants for disease severity. This review is informed both by current knowledge of the topics and our specific experiences accumulated from Cuba and Vietnam.

Introduction

Outbreaks of dengue haemorrhagic fever (DHF) and dengue shock syndrome (DSS) in Asia in the 1950s represented a change in the epidemiology of dengue disease at the global level (Guzman and Kouri, 2002; Halstead, 2007; Halstead *et al.*, 2007). In the following five decades, dengue has become the most widespread vector-borne viral disease in humans and a severe public health problem in tropical countries. Attack rates ranging as high as 40–50% amongst susceptible populations, an increase in the frequency of epidemics involving more than one serotype (hyperendemicity), and death rates between 2.5% and 5% have been documented (Stephenson, 2005). Rapid unplanned urbanization with substandard housing, demographic changes, migration, travel, and poor public health programmes all contribute to the current situation (Kroeger and Nathan, 2006). South-East Asia, the Western Pacific and the Americas report the highest number of cases and epidemics, but dengue also circulates in African and Mediterranean countries.

The four dengue serotypes (DENV-1–4) can produce infections that range in severity from undifferentiated febrile illness, to the acute febrile illness called dengue fever (DF), to the life-threatening severe illness, dengue haemorrhagic fever/dengue shock syndrome DHF/DSS (Gubler, 2002).

Classical DF is more commonly observed in adults and is characterized by sudden onset of high fever, with headache, myalgia, arthralgia, rashes and leucopenia. DF may be an incapacitating disease with severe muscle and joint pain. Minor haemorrhagic manifestations and occasionally thrombocytopenia can accompany the illness.

The fever rises rapidly to as high as 104°F, and may be accompanied by bradycardia. A maculopapular recovery rash appears 3–5 days after the onset of fever, and usually appears on the trunk first, before spreading peripherally. Symptoms usually persist for 7 days, hence one of the common names for the disease: seven-day fever.

Early symptoms of DHF are initially indistinguishable from DF, but patients progressively develop complications in the course of their illness. DHF is characterized by acute fever associated with haemorrhagic manifestations, plasma leakage and thrombocytopenia (<100,000 platelets/mm^3). Plasma leakage is characterised by haemoconcentration (haematocrit increase of 20%) or development of ascites or pleural effusion. Dengue shock syndrome (DSS) is distinguished from DHF by the presence of cardiovascular compromise as a consequence of significant plasma leakage and is typically detected as a narrowed or absent blood pressure. DSS is a life-threatening illness that requires urgent resuscitation with intravenous fluids to replace the lost intravascular volume. The mortality rate for DSS can be as high as 20%, but in most places with sufficient resources and clinical experience, mortality is less than 1%. Clinical warning signs for DSS include a rapidly rising haematocrit, intense abdominal pain, persistent vomiting, lethargy or agitation; patients with these symptoms require careful monitoring and judicious management of their intravascular fluid volume.

Recent studies suggest that dengue can produce long-term complications. During the DENV-3 epidemic of Havana, 2001–2002 some patients showed persistence of several dengue symptoms even 6 months after the acute illness. Asthenia (27.6%), headache (14.8%), and arthralgia (10.6%) were the most common symptoms (Gonzalez *et al.*, 2005). Similar observations were noted in Singapore, where a syndrome of post-infectious fatigue was described late after the acute illness (Seet *et al.*, 2007). Finally, after the DENV-2 epidemic of 1997 in Santiago de Cuba, some patients showed arthralgia, myalgia and fatigue related to dengue disease even after 5 years of the illness (B. Sierra, personal communication).

The WHO classification of dengue originated from descriptions of paediatric cases in South-East Asia in the 1970s. With the extension of the illness to other geographic areas, variations from the original description of dengue have been reported. The main problems identified with the WHO classification for DHF are the rigidity of the definitions, the low sensitivity and the assumption by some clinicians that DF means mild disease (Balasubramanian *et al.*, 2006; Bandyopadhyay *et al.*, 2006; Kroeger and Nathan, 2006a; Rigau-Perez, 2006). A revised and simplified categorization for dengue case classification directed to rapidly identify and adequately treat the most severe form of dengue and to simplify case reporting and surveillance has been proposed by TDR/WHO (UNICEF/UNDP/World Bank/WHO Special Programme for Research and Training in Tropical Diseases (TDR), 29 September to 1 October 2008). This new classification based on a review of clinical data from studies developed in seven countries across South-East Asia, Western Pacific and Latin America is now under validation in dengue endemic regions.

The pathogenesis of dengue infection is complicated and, despite intensive research, not well understood. Data suggest that viral, immunopathogenic and other host factors have a role in disease severity. The infecting viral strain, previous infection with a heterotypic serotype, age and genetic background of the individual are the main risk factors for severe disease (Guzman and Kouri, 2008; Halstead, 2007; Kouri *et al.*, 1987). A role for T-cells and a 'cytokine storm' have also been suggested in patients with secondary infections (Lin *et al.*, 2006; Rothman, 2003; Simmons *et al.*, 2005).

Why one individual develops an asymptomatic dengue infection while another in an apparently similar situation dies of DSS is a crucial point of research with implications for the control of the illness determinants for disease severity, based on current knowledge and our specific experiences accumulated from Cuba and Vietnam.; see also chapter 7). A limiting factor in dengue research is the absence of robust animal models of disease that reproducibly generate a vasculopathy that mimics that which occurs in patients. In this chapter, we present a general review of the burden of disease and the host (age, ethnicity, co-morbidities, immune and genetic factors) and viral (epidemiological and genetic factors

Burden of disease

With a total population of approximately three billion people living in endemic areas, and with roughly 120 million travellers to these regions each year, a large share of the world's population is at risk of dengue infection. Reliable estimates suggest that there are approximately 100 million cases of DF annually (Guzman *et al.*, 2006). As dengue spreads around the world, epidemics are often explosive (as recently experienced in Brazil and India (Coelho *et al.*, 2008; Kukreti *et al.*, 2008; Luz *et al.*, 2008; Weissmann, 2008). The successful care of patients with severe dengue is labour intensive and places an enormous strain on health care systems. Quantifying the dengue disease burden (DDB) is key to formulating policy decisions on research priorities, prevention programmes, clinical training for management of the disease, allocation of resources in the face or intermittent explosive epidemics and the introduction of new technologies (Mathers *et al.*, 2007).

It is likely that the twenty-first century will be characterized as a period of globalization, mass migration, urbanization, climate change and changing demographics, all factors that encourage further spread of dengue. The life style of the major dengue vector, *Aedes aegypti*, is ideally suited to the huge conurbations that exist already in Asia and Latin America and may appear in the future in Africa. Moreover, climate change may alter the geographic distribution of *Ae. aegypti*, allowing entry of degue into new areas. Human migration will also play a significant role in dengue epidemiology, as movement among locations will increase exposure to multiple dengue serotypes, a risk factor for severe disease (see below).

Dengue is endemic in all World Health Organization (WHO) regions except the European Region (EUR). Dengue disease burden can be expressed as a single health indicator such as quality-adjusted life years (QALYs) or disability-adjusted life years (DALYs) (Clark *et al.*, 2005; Mathers *et al.*, 2007; Shepard *et al.*, 2004). Internationally, DALY computations are most often employed. The disease burden imposed by dengue and the potential benefits of any intervention, such as vaccination, can then be expressed in terms of DALYs lost or saved and cost per DALY lost or saved (Meltzer *et al.*, 1998).

The DDB was estimated in Puerto Rico for the period 1984 to 1994 (Meltzer *et al.*, 1998) to be an average loss of 658 DALYs per million people per year, comparable to that attributed to meningitis, hepatitis, malaria, tuberculosis, the childhood cluster of diseases (polio, measles, pertussis, diphtheria, and tetanus), or intestinal helminthiasis. This is of the same order of magnitude as tuberculosis in Latin America and the Caribbean (Torres and Castro, 2007). A study in South-East Asia estimated a loss of 420 DALYs per million people per year, comparable to that of meningitis (390 DALYs per million population per year), twice the burden of hepatitis, and one-third of the burden imposed by HIV-AIDS in the region (Cho Min, 2000; Suaya *et al.*, 2007). A study of Thailand estimated that country's DDB at a loss of 427 DALYs per million people per year

for 2001 (Clark *et al.*, 2005). It has been estimated the global DDB as 528,000 DALYs for the year 2001 (Suaya *et al.*, 2007). This corresponds to a burden of 264 DALYs per million populations per year for two billion people living worldwide in areas at risk of dengue. However there are many problems with estimating the disease burden due to dengue, including lack of uniform application of WHO case definition, limited capabilities and standards of dengue laboratories, limited accuracy of rapid tests, misdiagnosis, lack of uniform criteria to report dengue cases to WHO, incomplete surveillance and reporting systems and misclassification in dengue case reporting. Better dengue disease burden estimates would be invaluable in planning cost-effectiveness analyses of new diagnostic, preventive, and therapeutic technologies and would facilitate a better understanding of disease pathogenesis by shedding light on the complete spectrum of infection, disease and immunity.

Host determinants in dengue severity

Host factors are a key issue in susceptibility to DHF. Among these, secondary infection by a different dengue serotype plays a major role, although age, chronic disease, ethnicity and host genotype of the individual are also important.

Humoral immune response and secondary infection

The humoral immune response to dengue infection is important for control of viral infection and dissemination. Neutralizing antibodies directed to the envelope (Env) glycoprotein of the virus inhibit viral attachment, internalization, and/or replication within cells. After a dengue infection, neutralizing antibodies are raised to specific and cross-reactive epitopes of Env protein. Studies in volunteers suggest that infection by a dengue serotype results in long-lived immunity to the homologous serotype with short-lived cross-protection (3–4 months), called heterotypic immunity, against the other serotypes (Sabin, 1952). Following a primary infection, after an initial period of cross-protective immunity, antibodies persist but lose their capacity to neutralize heterologous serotypes. During a secondary infection with a heterologous serotype, these pre-existing cross-reactive antibodies are hypothesized to bind to the infecting heterologous virus and facilitate viral infection of Fc receptor bearing cells through a phenomenon known as Antibody Dependent Enhancement (ADE) (Green and Rothman, 2006; Kurane, 2007; Kurane *et al.*, 1991b; Littaua *et al.*, 1990).

Antibody-dependent enhancement – a hypothesis to explain DHF

The ADE phenomena are believed to result in a greater viral biomass '*in vivo*' as a consequence of more cells being infected. This higher viral burden elicits a substantially greater host inflammatory response. Increased plasma levels of several soluble mediators, some directly linked to plasma leakage, have been detected in patients with severe disease (Mathew and Rothman, 2008). '*In vitro*' studies indicate that heterologous anti-Envelope antibodies mediate enhancement of virus infection through FcRII, while anti-pre-membrane antibodies enhance dengue infection of both FcRII and non FcRII-bearing cells (Huang *et al.*, 2006; Littaua *et al.*, 1990). According to the ADE hypothesis, severe disease is a direct consequence of increased viral burden. Consistent with this hypothesis, high initial viraemia titres have been associated with disease severity in children experiencing secondary dengue (Halstead *et al.*, 2005; Libraty *et al.*, 2002b; Vaughn *et al.*, 2000).

Low-affinity heterotypic antibodies could assist the viral infection of monocytes early after the binding of virus-antibody complex to FcR. Conversely, immunocomplexes with higher affinity antibodies could favour the monocyte's uptake and digestion by endolysosomal enzymes. That could explain the different evolution of the infection between individuals with a dengue secondary infection: those who are asymptomatic generally would have elevated levels of higher affinity neutralizing antibodies than those who suffer the disease with different ranges of severity. Then, antibody–virus immunocomplex affinity could play an important role modulating the infection evolution of different individuals with similar levels of cross-reacting antibodies.

The ADE hypothesis is supported by several direct and indirect lines of evidence. First, epidemiological studies in Thailand and elsewhere suggest that DHF occurs predominantly in a second infection and that up to 99% of DHF cases were associated with pre-existing heterotypic dengue antibodies (Halstead, 1993; Kurane, 2007; Kurane and Ennis, 1997). The role of secondary dengue infection as a risk factor for more severe disease has been supported in several prospective studies performed in Thailand, Burma and Indonesia (Burke *et al.*, 1988; Endy *et al.*, 2002; Kyle and Harris, 2008; Sangkawibha *et al.*, 1984; Thein, 1975).

The epidemiology of dengue in Cuba also supports the association of secondary infection with DHF/DSS. Well-spaced dengue outbreaks, caused by different serotypes, in Cuba have provided unique insights into this phenomenon. In three separate dengue epidemics, DHF has been significantly associated with secondary dengue infection. In 1977–79, a DENV-1 epidemic affected the whole country, producing more than 500,000 mild cases. According to Cantelar *et al.*, 44.46% of the Cuban population developed anti DENV-1 antibody (Cantelar de Francisco *et al.*, 1981). Four years later, DENV-2 entered the country producing more than 300 000 reports and 10,000 severe and very severe cases with 158 fatalities. DHF/DSS cases were observed both in children and adults (Kouri *et al.*, 1989). After a 15 year period without dengue transmission, DENV-2 entered Santiago de Cuba municipality, producing an estimated 5 208 clinical cases with 205 DHF/DSS and 12 fatalities, all of them in adults (Kouri *et al.*, 1998; Valdes *et al.*, 1999). Finally in 2001–02, DENV-3 affected Havana, the capital city of the country, with 12,889 cases, 78 DHF/DSS and three fatalities, all severe cases in adults (Guzman *et al.*, 2006; Pelaez *et al.*, 2004).

Considering the particular dengue situation in this country, the neutralizing antibody response to the four serotypes was retrospectively determined in several groups of individuals with the antecedent of DHF during the 1981, 1997 and 2001 epidemics (Alvarez *et al.*, 2006; Guzman *et al.*, 1984, 1990, 1997, 1999, 2000). These studies, in conjunction with seroepidemiological surveys, allowed a determination of the serotype and the sequence of the infections. In three groups of DENV-2 and DENV-3 DHF cases, secondary infection was observed in a large number of individuals supporting the role of secondary infection as a risk factor for DHF/DSS (Table

5.1). No single DHF/DSS case was reported in children in the 1997 and 2001 epidemics, as children were at risk of a primary DENV infection only. An indirect observation supporting the role of the secondary infection as risk factor for DHF/DSS was the mild disease that characterized the 1977–79 epidemic of DENV-1. At that time, most of the Cuban population was at risk of a primary dengue infection and no single DHF case was reported.

Experimental evidence for ADE

The ADE model attempts to explain the association of secondary infection with DHF/DSS. The ADE process is believed to increase viral infection resulting in increased viral burdens in the host. Evidence for ADE in humans are indirect. Initial reports by Kliks *et al.*, 1989 suggested that the ability of pre-illness plasma to enhance dengue virus infection in monocytes correlated with disease severity (Kliks *et al.*, 1989). More recently, reduced viraemia levels and decreased disease severity were observed in individuals secondarily infected by DENV-3 with documented neutralizing anti DENV-3 antibodies in pre-illness plasma (Endy *et al.*, 2004). However, similar observations were not observed in the same study for secondary DENV-1 or DENV-2 infections. Laoprasopwattana *et al.* reported that enhancing antibody activity in plasma does not predict subsequent disease severity or viraemia in secondary DENV infection (Laoprasopwattana *et al.*, 2005). These results have been discussed and criticized by Halstead (2006) and Burke *et al.* (2006) (Burke and Kliks, 2006; Halstead, 2006). More recently, Goncalvez *et al.* (2007) demonstrated a significant increase of DENV-4 viraemia titres in juvenile rhesus monkeys immunized with passively transferred monoclonal antibody supporting the establishment of an animal model

for ADE (see Chapter 14). Finally, Guzman *et al.* (manuscript in preparation) has observed long-lasting ADE activity *in vitro* in sera collected from DENV-immune individuals. Better experimental design and further studies are needed to elucidate the role of ADE.

Multiple heterotypic infections

Primary, secondary, tertiary and even quaternary dengue infections have been documented in endemic settings. Tertiary and quaternary DENV infections have been reported in seroepidemiological studies, however their relationship with disease severity is unknown (Halstead, 2007). Studies performed in Thailand documented that 0.08–0.8% of dengue hospitalizations may be caused by third or fourth infections (Gibbons *et al.*, 2007). In addition a report from Alvarez *et al.* (2006) suggested that DHF resulting from tertiary infections was a surprisingly common event in Cuba (17.5% of the total cases). Further studies to determine the role of tertiary and quaternary DENV infections as contributors to symptomatic infection are needed.

Cellular immune response

There are robust epidemiological, observational, empirical and experimental data that support the participation of both humoral and cellular immune responses in the pathogenesis of the DHF, with a dual/paradoxical role oscillating between protection and damage. Recent observations by Mongkolsapaya *et al.* (2003) are consistent with the 'original antigenic sin phenomenon' for T-cell response, previously reported for the dengue antibody response by Halstead *et al.* (1983). The 'original antigenic sin phenomenon' is based on the fact that the DENV causing secondary disease is of a different serotype than the virus that induced immune response during the earlier

Table 5.1 Secondary infection in individuals with DHF during 1981, 1997 and 2001–02 Cuban epidemics

	1981		1997		2001–02
	Children	Adults	Adults		Adults
Epidemics	DHF		DHF	Fatal	DHF
Sample	124	104	111	12	51
2d infection	98%	98%	98%	91%	97%

DENV infection. Therefore the antibodies and memory T lymphocytes induced by the primary DENV infection typically encounter antigens that differ in sequence from their original target antigen. The magnitude of these sequence differences is epitope dependent, but all sequence differences have the potential to affect the quality of the effector response and also modify the immunological repertoire. The pattern of antibody/T-cell responses in convalescence, therefore, is influenced heavily by the serotype of the primary DENV infection (Halstead *et al.*, 1983). It has been speculated that the immune response to these heterologous sequences has the net effect of altering the balance between a protective and pathological outcome.

T regulatory (T reg) cells may also play role in regulating immune responses during dengue. The function of these cells in chronic and persistent infectious diseases has been an exciting topic of research in recent years (Belkaid and Rouse, 2005; Mills, 2004). However, there is little information on how they control the immune response to acute viral pathogens. These studies are in fact complicated by the small antigen specific population in peripheral blood, which likely under-represents the effect of T reg cells at the site of viral infection (Keynan *et al.*, 2008). The role of the T reg cell response in acute infection may be inadequate to control immune activation. However, it was recently reported that T reg cells are fully functional and expand in acute dengue infection. Moreover they are able to suppress the production of vasoactive cytokines such as IFN-γ, TNF-α, and IL-6 in response to dengue antigens, but in some cases they might be insufficient to control the immune response in patients with severe disease (Luhn *et al.*, 2007). Since T reg cells regulate innate immunity also, they could influence antibody-dependent enhancement and innate immune cell–driven vasoactive cytokine release.

Cytokines and DHF

Elevated serum levels of cytokines and chemokines such as IL-2, IL-6, IL-8, IL-10, IL-13, IL-18, IFN γ, TNFα, MCP-1 are associated with secondary infections and severe disease in DHF/DSS patients (Azeredo *et al.*, 2001; Chakravarti and Kumaria, 2006; Hung *et al.*, 2004; Mustafa *et al.*,

2001; Perez *et al.*, 2004; Pinto *et al.*, 1999; Yang *et al.*, 1995), suggesting the immune response is a contributor to the pathogenesis of DHF/DSS.; see also Chapter 7). The contribution of the immune response includes circulating antibody from previous infections, the anamnestic B-cell and T-cell responses, with the effects of cytokines on both infected and bystander immune cells, hepatocytes, and endothelial cells (Anderson *et al.*, 1997; Green and Rothman, 2006; Halstead, 1989; Kurane *et al.*, 1991a; Lei *et al.*, 2001

Pro-inflammatory cytokines released by DENV serotype cross-reactive memory T-cells could explain the severe course in some patients after DENV re-infection, since it has been suggested that plasma leakage could be related to malfunction of vascular endothelial cells induced by cytokines or chemical mediators rather than by destruction of the small vessels (Avirutnan *et al.*, 1998; Beynon *et al.*, 1993; Bosch *et al.*, 2002; Chaturvedi *et al.*, 2000; Green *et al.*, 1999; Kurane *et al.*, 1994; Maruo *et al.*, 1992; Talavera *et al.*, 2004). In fact, endothelial cell monolayers have been rendered permeable by treatment with cytokines generated in various ways, including by antibody-mediated dengue virus infection of monocytes (Anderson *et al.*, 1997; Burke-Gaffney and Keenan, 1993), release of cytotoxic factors from dengue-infected monocytes, macrophages, or dendritic cells (Carr *et al.*, 2003; Palmer *et al.*, 2005) or endogenous interleukin 1β produced by peripheral blood mononuclear cells stimulated with interferon and tumour necrosis factor (Cardier *et al.*, 2005; Seynhaeve *et al.*, 2006). However, it is not clearly understood how these cytokines cause malfunction of vascular endothelial cells leading to plasma leakage.

In vitro evidence suggests that ADE results in decreased nitric oxide (NO) production, and increased IL-10 and virion production. Mirroring these *in vitro* data, reduced NO and enhanced IL-10 blood concentrations have been observed in DHF patients (Chareonsirisuthigul *et al.*, 2007; Simmons *et al.*, 2007b). There is strong evidence that NO and superoxide contribute to immune regulation including the development of inflammatory processes like IL-1β or TNFα expression, as well as the expression of IL-6 and IFNγ, all of them critically involved in dengue pathogenesis (Guzik *et al.*, 2003; Khare and Chaturvedi, 1997).

IL-10 is suggested to play a role in DHF pathogenesis. The increase in IL-10 levels in acute dengue has been correlated with platelet decay (Libraty et al., 2002a). IL-10 may be down-regulating lymphocyte and platelet function (Azeredo et al., 2001) and could be involved in the induction of T-cell apoptosis described in secondary virus infection (Mongkolsapaya et al., 2003). On the other hand, IL-10 could modulate the synthesis of platelet-activating factor (PAF), a potent inflammatory mediator of vascular injury (Bussolati et al., 2000). More recent in vivo studies, by Pérez et al. (2004) have shown higher levels of IL-10 in DHF patients compared to controls, being consistently higher in those patients with secondary infections. Differences in the levels of released soluble mediators like IL-10 between primary or secondary infections could be explained by differential activating mechanisms. Alternatively, different cytokine patterns could be released by activated naive or memory T-cell clones in a primary or sequential infection, since during a secondary infection the kinetics of cellular and molecular events may be different. In the presence of heterotypic antibodies, skin monocytes could be more efficiently infected than DCs, and be able to present antigens efficiently to memory T-cells (Perez et al., 2004).

TNFα is another cytokine that has attracted attention because of its well-known activity in inducing endothelial permeability. It was reported that dengue virus-infected monocytes and endothelial cells produce TNFα (Green and Rothman, 2006). It was also reported that dengue infected monocytes produce TNFα when they were infected with dengue virus in the presence of enhancing antibody and that the produced TNFα-induced plasma leakage in an in vitro experimental system using endothelial cells (Anderson et al., 1997).

A unifying model of DHF pathogenesis?

Efforts to identify the mechanisms that underlie the sudden onset of vascular permeability and of bleeding in DHF include the study of the T-cell response and infection-enhancing antibody. T-cells may contribute to pathogen clearance via help for antibody production by mediating lysis of virally infected cells and/or via production of macrophage-activating cytokines. Paradoxically, the T-cell response may also contribute to capillary permeability via release of vasoactive cytokines. There is, however, an important patient subgroup for which memory T-cell responses cannot explain pathogenesis – cases of DHF in infants – the vast majority of which is caused by primary dengue virus infections. The occurrence of DHF during primary DENV infection in infants born to DENV-immune mothers, who acquire anti DENV antibody transplacentally, supports the idea of an in vivo role of ADE, but not memory T-cells (Chau et al., 2008; Simmons et al., 2007b). There are, however, reasons to believe that the pathogenesis of dengue in infants versus children can have a different underlying basis. For example, the pathogenesis of dengue in infants is complicated by their inherent immunological immaturity, the process of massive angiogenesis (resulting in a very large vascular endothelia surface area per kg weight) and poor compensatory reserve. Dengue in infants does not necessarily mean that memory T-cell responses are not important in secondary infections in older children, simply that they cannot explain DHF in infants (Simmons et al., 2007a).

A unifying model of DHF in infants (primary infections) and older children/adults (secondary infections) proposes that enhancing antibodies lead to an increased infected cell mass, and in secondary infections, T-cell and cytokine responses are then proportionate to the antigenic stimulus. High titres of circulating virus in blood collected early during early secondary infection and high concentrations in blood of dengue viral RNA and dengue non-structural protein 1 (NS1) are consistent with this hypothesis (Libraty et al., 2002a; Libraty et al., 2002b; Vaughn et al., 2000; Wang et al., 2003). Although evidence that viral burdens or NS1 concentrations are higher in infants with DHF has not been forthcoming (Simmons et al, 2007), there does exist a temporal relationship between the DENV infection-enhancing activity of neat plasma from Vietnamese infants and the age-related epidemiology of DHF in this population. Cellular immune responses are also greater in infants with DSS compared to DHF and this is assumed to be driven by a greater viral burden in these infants (Chau et al., 2008). Further studies are needed to understand whether the pathogenesis of dengue in infants has a similar basis to that in older children.

Autoimmune responses in DHF

Autoimmune events could be also involved in dengue pathogenesis. The correlation of disease severity with immune activation markers (e.g. IL-6 and IL-8, TNF α, IFN-α and -γ, soluble TNFα receptor, complement components 3a and 5a) together with altered platelet, dendritic cell, monocyte, and T-cell functions suggest that immune responses to components of dengue viruses could contribute to autoimmune processes resulting in DHF/DSS (Falconar, 1997; Lin *et al.*, 2002, 2003, 2004). The autoimmune hypothesis posits that endothelial dysfunction is the result of cross-reactivity between anti-DENV NS1 to host proteins and endothelial cells (Lin *et al.*, 2006) Immune complex formation '*in vivo*' in association with complement activation has been detected in patients with severe disease (Falconar, 1997). Anti Env dengue protein antibodies cross-reactive with plasminogen have been associated with bleeding in acute DENV infection, but not with DHF (Chungue *et al.*, 1994). However, the autoimmune NS1 hypothesis is contradicted by the durability of NS1 antibodies in contrast to the transient nature of vascular permeability and haemostatic disorders.

Further immunological, virological and epidemiological studies will enlarge our understanding of the complex mechanisms of immunopathogenesis of dengue haemorrhagic fever and contribute to the development of therapeutics as well as safe and efficacious dengue vaccines and antiviral therapies.; see also Chapters 6, 7, 12 and 24

The sequence and interval of infection by different DENV serotypes

As heterotypic secondary DENV infection is a main risk factor for DHF/DSS, two related parameters, the sequence of the infecting serotypes and the interval between infections, deserve to be studied.

DHF/DSS has been observed in the course of a secondary infection by any of the four serotypes, but apparently at different frequencies (Alvarez *et al.*, 2006). This phenomenon could depend not only on the serotype of the second infection but also on the sequence of primary and secondary infections and the interval between them. In hyperendemic areas, it is not easy to clearly define the serotype of the primary infection and the time elapsed between infections. Since it experiences periodic waves of dengue infection, the Cuban dengue epidemiological situation allows an analysis of these factors. This population was initially infected with DENV-1 in 1977–78 (all primary infections) and then infected 4 and 20 years later with DENV-2 and 24 years later by DENV-3. This epidemiological situation allows insight into the impact of the first/second serotype combinations DENV-1/DENV-2; DENV-1/DENV-3 and DENV-2/DENV-3 on disease severity. The case fatality rate attributable to the DENV-1/DENV-2 sequence in the 1997 epidemic was 0.39%, significantly higher than case fatality rate due to DENV-1/DENV-3 in the 2001–02 epidemic (0.02%) (Guzman *et al.*, 1999, 2006; Pelaez *et al.*, 2004; Valdes *et al.*, 1999). These data and the higher ratio of DHF cases in the 1997 (6.8%) compared to the 2001–02 epidemic (0.6%) suggest a higher risk of severe disease in the DENV-1/DENV-2 sequence of infection.

Moreover, Alvarez *et al.* (2006) studied 69% of the total DHF cases during the 2001–02 DENV-3 epidemic. No single case in the sequence DENV-2/DENV-3 was observed while 77% of DHF cases were in the sequence DENV-1/DENV-3. Of interest is that 55% of DF cases were in the sequence DENV-1/DENV-3 while 15% were in the sequence of DENV-2/DENV-3. Seroepidemiological studies performed after the end of the epidemic support the conclusion that disease severity was significantly greater in the sequence DENV-1/DENV-3 compared to DENV-2/DENV-3 (M.G. Guzman, manuscript in preparation). Since the Cuban health care system is accessible to all strata of the population, the competent supportive treatment given to DHF/DSS in each epidemic assured that similar high-quality care was provided during these outbreaks and that variation in disease severity and case fatality rates are unlikely to be attributable to differences in clinical management or case detection. Outside of Cuba, DHF has also been reported in the sequences of DENV-3/DENV-2, DENV-4/ DENV-2, DENV-1/DENV-4, DENV-3/DENV-1, DENV-3/DENV-4 (Gibbons *et al.*, 2007; Sangkawibha *et al.*, 1984).

The 1997 Cuban epidemic allowed investigation of the role of the time interval between infections as risk factor for disease severity. For the first time, DHF case fatality rate and DHF rate by secondary DENV-2 infection were studied in individuals infected initially by DENV-1 at intervals of 4 and 20 years previously. Significantly higher DHF/DSS ratios were observed after the longer interval (Guzman *et al.*, 2002b). In particular, death and hospitalization rates for DHF/DSS per 10,000 secondary DEN-2 infections were compared in the same age group affected in the DENV-2 epidemics that occurred in all of Cuba and Havana in 1981 and in just Santiago de Cuba in 1997. Among adults aged 15–39 years the death rate per 10,000 secondary DENV-2 infections was 39.2 times higher in Santiago de Cuba in 1997 than in Havana in 1981. DENV-2 strains recovered in 1981 and in 1997 epidemics belong to the Asian genotype of DENV-2 (see Chapter 9). It is not known whether the nucleotide and amino acid differences observed between the 1981 and the 1997 DENV-2 strains led to the observed variation in the pathogenic consequences (Guzman *et al.*, 2002b). However, a second explanation for this phenomenon could be a progressive loss in time of heterotypic neutralizing antibodies to DENV-2 in DENV-1 immune individuals, a phenomenon that has been recently documented (Guzman *et al.*, 2007).

The above-mentioned situation allowed us to explore the long-term memory T-cell response to DENV after a natural primary infection. It was shown the existence of memory T-cell response that exhibits serotype cross-reactive proliferative response to DENV 20 years after the primary infection (Sierra *et al.*, 2002). Taking into account the reported role (Mathew and Rothman, 2008) of the cellular immune response in the DHF pathogenesis, the existence of memory cross-reactive T lymphocytes 20 years post primary infection could contribute to the 205 classical DHF/ DSS cases with 12 deaths in persons 18 years and older observed in Santiago de Cuba in the 1997 DENV-2 outbreak (Guzman *et al.*, 2002b).

Age

Most primary infections in children are asymptomatic or very mild while in adults classical DF can be observed. Recently. Egger and Coleman (2007) documented that the relative risk of classical DF during a primary DENV infection increases with age. However, silent primary DENV-2 infection was documented in 94% of both infected children and adults during the Santiago de Cuba 1997 epidemic (Guzman *et al.*, 2000). Moreover, the pattern of severe disease may be different in secondary infections.Today, DHF/ DSS is observed both in children and adults. In South-East Asia, DHF/DSS remains predominantly an illness of children while in the American region adults are the most affected (Rigau-Perez *et al.*, 2001). Classical DHF/DSS in Cuban adults was recognized during the DENV-2 epidemic in 1981, The observation of DHF both in children and in adults in this epidemic allowed an estimation of the DHF ratio of secondary DENV-2 infection by age groups. A higher DHF susceptibility was observed in young ages (Guzman *et al.*, 2002a). This observation is in agreement with the inherent higher vascular permeability of children relative to adults (Gamble *et al.*, 2000).

Preceding host conditions

Few studies have investigated associations between chronic illness and dengue severity. In the Cuban epidemics, an increased association between DHF and bronchial asthma, diabetes mellitus and sickle cell anaemia was reported (Bravo *et al.*, 1987; Diaz *et al.*, 1988; Gonzalez *et al.*, 2005a; Guzman *et al.*, 1999, 2006; Limonta *et al.*, 2008; Valdes *et al.*, 1999). In addition, peptic ulcer may be a risk factor for severe bleeding.

During the 1981, 1997 and 2001–2002 Cuban epidemics, bronchial asthma was reported in 23%, 8% and 22% of children (1981) and adult fatal cases respectively. On the other hand, sickle cell anaemia was associated with DHF in the 17%, 8% and 11% of adult fatal cases in the same epidemics (1981, 1997 and 2001–2002). These figures were significantly higher than figures observed in the general population.

Diabetes mellitus was reported in 4% of fatal cases during the 1981 Cuban epidemic. Similar observations have been recently made in Taiwan during a DENV-2 epidemic (Lee *et al.*, 1992). These authors report diabetes mellitus in 16.8% of DHF cases. Finally, Rigau-Perez (2006) reported some chronic illness in 70% of adult fatal case .

Case–control studies need to be developed in dengue-endemic areas to further elucidate the role of these and other co-morbidities in susceptibility to severe disease. Also studies need to be developed to learn more about the impact of dengue on the disease course of other pathogens. In this sense, HIV is of particular interest. Recently, Silva-Nunes *et al.* (2006) reported DHF in a patient with acquired immunodeficiency syndrome. No significant alteration in his illness was noted. Similar observations have been noted by Gonzalez *et al.* during the 2001–02 DENV-3 Cuban epidemic. This is particularly interesting in light of a recent report by McLinden *et al.* of the inhibition of HIV replication by the NS5 protein of DENV (McLinden *et al.*, 2008; Gonzalez *et al.*, 2008).

Host genetic background

Genetic polymorphisms include single nucleotide polymorphisms, gene copy number variants and deletions. These subtle changes might very well have important consequences for susceptibility to disease (Casanova and Abel, 2007). Investigations into the genetic basis of susceptibility to disease have made enormous strides since the completion of the human genome sequencing project. Until recently, this field of research was populated with many case–control association studies of small sample size, uncertain case ascertainment and unknown levels of population stratification (genetic admixture in one population that obscures any meaningful study of genetic association). Typically, these studies examined polymorphisms in a limited number of candidate genes for which there was some *a priori* rationale for their investigation. For the vast majority of case–control association studies, the genetic association has never been replicated in either the same or a different population. When this has been attempted, the same association has been observed only very rarely. Improvements in technology that permit high throughput genotyping of genetic polymorphisms have allowed a genome-wide approach to investigating host genetic susceptibility. Essential to this approach are large sample sizes (>5000 cases and controls), in order to make statistical correction of the number of comparisons being made. Replication of the genetic association in a second population is now considered an essential validation step.

Compared to these criteria, published studies of dengue disease susceptibility are underpowered and, with few exceptions, have never been replicated (Bashyam *et al.*, 2006; Stephens *et al.*, 2002; Loke *et al.*, 2001; Sakuntabhai *et al.*, 2005). Moreover, most have not attempted any functional studies to try to link the genetic association with any process in disease pathogenesis. Future genetic association studies in dengue should consider assembling, analysing and interpreting large case-control, genome-wide genetic data. Despite the fact the accepted 'gold standard' for performing case–control association studies has changed in recent years, it is nonetheless worthwhile to make some observations on the dengue literature in this field.

Indirect evidence implicating host genetic background in DHF/DSS has been observed in Cuban dengue epidemics where a reduced risk for DHF/DSS was observed in those with an African ancestry compared to those with European ancestry (Agramonte, 1906; Bravo *et al.*, 1987; Gonzalez *et al.*, 2005a; Guzman, 2005; Guzman *et al.*, 1990, 1999; Guzman and Kouri, 2002; Sierra *et al.*, 2007). In support of these observations, Agramonte in 1906 stated that 'Black people seem to have a remarkable degree of resistance to dengue disease' (Agramonte, 1906). These Cuban observations are of significant epidemiological interest, as the differences in susceptibility to DHF among racial groups in Cuba coincide with the low susceptibility reported in African and Black Caribbean populations (Halstead *et al.*, 2001; Roche *et al.*, 1983; Saluzzo *et al.*, 1986a).

Although dengue virus has been repeatedly isolated in Africa, and DF is known to be present in 19 countries in this continent, there are only sporadic reports of DHF cases (Gubler *et al.*, 1986; Saluzzo *et al.*, 1986b; Sharp *et al.*, 1995). An additional observation is that in Haiti, despite the presence of viral risk factors, no DHF/DSS cases have been reported (Halstead *et al.*, 2001). The current Cuban population principally originated from two well-defined ethnic groups – the Spanish colonizers and the African slaves – as the native population, the Cuban aborigines, was almost totally exterminated (Guerra, 1964; Ulloa, 2002).

The origins of the Cuban population and the dengue epidemiological observations related to ethnicity in Africa and Haiti could indicate the possible sharing of a common gene pool that moderates the clinical outcome of dengue infection in individuals who have African genetic background. Bearing in mind the central role of immunological mechanisms in the pathogenesis of DHF the genes associated with the immune response must be considered carefully in the context of regulation of dengue disease severity, since there may be considerable variation in the alleles present in individuals as a consequence of their ethnic ancestry.

In particular the polymorphic HLA genes have been among the most studied candidates for genetic associations with dengue disease severity. Several serological studies of HLA class I alleles have been performed in ethnically and geographically distinct populations, and positive associations of various HLA class I alleles with susceptibility to DHF have been found (Paradoa Perez *et al.*, 1987) (Chiewsilp *et al.*, 1981; Loke *et al.*, 2001). More recently a case–control study in ethnic Thai cases reported the association of classical HLA class I alleles (A2, A*0207, B46, B51) with different clinical outcomes, DF or DHF in dengue immunologically primed individuals (Stephens *et al.*, 2002).

Taking advantage the unique epidemiological Cuban dengue situation a molecular approach to HLA case–control study was recently conducted. A significantly higher frequency of HLA I alleles A*31 and B*15 was found in Cuban individuals with a history of symptomatic DENV dengue virus infection compared to controls. HLA II alleles DRB1*07 and DRB1*04, on the other hand, showed an elevated frequency in controls compared with dengue cases.

These results are consistent with the report of Loke *et al.* (2001) in a Vietnamese population, who found that the polymorphism of the HLA class I loci was significantly associated with DHF disease susceptibility, but the polymorphism in the HLA-DRB1 was not. (Loke *et al.*, 2001; Stephens *et al.*, 2002). Notably, HLA II results also match the report by Lafleur and colleagues

(LaFleur *et al.*, 2002), who showed DRB1*04, the most frequent allele in the Mexican Mestizo and Amerindian populations of the Americas, was associated with resistance to DHF in a Mexican population. This finding is extremely interesting considering the Cuban ethnographic history. (Centro, 1976; Rivero de la Calle, 1984).

The Cuban population studied here belong to Santiago de Cuba, a municipality located in the eastern part of the island, the place where aboriginal genetic influences persisted longer (Guanche, 1983) and where, in fact, the phenotypic difference between western and eastern Cuban people are perfectly distinguishable, with a predominance of some Amerindian features in the latter. Thus, the identification of the same HLA class II allotype associated to dengue disease protection in Mexican Mestizo population and in the Cuban population of Santiago de Cuba may be explained by their shared Amerindian genetic background.

Genes in the class III region encode a number of proteins, including complement proteins (C4A, C4B, C2), TNFα and TNFβ and heat shock proteins (Cooke and Hill, 2001). Loke *et al.* (2002) studied promoter polymorphisms in the TNFα gene but did not find an association with DHF, whereas Fernandez-Mestre *et al.* (2004) studied a single-nucleotide polymorphism and reported a significant increase of the TNF-308A allele in patients with DHF.

The number of studies on polymorphisms within non-HLA genes remains low. Loke and colleagues investigated the association between susceptibility to DHF and polymorphic non-HLA alleles: vitamin D receptor (VDR), Fcγ receptor II (FcγRII), IL-4, IL-1RA, and mannose-binding. The less frequent T allele of a variant at position 352 of the vitamin D receptor (VDR) gene was associated with resistance to severe dengue ($P = 0.03$). Homozygotes for the arginine variant at position 131 of the Fc gammaRIIA gene, who have less capacity to opsonize IgG2 antibodies, may also be protected from DHF (one-tailed $P = 0.03$). No associations were found with polymorphisms in the mannose binding lectin, interleukin-1 (IL-4), and IL-1 receptor antagonist genes. (Loke *et al.*, 2002).

Additionally, an allelic variant of a DC-SIGN1 coding gene CD209, DCSIGN1–336, shows a strong association with the risk of DF compared to with DHF or population controls. DC-SIGN1 is a dendritic cell-specific ICAM-3 grabbing non-integrin essential for productive infection of dendritic cells (Despres *et al.*, 2005).

Recently associations on TAP and HPA gene polymorphism in dengue were also described, suggesting that the heterozygous pattern at the TAP1 333 locus and HPA1a/1a and HPA2a/2b genotypes confer susceptibility to DHF. The HPA1a/1b genotype was determined to be a genetic risk factor for DSS (Soundravally and Hoti, 2007).

Viral determinants in dengue severity

Dengue virology and epidemiology

As described in Chapter 1, DENV is a single-stranded, positive-sense, RNA virus (genus Flavivirus, family Flaviviridae) with a genome of approximately 11,000 nucleotides (nt) that is translated as a single polyprotein and then cleaved into its individual components. The viral genome is comprised of 3 genes encoding structural proteins – the capsid (C), membrane (M) and Env proteins, and 7 non-structural (NS) genes – NS1, NS2A, NS2B, NS3, NS4A, NS4B and NS5. Functions of the non-structural proteins include an RNA-dependent RNA polymerase and methyltransferase (NS5) and a helicase/protease (NS3). The coding region is flanked by a 5′-UTR (untranslated region), of approximately 98 nt, and a 3′-UTR of 388–462 nt (Chambers *et al.*, 1990).

Infection of target cells is mediated through binding of the Env protein to a cell surface receptor, followed by internalization of the viral particle through receptor-mediated endocytosis and, finally, fusion of viral and cellular membranes following acidification of the endosome. Viral proteins are translated and either remains in the cytoplasm or are co-translationally threaded through the endoplasmic reticulum membrane (see Chapter 2). Genome replication occurs at vesicular sites where genomes are packaged and virions are assembled for secretion via secretory vesicles through the Golgi (Clyde and Harris, 2006) (see Chapter 3).

The four antigenically distinct DENV serotypes diverge at ~30% across their polyprotein. Each serotype also harbours phylogenetically distinct 'subtypes' or 'genotypes', which have differing geographical distributions (see Chapters 9 and 11). Humans are the only mammalian host of epidemic DENV, with *Aedes* mosquitoes the principal vectors, particularly the peri-domestic species *Ae. aegypti,* which also transmits yellow fever virus. *Ae. aegypti* is closely associated with human habitation, and larvae are often found in artificial water containers such as discarded tires, buckets, and water storage jars. Increases in dengue incidence and geographic range are largely due to the introduction of *Ae. aegypti,* followed by the virus, to non-endemic regions of the world. Given that breeding of *Ae. aegypti* is closely associated with standing water, it is no surprise that DENV transmission in the tropics is often strongly seasonal, with peaks during each rainy season.

The first well-documented outbreak of DHF/DSS occurred in Manila during 1953/54, and was followed by a larger outbreak in Bangkok in 1958 (Halstead, 1980). Since this time, DHF/DSS has become endemic in all countries in South-East Asia, with dramatic increases in case numbers, so much so that dengue is considered an archetypal 'emerging' disease. At the same time, the geographic range of DHF/DSS has expanded considerably, and it is now reported in over 100 countries (PAHO, 2002). The precise causes of DHF/DSS remain elusive despite intense research efforts. In addition to epidemiological studies that have shown that secondary infections with a different DENV serotype correlate with severe disease (Halstead, 2002; Halstead and Yamarat, 1965) recent work has demonstrated that Asian, but not American, genotypes of DENV-2 are associated with severe disease (Rico-Hesse *et al.*, 1997). Therefore, the specific genetic make-up of the infecting strain does indeed appear to play a role in determining disease outcome. One hypothesis is that this is due to the fixation of mutations that result in increased replication of the virus within host cells (or certain cell types) – in turn increasing fitness (Cologna *et al.*, 2005; Cologna and Rico-Hesse, 2003; Pryor *et al.*, 2001) In addition, the virus may also be less sensitive to targeting by the host immune system, including altered sensitivity to neutralization by host antibody responses (Kochel *et al.*, 2005).

As noted above, one of the most remarkable features of DENV is that it exists as four distinct serotypes that exhibit complex immunological and epidemiological patterns, including ADE. For example, the pattern of DHF/DSS incidence within endemic populations such as Thailand exhibits complex wave-like dynamics, with the dominant serotype changing on an 8–10 year cycle (Cummings *et al.*, 2004).

Similarly, complex oscillations in serotype frequency are observed within individual populations, most likely stemming from different levels of cross-protection (and perhaps enhancement) among the antigenically distinct serotypes (Endy *et al.*, 2004). In Bangkok, for example, DENV-1, -2 and -3 appear to be cycling in-phase with each other (i.e. on the same epidemic cycle), while DENV-4 is consistently out-of-phase with the other serotypes. (Adams *et al.*, 2006; Zhang *et al.*, 2005). However, the underlying cause of this complex dynamical behaviour, particularly the roles played by immune cross-protection and enhancement, is unclear and constitutes a major goal in studies of dengue epidemiology (see Chapter 10).

The molecular epidemiology of DENV

The growing availability of comparative genome sequence data has provided important insights into the molecular evolution and epidemiology of DENV. Even before gene sequence data was available, it was known that genetic variation was apparent within each serotype (Monath *et al.*, 1986; Trent *et al.*, 1989).

Since the early 1990s, the scope of genotyping studies has greatly expanded, particularly through the use of complete Env gene sequences (~1485 nt) as a phylogenetic marker (Chungue *et al.*, 1993; Goncalvez *et al.*, 2002; Klungthong *et al.*, 2004; Lanciotti *et al.*, 1997; Lanciotti *et al.*, 1994; Lewis *et al.*, 1993; Rico-Hesse *et al.*, 1998; Rico-Hesse *et al.*, 1997; Twiddy *et al.*, 2002) and has now been extended to the analysis of complete viral genomes (Zhang *et al.*, 2005; Zhou, 2006). The most obvious result stemming from these studies is that viral isolates form distinct clusters of sequences – subtypes or genotypes – within each serotype. Currently, 3 subtypes can be identified in DENV-1, 6 in DENV-2 (one of which is only found in non-human primates, see

Chapter 11), 4 in DENV-3 and 4 in DENV-4, (with another exclusive to non-human primates). Further, phyogenetically discrete groups of isolates are observed within individual genotypes, which can be considered viral 'clades'. For example, two clades are observed in genotype I of DENV-1 which differ in their temporal distribution (Zhang *et al.*, 2005).

Although there is an arbitrary component to genotypic classifications, the detailed analysis of the phylogenetic structure of DENV provides important insights into the evolutionary and epidemiological processes that shape diversity. For example, subtypes often have differing spatial distributions, with some more widespread (i.e. 'cosmopolitan') than others. This is best documented in DENV-2, where two subtypes are apparently restricted to South-East Asia and another to the Americas (Twiddy *et al.*, 2002). In contrast, a Cosmopolitan subtype has been sampled from a wider range of geographic localities. Other important conclusions that can be drawn from the phylogenetic analysis of DENV genetic diversity are that (i) subtypes frequently co-circulate within the same locality, (ii) South east Asia harbours the greatest degree of genetic diversity, suggesting that it acts as a global 'source' population, generating strains that then ignite epidemics elsewhere, (iii) there is a distinction between 'endemic' subtypes that have circulated within particular localities for extended time periods and 'epidemic' subtypes that seem to spread rapidly through multiple populations, and (iv) there is a relatively high rate of clade (including subtype) replacement, generating periodic fluctuations in genetic diversity which can have a major impact on phylogenetic structure (Klungthong *et al.*, 2004; Sittisombut *et al.*, 1997; Wittke *et al.*, 2002). The occurrence of clade replacement is one the most intriguing, and potentially important aspect of DENV molecular epidemiology, particularly as its mechanistic basis is unclear. For example, in Thailand a turnover of DENV-2 lineages was observed between 1980 and 1987 (Sittisombut *et al.*, 1997) and of DENV-3 lineages in the 1990s (Wittke *et al.*, 2002).

Similarly, in Vietnam, a lineage shift in DENV-2 occurred in 2005/6, with the prevalent Asian/American genotype being entirely replaced by an Asian 1 virus (Simmons *et al.*, unpublished observations). Interestingly, this clade replacement occurred at a time when a serotype shift was occurring in the viral population, from DENV-2 to DENV-1. Such major, and abrupt, changes in genetic diversity through time could be explained by either (a) random population bottlenecks, for example caused by large-scale declines in mosquito numbers during the annual dry season or (b) because DENV clades differ in fitness so that one is able outcompete another – a process that is perhaps in part mediated by the extent of cross-protective immune responses. Indeed, it is likely that clade replacement, even involving entire genotypes, has been a regular occurrence in DENV evolution. Choosing between these two models is a key question in DENV evolutionary genetics; for example, frequent genetic drift will mean that the fate of a strain in a population will not always reflect its fitness. Given its major role in shaping the structure of genetic diversity within DENV populations, determining the biological processes that underlie clade replacement is a research area requiring more attention.

Epidemiologically, there have been specific geographic examples of the appearance of new strains correlating with DHF/DSS epidemics. The introduction/emergence of a South-East Asian strain of DENV-2 in the Americas in 1981 (Rico-Hesse, 1990; Rico-Hesse *et al.*, 1997) and of a Group B subtype III DENV-3 strain in Sri Lanka in 1989 (Messer *et al.*, 2003; Messer *et al.*, 2002) are two relevant examples of clade replacement correlating with an increase in DHF/DSS. Determining whether viral clades (including whole genotypes) differ in virulence, as well as the genetic basis of such differences, is therefore of fundamental importance. Two non-mutually exclusive theories have been proposed to explain the association between clade replacement and increased viral pathogenicity (Rico-Hesse, 2007). The first hypothesis suggests that highly pathogenic strains are inherently able to produce more viral output per infected human cell, ultimately resulting in higher viraemia than low pathogenicity strains (Cologna and Rico-Hesse, 2003; Leitmeyer *et al.*, 1999) The second proposes that the difference in virulence between strains associated with DF and those associated with severe DHF/DSS outbreaks may result from the improved ability of the latter strains to avoid

neutralization by serotype cross-reactive antibodies present in the semi-immune host (Kochel *et al.*, 2005). The introduction and ultimate replacement of the less pathogenic strains by the highly pathogenic strains implies an increase in viral fitness associated with this reduced neutralizability (Cologna *et al.*, 2005).

Despite the growing database of DENV sequences (at the time of writing, >5000 DENV sequences are available on GenBank), there are critical and fundamental gaps in our understanding of the epidemiological and evolutionary dynamics of DENV. In particular, there is a marked absence of data that links the epidemiological and genomic scales. At present, the best-characterized dataset comes from the Queen Sirikit National Hospital for Children (QSNICH) in Bangkok, Thailand. This dataset is remarkable in that sample collecting started in 1973 and has continues to this day, providing a unique glimpse into the changing frequency of the four DENV serotypes through time, as well as the incidence of DF and DHF/DSS (Nisalak *et al.*, 2003).

To date, some 250 Env genes and 40 complete genomes have been sampled across the 30-year study period which, in turn, have led to fundamental insights into the molecular evolution of DENV, most notably a high frequency of clade extinction and replacement and how this relates to patterns of changing serotype prevalence (Adams *et al.*, 2006; Zhang *et al.*, 2005). However, in reality, such a data set is still both small and potentially biased in structure; the total of 250 Env genes equates to approximately two gene sequences per serotype per year, with very little information on how the rest of the viral genome changes during this period or over the course of a single DENV season. To obtain a more accurate picture of the evolutionary and epidemiological dynamics of dengue, it is essential to undertake a far larger and intensive sampling scheme, considering a single population restricted in both time and space, to sequence complete viral genomes, many components of which will undoubtedly contribute to viral fitness.

Vector and virus biology and evolution

DENV evolution may be influenced just as much by selective pressure in the mosquito as in the human host. DENV infects the mosquito by entering and multiplying within cells of the midgut epithelium before disseminating out of the midgut for a second round of amplification in the salivary gland cells and surrounding tissues (Salazar *et al.*, 2007). The virus is transmitted to the vertebrate host when the mosquito salivates during blood feeding. Selective pressure on the viral population could be exerted at any of these stages. For example, viral variants that are better able to infect and disseminate in the mosquito host, or which reach higher titres in the salivary glands, or which confer a survival advantage to the mosquito host, could be selected for over other variants (see Chapter 8).

Earlier studies have demonstrated inherent differences in vector infectivity among distinct DENV serotypes and strains within serotypes (Gubler *et al.*, 1979; Rosen *et al.*, 1985). For example, DENV-2 and DENV-3 viruses were found to be more infectious for mosquitoes than DENV-1 and DENV-4 viruses at comparable viraemias. Efforts to better understand the virus–vector relationship have in part be driven by an improved understanding of DENV evolution and the knowledge that some virus lineages have measurable and distinctive biological characteristics (Rico-Hesse, 2003). For example, the sylvatic genotype of DENV-2 virus comprises strains from West Africa and South-East Asia that are phylogenetically distinct from the DENV-2 viruses involved in human outbreaks (Rico-Hesse, 1990; Wang *et al.*, 2000). These viruses are transmitted in a cycle involving non-human primates and canopy-dwelling mosquitoes but still retain the ability to infect the primary vectors of human transmission, the peridomestic mosquitoes *Ae. aegypti* and *Ae. albopictus* (Diallo *et al.*, 2003), as well as human hosts (Zeller *et al.*, 1992). Sylvatic strains are, however, less infectious for *Ae. aegypti* and *Ae albopictus* mosquitoes than endemic/epidemic strains of the DENV-2 virus (Moncayo *et al.*, 2004). Potentially, the selection of sylvatic DENV lineages that had improved fitness in *Ae. aegypti* or *Ae albopictus* mosquitoes represented an important step in the emergence of DENV from the sylvatic cycle to a peridomestic/urban cycle involving human hosts instead of primates. However, recent studies contradict these previous observations as no difference in the infectivity of sylvatic and endemic isolates for *Aedes aegypti* or

albopictus was noted (Hanley *et al.*, manuscript in preparation)

The spread of the South-East Asian genotype of DENV-2 in the Americas and displacement of 'indigenous' lineages represents an example of a virus having an apparent fitness advantage in both human and mosquito hosts (Cologna *et al.*, 2005). Viruses belonging to this lineage replicate to higher titres in human dendritic cells than American genotype viruses. Furthermore, South-East Asian DENV-2 genotype were found to infect and disseminate to the head tissue in a greater proportion of *Ae. aegypti* mosquitoes, and do so more rapidly, than American genotype viruses, and this pattern of differential infectivity was maintained in mosquito populations sampled from Perú, Mexico, and Texas (Anderson and Rico-Hesse, 2006; Armstrong and Rico-Hesse, 2001; Armstrong and Rico-Hesse, 2003; Cologna *et al.*, 2005). This is an important consideration because *Ae. aegypti* are short-lived organisms under natural conditions and few survive the incubation period required to become capable of transmitting the virus (Clements and Patterson, 1981; Sheppard, 1969). In another comparison, invasive and native DENV-3 strains from Sri Lanka were shown to infect a similar proportion of mosquitoes, but the invasive strains replicated to higher levels in mosquitoes and disseminated to the head tissue more readily than the displaced, native strains (Hanley *et al.*, 2008). Overall, these observations are consistent with the notion that viruses that have a selective advantage in either the mosquito or human host are more likely to become the dominant viral lineage in a given setting.

Considerable progress has been made in describing how DENV evolution is occurring on a macro scale. Thus it is clear that the relative abundance of different serotypes and viral lineages is continually changing, possibly in the face of changing thresholds of host immunity, and result in the local extinction and emergence of viral clades, as well as due to the introduction of new viruses from different geographic regions. However, our understanding of the functional determinants that result in selection of certain viral serotypes or lineages remains poor. Thus further prospective studies are needed to understand changing population immunity and in parallel, functional studies of viruses and their biological characteristics in humans and mosquitoes. Advances in these areas will assist in understanding how the introduction of live DENV vaccines might impact the viral evolutionary process, and whether wild-type viruses might evolve to 'escape' vaccine elicited immune responses.

Perspectives

Dengue affects millions of people in predominantly developing countries and causes thousands of deaths each year. On a global perspective, vector control as a mechanism for reducing transmission and the disease burden has clearly failed. The enormous disease burden of dengue is justifiably the catalyst for an invigorated agenda of basic research in pathogenesis and viral determinants of disease severity. The hope is that a better understanding of host factors could help guide rational disease interventions to modulate immunopathogenesis and reduce the severity of disease. Similarly, a better understanding of the sequence or functional characteristics of viruses associated with symptomatic or even severe disease could inform rational drug discovery efforts and vaccine development, leading to improved disease control. However, large gaps exist in the current body of knowledge, and these need to be addressed. Important but poorly understood areas include:

- At a mechanistic level, what drives endothelial cell dysfunction and allows capillary leakage to occur?
- To what extent and by what mechanism does host genotype influence susceptibility to severe dengue disease?
- What are the immune correlates of host immunity and disease susceptibility?
- How do antibodies and T-cells contribute to either immunity or pathogenesis?
- What are the mechanistic factors driving viral evolution in endemic regions and could these influences the long-term efficacy of live dengue vaccines?
- How does vector biology and host immunity influence viral evolution?
- How does virus serotype and strain influence the infection outcome?
- How do co-morbidities impact in the severity of dengue infection?

- How does vector–host interaction influence in infection outcome?

The answers to these questions will only be obtained by careful prospective studies in dengue endemic regions and through collaborations between clinicians and basic scientists. Such studies are urgently needed if we are going to make the discoveries that lead to the next generation of clinical interventions, vaccines or vector control systems.

Acknowledgements

We thank Eva Harris, Eddie Holmes and Phil Armstrong for their comments and contributions to this chapter.

References

Adams, B., E. C., Holmes, C., Zhang, M.P., Mammen, J., Nimmannitya, S., Kalayanarooj, S., and Boots, M. (2006). Cross-protective immunity can account for the alternating epidemic pattern of dengue virus serotypes circulating in Bangkok. Proc. Natl. Acad. Sci. U.S.A. *103*, 14234–14239.

Agramonte, A. (1906). Notas clínicas sobre una epidemia reciente de dengue. Revista de Medicina y Cirugía de la Habana, Cuba Enero(222–226).

Alvarez, M., Rodriguez-Roche, R., Bernardo, L., Vázquez, S., Morier, L., Gonzalez, D., Castro, O., Kouri, G., Halstead, S.B., and Guzman, M.G. (2006). Dengue hemorrhagic Fever caused by sequential dengue 1–3 virus infections over a long time interval: Havana epidemic, 2001–2002. Am. J. Trop. Med. Hyg. *75*, 1113–1117.

Anderson, R., Wang, S., Osiowy, C., and Issekutz, A.C. (1997). Activation of endothelial cells via antibody-enhanced dengue virus infection of peripheral blood monocytes. J. Virol. *71*, 4226–4232.

Avirutnan, P., Malasit, P., Seliger, B., Bhakdi, S., and Husmann, M. (1998). Dengue virus infection of human endothelial cells leads to chemokine production, complement activation, and apoptosis. J. Immunol. *161*, 6338–6346.

Azeredo, E.L., Zagne, S.M., Santiago, M.A., Gouvea, A.S., Santana, A.A., Neves-Souza, P.C., Nogueira, R.M., Miagostovich, M.P., and Kubelka, C.F. (2001). Characterisation of lymphocyte response and cytokine patterns in patients with dengue fever. Immunobiology *204*, 494–507.

Bashyam, H.S., Green, S., and Rothman, A.L. (2006). Dengue virus-reactive CD8+ T-cells display quantitative and qualitative differences in their response to variant epitopes of heterologous viral serotypes. J. Immunol. *176*, 2817–2824.

Belkaid, Y., and Rouse, B.T. (2005). Natural regulatory T-cells in infectious disease. Nat. Immunol. *6*, 353–360.

Beynon, H.L., Haskard, D.O., Haroutunian, R., Walport, M.J., and Davies, K.A. (1993). Combinations of low concentrations of cytokines and acute agonists synergize in increasing the permeability of endothelial monolayers. Clin. Exp. Immunol. *91*, 314–319.

Bosch, I., Xhaja, K., Estevez, L., Raines, G., Melichar, H., Warke, R.V., Fournier, M.V., Ennis, F.A., and Rothman, A.L. (2002). Increased production of interleukin-8 in primary human monocytes and in human epithelial and endothelial cell lines after dengue virus challenge. J. Virol. *76*, 5588–97.

Burke-Gaffney, A., and Keenan, A.K. (1993). Modulation of human endothelial cell permeability by combinations of the cytokines interleukin-1, tumor necrosis factor-alpha and interferon-gamma. Immunopharmacology *25*, 1–9.

Burke, D.S., and Kliks, S. (2006). Antibody-dependent enhancement in dengue virus infections. J. Infect. Dis. *193*, 601–603; author reply 603–604.

Burke, D.S., Nisalak, A., Johnson, D.E., and Scott, R.M. (1988). A prospective study of dengue infections in Bangkok. Am. J. Trop. Med. Hyg. *38*, 172–180.

Bussolati, B., Rollino, C., Mariano, F., Quarello, F., and Camussi, G. (2000). IL-10 stimulates production of platelet-activating factor by monocytes of patients with active systemic lupus erythematosus (SLE). Clin. Exp. Immunol. Cell Biol. *122*, 471–476.

Cantelar de Francisco, N., Fernandez, A., Albert Molina, L., and Perez Balbis, E. (1981). [Survey of dengue in Cuba. 1978–1979]. Rev. Cubana Med. Trop. *33*, 72–78.

Cardier, J.E., Marino, E., Romano, E., Taylor, P., Liprandi, F., Bosch, N., and Rothman, A.L. (2005). Proinflammatory factors present in sera from patients with acute dengue infection induce activation and apoptosis of human microvascular endothelial cells: possible role of TNF-alpha in endothelial cell damage in dengue. Cytokine *30*, 359–365.

Carr, J.M., Hocking, H., Bunting, K., Wright, P.J., Davidson, A., Gamble, J., Burrell, C.J., and Li, P. (2003). Supernatants from dengue virus type-2 infected macrophages induce permeability changes in endothelial cell monolayers. J Med Virol *69*, 521–528.

Casanova, J.L., and Abel, L. (2007). Human genetics of infectious diseases: a unified theory. EMBO J. *26*, 915–922.

Centro, d.E.D., Cubanos. (1976). La Población de Cuba. Editorial Ciencias Sociales La Habana.

Chakravarti, A., and Kumaria, R. (2006). Circulating levels of tumour necrosis factor-alpha and interferon-gamma; in patients with dengue and dengue haemorrhagic fever during an outbreak. Indian J. Med. Res. *123*, 25–30.

Chambers, T.J., Hahn, C.S., Galler, R., and Rice, C.M. (1990). Flavivirus genome organization, expression, and replication. Annu. Rev. Microbiol. *44*, 649–688.

Chareonsirisuthigul, T., Kalayanarooj, S., and Ubol, S. (2007). Dengue virus (DENV) antibody-dependent enhancement of infection up-regulates the production of anti-inflammatory cytokines, but suppresses anti-DENV free radical and pro-inflammatory cytokine production, in THP-1 cells. J. Gen. Virol. *88*, 365–375.

Chaturvedi, U.C., Agarwal, R., Elbishbishi, E.A., and Mustafa, A.S. (2000). Cytokine cascade in dengue hemorrhagic fever: implications for pathogenesis. FEMS Immunol Med Microbiol *28*, 183–188.

Chau, T.N., N.T., Q., Thuy, T.T., Tuan, N.M., Hoang, D.M., Dung, N.T., L.B., L., Quy, N.T., Hieu, N.T., Hieu, L.T., Hien, T.T., Hung, N.T., Farrar, J., and Simmons, C.P. (2008). Dengue in Vietnamese infants – results of infection-enhancement assays correlate with age-related disease epidemiology, and cellular immune responses correlate with disease severity. J. Infect. Dis. *198*, 516–524.

Chiewsilp, P., Scott, R.M., and Bhamarapravati, N. (1981). Histocompatibility antigens and dengue hemorrhagic fever. Am. J. Trop. Med. Hyg. *30*, 1100–5.

Cho Min, N. (2000). Assessment of dengue hemorrhagic fever in Myanmar. South-East Asian J. Trop. Med. Public Health *31*, 636–461.

Chungue, E., Deubel, V., Cassar, O., Laille, M., and Martin, P.M. (1993). Molecular epidemiology of dengue 3 viruses and genetic relatedness among dengue 3 strains isolated from patients with mild or severe form of dengue fever in French Polynesia. J. Gen. Virol. *74*, 2765–2770.

Chungue, E., Poli, L., Roche, C., Gestas, P., Glaziou, P., and Markoff, L.J. (1994). Correlation between detection of plasminogen cross-reactive antibodies and hemorrhage in dengue virus infection. J. Infect. Dis. *170*, 1304–1307.

Clark, D.V., Mammen, M.P., Jr., Nisalak, A., Puthimethee, V., and Endy, T.P. (2005). Economic impact of dengue fever/dengue hemorrhagic fever in Thailand at the family and population levels. Am. J. Trop. Med. Hyg. *72*, 786–791.

Clements, A.N., and Patterson, G.D. (1981). The analysis of mortality and survival rates in wild populations of mosquitoes. J. Appl. Ecol. *18*, 373–399.

Clyde, K., and Harris, E. (2006). RNA secondary structure in the coding region of dengue virus type 2 directs translation start codon selection and is required for viral replication. J. Virol. *80*, 2170–2182.

Coelho, G.E., Burattini, M.N., Teixeira, M.G., Coutinho, F.A., and Massad, E. (2008). Dynamics of the 2006/2007 dengue outbreak in Brazil. Mem. Inst. Oswaldo Cruz *103*, 535–9.

Cologna, R., Armstrong, P.M., and Rico-Hesse, R. (2005). Selection for virulent dengue viruses occurs in humans and mosquitoes. J. Virol. *79*, 853–859.

Cologna, R., and Rico-Hesse, R. (2003). American genotype structures decrease dengue virus output from human monocytes and dendritic cells. J. Virol. *77*, 3929–3938.

Cooke, G.S., and Hill, A.V. (2001). Genetics of susceptibility to human infectious disease. Nat. Rev. Genet. *2*, 967–977.

Cummings, D.A., Irizarry, R.A., Huang, N.E., Endy, T.P., Nisalak, A., Ungchusak, K., and Burke, D.S. (2004). Travelling waves in the occurrence of dengue haemorrhagic fever in Thailand. Nature *427*, 344–347.

Despres, P., Sakuntabhai, A., and Julier, C. (2005). [A variant in the CD209 promoter is associated with severity of dengue disease]. Med. Sci. (Paris) 21, 905–6.

Diallo, M., Ba, Y., Sall, A.A., Diop, O.M., Ndione, J.A., Mondo, M., Girault, L., and Mathiot, C. (2003). Amplification of the sylvatic cycle of dengue virus type 2, Senegal, 1999–2000: entomologic findings and epidemiologic considerations. Emerg. Infect. Dis. *9*, 362–367.

Egger, J.R., and Coleman, P.G. (2007). Age and clinical dengue illness. Emerg. Infect. Dis. *13*, 924–925.

Endy, T.P., Nisalak, A., Chunsuttitwat, S., Vaughn, D.W., Green, S., Ennis, F.A., Rothman, A.L., and Libraty, D.H. (2004). Relationship of preexisting Dengue virus (DV) neutralizing antibody levels to viremia and severity of disease in a prospective cohort study of DV infection in Thailand. J. Infect. Dis. *189*, 990–1000.

Endy, T.P., Nisalak, A., Chunsuttitwat, S., Libraty, D.H., Green, S., Rothman, A.L., Vaughn, D.W., and Ennis, F.A. (2002). Spatial and temporal circulation of dengue virus serotypes: a prospective study of primary school children in Kamphaeng Phet, Thailand. Am. J. Epidemiol. *156*, 52–95.

Falconar, A.K. (1997). The dengue virus nonstructural-1 protein (NS1) generates antibodies to common epitopes on human blood clotting, integrin/adhesin proteins and binds to human endothelial cells: potential implications in haemorrhagic fever pathogenesis. Arch Virol *142*, 897–916.

Fernandez-Mestre, M.T., Gendzekhadze, K., Rivas-Vetencourt, P., and Layrisse, Z. (2004). TNF-alpha-308A allele, a possible severity risk factor of hemorrhagic manifestation in dengue fever patients. Tissue Antigens *64*, 469–472.

Gamble, J., Bethell, D., Day, N.P., Loc, P.P., Phu, N.H., Gartside, I.B., Farrar, J.F., and White, N.J. (2000). Age-related changes in microvascular permeability: a significant factor in the susceptibility of children to shock? Clin. Sci. (Lond.) *98*, 211–216.

Gibbons, R.V., Kalanarooj, S., Jarman, R.G., Nisalak, A., Vaughn, D.W., Endy, T.P., Mammen, M.P., and Srikiatkhachorn, A. (2007). Analysis of repeat hospital admissions for dengue to estimate the frequency of third or fourth dengue infections resulting in admissions and dengue hemorrhagic fever, and serotype sequences. Am. J. Trop. Med. Hyg. *77*, 910–913.

Goncalvez, A.P., Engle, R.E., St Claire, M., Purcell, R.H., and Lai, C.J. (2007). Monoclonal antibody-mediated enhancement of dengue virus infection *in vitro* and *in vivo* and strategies for prevention. Proc. Natl. Acad. Sci. U.S.A. *104*, 9422–9427.

Goncalvez, A.P., Escalante, A.A., Pujol, F.H., Ludert, J.E., Tovar, D., Salas, R.A., and Liprandi, F. (2002). Diversity and evolution of the envelope gene of dengue virus type 1. Virology *303*, 110–119.

Gonzalez, D., Limonta, D., Bandera, J.R., Pérez, J., Kouri, G., and Guzmán, M.G. (2008). Dual infection with dengue virus 3 and human immunodeficiency virus 1 in Havana, Cuba. J. Infect. Developing Countries *3*, 318–320.

Gonzalez, D., Martinez, R., Castro, O., Serrano, T., Portela, D., Vazquez, S., Perez, J., and Guzman, M.G. (2005). Evaluation of some clinical, humoral and imagenological parameters in patients of dengue haemorrhagic fever six months after acute illness. Dengue Bull. *29*, 79–84.

Green, S., and Rothman, A. (2006). Immunopathological mechanisms in dengue and dengue hemorrhagic fever. Curr. Opin. Infect. Dis. *19*, 429–436.

Green, S., Vaughn, D.W., Kalayanarooj, S., Nimmannitya, S., Suntayakorn, S., Nisalak, A., Lew, R., Innis, B.L., Kurane, I., Rothman, A.L., and Ennis, F.A. (1999). Early immune activation in acute dengue illness is related to development of plasma leakage and disease severity. J. Infect. Dis. *179*, 755–762.

Guanche, J. (1983). Procesos Etno-Culturales de Cuba. Editorial de Letras Cubanas, La Habana.

Gubler, D.J. (2002). Epidemic dengue/dengue hemorrhagic fever as a public health, social and economic problem in the 21st century. Trends Microbiol. *10*, 100–103.

Gubler, D.J., Nalim, S., Tan, R., Saipan, H., and Sulianti Saroso, J. (1979). Variation in susceptibility to oral infection with dengue viruses among geographic strains of *Aedes aegypti*. Am. J. Trop. Med. Hyg. *28*, 1045–1052.

Gubler, D.J., Sather, G.E., Kuno, G., and Cabral, J.R. (1986). Dengue 3 virus transmission in Africa. Am. J. Trop. Med. Hyg. *35*, 1280–1284.

Guerra, R. (1964). Manual de Historia de Cuba desde su Descubrimiento. Colección Histórica. Editorial Nacional de Cuba, C. Habana.

Guzik, T.J., Korbut, R., and Adamek-Guzik, T. (2003). Nitric oxide and superoxide in inflammation and immune regulation. Journal of Physiology and Pharmacology *54*, 469–487.

Guzman, M.G., Alvarez, M., Rodriguez-Roche, R., Bernardo, L., Montes, T., Vazquez, S., Morier, L., Alvarez, A., Gould, E.A., Kouri, G., and Halstead, S.B. (2007). Neutralizing antibodies after infection with dengue 1 virus. Emerg. Infect. Dis. *13*, 282–286.

Guzman, M.G., Alvarez, M., Rodriguez, R., Rosario, D., Vazquez, S., Valdes, L., Cabrera, M.V., and Kouri, G. (1999). Fatal dengue hemorrhagic fever in Cuba, 1997. Int. J. Infect. Dis. *3*, 130–135.

Guzman, M.G., Garcia, G., and Kouri, G. (2006). [Dengue and dengue hemorrhagic fever: research priorities]. Rev. Panam. Salud Publica *19*, 204–215.

Guzman, M.G., and Kouri, G. (2002). Dengue: an update. Lancet Infect. Dis. *2*, 33–42.

Guzman, M.G., and Kouri, G. (2008). Dengue haemorrhagic fever integral hypothesis: confirming observations, 1987–2007. Trans. R. Soc. Trop. Med. Hyg. *102*, 522–523.

Guzman, M.G., Kouri, G., Bravo, J., Valdes, L., Vazquez, S., and Halstead, S.B. (2002a). Effect of age on outcome of secondary dengue 2 infections. Int. J. Infect. Dis. *6*, 118–124.

Guzman, M.G., Kouri, G., Martinez, E., Bravo, J., Riveron, R., Soler, M., Vazquez, S., and Morier, L. (1987). Clinical and serologic study of Cuban children with dengue hemorrhagic fever/dengue shock syndrome (DHF/DSS). Bull. Pan Am. Health Org. *21*, 270–279.

Guzman, M.G., Kouri, G., Valdes, L., Bravo, J., Alvarez, M., Vazques, S., Delgado, I., and Halstead, S.B. (2000). Epidemiologic studies on Dengue in Santiago de Cuba, 1997. Am. J. Epidemiol. *152*, 793–799.

Guzman, M.G., Kouri, G., Valdes, L., Bravo, J., Vazquez, S., and Halstead, S.B. (2002b). Enhanced severity of secondary dengue-2 infections: death rates in 1981 and 1997 Cuban outbreaks. Rev Panam Salud Publica *11*, 223–227.

Guzman, M.G., Kouri, G.P., Bravo, J., Calunga, M., Soler, M., Vazquez, S., and Venereo, C. (1984). Dengue haemorrhagic fever in Cuba. I. Serological confirmation of clinical diagnosis. Trans. R. Soc. Trop. Med. Hyg. *78*, 235–238.

Guzman, M.G., Kouri, G.P., Bravo, J., Soler, M., Vazquez, S., and Morier, L. (1990). Dengue hemorrhagic fever in Cuba, 1981: a retrospective seroepidemiologic study. Am. J. Trop. Med. Hyg. *42*, 179–184.

Guzmán, M.G., Peláez, O., Kourí, G., Quintana, I., Vázquez, S., Pentón, M., and Avila, L.C. (2006). Grupo Multidisciplinario para el Control de la Epidemia de Dengue 2001–2002.[Final characterization of and lessons learned from the dengue 3 epidemic in Cuba, 2001–2002]. Rev. Panam Salud Publica. *19*, 282–289.

Halstead, S. (1993). Pathophysiology and pathogenesis of dengue hemorrhagic fever. In: W.H. Organization (Ed), Thongcharoen P (Ed),, pp. 80–103. SEARO No. 22, Regional Publication.

Halstead, S.B. (1980). Dengue haemorrhagic fever – a public health problem and a field for research. Bull. World Health Org. *58*, 1–21.

Halstead, S.B. (1989). Antibody, macrophages, dengue virus infection, shock, and hemorrhage: a pathogenetic cascade. Rev. Infect. Dis. 11 Suppl. 4, S830–839.

Halstead, S.B. (2002). Dengue. Curr. Opin. Infect. Dis. *15*, 471–476.

Halstead, S.B. (2006). Measuring dengue enhancing antibodies: caveats. J. Infect. Dis. *193*, 601.

Halstead, S.B. (2007). Dengue. Lancet *370*, 1644–1652.

Halstead, S.B., Rojanasuphot, S., and Sangkawibha, N. (1983). Original antigenic sin in dengue. Am. J. Trop. Med. Hyg. *32*, 154–156.

Halstead, S.B., Streit, T.G., Lafontant, J.G., Putvatana, R., Russell, K., Sun, W., Kanesa Thasan, N., Hayes, C.G., and Watts, D.M. (2001). Haiti: absence of dengue hemorrhagic fever despite hyperendemic dengue virus transmission. Am. J. Trop. Med. Hyg. *65*, 180–183.

Halstead, S.B., Heinz, F.X., Barrett, A.D., and Roehrig, J.T. (2005). Dengue virus: molecular basis of cell entry and pathogenesis, 25–27 June 2003, Vienna, Austria. Vaccine 23, 849–856.

Halstead, S.B., Suaya, J.A., and Shepard, D.S. (2007). The burden of dengue infection. Lancet. *369*, 1410–1411.

Halstead, S.B., and Yamarat, C. (1965). Recent Epidemics of Hemorrhagic Fever in Thailand. Observations Related to Pathogenesis of a 'New' Dengue Disease. Am. J. Public Health Nations Health *55*, 1386–9135.

Hanley, K.A., Nelson, J.T., Schirtzinger, E.E., Whitehead, S.S., and Hanson, C.T. (2008). Superior infectivity for mosquito vectors contributes to competitive displacement among strains of dengue virus. BMC Ecol. 8, 1.

Huang, K.J., Yang, Y.C., Lin, Y.S., Huang, J.H., Liu, H.S., Yeh, T.M., Chen, S.H., Liu, C.C., and Lei, H.Y. (2006). The dual-specific binding of dengue virus and target cells for the antibody-dependent enhancement of dengue virus infection. J. Immunol. *176*, 2825–2832.

Hung, N.T., Lei, H.Y., Lan, N.T., Lin, Y.S., Huang, K.J., Lien le, B., Lin, C.F., Yeh, T.M., Ha do, Q., Huong, V.T., Chen, L.C., Huang, J.H., My, L.T., Liu, C.C., and

Halstead, S.B. (2004). Dengue hemorrhagic Fever in infants: a study of clinical and cytokine profiles. J. Infect. Dis. *189*, 221–232.

Keynan, Y., Card, C.M., McLaren, P.J., Dawood, M.R., Kasper, K., and Fowke, K.R. (2008). The role of regulatory T-cells in chronic and acute viral infections. Clin. Infect. Dis. *46*, 1046–1052.

Khare, M., and Chaturvedi, U.C. (1997). Role of nitric oxide in transmission of dengue virus specific suppressor signal. Indian J Exp Biol 35, 855–60.

Kliks, S.C., Nisalak, A., Brandt, W.E., Wahl, L., and Burke, D.S. (1989). Antibody-dependent enhancement of dengue virus growth in human monocytes as a risk factor for dengue hemorrhagic fever. Am. J. Trop. Med. Hyg. *40*, 444–451.

Klungthong, C., Zhang, C., Mammen, M.P., Jr., Ubol, S., and Holmes, E.C. (2004). The molecular epidemiology of dengue virus serotype 4 in Bangkok, Thailand. Virology 329, 168–179.

Kochel, T.J., Watts, D.M., Gozalo, A.S., Ewing, D.F., Porter, K.R., and Russell, K.L. (2005). Cross-Serotype Neutralization of Dengue Virus in Aotus nancymae Monkeys. J. Infect. Dis. *191*, 1000–1004.

Kouri, G., Guzman, M.G., Valdes, L., Carbonel, I., del Rosario, D., Vazquez, S., Laferte, J., Delgado, J., and Cabrera, M.V. (1998). Reemergence of dengue in Cuba: a 1997 epidemic in Santiago de Cuba. Emerg. Infect. Dis. *4*, 89–92.

Kouri, G.P., Guzman, M.G., and Bravo, J.R. (1987). Why dengue haemorrhagic fever in Cuba? 2. An integral analysis. Trans. R. Soc. Trop. Med. Hyg. *81*, 821–823.

Kouri, G.P., Guzman, M.G., Bravo, J.R., and Triana, C. (1989). Dengue haemorrhagic fever/dengue shock syndrome: lessons from the Cuban epidemic, 1981. Bull. World Health Org. *67*, 375–80.

Kroeger, A., and Nathan, M.B. (2006). Dengue: setting the global research agenda. Lancet *368*, 2193–2195.

Kukreti, H., Chaudhary, A., Rautela, R.S., Anand, R., Mittal, V., Chhabra, M., Bhattacharya, D., Lal, S., and Rai, A. (2008). Emergence of an independent lineage of dengue virus type 1 (DENV-1) and its co-circulation with predominant DENV-3 during the 2006 dengue fever outbreak in Delhi. Int. J. Infect. Dis. *12*, 542–549.

Kurane, I. (2007). Dengue hemorrhagic fever with special emphasis on immunopathogenesis. Comp Immunol Microbiol Infect Dis 30(5–6), 329–40.

Kurane, I., and Ennis, F.A. (1997). Immunopathogenesis of dengue virus infection. In: D. Gubler and G. Kuno (Eds), Dengue and dengue hemorrhagic fever, pp. 273–290. CAB International, New York.

Kurane, I., Innis, B.L., Nimmannitya, S., Nisalak, A., Meager, A., Janus, J., and Ennis, F.A. (1991a). Activation of T lymphocytes in dengue virus infections. High levels of soluble interleukin 2 receptor, soluble CD4, soluble CD8, interleukin 2, and interferon-gamma in sera of children with dengue. J. Clin. Invest. 88, 1473–80.

Kurane, I., Mady, B.J., and Ennis, F.A. (1991b). Antibody-dependent enhancement of dengue virus infection. Rev. Med. Virol. *1*, 211–221.

Kurane, I., Rothman, A.L., Livingston, P.G., Green, S., Gagnon, S.J., Janus, J., Innis, B.L., Nimmannitya,

S., Nisalak, A., and Ennis, F.A. (1994). Immunopathologic mechanisms of dengue hemorrhagic fever and dengue shock syndrome. Arch. Virol. Suppl. *9*, 59–64.

Kyle, J.L., and Harris, E. (2008). Global Spread and Persistence of Dengue. Annu. Rev. Microbiol. Apr 22.

LaFleur, C., Granados, J., Vargas-Alarcon, G., Ruiz-Morales, J., Villarreal-Garza, C., Higuera, L., Hernandez-Pacheco, G., Cutino-Moguel, T., Rangel, H., Figueroa, R., Acosta, M., Lazcano, E., and Ramos, C. (2002). HLA-DR antigen frequencies in Mexican patients with dengue virus infection: HLA-DR4 as a possible genetic resistance factor for dengue hemorrhagic fever. Hum. Immunol. *63*, 1039–1044.

Lanciotti, R.S., Gubler, D.J., and Trent, D.W. (1997). Molecular evolution and phylogeny of dengue-4 viruses. J. Gen. Virol. *78*, 2279–2284.

Lanciotti, R.S., Lewis, J.G., Gubler, D.J., and Trent, D.W. (1994). Molecular evolution and epidemiology of dengue-3 viruses. J. Gen. Virol. *75*, 65–75.

Laoprasopwattana, K., Libraty, D.H., Endy, T.P., Nisalak, A., Chunsuttiwat, S., Vaughn, D.W., Reed, G., Ennis, F.A., Rothman, A.L., and Green, S. (2005). Dengue virus (DV) enhancing antibody activity in preillness plasma does not predict subsequent disease severity or viremia in secondary DV infection. J. Infect. Dis. *192*, 510–519.

Lee, E., Nestorowicz, A., Marshall, I.D., Weir, R.C., and Dalgarno, L. (1992). Direct sequence analysis of amplified dengue virus genomic RNA from cultured cells, mosquitoes and mouse brain. J. Virol. Methods *37*, 275–288.

Lei, H.Y., Yeh, T.M., Liu, H.S., Lin, Y.S., Chen, S.H., and Liu, C.C. (2001). Immunopathogenesis of dengue virus infection. J. Biomed. Sci. *8*, 377–388.

Leitmeyer, K.C., Vaughn, D.W., Watts, D.M., Salas, R., Villalobos, I., de, C., Ramos, C., and Rico Hesse, R. (1999). Dengue virus structural differences that correlate with pathogenesis. J. Virol. *73*, 4738–4747.

Lewis, J.A., Chang, G.J., Lanciotti, R.S., Kinney, R.M., Mayer, L.W., and Trent, D.W. (1993). Phylogenetic relationships of dengue-2 viruses. Virology *197*, 216–224.

Libraty, D.H., Endy, T.P., Houng, H.S., Green, S., Kalayanarooj, S., Suntayakorn, S., Chansiriwongs, W., Vaughn, D.W., Nisalak, A., Ennis, F.A., and Rothman, A.L. (2002a). Differing influences of virus burden and immune activation on disease severity in secondary dengue-3 virus infections. J. Infect. Dis. *185*, 1213–1221.

Libraty, D.H., Young, P.R., Pickering, D., Endy, T.P., Kalayanarooj, S., Green, S., Vaughn, D.W., Nisalak, A., Ennis, F.A., and Rothman, A.L. (2002b). High circulating levels of the dengue virus nonstructural protein NS1 early in dengue illness correlate with the development of dengue hemorrhagic fever. J. Infect. Dis. *186*, 1165–1168.

Lin, C.F., Lei, H.Y., Shiau, A.L., Liu, C.C., Liu, H.S., Yeh, T.M., Chen, S.H., and Lin, Y.S. (2003). Antibodies from dengue patient sera cross-react with endothelial cells and induce damage. J. Med. Virol. *69*, 82–90.

Lin, C.F., Lei, H.Y., Shiau, A.L., Liu, H.S., Yeh, T.M., Chen, S.H., Liu, C.C., Chiu, S.C., and Lin, Y.S. (2002). Endothelial cell apoptosis induced by antibodies against dengue virus nonstructural protein 1 via production of nitric oxide. J. Immunol. *169*, 657–664.

Lin, C.F., Wan, S.W., Cheng, H.J., Lei, H.Y., and Lin, Y.S. (2006). Autoimmune pathogenesis in dengue virus infection. Viral Immunol. *19*, 127–132.

Lin, Y.S., Lin, C.F., Lei, H.Y., Liu, H.S., Yeh, T.M., Chen, S.H., and Liu, C.C. (2004). Antibody-mediated endothelial cell damage via nitric oxide. Curr. Pharm. Des. *10*, 213–221.

Littaua, R., Kurane, I., and Ennis, F.A. (1990). Human IgG Fc receptor II mediates antibody-dependent enhancement of dengue virus infection. J. Immunol. *144*, 3183–3186.

Loke, H., Bethell, D., Phuong, C.X., Day, N., White, N., Farrar, J., and Hill, A. (2002). Susceptibility to dengue hemorrhagic fever in vietnam: evidence of an association with variation in the vitamin d receptor and Fc gamma receptor IIa genes. Am. J. Trop. Med. Hyg. *67*, 102–106.

Loke, H., Bethell, D.B., Phuong, C.X., Dung, M., Schneider, J., White, N.J., Day, N.P., Farrar, J., and Hill, A.V. (2001). Strong HLA class I – restricted T-cell responses in dengue hemorrhagic fever: a double-edged sword? J. Infect. Dis. *184*, 1369–1373.

Luhn, K., Simmons, C.P., Moran, E., Dung, N.T., Chau, T.N., Quyen, N.T., Thao le, T.T., Van Ngoc, T., Dung, N.M., Wills, B., Farrar, J., McMichael, A.J., Dong, T., and Rowland-Jones, S. (2007). Increased frequencies of CD4+ CD25(high) regulatory T-cells in acute dengue infection. J. Exp. Med. *204*, 979–985.

Luz, P.M., Mendes, B.V., Codeço, C.T., Struchiner, C.J., and Galvani, A.P. (2008). Time Series Analysis of Dengue Incidence in Rio de Janeiro, Brazil. Am. J. Trop. Med. Hyg. *79*, 933–939.

Maruo, N., Morita, I., Shirao, M., and Murota, S. (1992). IL-6 increases endothelial permeability *in vitro*. Endocrinology *131*, 710–714.

Mathers, C.D., Ezzati, M., and Lopez, A.D. (2007). Measuring the burden of neglected tropical diseases: the global burden of disease framework. PLoS. Negl. Trop. Dis. *1*, 114.

Mathew, A., and Rothman, A.L. (2008). Understanding the contribution of cellular immunity to dengue disease pathogenesis. Immunol. Rev. *225*, 300–313.

McLinden, J.H., Stapleton, J.T., Chang, Q., and Xiang, J. (2008). Expression of the Dengue virus type 2 NS5 protein in a CD4 T-cell line inhibits HIV replication. J. Infect. Dis. *198*, 860–863.

Meltzer, M.I., Rigau-Perez, J.G., Clark, G.G., Reiter, P., and Gubler, D.J. (1998). Using disability-adjusted life years to assess the economic impact of dengue in Puerto Rico: 1984–1994. Am. J. Trop. Med. Hyg. *59*, 265–271.

Messer, W.B., Gubler, D.J., Harris, E., Sivananthan, K., and de Silva, A.M. (2003). Emergence and global spread of a dengue serotype 3, subtype III virus. Emerg. Infect. Dis. *9*, 800–809.

Messer, W.B., Vitarana, U.T., Sivananthan, K., Elvtigala, J., Preethimala, L.D., Ramesh, R., Withana, N., Gubler, D.J., and De Silva, A.M. (2002). Epidemiology of dengue in Sri Lanka before and after the emergence of epidemic dengue hemorrhagic fever. Am. J. Trop. Med. Hyg. *66*, 765–773.

Mills, K.H. (2004). Regulatory T-cells: friend or foe in immunity to infection? Nat. Rev. Immunol. *4*, 841–855.

Monath, T.P., Wands, J.R., Hill, L.J., Brown, N.V., Marciniak, R.A., Wong, M.A., Gentry, M.K., Burke, D.S., Grant, J.A., and Trent, D.W. (1986). Geographic classification of dengue-2 virus strains by antigen signature analysis. Virology *154*, 313–324.

Moncayo, A.C., Fernandez, Z., Ortiz, D., Diallo, M., Sall, A., Hartman, S., Davis, C.T., Coffey, L., Mathiot, C.C., Tesh, R.B., and Weaver, S.C. (2004). Dengue emergence and adaptation to peridomestic mosquitoes. Emerg. Infect. Dis. *10*, 1790–1796.

Mongkolsapaya, J., Dejnirattisai, W., Xu, X.N., Vasanawathana, S., Tangthawornchaikul, N., Chairunsri, A., Sawasdivorn, S., Duangchinda, T., Dong, T., Rowland-Jones, S., Yenchitsomanus, P.T., McMichael, A., Malasit, P., and Screaton, G. (2003). Original antigenic sin and apoptosis in the pathogenesis of dengue hemorrhagic fever. Nat. Med. *9*, 921–927.

Mustafa, A.S., Elbishbishi, E.A., Agarwal, R., and Chaturvedi, U.C. (2001). Elevated levels of interleukin-13 and IL-18 in patients with dengue hemorrhagic fever. FEMS Immunol Med. Microbiol. *30*, 229–233.

Nisalak, A., Endy, T.P., Nimmannitya, S., Kalayanarooj, S., Thisayakorn, U., Scott, R.M., Burke, D.S., Hoke, C.H., Innis, B.L., and Vaughn, D.W. (2003). Serotype-specific dengue virus circulation and dengue disease in Bangkok, Thailand from 1973 to 1999. Am. J. Trop. Med. Hyg. *68*, 191–202.

PAHO. (2002). Framework: New Generation of Dengue Prevention and Control Programs in the Americas. Pan American Health Organization Washington D.C.

Palmer, D.R., Sun, P., Celluzzi, C., Bisbing, J., Pang, S., Sun, W., Marovich, M.A., and Burgess, T. (2005). Differential effects of dengue virus on infected and bystander dendritic cells. J. Virol. *79*, 2432–2439.

Paradoa Perez, M.L., Trujillo, Y., and Basanta, P. (1987). Association of dengue hemorrhagic fever with the HLA system. Haematologia (Budap) 20, 83–87.

Pelaez, O., Guzman, M.G., Kouri, G., Perez, R., San Martin, J.L., Vazquez, S., Rosario, D., Mora, R., Quintana, I., Bisset, J., Cancio, R., Masa, A.M., Castro, O., Gonzalez, D., Avila, L.C., Rodriguez, R., Alvarez, M., Pelegrino, J.L., Bernardo, L., and Prado, I. (2004). Dengue 3 epidemic, Havana, 2001. Emerg. Infect. Dis. *10*, 719–722.

Perez, A.B., Garcia, G., Sierra, B., Alvarez, M., Vazquez, S., Cabrera, M.V., Rodriguez, R., Rosario, D., Martinez, E., Denny, T., and Guzman, M.G. (2004). IL-10 levels in Dengue patients: some findings from the exceptional epidemiological conditions in Cuba. J. Med. Virol. *73*, 230–234.

Pinto, L.M., Oliveira, S.A., Braga, E.L., Nogueira, R.M., and Kubelka, C.F. (1999). Increased pro-inflammatory cytokines (TNF-alpha and IL-6) and anti-inflammatory compounds (sTNFRp55 and sTNFRp75) in Brazilian patients during exanthematic dengue fever. Mem Inst Oswaldo Cruz *94*, 387–394.

Pryor, M.J., Carr, J.M., Hocking, H., Davidson, A.D., Li, P., and Wright, P.J. (2001). Replication of dengue virus type 2 in human monocyte-derived macrophages: comparisons of isolates and recombinant viruses with substitutions at amino acid 390 in the envelope glycoprotein. Am. J. Trop. Med. Hyg. *65*, 427–434.

Rico-Hesse, R. (1990). Molecular evolution and distribution of dengue viruses type 1 and 2 in nature. Virology *174*, 479–493.

Rico-Hesse, R. (2003). Microevolution and virulence of dengue viruses. Adv. Virus Res. *59*, 315–341.

Rico-Hesse, R. (2007). Dengue virus evolution and virulence models. Emerg. Infect. *44*, 1462–1466.

Rico-Hesse, R., Harrison, L.M., Nisalak, A., Vaughn, D.W., Kalayanarooj, S., Green, S., Rothman, A.L., and Ennis, F.A. (1998). Molecular evolution of dengue type 2 virus in Thailand. Am. J. Trop. Med. Hyg. *58*, 96–101.

Rico-Hesse, R., Harrison, L.M., Salas, R.A., Tovar, D., Nisalak, A., Ramos, C., Boshell, J., de Mesa, M.T., Nogueira, R.M., and da Rosa, A.T. (1997). Origins of dengue type 2 viruses associated with increased pathogenicity in the Americas. Virology *230*, 244–51.

Rigau-Perez, J.G. (2006). Severe dengue: the need for new case definitions. Lancet Infect. Dis. *6*, 297–302.

Rigau-Perez, J.G., Vorndam, A.V., and Clark, G.G. (2001). The dengue and dengue hemorrhagic fever epidemic in Puerto Rico, 1994–1995. Am. J. Trop. Med. Hyg. *64*, 67–74.

Rivero de la Calle, M. (1984). Antropología de la Población Adulta Cubana.. Editorial Científico Técnica, Ciudad de La Habana 44.

Roche, J.C., Cordellier, E., Hervy, J.P., Digoutte, J.P., and Monteny, N. (1983). Isolement de 96 souchess de virus dengue a partir de moustiques captures en Cote D'Ivoire et Haute-Volta. Ann. Virol. (Institute Pasteur) *134*, 233–244.

Rosen, L., Roseboom, L.E., Gubler, D.J., Lien, J.C., and Chaniotis, B.N. (1985). Comparative susceptibility of mosquito species and strains to oral and parenteral infection with dengue and Japanese encephalitis viruses. Am. J. Trop. Med. Hyg. *34*, 603–615.

Rothman, A.L. (2003). Immunology and immunopathogenesis of dengue disease. Adv Virus Res. *60*, 397–419.

Sabin, A.B. (1952). Research on dengue during World War I.I. Am. J. Trop. Med. Hyg. *1*, 30–50.

Sakuntabhai, A., Turbpaiboon, C., Casademont, I., Chuansumrit, A., Lowhnoo, T., Kajaste-Rudnitski, A., Kalayanarooj, S.M., Tangnararatchakit, K., Tangthawornchaikul, N., Vasanawathana, S., Chaiyaratana, W., Yenchitsomanus, P.T., Suriyaphol, P., Avirutnan, P., Chokephaibulkit, K., Matsuda, F., Yoksan, S., Jacob, Y., Lathrop, G.M., Malasit, P., Despres, P., and Julier, C. (2005). A variant in the CD209 promoter is associated with severity of dengue disease. Nat. Genet. *37*, 507–513.

Salazar, M.I., Richardson, J.H., Sanchez-Vargas, I., Olson, K.E., and Beaty, B.J. (2007). Dengue virus type 2: replication and tropisms in orally infected *Aedes aegypti* mosquitoes. BMC Microbiol 7, 9.

Saluzzo, J.F., Cornet, M., Castagnet, P., Rey, C., and Digoutte, J.P. (1986a). Isolation of dengue 2 and dengue 4 viruses from patients in Senegal. Trans. R. Soc. Trop. Med. Hyg. *80*, 5.

Saluzzo, J.F., Cornet, M., Castaneg, P., Rey, C., and Digoutte, J.P. (1986b). Isolation of dengue 2 and dengue 4 viruses from patients in Senegal. Trans. R. Soc .Trop. Med. Hyg. 80, 5.

Sangkawibha, N., Rojanasuphot, S., Ahandrik, S., Viriyapongse, S., Jatanasen, S., Salitul, V., Phanthumachinda, B., and Halstead, S.B. (1984). Risk factors in dengue shock syndrome: a prospective epidemiologic study in Rayong, Thailand. I. The 1980 outbreak. Am. J. Epidemiol. 120, 653–669.

Seet, R.C.S., Quekb, A.M.L., and Lima, E.C.H. (2007). Post-infectious fatigue syndrome in dengue infection. J. Clin. Virol. 38, 1–6.

Seynhaeve, A.L., Vermeulen, C.E., Eggermont, A.M., and T.L., H. (2006). Cytokines and vascular permeability: an *in vitro* study on human endothelial cells in relation to tumor necrosis factor-alpha primed peripheral blood mononuclear cells. Cell Biochem. Biophys. 44, 157–169.

Sharp, T.W., Wallace, M.R., Hayes, C.G., Sanchez, J.L., DeFraites, R.F., Arthur, R.R., Thornton, S.A., Batchelor, R.A., Rozmajzl, P.J., Hanson, R.K., *et al.* (1995). Dengue fever in U.S. troops during Operation Restore Hope, Somalia, 1992–1993. Am. J. Trop. Med. Hyg. 53, 89–94.

Shepard, D.S., Suaya, J.A., Halstead, S.B., Nathan, M.B., Gubler, D.J., Mahoney, R.T., Wang, D.N., and Meltzer, M.I. (2004). Cost-effectiveness of a pediatric dengue vaccine. Vaccine 22, 1275–1280.

Sheppard, P.M., W. W. MacDonald, R. J. Tonn, and B. Grab. (1969). The dynamics of an adult population of *Aedes aegypti* in relation to dengue haemorrhagic fever in Bangkok. J Anim Ecol. 38, 661–702.

Sierra, B., Garcia, G., Perez, A.B., Morier, L., Rodriguez, R., Alvarez, M., and Guzman, M.G. (2002). Long-term memory cellular immune response to dengue virus after a natural primary infection. Int. J. Infect. Dis. 6, 125–128.

Simmons, C.P., Chau, T.N., Thuy, T.T., Tuan, N.M., Hoang, D.M., Thien, N.T., Lien le, B., Quy, N.T., Hieu, N.T., Hien, T.T., McElnea, C., Young, P., Whitehead, S., Hung, N.T., and Farrar, J. (2007a). Maternal antibody and viral factors in the pathogenesis of dengue virus in infants. J. Infect. Dis. 196, 416–24.

Simmons, C.P., Dong, T., Chau, N.V., Dung, N.T., Chau, T.N., Thao le, T.T., Hien, T.T., Rowland-Jones, S., and Farrar, J. (2005). Early T-cell responses to dengue virus epitopes in Vietnamese adults with secondary dengue virus infections. J. Virol. 79, 5665–5675.

Simmons, C.P., Popper, S., Dolocek, C., Chau, T.N., Griffiths, M., Dung, N.T., Long, T.H., Hoang, D.M., Chau, N.V., Thao le, T.T., Hien, T.T., Relman, D.A., and Farrar, J. (2007b). Patterns of host genome-wide gene transcript abundance in the peripheral blood of patients with acute dengue hemorrhagic fever. J. Infect. Dis. 195, 1097–1107.

Sittisombut, N., Sistayanarain, A., Cardosa, M.J., Salminen, M., Damrongdachakul, S., Kalayanarooj, S., Rojanasuphot, S., Supawadee, J., and Maneekarn, N. (1997). Possible occurrence of a genetic bottleneck in dengue serotype 2 viruses between the 1980 and 1987 epidemic seasons in Bangkok, Thailand. Am. J. Trop. Med. Hyg. 57, 100–108.

Soundravally, R., and Hoti, S.L. (2007). Immunopathogenesis of dengue hemorrhagic fever and shock syndrome: role of TAP and HPA gene polymorphism. Hum. Immunol. 68, 973–979.

Stephens, H.A., Klaythong, R., Sirikong, M., Vaughn, D.W., Green, S., Kalayanarooj, S., Endy, T.P., Libraty, D.H., Nisalak, A., Innis, B.L., Rothman, A.L., Ennis, F.A., and Chandanayingyong, D. (2002). HLA-A and -B allele associations with secondary dengue virus infections correlate with disease severity and the infecting viral serotype in ethnic Thais. Tissue Antigens 60, 309–318.

Stephenson, J.R. (2005). The problem with dengue. Trans. R. Soc. Trop. Med. Hyg. 99, 643–646.

Suaya, J.A., Shepard, D.S., Chang, M.S., Caram, M., Hoyer, S., Socheat, D., Chantha, N., and M.B., N. (2007). Cost-effectiveness of annual targeted larviciding campaigns in Cambodia against the dengue vector *Aedes aegypti*. Trop. Med. Int. Health 12, 1026–1036.

Talavera, D., Castillo, A.M., Dominguez, M.C., Gutierrez, A.E., and Meza, I. (2004). IL8 release, tight junction and cytoskeleton dynamic reorganization conducive to permeability increase are induced by dengue virus infection of microvascular endothelial monolayers. J. Gen. Virol. 85, 1801–1813.

Thein, S. (1975). Haemorrhagic manifestations of influenza A infection in children. Am. J. Trop. Med. Hyg. 78, 78–80.

Torres, J.R., and Castro, J. (2007). The health and economic impact of dengue in Latin America. Cad Saude Publica 23(Suppl. 1), S23–31.

Trent, D.W., Grant, J.A., Monath, T.P., Manske, C.L., Corina, M., and Fox, G.E. (1989). Genetic variation and microevolution of dengue 2 virus in South-East Asia. Virology 172, 523–535.

Twiddy, S.S., Farrar, J.J., Vinh Chau, N., Wills, B., Gould, E.A., Gritsun, T., Lloyd, G., and Holmes, E.C. (2002). Phylogenetic relationships and differential selection pressures among genotypes of dengue-2 virus. Virology 298, 63–72.

Ulloa, J. (2002). Archaeology and Rescue of the Aboriginal Presence in Cuba and the Caribbean. KACIKE: The Journal of Caribbean Amerindian History and Anthropology [On-line Journal], Special Issue (Lynne Guitar, Ed), Available at: http://www.kacike.org/UlloaEnglish.html.

UNICEF/UNDP/World Bank/WHO Special Programme for Research and Training in Tropical Diseases (TDR), G., WHO Headquarters (29 September- 1 October 2008.). Expert meeting on dengue classification and case management.

Valdes, L., Guzman, M.G., Kouri, G., Delgado, J., Carbonell, I., Cabrera, M.V., Rosario, D., and Vazquez, S. (1999). [Epidemiology of dengue and hemorrhagic dengue in Santiago, Cuba 1997]. Rev. Panam Salud Publica 6, 16–25.

Vaughn, D.W., Green, S., Kalayanarooj, S., Innis, B.L., Nimmannitya, S., Suntayakorn, S., Endy, T.P., Raengsakulrach, B., Rothman, A.L., Ennis, F.A., and Nisalak, A. (2000). Dengue viremia titer, antibody response pattern, and virus serotype correlate with disease severity. J. Infect. Dis. 181, 2–9.

Wang, E., Ni, H., Xu, R., Barrett, A.D., Watowich, S.J., Gubler, D.J., and Weaver, S.C. (2000). Evolutionary relationships of endemic/epidemic and sylvatic dengue viruses. J. Virol. *74*, 3227–34.

Wang, W.K., Chao, D.Y., Kao, C.L., Wu, H.C., Liu, Y.C., Li, C.M., Lin, S.C., Ho, S.T., Huang, J.H., and King, C.C. (2003). High levels of plasma dengue viral load during defervescence in patients with dengue hemorrhagic fever: implications for pathogenesis. Virology *305*, 330–8.

Weissmann, G. (2008). Dengue fever in Rio: Macumba versus Voltaire. FASEB J. *22*, 2109–2112.

Wittke, V., Robb, T., Thu, H., Nisalak, A., Nimmannitya, S., Kalayanrooj, S., Vaughn, D., Endy, T., Holmes, E., and Aaskov, J. (2002). Extinction and Rapid Emergence of Strains of Dengue 3 Virus during an interepidemic period. Virology *301*, 148.

Yang, K.D., Wang, C.L., and Shaio, M.F. (1995). Production of cytokines and platelet activating factor in secondary dengue virus infections. J. Infect. Dis. *172*, 604–605.

Zeller, H.G., Traore-Lamizana, M., Monlun, E., Hervy, J.P., Mondo, M., and Digoutte, J.P. (1992). Dengue-2 virus isolation from humans during an epizootic in southeastern Senegal in November, 1990. Res. Virol. *143*, 101–102.

Zhang, C., Mammen, M.P., Jr., Chinnawirotpisan, P., Klungthong, C., Rodpradit, P., Monkongdee, P., Nimmannitya, S., Kalayanarooj, S., and Holmes, E.C. (2005). Clade replacements in dengue virus serotypes 1 and 3 are associated with changing serotype prevalence. J. Virol. *79*, 15123–15130.

Zhou, Y., M. P. J. Mammen, P. Chinnawirotpisan, C. Klungthong, D. W. Vaughn, S. Nimmannitya, S. Kalayanarooj, E. C. Holmes, and C. Zhang. (2006). Comparative analysis reveals no consistent association between the secondary structure of the 3'-UTR of dengue viruses and disease syndrome. J. Gen.Virol. *87*, 2595–2603.

Animal Models of Dengue Virus Infection and Disease: Applications, Insights and Frontiers

6

Scott J. Balsitis and Eva Harris

Abstract

Many important questions in dengue pathogenesis are difficult to address without appropriate animal models of infection and disease. However, animal models of dengue virus (DENV) infection have been hampered by the fact that no non-human species naturally exhibits disease similar to human dengue fever or dengue shock syndrome following natural or experimental infection. Recently, though, many aspects of human DENV infection and disease have been reproduced in mice by manipulating the host and/or the virus to facilitate susceptibility to infection, and judicious use of these mouse models in combination with non-human primate models of DENV infection has resulted in substantial progress. This chapter briefly reviews the animal models for DENV infection currently available, and surveys the contributions each has made to four key areas in dengue research: pathogenesis, immunity and immunopathogenesis, therapeutic drug development, and vaccine development.

Introduction

Animal models of dengue virus (DENV) infection and disease are crucial for many areas of dengue research, including pathogenesis, immunity, drug development, and vaccine design and testing. However, early attempts to induce dengue-like disease in a wide variety of animals failed to identify any non-human species that naturally develops disease following virus inoculation into peripheral sites (reviewed in Bente and Rico-Hesse, 2006). Consequently, animal models of DENV infection were initially limited to non-human primates, which develop a low level of viraemia but no severe disease, and intracranially inoculated mouse models, which succumb to infection but with a neurovirulent phenotype quite unlike human dengue haemorrhagic fever (DHF) or dengue shock syndrome (DSS).

Fortunately, subsequent efforts have produced a stream of mouse models for DENV infection, each with distinct advantages and disadvantages, and with increasing relevance to human infection. Viral replication has been established in transplanted tumour cell lines, xenografted human haematopoietic cells, and relevant endogenous mouse cells. These advances provide useful tools for analysing viral infection *in vivo* and developing antiviral countermeasures, and further progress has been made recently by several groups in modelling important disease features of dengue. Recent mouse models have succeeded in reproducing key features of human dengue disease in mice, including viraemia, serum nonstructural protein 1 (NS1) circulation, fever, rash, thrombocytopenia, haemorrhage, and death associated with increased vascular permeability and elevated haematocrit. In addition, primate models continue to play indispensable roles in dengue animal modelling due to their natural susceptibility to infection and close relatedness to humans.

Applications of these models have begun to generate useful insights into pathogenesis and immunity and to provide more relevant models for drug and vaccine development. The history of dengue animal models has been reviewed recently in some detail (Bente and Rico-Hesse, 2006; Yauch and Shresta, 2008). In this chapter, we introduce the various models briefly and then focus on the insights and advances gained through dengue animal models, as well as the challenges remaining for the future of dengue animal research.

Animal models of dengue virus infection and disease

Dengue animal models can be divided into five groups: non-human primate (NHP) models, immunocompetent mouse models, severe combined immunodeficient (SCID)-tumour transplant mouse models, humanized mouse models, and interferon-deficient mouse models (Table 6.1).

Non-human primates

DENV infections in rhesus monkeys and humans are acute, self-limited and of similar duration, although no NHP species that has been experimentally infected with DENV consistently develops DF or DHF/DSS-like disease. However, a low level of viraemia and a neutralizing antibody response occur regularly in most NHP species inoculated with wild-type DENV (reviewed in Bente and Rico-Hesse, 2006). These features, combined with the close relatedness of NHPs to humans, have made NHPs the most widely used model in vaccine testing. Additionally, in spite of the absence of severe disease in NHPs, several key insights into pathogenesis and immunity have come from NHP studies. Furthermore, NHPs are also useful as a basis of comparison to validate results in dengue mouse models, which are more tractable but potentially less relevant to human infection.

Immunocompetent mouse models

Immunocompetent mice are inexpensive and readily available, and the unimpaired nature of their immune system makes them valuable for assessing vaccine immunogenicity. However, they are poorly susceptible to DENV infection, although neurologic disease can be induced by intracranial and/or extremely high-dose infection. Thus, immunocompetent mouse models of DENV infection have been a mainstay for drug and vaccine development studies, although only limited insights into pathogenesis or immunity have been drawn from these models. Importantly, though, one group has recently reproduced a

DHF-like disease in BALB/c mice using a novel mouse-adapted DENV strain (Atrasheuskaya et al., 2003), demonstrating that better immunocompetent mouse models remain a possibility.

Immunodeficient mouse models

Severe combined immunodeficient (SCID) mice lack an adaptive immune system, and so can be transplanted with cultured, DENV-susceptible tumour cells such as Huh7 or HepG2 (hepatocarcinoma) or K562 (erythroleukaemia) (An et al., 1999; Blaney et al., 2002; Lin et al., 1998). The transplanted tumour mass provides a replication site for injected DENV viruses, resulting in abundant viral replication, viraemia, and some signs of illness. SCID mice have also been transplanted with primary human peripheral blood leukocytes, but only a minority of mice transplanted in this manner exhibited detectable DENV infection, and this model has not been pursued further (Wu et al., 1995). SCID models have been successful in establishing DENV infection within the transplanted cells and even reproducing some dengue disease features, but since viral replication occurs primarily in the transplanted cells, it is unclear how any insights into pathogenesis gained from SCID-tumour transplant mice might apply to human disease pathogenesis. Nevertheless, some antiviral drug testing has been performed in SCID-K562 mice (Lin et al., 1998), and vaccine testing has been conducted in SCID-Huh7 mice (Blaney et al., 2002; Blaney et al., 2007).

Researchers have improved upon the SCID-tumour models by creating 'humanized' SCID mice (Bente et al., 2005; Kuruvilla et al., 2007). In these models, SCID mice are irradiated to destroy the haematopoietic progenitors in the bone marrow and then transplanted with human CD34+ haematopoietic stem cells. Over a period of several weeks, the transplanted cells engraft into the mouse bone marrow and reconstitute a variety of haematopoietic cell types, resulting in an adaptive immune system consisting exclusively of human cells, with certain parts of the innate immune system being humanized as well. Haematopoietic-lineage cells such as monocytes are key targets of DENV infection, and consequently these mice can be infected with clinical DENV isolates without adaptation to mice, and can display some signs of human disease. These

Table 6.1 Animal models of DENV infection

Model	Serotypes	Endpoints	Reference
Newborn mouse	1–4	death (neurovirulence)	Cole and Wisseman, (1969), Schlesinger and Frankel (1952)
Immunocompetent models			
A/J	2	Paralysis, thrombocytopenia, increased haematocrit	Huang *et al.* (2000), Shresta *et al.* (2004a)
C57/BL6	2	Haemorrhage	Chen *et al.* (2007a)
Balb/c	2, mouse-neuroadapted	Viraemia, thrombocytopenia, haemorrhage, paralysis	Atrasheuskaya *et al.* (2003)
SCID-tumour models			
SCID-K562	1–4	Paralysis	Lin *et al.* (1998)
SCID-HepG2	1,2	Paralysis, thrombocytopenia, increased haematocrit	An *et al.* (1999)
SCID-Huh7	4	Viraemia	Blaney *et al.* (2002)
Immunodeficient-humanized models			
Humanized NOD/SCID	2	Rash, thrombocytopenia, viraemia	Bente *et al.* (2005)
Rag-Hu	2	Viraemia, fever	Kuruvilla *et al.* (2007)
Interferon-deficient models			
AG129	2, mouse – neuroadapted	Paralysis	Johnson and Roehrig (1999), Lee *et al.* (2006b)
AG129	1, 2, 4	Tissue viral burden	Kyle *et al.* (2008), Kyle *et al.*, (2007)
AG129	2	viraemia	Schul *et al.* (2007)
AG129	2, mouse – peripherally passaged	Viraemia, vascular leak, death	Prestwood *et al.* (2008), Shresta *et al.* (2006)
STAT1$^{-/-}$129	1, 2	paralysis	Shresta *et al.* (2005)
STAT1$^{-/-}$ B6	2, mouse – neuroadapted	haemorrhage, vascular leak, paralysis, death	Chen *et al.* (2008)
Non-human primates			
Multiple NHP species	1–4	Viraemia	Reviewed in Bente and Rico-Hesse (2006)

new models have potential to answer important basic questions surrounding infection of human cells *in vivo*, but difficulty in generating sufficient numbers of humanized mice, genetic variation in stem cell donors, and variability in the degree of human cell engraftment cause humanized mouse models to be slow and labour-intensive, and thus unsuitable for most drug and vaccine develop-

ment purposes. Efforts to improve humanized mouse models of DENV infection continue, and may be able to address some of these limitations.

Lastly, because the interferon (IFN) system is a potent suppressor of DENV replication (Diamond and Harris, 2001; Diamond *et al.*, 2000), mice deficient in the IFN pathway are much more susceptible to DENV infection than wild-type mice (Johnson and Roehrig, 1999; Shresta *et al.*, 2004b). Two types of IFN deficiency have been used in dengue mouse models: IFN-receptor deficiency, and STAT1 deficiency. Mice of the 129 strain deficient in both the IFN-α/β and -γ receptors (AG129 mice) were initially used to provide a model in which adult mice could be both vaccinated and lethally challenged without intracranial inoculation (Johnson and Roehrig, 1999). Since then, AG129 mice have proven to be one of the most versatile dengue mouse models, as they exhibit viral replication in relevant peripheral cell types and can be infected with a variety of DENV isolates. Depending on the viral isolate and route of inoculation, they can develop either neurologic disease or a DSS-like vascular permeability syndrome (Huang *et al.*, 2003; Johnson and Roehrig, 1999; Kyle *et al.*, 2007; Lee *et al.*, 2006b; Schul *et al.*, 2007; Shresta *et al.*, 2006; Stein *et al.*, 2008). AG129 mice have been used for numerous studies including tropism, pathogenesis, immune protection and enhancement, and antiviral drug and vaccine testing (Balsitis *et al.*, submitted; Johnson and Roehrig, 1999; Kyle *et al.*, 2008; Kyle *et al.*, 2007; Schul *et al.*, 2007; Shresta *et al.*, 2004b, 2006; Prestwood *et al.*, 2008). However, AG129 mice are difficult to use for genetic studies, since many genetically deficient strains are in the C57BL/6 background, and breeding a third knock-out gene into the AG129 strain would be time-consuming. STAT1$^{-/-}$ mice in the C57BL/6 background may provide a useful alternative because they are susceptible to DENV infection (Shresta *et al.*, 2005), and a combination of vascular leakage and paralysis can be induced if mice are simultaneously inoculated both intraperitoneally (i.p.) and intracranially (i.c.) (Chen *et al.*, 2008).

Each animal model of DENV infection carries with it clear advantages and disadvantages (Table 6.1), and consequently the various models have made differing contributions to our understanding of DENV pathogenesis and immunity or have aided in the testing and development of antiviral drugs and vaccines.

Pathogenesis: insights from dengue animal models

Animal models of DENV infection have progressed beyond the point of simply allowing virus replication *in vivo* to examining and illuminating the pathogenic mechanisms responsible for disease features of DF and DHF/DSS.

Tropism

Understanding viral tropism is key to viral pathogenesis; accordingly, one of the most important features of any animal model is the degree to which viral tropism in the animal reflects that in humans. In rhesus macaques, subcutaneous (s.c.) inoculation of DENV was performed, followed by attempted virus isolation from serum and a variety of tissues at necropsy on days 1–12 postinfection (Marchette *et al.*, 1973). Virus spread from the inoculation site first to regional lymph nodes, then to systemic lymphatic tissues and disseminated skin sites, while virus was rarely recovered from other organs. These data evidenced a preferential replication in leukocyte-rich tissues, with limited and inconsistent dissemination to solid organs. Notably, tissue infection peaked 3–4 days later than the peak of viraemia (Halstead, 1980; Marchette *et al.*, 1973). Since fatal dengue shock syndrome in humans also occurs after peak viraemia around the time of defervescence (Gubler, 1998; Vaughn *et al.*, 2000), this observation suggests that disease severity may correlate with high tissue viral burdens, even if viraemia is reduced. Consistent with the observed tropism of DENV in macaques, infected monkeys developed lymphadenomegaly, lymphocytosis, and leucopenia. However, no abnormalities were detected in haematocrit or prothrombin time, and only small to moderate decreases in platelet counts occurred in a minority of infected animals (Halstead *et al.*, 1973b; Halstead *et al.*, 1973c). In sum, DENV infection in macaques is similar to but more limited than that in humans, so while macaques remain an important infection model, thus far they have not been able to offer major insights into molecular mechanisms of pathogenesis.

In most mouse models of DENV infection, viral tropism is either clearly different from human infection, as is the case with mouse brain-adapted DENV strains and SCID-tumour transplant models, or is poorly characterized. In the humanized mouse models, human haematopoietic cells are the presumed targets of infection, and viral RNA is detectable by PCR in spleen, liver, and skin of humanized NOD-SCID mice (Bente *et al.*, 2005). All of these sites are known targets of DENV infection in humans as well (Balsitis *et al.*, 2008; Bhoopat *et al.*, 1996; Jessie *et al.*, 2004; Miagostovich *et al.*, 1997; Wu *et al.*, 2000), but the exact cell types in which virus replicates in humanized mice have not been identified. In contrast, tropism in the AG129 model is quite well characterized. Clinical isolates of all four DENV serotypes replicate in AG129 mice following subcutaneous (s.c.) inoculation, and virus is consistently detected in spleen, lymph nodes, bone marrow, serum, and to a lesser degree peripheral blood cells and liver, in some cases followed by spread into the central nervous system (Balsitis *et al.*, 2008; Balsitis *et al.*, submitted; Kyle *et al.*, 2007; Schul *et al.*, 2007; Shresta *et al.*, 2006, S. Balsitis and E. Harris, unpublished data). Cellular tropism has also been studied in some detail. Negative-strand viral RNA and structural and non-structural proteins of DENV-2 were detected in macrophages and dendritic cells in spleen and lymph node shortly after s.c. inoculation (Kyle *et al.*, 2007), and non-structural protein 3 (NS3)-specific immunohistochemistry also detected infected macrophages and dendritic cells in spleen and lymph node, plus hepatocytes in spleen and myeloid cells in bone marrow (Balsitis *et al.*, 2008). All of these cell types have also been implicated in human DENV pathogenesis (Balsitis *et al.*, 2008; Bhoopat *et al.*, 1996; Jessie *et al.*, 2004; Miagostovich *et al.*, 1997; Rothwell *et al.*, 1996), indicating that viral infection in AG129 mice at early time-points post-infection occurs primarily or exclusively in relevant cell types.

Viral virulence

Building upon these strengths of the AG129 model, two groups have observed that the affinity of DENV virions for glycosaminoglycans (GAG) is an important determinant of viral virulence. Lee *et al.* (2006b) compared the ability of clinical and mouse-adapted strains of DENV-2 to disseminate from peripheral tissues into the central nervous system (CNS), and discovered that high GAG affinity decreases the half-life of virions in the bloodstream, resulting in a lower viraemia and decreased CNS invasion.

The relevance of this observation was extended using a peripherally passaged, mouse-adapted strain of DENV-2. This strain, called D2S10, was adapted to replication in peripheral mouse tissues by passaging a DENV-2 clinical strain, PL046, in AG129 mice and then inoculating C6/36 mosquito cells with mouse serum and using the resulting virus to infect new mice; this cycle was repeated a total of ten times. The resulting D2S10 strain induces a fatal vascular leak syndrome bearing many similarities to human DSS (Shresta *et al.*, 2006). Two mutations in the envelope (E) protein, K124E and N128D, were found to be responsible for the vascular leak phenotype (Prestwood *et al.*, 2008), and both mutations occur in the GAG-binding region previously identified by Lee *et al.* (2006b). Furthermore, the mechanism of increased neurovirulence described by Lee *et al.* (2006b) is consistent with the increased virulence of D2S10. The D2S10 virus has lower GAG affinity than the parental PL046 strain and thus exhibits a longer serum-half life and consequently establishes higher viraemia and higher tissue viral loads in mice (Prestwood *et al.*, 2008).

Combined, these observations demonstrate that E protein mutations in the GAG binding region can modulate the level of viraemia attained by DENV, and thus act as virulence determinants. More work is needed to determine whether similar changes are responsible for virulence differences between naturally occurring DENV strains in humans, or whether GAG-binding region mutations can be deliberately introduced into other DENV strains or serotypes to improve viral replication in animal models.

Mechanisms of haemorrhage and vascular leak

In vitro and in human studies, a broad variety of cytokines, complement components, and other factors have been correlated with vascular leakage in severe dengue (Pang *et al.*, 2007; Rothman and Ennis, 1999), but it remains unclear which factors contribute significantly to disease severity and which are effects of increased disease severity. Fortunately, animal models of DENV have begun to shed light on this question.

Three studies using independent DENV mouse models have all shown that TNF-α is an important mediator of vascular leakage during DENV infection. In the AG129/D2S10 model, lethality due to vascular leak was completely blocked by treatment with anti-TNF-α antibody (Shresta *et al.*, 2006). Similarly, two groups have implicated TNF-α in immunocompetent models of haemorrhagic dengue. Chen *et al.* (2007a) inoculated C57BL/6 mice intradermally with 3×10^9 pfu of DENV-2 16681, and while no overt signs of illness occurred, focal haemorrhage occurred in subcutaneous tissue and intestines. Notably, platelet levels at the time of haemorrhage were only slightly reduced and were not different between mice with and without haemorrhage, implying another mechanism driving the haemorrhagic phenotype. Interestingly, sites of subcutaneous haemorrhage contained DENV antigen-positive apoptotic endothelial cells and macrophages, accompanied by macrophage infiltrates that stained positive for TNF-α. The role of TNF-α was confirmed by showing that its deficiency greatly reduced the susceptibility of mice to DENV-induced haemorrhage, while complement C5 deficiency had no effect. This emphasized an important role for TNF-α in haemorrhage, which in this model may be due to TNF-α-induced endothelial cell apoptosis. Atrasheuskata *et al.* (2003) created a mouse-adapted DENV-2 strain by a series of passages through suckling mouse brain followed by passages in mice of increasing age via i.p. injection. The resulting mouse-adapted strain produces severe disease in i.p.-infected mice consisting of viraemia, thrombocytopenia, haemorrhage, and paralysis. Blocking TNF-α function with anti-TNF-α antibodies reduced mortality in this model by 60% without reducing viraemia.

Another recent study has shed light on the molecular signalling pathways involved in DENV-induced TNF-α production. In a series of elegant *in vitro* experiments, DENV binding to the C-type lectin family member CLEC5A on the surface of macrophages was shown to be dispensable for infection of macrophages by DENV, but was important for inducing secretion of pro-inflammatory cytokines including TNF-α (Chen *et al.*, 2008). The researchers then induced haemorrhage, vascular leakage, and death accompanied by high levels of TNF-α in STAT1$^{-/-}$ mice via simultaneous i.p. and i.c. inoculations with mouse-brain adapted DENV-2 strain New Guinea C. Adding anti-CLEC5A antibodies

reduced DENV-induced TNF-α production and concomitantly reduced haemorrhage, vascular leak, and mortality.

Combined, these four distinct models offer strong evidence that TNF-α is a key pathologic mediator during severe dengue and that macrophages or other phagocytes are important sources of TNF-α. These results raise the possibility that anti-TNF-α therapies may alleviate the severity of human dengue disease.

Pathogenic role of anti-NS1 antibodies

Several *in vitro* studies have demonstrated that antibodies elicited by DENV NS1 proteins cross-react with endothelial cells, platelets, or blood clotting proteins (Falconar, 1997; Lin *et al.*, 2002, 2003, 2005). Thus, anti-NS1 antibody-mediated platelet destruction, endothelial cell damage, and coagulopathy may contribute to dengue disease. It has been unclear if any of these pathogenic mechanisms is active *in vivo*, but recently anti-NS1 antibodies have been shown to bind to endothelium in mouse liver (Lin *et al.*, 2008). This effect was associated with endothelial apoptosis, inflammatory infiltration into the liver, and evidence of liver damage indicated by increased serum levels of liver enzymes (Lin *et al.*, 2008). In another study, polyclonal anti-NS1 IgG was able to bind platelets *in vitro*, induce platelet depletion after intravenous administration into mice, and contribute to clotting system dysfunction in mice (Sun *et al.*, 2007). Notably, pre-absorption of the anti-NS1 IgG with platelets eliminated the latter effect, connecting anti-NS1 antibody binding of platelets to a disease phenotype *in vivo* for the first time. However, neither of these studies examined the effects of anti-NS1 antibodies in the context of DENV infection. Consequently, the effects of anti-NS1 antibodies on DENV infection require further examination, preferably in mouse models that develop thrombocytopenia and haemorrhage or vascular leak.

Pathogenesis: future directions

Recent years have witnessed the development of multiple animal models able to reproduce key features of dengue pathogenesis. It will be crucial to utilize these advances to understand molecular mechanisms of pathogenesis in sufficient detail to design and test new interventions to disrupt pathogenic processes.

The multiple mouse models of DENV-induced vascular leak and haemorrhage provide the necessary tools to continue to identify factors responsible for these crucial features of human disease. Many factors other than TNF-α have been proposed to affect the vascular system in human dengue. Thus, more work is required to determine which factors other than TNF-α may contribute to disease in mouse models of haemorrhage and vascular leak and to understand in greater detail the molecular pathways by which DENV triggers vascular dysfunction.

Thrombocytopenia is one of the most common characteristics of human dengue. Multiple competing hypotheses have been advanced to explain the loss of platelets that occurs during human DENV infection, including reduced platelet synthesis due to bone marrow suppression, platelet consumption at sites of haemorrhage, inappropriate platelet activation due to the release of an unidentified signalling factor, or autoimmune platelet destruction mediated by cross-reactive antibodies raised against dengue virions or NS1 protein (Falconar, 1997, 2007; Funahara et al., 1987; Lin et al., 2001; Nelson et al., 1964; Oishi et al., 2003; Saito et al., 2004). Five mouse models have now been shown to develop significant platelet depletion correlated with the magnitude of DENV infection, including SCID-HepG2 mice (An et al., 1999), AG129 mice (Balsitis et al.), A/J mice (Huang et al., 2000), BALB/c mice with mouse-adapted DENV-2 infection (Atrasheuskaya et al., 2003), and humanized NOD-SCID mice (Bente et al., 2005). These advances set the stage for mechanistic studies of the origins of thrombocytopenia to be performed, and for observations and interventions to be confirmed in multiple distinct models.

A wide variety of additional questions are awaiting further advances in dengue animal models. First, the only mouse model demonstrated to exhibit clotting system abnormalities is the SCID-HepG2 mouse, where irrelevant tropism hampers interpretation of the data. Similarly, extensive epidemiologic data indicates that viral strain variation is an important pathogenic determinant, but no animal model has yet reproduced strain variation in pathogenesis using clinical DENV isolates. Thirdly, the secreted form of the viral NS1 protein circulates at high levels during infection and is likely to play an important pathogenic role, but its in vivo function remains unclear. Secreted NS1 has been ascribed several possible functions, including complement consumption, complement inhibition, endothelial cell binding, and facilitation of viral infection (Alcon-LePoder et al., 2005; Avirutnan et al., 2006, 2007; Chung et al., 2006; Kurosu et al., 2007). Schul et al. (2007) and our group (S.B., and E.H., unpublished results) have found that infected AG129 mice develop NS1 antigenaemia with levels and kinetics reflective of human DENV infection, providing an in vivo infection model for NS1 function. Lastly, perhaps the most pressing questions in dengue research are focused on immune protection and immunopathogenesis of DENV infection.

Immunity and immunopathogenesis

In addition to their role in viral clearance and protection from disease, immunologic factors are major determinants of human dengue pathogenesis. Some studies have identified potential differences in innate immunity underlying differential susceptibility to disease (Simmons et al., 2007; Ubol et al., 2008), but adaptive immune factors appear to have the strongest effect on determining disease outcome. Infection with one DENV serotype is widely believed to provide lifelong protection against re-infection with the same serotype, but prior immunity to a heterologous serotype is the greatest risk factor for the development of severe disease (Halstead, 2003; Halstead, 2007; Rothman, 2004). Despite this latter observation, human challenge studies and epidemiological models also indicate that serotype cross-reactive immune protection is common (Adams et al., 2006; Sabin, 1952; Wearing and Rohani, 2006). This leads to the most urgent question in dengue research today: what factors distinguish protective from pathogenic immunity to DENV infections?

NHP models of immune protection and immunopathogenesis

Antibody responses to DENV infection in several different primate species are broadly similar to those observed in humans. Marchette *et al.* (Marchette *et al.*, 1973) showed that primary infection of rhesus macaques with any of the four DENV serotypes elicits an antibody response that is broadly flavivirus cross-reactive, but with highest titre against the infecting serotype. Similar results were observed following infection of owl monkeys (*Aotus nancymae*) or cynomologous macaques (Kochel *et al.*, 2005; Koraka *et al.*, 2007). In the owl monkey study, primary DENV-1 infection elicited antibodies that neutralized American lineage DENV-2 strains much more effectively than Asian lineage strains, a phenomenon also observed in humans that may contribute to the much greater incidence of severe disease observed during secondary infections with Asian lineage DENV-2 strains (Kochel *et al.*, 2005; Kochel *et al.*, 2002). In all three of these NHP models, secondary infection with a different DENV serotype elicited a broadly serotype cross-reactive antibody response, with the highest titres in many cases against the primary, rather than secondary, infecting serotypes (Halstead *et al.*, 1973a; Halstead *et al.*, 1973c; Kochel *et al.*, 2005; Koraka *et al.*, 2007).

The effects of antibodies on DENV infection in NHPs are variable and appear to be governed by factors that are poorly understood, as is the case in humans. In sequential infections of rhesus macaques, previous exposure to one DENV serotype reduced viraemia upon subsequent challenge when DENV-1 or DENV-4 were the secondary infection viruses, yet viraemia was enhanced if DENV-2 was used for the secondary infection (Halstead *et al.*, 1973c). Increased DENV-2 viraemia was accompanied by a greater incidence of mild thrombocytopenia and complement consumption, and one monkey developed pronounced thrombocytopenia, increased prothrombin time, and increased haematocrit. Thus, cross-reactive immunity can be either protective (Halstead and Palumbo, 1973; Halstead *et al.*, 1973c) or pathogenic (Halstead *et al.*, 1973c) in rhesus macaques. In contrast, sequential infections of owl monkey and cynomologous macaques have only generated serotype cross-protection during DENV secondary infection, although many fewer conditions have been tested in these species than in rhesus macaques (Kochel *et al.*, 2005; Koraka *et al.*, 2007). Passive antibody

transfer experiments in rhesus macaques have further demonstrated that this species of NHP is capable of modelling antibody-dependent enhancement of DENV infection *in vivo*. When two groups of rhesus monkeys were administered either dengue-naïve human serum or dengue-immune serum and infected with DENV-2, viraemia in dengue-immune serum recipients was enhanced 3–50 fold (Halstead, 1979). Notably, the dose of dengue-immune serum was chosen such that the neutralizing antibody titre after transfer into macaques was below the limit of detection. Nearly 30 years later, Goncalvez *et al.* (Goncalvez *et al.*, 2007) achieved similar results using passive transfer of a serotype cross-reactive monoclonal antibody followed by DENV-4 infection. Antibody doses of 0.22–6.0 mg/kg enhanced viraemia 22- to 165-fold, with 18 mg/kg of the same monoclonal antibody needed to prevent antibody-dependent enhancement (ADE). These results demonstrate that ADE is mediated by both polyclonal and monoclonal antibodies *in vivo*, and, at least with monoclonal antibody, occurs over a broad concentration range.

While much less work has been done to characterize T-cell responses in infected NHPs, T-cell responses have been characterized in cynomologous macaques (Koraka *et al.*, 2007). T-cells elicited by primary DENV infection are largely serotype cross-reactive, and secondary infection results in activation of T-cells principally directed against the primary infecting virus.

In summary, NHPs reproduce many critical features of human immune protection and immunopathogenesis, including broad serotype cross-reactive antibody and T-cell responses. These cross-reactive responses reduce the magnitude or duration of heterologous secondary infection viraemia in many cases. On the other hand, haemostatic abnormalities have been observed during secondary infections, albeit relatively rarely, and infection of animals circulating passively transferred anti-DENV antibody resulted in enhanced viraemia. However, in none of these studies did infected primates develop DHF-like vascular permeability. The high cost, absence of severe disease, and lack of genetic tractability have thus far limited practical applications of NHP models in understanding immune protection and pathogenesis of dengue.

Mouse models of DENV immune protection and pathogenesis

Modeling secondary DENV infection in most mouse models is problematic for several reasons. In some cases, only a single serotype or strain of virus is able to infect the mouse, while others possess little or no adaptive immune system, and still others develop disease due to infection of cell types that lack Fcγ receptors, such as neurons, and are thus unable to model ADE.

AG129 mice, in contrast, exhibit infection of macrophages, dendritic cells, and myeloid cells in multiple tissues and can be infected with all four serotypes of DENV (Balsitis *et al.*, 2008; Kyle *et al.*, 2007); in addition, primary infection with any DENV serotype elicits a broadly sero-type cross-reactive response, with the strongest response against the infecting virus (S. Balsitis, K. Williams, and E. Harris, unpublished data). Consequently, we have utilized AG129 mice to model secondary infection by both sequential infections and passive antibody transfer. In both DENV-1/DENV-2 and DENV-2/DENV-4 sequences of infection, primary infection reduced the viral burden upon subsequent secondary infection (Kyle *et al.*, 2008). While substantial serotype cross-neutralizing antibody was elicited by primary infection, serotype cross-reactive T-cells were not detected when stimulated *ex vivo*. Consistent with this, passive transfer of antiserum from DENV-1-immune mice substantially reduced viral burden, while adoptive transfer of DENV-1-immune splenocytes only reduced viral burden when a lower challenge dose of DENV-2 was used (Kyle *et al.*, 2008). Thus, AG129 mice model serotype cross-protective immunity similar to that observed in primates and humans, and this protection appears to be predominantly antibody-mediated. However, it is difficult to interpret data on protective immunity against dengue unless conditions for immunopathogenesis have been established in the same model.

Fortunately, we have recently succeeded in reproducing ADE of dengue disease in AG129 mice using the D2S10 peripherally mouse-passaged DENV-2 strain (Balsitis *et al.*, submitted). In the absence of antibody, D2S10 induces fatal vascular leakage only when a dose of 10^7 pfu or higher is administered (Shresta *et al.*, 2006). However, passive transfer of DENV-1-immune

AG129 serum one day before DENV-2 infection reduced the viral dose required for lethality by 100-fold (Balsitis *et al.*, submitted), to within the likely range of mosquito-delivered doses (Styer *et al.*, 2007; Vanlandingham *et al.*, 2004). Comparison of naïve and anti-DENV-1 serum recipients challenged with the same DENV-2 dose showed that antibody-enhanced lethality is accompanied by a systemic increase in viral burden, with a 20-fold increase in viraemia accompanied by up to log increases in viral load in peripheral white blood cells, bone marrow, lymph node, liver, and small intestine. ADE also caused increased vascular leakage into visceral organs, increased serum levels of TNF-α and IL-10, and caused greater depletion of platelets. Notably, all of these ADE-induced phenotypes are reproduced in the absence of ADE by infection with a higher viral inoculum (10^7 pfu); thus, all detected effects of ADE could be attributed to increased viral burden. Consistent with this, no differences in infected cell types were observed between mice lethally infected via ADE and those lethally infected by a high viral inoculum. Thus, severe dengue disease in AG129 mice occurs when a sufficient viral burden is achieved, regardless of whether that burden was reached via ADE or via a high viral dose. These data represent the first experimental demonstration that antibodies are sufficient to induce lethal disease in the absence of serotype cross-reactive T-cells; this is consistent with infant DHF in humans, where infants receiving antibody but not T-cells from dengue-immune mothers are at increased risk of DHF (Chau *et al.*, 2008; Halstead, 2003; Kliks *et al.*, 1988).

The ability of serotype-specific and cross-reactive antibodies to mediate ADE *in vivo* had never been directly compared, so we examined the ability of both DENV-1- and DENV-2-immune sera to protect against or enhance D2S10 infection (Balsitis *et al.*, submitted). A high dose of anti-DENV-2 serum reduced viraemia, but lower doses caused lethal enhancement of infection. This parallels observations of antibody-mediated neutralization and enhancement *in vitro*, where all virion-binding antibodies, regardless of epitope, are capable of enhancing infection when diluted below the threshold of neutralization (Pierson *et al.*, 2007). DENV-1-immune sera, DENV-2-

immune sera, and mouse pan-flavivirus MAb 4G2 all enhanced infection *in vivo* even when sera taken from mice after transfer but before infection had detectable *in vitro* neutralizing titres as measured by the traditional $PRNT_{50}$ assay (Balsitis *et al.*, submitted), a phenomenon also observed in human infections (Endy *et al.*, 2004). This observation shows that low levels of serotype-specific antibodies can enhance secondary DENV infection as can serotype cross-reactive antibodies. However, it is important to emphasize here that in these studies passive transfer recipients do not have memory immune responses, and conditions for ADE are likely to be much more restricted when memory immunity is present.

Future directions in dengue immunity and immunopathogenesis

With the development of a mouse model of ADE, many questions surrounding dengue immunity and immunopathology can be addressed experimentally for the first time. One priority must be an analysis of how the epitope reactivity and isotype of antibodies affect their ability to protect against or enhance infection. Such an analysis could lead both to the development of improved dengue vaccines and to the development of *in vitro* tests to identify individuals at risk for severe disease. As mentioned above, an understanding of the role of memory immunity in ADE is also sorely needed to clarify when naturally DENV-infected patients or vaccinees are potentially at risk for severe disease.

Another key question facing dengue researchers is how viral strain and serotype variation and infection sequence affect the conditions for immune protection and enhancement. Without such an understanding, correlates of protection and enhancement identified in controlled laboratory settings may not transfer into more complex field settings. In this regard, it will be important to establish mouse models for ADE with multiple serotypes, either by identifying conditions for enhancement of other serotypes in the AG129 model or by developing additional ADE models that are permissive for a wide variety of DENV strains and serotypes.

Regardless of the progress made in the AG129 model, though, additional mouse models of ADE are needed. These models would allow for comparisons of ADE in multiple backgrounds, and additional models may have strengths that complement the shortcomings of the AG129 model. On that note, it is unclear how interferon deficiency or the rapid onset of lethal vascular permeability in AG129 mice may influence the applicability of conclusions drawn from the AG129 model to human infection. Additionally, establishing ADE in STAT1$^{-/-}$ or immunocompetent mice would greatly ease genetic studies of ADE, since such studies in AG129 mice would require triple-knockout mice to be generated. ADE models in humanized mice would also be powerful research tools, as features unique to human cells could be studied *in vivo* and observations made in mouse cells could be confirmed.

A better understanding of the role of T-cells in primary and secondary infections is also needed, as T-cells could protect against infection or exacerbate disease in ways that are not currently understood. A limited number of studies have begun to address this issue in mice. The chemokine interferon-inducible protein 10 (IP-10) and its receptor CXCR3 promote CD8$^+$ T-cell recruitment into the brains of mice challenged intracerebrally with neuroadapted DENV-2, and CXCR3$^{-/-}$ and IP-10$^{-/-}$ mice exhibit higher CNS viral burden and greater mortality than wild-type mice (Hsieh *et al.*, 2006). The relevance of T-cell recruitment into mouse brain to human DENV infection in peripheral tissues is unclear, but since IP-10 is up-regulated in the blood of human dengue patients (Fink *et al.*, 2007), the role of IP-10 in T-cell function during DENV infection merits further study in other, more relevant animal models. Other recent work demonstrated that T-cells elicited by exposure of immunocompetent mice to DENV are serotype cross-reactive and are activated more rapidly and robustly during secondary DENV infection than newly activated serotype-specific T-cells, and they produce more TNF-α (Beaumier *et al.*, 2008). Under some conditions, this resulted in memory T-cell responses during secondary DENV infection that were predominantly directed against the primary infection virus, as has been reported in humans (Dong *et al.*, 2007; Mongkolsapaya *et al.*, 2003, 2006). These observations indicate that mice may be able to model important features of T-cell responses to secondary DENV infection. However, since no significant viral replication or disease occurred in the mice, new models are needed that allow for in-depth analysis of the effects of T-cells in the context of robust viral infection and dengue-like disease.

Returning to the NHP model, one factor that can only be properly assessed in NHPs is the waning of immunity over time and the effects of increasing time intervals between primary and secondary DENV infections. Sabin demonstrated several decades ago that immunity generated against a primary DENV infection changes over time; primary infections resulted in complete protection against secondary infections with another DENV serotype if the infections were less than 2 months apart, while shorter duration and reduced severity of disease was observed with infections up to 9 months apart (Sabin, 1950, 1952). While primary infection protected against secondary infection in AG129 mice at intervals of up to 1 year (Kyle *et al.*, 2008), much longer intervals cannot be addressed using mice. Secondary infection of NHPs has thus far been performed from 2 weeks to 6 months post primary infection in Rhesus macaques (Halstead *et al.*, 1973a; Halstead *et al.*, 1973b) or one year post-primary infection in cynomologous macaques (Koraka *et al.*, 2007), but the longevity of NHPs allow for future investigations to conduct a more thorough analysis of the waning of immunity over time and the consequences for immune protection and pathogenesis.

Antiviral drug development and testing

Intensive global research efforts are under way to identify effective therapeutics for DENV infection, and these efforts will require candidate therapies to be tested in a combination of animal models. Surprisingly, though, only a handful of anti-DENV therapeutics has been tested in animal models of DENV infection thus far (Table 6.2). Much antiviral drug testing has been performed in dengue encephalitis mouse models, which may not be representative of a drug's effects on peripheral virus infection. As the next generation of anti-dengue drugs progresses into animal testing, there is a need for models that are easy to use, well-characterized, and have endpoints relevant to human infection. To this end, Schul *et al.* (Schul

et al., 2007) recently adapted the AG129 model to drug testing by identifying a clinical isolate of DENV-2, TSV01, that produces substantial viraemia following i.p. injection, and demonstrated that three separate lead compounds were able to reduce viraemia *in vivo*. This represents a substantial step forward, although the TSV01 strain does not produce dengue-like disease.

However, antibody-enhanced infection of AG129 mice with DENV-2 strain D2S10 produces both viraemia and fatal vascular leakage, so viraemia, vascular permeability, and mortality can all be used as endpoints. Indeed, this model was recently used to show that an engineered antibody unable to bind to Fcγ receptors has prophylactic and therapeutic efficacy against lethal DENV-induced vascular leak (Balsitis *et al.*). Due to these advances, AG129 mice now represent a robust model for testing drugs that aim to treat dengue, by interfering with the interaction between anti-DENV antibodies and Fcγ receptors, or by inhibiting host mediators of pathogenesis such as TNF-α. The latter category of drugs would have the advantages of being effective against the disease, rather than the virus, and thus may be effective against any dengue virus isolate and preclude the possibility of drug-resistant virus emergence. Furthermore, since multiple mouse models of DENV-induced vascular leak and haemorrhage are now available, investigators also have the opportunity to test promising therapeutics in two or more models to ensure that the shortcomings of each model do not give rise to misleading results. In sum, then, recent advances in dengue mouse modelling have provided the necessary tools for testing promising therapeutics *in vivo* with multiple readouts relevant to human infection.

Vaccine development and testing

Numerous dengue vaccines are currently at various stages of discovery, development, and clinical testing. The status of dengue vaccine development has been reviewed recently (Edelman, 2007; Guy and Almond, 2008; Hombach, 2007; Whitehead *et al.*, 2007) and see Chapter 12. We focus here on the advantages and limitations of the animal models available to aid in pre-clinical immunogenicity, safety, and efficacy testing. No

Table 6.2 Therapeutic drug testing in dengue mouse models

Drug	Mechanism of action	Mouse model used	Result	Reference
Neem leaf extract	Unknown	suckling mouse	Absence of disease symptoms	Parida et al. (2002)
Castanospermine	α-Glucosidase inhibitor	A/J	Increased survival	Whitby et al. (2005)
PI-88	Heparan sulphate mimetic	AG129	Increased survival time	Lee et al. (2006a)
7-DMA	RdRP inhibitor	AG129	70% reduction in viraemia	Schul et al. (2007)
NN-DNJ	α-Glucosidase inhibitor	AG129	90% reduction in viraemia	Schul et al. (2007)
6-O-butanoyl castanospermine	α-Glucosidase inhibitor	AG129	90% reduction in viraemia	Schul et al. (2007)
PMO	Antisense oligonucleotide	AG129	Increased survival time post-neurovirulent challenge	Stein et al. (2008)
E60-N297Q	Engineered antibody	AG129	Increased survival	Balsitis et al. (submitted)

Abbreviations: 7-DMA, 7-deaza-2′-C-methyl-adenosine; RdRp, RNA-dependent RNA polymerase; NN-DNJ, N-nonyl-deoxynojirimycin; PMO, phosphorodiamidate morpholino oligomers.

single dengue animal model is ideal for all of these purposes and for all vaccine formulations, but due to the cost of non-human primates, testing usually begins with mouse models.

Model selection for initial testing for immunogenicity and efficacy is dependent upon the nature of the vaccine. For subunit or DNA vaccines, initial immunogenicity testing is best performed in immunocompetent mice, as immunocompromised models of DENV infection may exhibit poor responses to vaccination. However, this implies that early efficacy testing must often be performed in various dengue encephalitis models. A wide variety of subunit and DNA vaccines targeting the viral E, pre-membrane (prM), NS1, and capsid genes have been tested in this manner (Babu et al., 2008; Chen et al., 2007b; Costa et al., 2006; Henchal et al., 1988; Lazo et al., 2007; Lazo et al., 2008; Lu et al., 2003; Mota et al., 2005; Schlesinger et al., 1987; White et al., 2007; Wu et al., 2003). Such studies serve as useful proof-of-concept studies to support further development, but due to the many limitations of dengue encephalitis models, subsequent testing in more relevant models is required.

Live-attenuated vaccines have been tested in immunocompetent mice (Chambers et al., 2003; van Der Most et al., 2000), but due to the extremely low level of DENV replication that occurs in these mice, immunogenicity may be underestimated and attenuation may be overestimated. Consequently, the AG129 model was originally developed to provide a better model for testing of live-attenuated DENV vaccines (Johnson and Roehrig, 1999). This model has indeed proven useful for live-attenuated vaccine testing (Huang et al., 2003), and has also been used to examine the protective efficacy of immune responses targeted against the viral non-structural proteins (Calvert et al., 2006). Nevertheless, the same interferon deficiency that allows live attenuated vaccine candidates to replicate in this model also can pose issues in the interpretation of such studies, as both humoral and cellular immunity may be partially impaired.

Several attempts have been made to gauge the degree of attenuation of live attenuated dengue vaccines and identify attenuation markers using mouse models, including SCID-Huh7 and intracranial inoculation models (Blaney et al.,

2003; Blaney *et al.*, 2007; Butrapet *et al.*, 2000; Hanley *et al.*, 2002; Monath *et al.*, 2005). Loss of neurovirulence in infant mice appears to be a useful attenuation marker, as infant mice are more sensitive at detecting viral virulence than are neurovirulence safety tests in non-human primates, which are currently required for safety testing of live attenuated vaccines derived from potentially neurovirulent viruses (Monath *et al.*, 2005). However, attenuation in other mouse models may or may not be meaningful; while some candidate vaccines that were attenuated in SCID-HepG2 mice were also attenuated in rhesus macaques, others candidates were attenuated in mice but not in macaques (Blaney *et al.*, 2007).

Mouse models that exhibit relevant disease phenotypes following infection with multiple dengue serotypes and with demonstrated ability to exhibit ADE would be very helpful in vaccine development. While the AG129 model is closest to this goal, it is not ideal and will not be appropriate for testing some vaccine formulations. Other models require further characterization in terms of tropism, ADE, and susceptibility to multiple virus isolates, and some models, such as humanized SCID mice, are simply too labour-intensive to be of widespread use for vaccine testing. Thus, in spite of the substantial progress in dengue mouse models in recent years, there is yet room for improvement.

Ultimately, testing in NHPs is a necessary step for any dengue vaccine, as no other model has the advantages of being (1) closely related to humans; (2) immunocompetent yet susceptible to infection with clinical DENV isolates of all serotypes, with appropriate viraemia kinetics as an endpoint relevant to human disease; and (3) demonstrated to exhibit both protective and pathogenic immunity. Rhesus macaques have been the species of choice for such studies, as they are by far the best characterized NHP model for dengue. Evaluation of some live attenuated vaccines in Rhesus macaques has been predictive of vaccine attenuation, viraemia, and immunogenicity in early human trials (Blaney *et al.*, 2007; Durbin *et al.*, 2001, 2006a,b; Hoke *et al.*, 1990; Marchette *et al.*, 1990; Men *et al.*, 1996; Whitehead *et al.*, 2003), and Rhesus macaques have also been used for optimizing vaccine formulations to promote the development of a balanced tetravalent

response (Guirakhoo *et al.*, 2002). However, in two cases live attenuated vaccine candidates that were attenuated in Rhesus macaques were found to be under-attenuated in humans (Innis *et al.*, 1988; Sanchez *et al.*, 2006), demonstrating that even macaque studies are not always predictive of vaccine performance in humans.

In summary, both mouse and NHP models of DENV infection can be very useful for vaccine development and testing, provided that the strengths and limitations of each model are kept in mind, experiments are designed and interpreted accordingly, and no single model is relied upon too heavily.

In conclusion, enormous progress in dengue animal models has been made in the last ten years. This has enabled researchers to begin to dissect the molecular mechanisms of pathogenesis, better understand the nature of protective and pathogenic immune responses to DENV, and develop and test vaccines and antivirals in increasingly useful and relevant animal models. These models provide an important tool set that is helping to set the stage for continued rapid progress in understanding how dengue viruses cause human disease and how that disease can be treated and prevented. Still, progress to date has also revealed limitations of current animal models, so efforts to improve dengue animal models will continue for some time to come.

Acknowledgements

We thank Scott Halstead and Anna Durbin for helpful editorial comments. Support was provided by grants from the Pediatric Dengue Vaccine Initiative (#CRA-14) and the National Institutes of Health (#AI65359) to E.H., and NIH grant #AI72933 to S.B.

References

Adams, B., Holmes, E.C., Zhang, C., Mammen, M.P., Jr., Nimmannitya, S., Kalayanarooj, S., and Boots, M. (2006). Cross-protective immunity can account for the alternating epidemic pattern of dengue virus serotypes circulating in Bangkok. Proc. Natl. Acad. Sci. U.S.A. *103*, 14234–14239.

Alcon-LePoder, S., Drouet, M.T., Roux, P., Frenkiel, M.P., Arborio, M., Durand-Schneider, A.M., Maurice, M., Le Blanc, I., Gruenberg, J., and Flamand, M. (2005). The secreted form of dengue virus nonstructural protein NS1 is endocytosed by hepatocytes and accumulates in late endosomes: implications for viral infectivity. J. Virol. 79, 11403–11411.

An, J., Kimura-Kuroda, J., Hirabayashi, Y., and Yasui, K. (1999). Development of a novel mouse model for dengue virus infection. Virology *263*, 70–77.

Atrasheuskaya, A., Petzelbauer, P., Fredeking, T.M., and Ignatyev, G. (2003). Anti-TNF antibody treatment reduces mortality in experimental dengue virus infection. FEMS Immunol. Med. Microbiol. *35*, 33–42.

Avirutnan, P., Punyadee, N., Noisakran, S., Komoltri, C., Thiemmeca, S., Auethavornanan, K., Jairungsri, A., Kanlaya, R., Tangthawornchaikul, N., Puttikhunt, C., Pattanakitsakul, S.N., Yenchitsomanus, P.T., Mongkolsapaya, J., Kasinrerk, W., Sittisombut, N., Husmann, M., Blettner, M., Vasanawathana, S., Bhakdi, S., and Malasit, P. (2006). Vascular leakage in severe dengue virus infections: a potential role for the nonstructural viral protein NS1 and complement. J. Infect. Dis. *193*, 1078–1088.

Avirutnan, P., Zhang, L., Punyadee, N., Manuyakorn, A., Puttikhunt, C., Kasinrerk, W., Malasit, P., Atkinson, J. P., and Diamond, M. S. (2007). Secreted NS1 of dengue virus attaches to the surface of cells via interactions with heparan sulfate and chondroitin sulfate E. PLoS Pathog. 3, e183.

Babu, J. P., Pattnaik, P., Gupta, N., Shrivastava, A., Khan, M., and Rao, P.V. (2008). Immunogenicity of a recombinant envelope domain III protein of dengue virus type-4 with various adjuvants in mice. Vaccine *26*, 4655–4663.

Balsitis, S.J., Coloma, J., Castro, G., Alava, A., Flores, D., McKerrow, J.H., Beatty, P.R., and Harris, E. (2008). Tropism of dengue virus in mice and humans defined by viral nonstructural protein 3-specific immunostaining. Am. J. Trop. Med. Hyg. *80*, 416–424.

Balsitis, S.J., Williams, K.L., Flores, D., Kyle, J., Mehlhop, E., Johnson, S., Diamond, M.S., Beatty, P.R., and Harris, E. Lethal antibody enhancement of dengue disease in mice is prevented by Fc modification. Manuscript submitted (in revision).

Beaumier, C.M., Mathew, A., Bashyam, H.S., and Rothman, A.L. (2008). Cross-reactive memory CD8(+) T-cells alter the immune response to heterologous secondary dengue virus infections in mice in a sequence-specific manner. J. Infect. Dis. *197*, 608–617.

Bente, D.A., Melkus, M.W., Garcia, J.V., and Rico-Hesse, R. (2005). Dengue fever in humanized NOD/SCID mice. J. Virol. *79*, 13797–13799.

Bente, D.A., and Rico-Hesse, R. (2006). Models of dengue virus infection. Drug Discov. Today Dis Models 3, 97–103.

Bhoopat, L., Bhamarapravati, N., Attasiri, C., Yoksarn, S., Chaiwun, B., Khunamornpong, S., and Sirisanthana, V. (1996). Immunohistochemical characterization of a new monoclonal antibody reactive with dengue virus-infected cells in frozen tissue using immunoperoxidase technique. Asian Pac J Allergy Immunol *14*, 107–113.

Blaney, J.E., Jr., Johnson, D.H., Manipon, G.G., Firestone, C.Y., Hanson, C.T., Murphy, B.R., and Whitehead, S.S. (2002). Genetic basis of attenuation of dengue virus type 4 small plaque mutants with restricted replication in suckling mice and in SCID mice transplanted with human liver cells. Virology *300*, 125–139.

Blaney, J.E., Jr., Manipon, G.G., Murphy, B.R., and Whitehead, S.S. (2003). Temperature sensitive mutations in the genes encoding the NS1, NS2A., NS3, and NS5 nonstructural proteins of dengue virus type 4 restrict replication in the brains of mice. Arch. Virol. *148*, 999–1006.

Blaney, J.E., Jr., Sathe, N.S., Hanson, C.T., Firestone, C.Y., Murphy, B.R., and Whitehead, S.S. (2007). Vaccine candidates for dengue virus type 1 (DEN1) generated by replacement of the structural genes of rDEN4 and rDEN4Delta30 with those of DEN1. Virol. J. *4*, 23.

Butrapet, S., Huang, C.Y., Pierro, D.J., Bhamarapravati, N., Gubler, D.J., and Kinney, R.M. (2000). Attenuation markers of a candidate dengue type 2 vaccine virus, strain 16681 (PDK-53), are defined by mutations in the 5′ noncoding region and nonstructural proteins 1 and 3. J. Virol. *74*, 3011–3019.

Calvert, A.E., Huang, C.Y., Kinney, R.M., and Roehrig, J.T. (2006). Non-structural proteins of dengue 2 virus offer limited protection to interferon-deficient mice after dengue 2 virus challenge. J. Gen. Virol. *87*, 339–346.

Chambers, T.J., Liang, Y., Droll, D.A., Schlesinger, J.J., Davidson, A.D., Wright, P.J., and Jiang, X. (2003). Yellow fever virus/dengue-2 virus and yellow fever virus/dengue-4 virus chimeras: biological characterization, immunogenicity, and protection against dengue encephalitis in the mouse model. J. Virol. *77*, 3655–3668.

Chau, T.N., Quyen, N.T., Thuy, T.T., Tuan, N.M., Hoang, D.M., Dung, N.T., Lien, L.B., Quy, N.T., Hieu, N.T., Hieu, L.T., Hien, T.T., Hung, N.T., Farrar, J., and Simmons, C.P. (2008). Dengue in Vietnamese infants-results of infection-enhancement assays correlate with age-related disease epidemiology, and cellular immune responses correlate with disease severity. J. Infect. Dis. *198*, 516–524.

Chen, H.C., Hofman, F.M., Kung, J.T., Lin, Y.D., and Wu-Hsieh, B.A. (2007a). Both virus and tumor necrosis factor alpha are critical for endothelium damage in a mouse model of dengue virus-induced hemorrhage. J. Virol. *81*, 5518–5526.

Chen, S., Yu, M., Jiang, T., Deng, Y., Qin, C., and Qin, E. (2007b). Induction of tetravalent protective immunity against four dengue serotypes by the tandem domain III of the envelope protein. DNA Cell Biol. *26*, 361–367.

Chen, S.T., Lin, Y.L., Huang, M.T., Wu, M.., Cheng, S.C., Lei, H.Y., Lee, C.K., Chiou, T.W., Wong, C.H., and Hsieh, S.L. (2008). CLEC5A is critical for dengue-virus-induced lethal disease. Nature *453*, 672–676.

Chung, K.M., Liszewski, M.K., Nybakken, G., Davis, A.E., Townsend, R.R., Fremont, D.H., Atkinson, J.P., and Diamond, M.S. (2006). West Nile virus nonstructural protein NS1 inhibits complement activation by binding the regulatory protein factor H. Proc. Natl. Acad. Sci. U.S.A. *103*, 19111–19116.

Cole, G.A., and Wisseman, C.L., Jr. (1969). Pathogenesis of type 1 dengue virus infection in suckling, weanling and adult mice. 1. The relation of virus replication to interferon and antibody formation. Am. J. Epidemiol. *89*, 669–680.

Costa, S.M., Paes, M.V., Barreto, D.F., Pinhao, A.T., Barth, O.M., Queiroz, J.L., Armoa, G.R., Freire, M.S., and Alves, A.M. (2006). Protection against dengue type 2 virus induced in mice immunized with a DNA plasmid encoding the non-structural 1 (NS1) gene fused to the tissue plasminogen activator signal sequence. Vaccine 24, 195–205.

Diamond, M.S., and Harris, E. (2001). Interferon inhibits dengue virus infection by preventing translation of viral RNA through a PKR-independent mechanism. Virology 289, 297–311.

Diamond, M.S., Roberts, T.G., Edgil, D., Lu, B., Ernst, J., and Harris, E. (2000). Modulation of dengue virus infection in human cells by alpha, beta, and gamma interferons. J. Virol. 74, 4957–4966.

Dong T., Moran, E., Vinh Chau N., Simmons, C., Luhn, K., Peng, Y., Wills, B., Phuong Dung N., Thi Thu Thao L., Hien, T.T., McMichael A., Farrar, J., and Rowland-Jones S. (2007). High pro-inflammatory cytokine secretion and loss of high avidity cross-reactive cytotoxic T-cells during the course of secondary dengue virus infection. PLoS ONE 2, e1192.

Durbin A.P., Karron, R.A., Sun, W., Vaughn, D.W., Reynolds, M.J., Perreault, J.R., Thumar, B., Men, R., Lai, C.J., Elkins, W.R., Chanock, R.M., Murphy, B.R., and Whitehead, S.S. (2001). Attenuation and immunogenicity in humans of a live dengue virus type-4 vaccine candidate with a 30 nucleotide deletion in its 3′-untranslated region. 65:. Am. J. Trop. Med. Hyg. 65, 405–413.

Durbin, A.P., McArthur, J., Marron, J.A., Blaney, J.E. J., Thumar, B., Wanionek, K., Murphy, B.R., and Whitehead, S.S. (2006a). The live attenuated dengue serotype 1 vaccine rDEN1Delta30 is safe and highly immunogenic in healthy adult volunteers. Hum. Vaccin. 2, 167–173.

Durbin, A.P., McArthur, J.H., Marron, J.A., Blaney, J.E., Thumar, B., Wanionek, K., Murphy, B.R., and Whitehead, S.S. (2006b). rDEN2/4Delta30(ME), a live attenuated chimeric dengue serotype 2 vaccine is safe and highly immunogenic in healthy dengue-naive adults. Hum Vaccin 2, 255–260.

Edelman, R. (2007). Dengue vaccines approach the finish line. Clin. Infect. Dis. 45 Suppl. 1, S56–60.

Endy, T.P., Nisalak, A., Chunsuttitwat, S., Vaughn, D.W., Green, S., Ennis, F.A., Rothman, A.L., and Libraty, D.H. (2004). Relationship of preexisting dengue virus (DV) neutralizing antibody levels to viremia and severity of disease in a prospective cohort study of DV infection in Thailand. J. Infect. Dis. 189, 990–1000.

Falconar, A.K. (1997). The dengue virus nonstructural-1 protein (NS1) generates antibodies to common epitopes on human blood clotting, integrin/adhesin proteins and binds to human endothelial cells: potential implications in haemorrhagic fever pathogenesis. Arch. Virol. 142, 897–916.

Falconar, A.K. (2007). Antibody responses are generated to immunodominant ELK/KLE-type motifs on the nonstructural-1 glycoprotein during live dengue virus infections in mice and humans: implications for diagnosis, pathogenesis, and vaccine design. Clin. Vaccine Immunol. 14, 493–504.

Fink J., Gu, F., Ling, L., Tolfvenstam, T., Olfat, F., Chin, K.C., Aw, P., George, J., Kuznetsov, V.A., Schreiber, M., Vasudevan, S.G., and Hibberd, M.L. (2007). Host gene expression profiling of dengue virus infection in cell lines and patients. PLoS Negl. Trop. Dis. 1, e86.

Funahara, Y., Ogawa, K., Fujita, N., and Okuno, Y. (1987). Three possible triggers to induce thrombocytopenia in dengue virus infection. South-East Asian J. Trop. Med. Public Health 18, 351–355.

Goncalvez, A.P., Engle, R.E., St Claire, M., Purcell, R.H., and Lai, C.J. (2007). Monoclonal antibody-mediated enhancement of dengue virus infection in vitro and in vivo and strategies for prevention. Proc. Natl. Acad. Sci. U.S.A. 104, 9422–9427.

Gubler, D.J. (1998). Dengue and dengue hemorrhagic fever. Clin Microbiol Rev 11, 480–496.

Guirakhoo, F., Pugachev, K., Arroyo, J., Miller, C., Zhang, Z.X., Weltzin, R., Georgakopoulos, K., Catalan, J., Ocran, S., Draper, K., and Monath, T.P. (2002). Viremia and immunogenicity in nonhuman primates of a tetravalent yellow fever-dengue chimeric vaccine: genetic reconstructions, dose adjustment, and antibody responses against wild-type dengue virus isolates. Virology 298, 146–159.

Guy, B., and Almond, J.W. (2008). Towards a dengue vaccine: progress to date and remaining challenges. Comp Immunol Microbiol Infect Dis 31, 239–252.

Hall, W.C., Crowell, T.P., Watts, D.M., Barros, V.L., Kruger, H., Pinheiro, F., and Peters, C.J. (1991). Demonstration of yellow fever and dengue antigens in formalin-fixed paraffin-embedded human liver by immunohistochemical analysis. Am. J. Trop. Med. Hyg. 45, 408–417.

Halstead, S.B. (1979) In vivo enhancement of dengue virus infection in rhesus monkeys by passively transferred antibody. J. Infect. Dis. 140, 527–533.

Halstead, S.B. (1980). Immunological parameters of Togavirus disease syndromes, In The togaviruses, biology, structure, replication, R. W. Schlesinger, ed. (New York: Academic Press), pp. 107–173.

Halstead, S.B. (2003). Neutralization and antibody-dependent enhancement of dengue viruses. Adv. Virus Res. 60, 421–467.

Halstead, S.B. (2007). Dengue. Lancet 370, 1644–1652.

Halstead, S.B., Casals, J., Shotwell, H., and Palumbo, N. (1973a). Studies on the immunization of monkeys against dengue. I. Protection derived from single and sequential virus infections. Am. J. Trop. Med. Hyg. 22, 365–374.

Halstead, S.B., and Palumbo, N. (1973). Studies on the immunization of monkeys against dengue. I.I. Protection following inoculation of combinations of viruses. Am. J. Trop. Med. Hyg. 22, 375–381.

Halstead, S.B., Shotwell, H., and Casals, J. (1973b). Studies on the pathogenesis of dengue infection in monkeys. I. Clinical laboratory responses to primary infection. J. Infect. Dis. 128, 7–14.

Halstead, S.B., Shotwell, H., and Casals, J. (1973c). Studies on the pathogenesis of dengue infection in monkeys. I.I. Clinical laboratory responses to heterologous infection. J. Infect. Dis. 128, 15–22.

Hanley, K.A., Lee, J.J., Blaney, J.E., Jr., Murphy, B.R., and Whitehead, S.S. (2002). Paired charge-to-alanine mutagenesis of dengue virus type 4 NS5 generates mutants with temperature-sensitive, host range, and mouse attenuation phenotypes. J. Virol. 76, 525–531.

Henchal, E.A., Henchal, L.S., and Schlesinger, J.J. (1988). Synergistic interactions of anti-NS1 monoclonal antibodies protect passively immunized mice from lethal challenge with dengue 2 virus. J. Gen. Virol. 69, 2101–2107.

Hoke, C.H., Jr., Malinoski, F.J., Eckels, K.H., Scott, R.M., Dubois, D.R., Summers, P.L., Simms,T., Burrous, J., Hasty, S.E., and Bancroft, W.H. (1990). Preparation of an attenuated dengue 4 (341750 Carib) virus vaccine. I.I. Safety and immunogenicity in humans. Am. J. Trop. Med. Hyg. 43, 219–226.

Hombach, J. (2007). Vaccines against dengue: a review of current candidate vaccines at advanced development stages. Rev Panam Salud Publica 21, 254–260.

Hsieh, M. F., Lai, S.L., Chen, J.P., Sung, J.M., Lin, Y.L., Wu-Hsieh, B.A., Gerard, C., Luster, A., and Liao, F. (2006). Both CXCR3 and CXCL10/IFN-inducible protein 10 are required for resistance to primary infection by dengue virus. J. Immunol. 177, 1855–1863.

Huang, C.Y., Butrapet, S., Tsuchiya, K.R., Bhamarapravati, N., Gubler, D.J., and Kinney, R.M. (2003). Dengue 2 PDK-53 virus as a chimeric carrier for tetravalent dengue vaccine development. J. Virol. 77, 11436–11447.

Huang, K.J., Li, S.Y., Chen, S.C., Liu, H.S., Lin, Y.S., Yeh, T.M., Liu, C.C., and Lei, H.Y. (2000). Manifestation of thrombocytopenia in dengue-2-virus-infected mice. J. Gen. Virol. 81, 2177–2182.

Innis, B.L., Eckels, K.H., Kraiselburd, E., Dubois, D.R., Meadors, G.F., Gubler, D.J., Burke, D.S., and Bancroft, W.H. (1988). Virulence of a live dengue virus vaccine candidate: a possible new marker of dengue virus attenuation. J. Infect. Dis. 158, 876–880.

Jessie, K., Fong, M.Y., Devi, S., Lam, S.K., and Wong, K.T. (2004). Localization of dengue virus in naturally infected human tissues, by immunohistochemistry and in situ hybridization. J. Infect. Dis. 189, 1411–1418.

Johnson, A.J., and Roehrig, J.T. (1999). New mouse model for dengue virus vaccine testing. J. Virol. 73, 783–786.

Kliks, S.C., Nimmanitya, S., Nisalak, A., and Burke, D.S. (1988). Evidence that maternal dengue antibodies are important in the development of dengue hemorrhagic fever in infants. Am. J. Trop. Med. Hyg. 38, 411–419.

Kochel, T.J., Watts, D.M., Gozalo, A.S., Ewing, D.F., Porter, K.R., and Russell, K.L. (2005). Cross-serotype neutralization of dengue virus in Aotus nancymae monkeys. J. Infect. Dis. 191, 1000–1004.

Kochel, T.J., Watts, D.M., Halstead, S.B., Hayes, C.G., Espinoza, A., Felices, V., Caceda, R., Bautista, C.T., Montoya, Y., Douglas, S., and Russell, K.L. (2002). Effect of dengue-1 antibodies on American dengue-2 viral infection and dengue haemorrhagic fever. Lancet 360, 310–312.

Koraka, P., Benton, S., van Amerongen, G., Stittel aar, K.J., and Osterhaus, A.D. (2007). Characterization of humoral and cellular immune responses in cynomolgus macaques upon primary and subsequent heterologous infections with dengue viruses. Microbes Infect. 9, 940–946.

Kurosu, T., Chaichana, P., Yamate, M., Anantapreecha, S., and Ikuta, K. (2007). Secreted complement regulatory protein clusterin interacts with dengue virus nonstructural protein 1. Biochem. Biophys. Res. Commun. 362, 1051–1056.

Kuruvilla, J.G., Troyer, R.M., Devi, S., and Akkina, R. (2007). Dengue virus infection and immune response in humanized RAG2(-/-)gamma(c) (-/-) (RAG-hu) mice. Virology 369, 143–152.

Kyle, J.L., Balsitis, S.J., Zhang, L., Beatty, P.R., and Harris, E. (2008). Antibodies play a greater role than immune cells in heterologous protection against secondary dengue virus infection in a mouse model. Virology 380, 296–303.

Kyle, J.L., Beatty, P.R., and Harris, E. (2007). Dengue virus infects macrophages and dendritic cells in a mouse model of infection. J. Infect. Dis. 195, 1808–1817.

Lazo, L., Hermida, L., Zulueta, A., Sanchez, J., Lopez, C., Silva, R., Guillen, G., and Guzman, M.G. (2007). A recombinant capsid protein from Dengue-2 induces protection in mice against homologous virus. Vaccine 25, 1064–1070.

Lazo, L., Zulueta, A., Hermida, L., Blanco, A., Sanchez, J., Valdes, I., Gil, L., Lopez, C., Romero, Y., Guzman, M.G., and Guillen, G. (2008). The Dengue-4 envelope domain III fused twice within the meningococcal P64k protein carrier induces partial protection in mice. Biotechnol. Appl. Biochem. 52, 265–271.

Lee, E., Pavy, M., Young, N., Freeman, C., and Lobigs, M. (2006a). Antiviral effect of the heparan sulfate mimetic, PI-88, against dengue and encephalitic flaviviruses. Antiviral Res 69, 31–38.

Lee, E., Wright, P.J., Davidson, A., and Lobigs, M. (2006b). Virulence attenuation of Dengue virus due to augmented glycosaminoglycan-binding affinity and restriction in extraneural dissemination. J. Gen. Virol. 87, 2791–2801.

Lin, C.F., Chiu, S.C., Hsiao, Y.L., Wan, S.W., Lei, H.Y., Shiau, A.L., Liu, H.S., Yeh, T.M., Chen, S.H., Liu, C.C., and Lin, Y.S. (2005). Expression of cytokine, chemokine, and adhesion molecules during endothelial cell activation induced by antibodies against dengue virus nonstructural protein 1. J. Immunol. 174, 395–403.

Lin, C.F., Lei, H.Y., Liu, C.C., Liu, H.S., Yeh, T.M., Wang, S.T., Yang, T.I., Sheu, F.C., Kuo, C.F., and Lin, Y.S. (2001). Generation of IgM anti-platelet autoantibody in dengue patients. J Med Virol 63, 143–149.

Lin, C.F., Lei, H.Y., Shiau, A.L., Liu, C. C., Liu,H.S., Yeh, T.M., Chen, S.H., and Lin, Y.S. (2003). Antibodies from dengue patient sera cross-react with endothelial cells and induce damage. J. Med. Virol. 69, 82–90.

Lin, C.F., Lei, H.Y., Shiau, A.L., Liu, H.S., Yeh, T.M., Chen, S.H., Liu, C.C., Chiu, S.C., and Lin, Y.S. (2002). Endothelial cell apoptosis induced by antibodies against dengue virus nonstructural protein 1 via production of nitric oxide. J. Immunol. 169, 657–664.

Lin, C.F., Wan, S.W., Chen, M.C., Lin, S.C., Cheng, C.C., Chiu, S.C., Hsiao, Y.L., Lei, H.Y., Liu, H.S., Yeh, T.M., and Lin, Y.S. (2008). Liver injury caused by antibodies against dengue virus nonstructural protein 1 in a murine model. Lab. Invest. *88*, 1079–1089.

Lin, Y.L., Liao, C.L., Chen, L.K., Yeh, C.T., Liu, C.I., Ma, S.H., Huang, Y.Y., Huang, Y.L., Kao, C.L., and King, C.C. (1998). Study of Dengue virus infection in SCID mice engrafted with human K562 cells. J. Virol. *72*, 9729–9737.

Lu, Y., Raviprakash, K., Leao, I.C., Chikhlikar, P.R., Ewing, D., Anwar, A., Chougnet, C., Murphy, G., Hayes, C.G., August, T.J., and Marques, E.T., Jr. (2003). Dengue 2 PreM-E/LAMP chimera targeted to the MHC class II compartment elicits long-lasting neutralizing antibodies. Vaccine *21*, 2178–2189.

Marchette, N.J., Dubois, D.R., Larsen, L.K., Summers, P.L., Kraiselburd, E.G., Gubler, D.J., and Eckels, K.H. (1990). Preparation of an attenuated dengue 4 (341750 Carib) virus vaccine. I. Pre-clinical studies. Am. J. Trop. Med. Hyg. *43*, 212–218.

Marchette, N.J., Halstead, S.B., Falkler, W.A., Jr., Stenhouse, A., and Nash, D. (1973). Studies on the pathogenesis of dengue infection in monkeys. 3. Sequential distribution of virus in primary and heterologous infections. J. Infect. Dis. *128*, 23–30.

Men, R., Bray, M., Clark, D., Chanock, R.M., and Lai, C.-J. (1996). Dengue type 4 virus mutants containing deletions in the 3′ noncoding region of the RNA genome: analysis of growth restriction in cell culture and altered viremia pattern and immunogenicity in Rhesus monkeys. J. Virol. *70*, 3930–3937.

Miagostovich, M.P., Ramos, R.G., Nicol, A.F., Nogueira, R.M., Cuzzi-Maya, T., Oliveira, A.V., Marchevsky, R.S., Mesquita, R.P., and Schatzmayr, H.G. (1997). Retrospective study on dengue fatal cases. Clin. Neuropathol. *16*, 204–208.

Monath T.P., Myers, G.A., Beck, R.A., Knauber, M., Scappaticci, K., Pullano, T., Archambault, W.T., Catalan, J., Miller, C., Zhang, Z.X., Shin, S., Pugachev, K., Draper, K., Levenbook, I.S., and Guirakhoo, F. (2005). Safety testing for neurovirulence of novel live, attenuated flavivirus vaccines: infant mice provide an accurate surrogate for the test in monkeys. Biologicals *33*, 131–144.

Mongkolsapaya J., Dejnirattisai, W., Xu, X.N., Vasanawathana, S., Tangthawornchaikul, N., Chairunsri, A., Sawasdivorn, S., Duangchinda, T., Dong, T., Rowland-Jones S., Yenchitsomanus, P.T., McMichael, A., Malasit, P., and Screaton, G. (2003). Original antigenic sin and apoptosis in the pathogenesis of dengue hemorrhagic fever. Nat. Med. *9*, 921–927.

Mongkolsapaya J., Duangchinda, T., Dejnirattisai, W., Vasanawathana, S., Avirutnan, P., Jairungsri, A., Khemnu, N., Tangthawornchaikul, N., Chotiyarnwong, P., Sae-Jang K., Koch, M., Jones, Y., McMichael, A., Xu, X., Malasit, P., and Screaton, G. (2006). T-cell responses in dengue hemorrhagic fever: are cross-reactive T-cells suboptimal? J. Immunol. *176*, 3821–3829.

Mota, J., Acosta, M., Argotte, R., Figueroa, R., Mendez, A., and Ramos, C. (2005). Induction of protective antibodies against dengue virus by tetravalent DNA immunization of mice with domain III of the envelope protein. Vaccine *23*, 3469–3476.

Nelson, E.R., Bierman, H.R., and Chulajata, R. (1964). Hematologic findings in the 1960 hemorrhagic fever epidemic (dengue) in Thailand. Am. J. Trop. Med. Hyg. *13*, 642–649.

Oishi K., Inoue, S., Cinco, M.T., Dimaano, E.M., Alera, M.T., Alfon, J.A., Abanes, F., Cruz, D.J., Matias, R.R., Matsuura, H., Hasebe, F., Tanimura, S., Kumatori, A., Morita, K., Natividad, F.F., and Nagatake, T. (2003). Correlation between increased platelet-associated IgG and thrombocytopenia in secondary dengue virus infections. J Med Virol *71*, 259–264.

Pang, T., Cardosa, M.J., and Guzman, M.G. (2007). Of cascades and perfect storms: the immunopathogenesis of dengue haemorrhagic fever-dengue shock syndrome (DHF/DSS). Immunol Cell Biol *85*, 43–45.

Parida, M.M., Upadhyay, C., Pandya, G., and Jana, A.M. (2002). Inhibitory potential of neem (Azadirachta indica Juss) leaves on dengue virus type-2 replication. J Ethnopharmacol *79*, 273–278.

Pierson, T.C., Xu, Q., Nelson, S., Oliphant, T., Nybakken, G.E., Fremont, D.H., and Diamond, M.S. (2007). The stoichiometry of antibody-mediated neutralization and enhancement of West Nile virus infection. Cell Host Microbe *1*, 135–145.

Prestwood, T.R., Prigozhin, D.M., Sharar, K.L., Zellweger, R.M., and Shresta, S. (2008). A mouse-passaged dengue virus strain with reduced affinity for heparan sulfate causes severe disease in mice by establishing increased systemic viral loads. J Virol. *82*, 8411–8421.

Rothman, A.L. (2004). Dengue: defining protective versus pathologic immunity. J. Clin. Invest, *113*, 946–951.

Rothman, A.L., and Ennis, F.A. (1999). Immunopathogenesis of dengue hemorrhagic fever. Virology *257*, 1–6.

Rothwell, S.W., Putnak, R., and La Russa, V.F. (1996). Dengue-2 virus infection of human bone marrow: characterization of dengue-2 antigen-positive stromal cells. Am. J. Trop. Med. Hyg. *54*, 503–510.

Sabin, A.B. (1950). The dengue group of viruses and its family relationships. Bacteriol. Rev. *14*, 225–232.

Sabin, A.B. (1952). Research on dengue during World War I.I. Am. J. Trop. Med. Hyg. *1*, 30–50.

Saito M., Oishi, K., Inoue, S., Dimaano, E.M., Alera, M.T., Robles, A.M., Estrella, BD Jr, Kumatori, A., Moji, K., Alonzo, M.T., Buerano, C.C., Matias, R.R., Morita, K., Natividad, F.F., and Nagatake, T. (2004). Association of increased platelet-associated immunoglobulins with thrombocytopenia and the severity of disease in secondary dengue virus infections. Clin. Exp. Immunol. *138*, 299–303.

Sanchez, V., Gimenez, S., Tomlinson, B., Chan, P.K., Thomas, G.N., Forrat, R., Chambonneau, L., Deauvieau, F., Lang, J., and Guy, B. (2006). Innate and adaptive cellular immunity in flavivirus-naive human recipients of a live-attenuated dengue serotype 3 vaccine produced in Vero cells (VDV3). Vaccine *24*, 4914–4926.

Schlesinger, J.J., Brandriss, M.W., and Walsh, E.E. (1987). Protection of mice against dengue 2 virus encephalitis by immunization with the dengue 2 virus non-structural glycoprotein NS1. J. Gen. Virol. *68*, 853–857.

Schlesinger, W., and Frankel, J.W. (1952). Adaptation of the New Guinea B strain of dengue virus to suckling and to adult swiss mice; a study in viral variation. Am. J. Trop. Med. Hyg. *1*, 66–77.

Schul, W., Liu, W., Xu, H.Y., Flamand, M., and Vasudevan, S.G. (2007). A dengue fever viremia model in mice shows reduction in viral replication and suppression of the inflammatory response after treatment with antiviral drugs. J. Infect. Dis. *195*, 665–674.

Shresta, S., Kyle, J.L., Beatty, P.R., and Harris, E. (2004a). Early activation of natural killer and B-cells in response to primary dengue virus infection in A/J mice. Virology *319*, 262–273.

Shresta, S., Kyle, J.L., Snider, H.M., Basavapatna, M., Beatty, P.R., and Harris, E. (2004b). Interferon-dependent immunity is essential for resistance to primary dengue virus infection in mice, whereas T- and B-cell-dependent immunity are less critical. J. Virol. *78*, 2701–2710.

Shresta, S., Sharar, K.L., Prigozhin, D.M., Beatty, P.R., and Harris, E. (2006). A murine model for dengue virus-induced lethal disease with increased vascular permeability. J. Virol. *80*, 10208–10217.

Shresta, S., Sharar, K.L., Prigozhin, D.M., Snider, H.M., Beatty, P.R., and Harris, E. (2005). Critical roles for both STAT1-dependent and STAT1-independent pathways in the control of primary dengue virus infection in mice. J. Immunol. *175*, 3946–3954.

Simmons, C.P., Popper, S., Dolocek, C., Chau, T.N., Griffiths, M., Dung, N.T., Long, T.H., Hoang, D.M., Chau, N.V., Thao le, T.T., Hien, T.T., Relman, D.A., and Farrar, J. (2007). Patterns of host genome-wide gene transcript abundance in the peripheral blood of patients with acute dengue hemorrhagic fever. J. Infect. Dis. *195*, 1097–1107.

Stein, D.A., Huang, C.Y., Silengo, S., Amantana, A., Crumley, S., Blouch, R.E., Iversen, P.L., and Kinney, R.M. (2008). Treatment of AG129 mice with antisense morpholino oligomers increases survival time following challenge with dengue 2 virus. J. Antimicrob. Chemother. *62*, 555–565.

Styer, L.M., Kent, K.A., Albright, R.G., Bennett, C.J., Kramer, L.D., and Bernard, K.A. (2007). Mosquitoes inoculate high doses of West Nile virus as they probe and feed on live hosts. PLoS Pathog. *3*, 1262–1270.

Sun, D.S., King, C.C., Huang, H.S., Shih, Y.L., Lee, C.C., Tsai, W.J., Yu, C.C., and Chang, H.H. (2007). Antiplatelet autoantibodies elicited by dengue virus non-structural protein 1 cause thrombocytopenia and mortality in mice. J. Thromb. Haemost. *5*, 2291–2299.

Ubol, S., Masrinoul, P., Chaijaruwanich, J., Kalayanarooj, S., Charoensirisuthikul, T., and Kasisith, J. (2008). Differences in global gene expression in peripheral blood mononuclear cells indicate a significant role of the innate responses in progression of dengue fever but not dengue hemorrhagic fever. J. Infect. Dis. *197*, 1459–1467.

van Der Most, R.G., Murali-Krishna, K., Ahmed, R., and Strauss, J.H. (2000). Chimeric yellow fever/dengue virus as a candidate dengue vaccine: quantitation of the dengue virus-specific CD8 T-cell response. J. Virol. *74*, 8094–8101.

Vanlandingham, D.L., Schneider, B.S., Klingler, K., Fair, J., Beasley, D., Huang, J., Hamilton, P., and Higgs, S. (2004). Real-time reverse transcriptase-polymerase chain reaction quantification of West Nile virus transmitted by *Culex pipiens quinquefasciatus*. Am. J. Trop. Med. Hyg. *71*, 120–123.

Vaughn, D.W., Green, S., Kalayanarooj, S., Innis, B.L., Nimmannitya, S., Suntayakorn, S., Endy, T.P., Raengsakulrach, B., Rothman, A.L., Ennis, F.A., and Nisalak, A. (2000). Dengue viremia titer, antibody response pattern, and virus serotype correlate with disease severity. J. Infect. Dis. *181*, 2–9.

Wearing, H.J., and Rohani, P. (2006). Ecological and immunological determinants of dengue epidemics. Proc. Natl. Acad. Sci. U.S.A. *103*, 11802–11807.

Whitby, K., Pierson, T.C., Geiss, B., Lane, K., Engle, M., Zhou, Y., Doms, R.W., and Diamond, M.S. (2005). Castanospermine, a potent inhibitor of dengue virus infection *in vitro* and *in vivo*. J. Virol. *79*, 8698–8706.

White, L.J., Parsons, M.M., Whitmore, A.C., Williams, B.M., de Silva, A., and Johnston, R.E. (2007). An immunogenic and protective alphavirus replicon particle-based dengue vaccine overcomes maternal antibody interference in weanling mice. J. Virol. *81*, 10329–10339.

Whitehead, S.S., Blaney, J.E., Durbin, A.P., and Murphy, B.R. (2007). Prospects for a dengue virus vaccine. Nat. Rev. Microbiol. *5*, 518–528.

Whitehead, S.S., Hanley, K.A., Blaney, J.E. J., Gilmore, L.E., Elkins, W.R., and Murphy, B.R. (2003). Substitution of the structural genes of dengue virus type 4 with those of type 2 results in chimeric vaccine candidates which are attenuated for mosquitoes, mice, and rhesus monkeys. Vaccine *21*, 4307–4316.

Wu, S.F., Liao, C.L., Lin, Y.L., Yeh, C.T., Chen, L.K., Huang, Y.F., Chou, H.Y., Huang, J.L., Shaio, M.F., and Sytwu, H.K. (2003). Evaluation of protective efficacy and immune mechanisms of using a non-structural protein NS1 in DNA vaccine against dengue 2 virus in mice. Vaccine *21*, 3919–3929.

Wu, S.J., Grouard-Vogel G., Sun, W., Mascola, J.R., Brachtel, E., Putvatana, R., Louder, M.K., Filgueira, L., Marovich, M.A., Wong, H.K., Blauvelt, A., Murphy, G.S., Robb, M.L., Innes, B.L., Birx, D.L., Hayes, C.G., and Frankel, S.S. (2000). Human skin Langerhans cells are targets of dengue virus infection. Nat. Med. *6*, 816–820.

Wu, S.J., Hayes, C.G., Dubois, D.R., Windheuser, M.G., Kang, Y.H., Watts, D.M., and Sieckmann, D.G. (1995). Evaluation of the severe combined immunodeficient (SCID) mouse as an animal model for dengue viral infection. Am. J. Trop. Med. Hyg. *52*, 468–476.

Yauch, L.E., and Shresta, S. (2008). Mouse models of dengue virus infection and disease. Antiviral Res. *80*, 87–93.

Modulation of the Antiviral Response by Dengue Virus

7

Jorge L. Muñoz-Jordán and Irene Bosch

Abstract

Dengue virus (DENV) produces a wide range of human illness, ranging from asymptomatic infections to haemorrhagic and potentially fatal disease. Severe disease is associated with high viraemia, immune enhancement of sequential infections, and exacerbated inflammatory response. DENV is sensed in mammalian cells by endosomal and cytoplasmic receptors and stimulates the type-1 interferon (IFNα/β) response. Secreted IFNα/β stimulates JAK/STAT signalling, which results in the activation of IFNα/β-stimulated genes that lead the infected cells towards the establishment of an antiviral response. Genomic technology has enabled the identification of a remarkable list of genes induced in human host cells in response to DENV infection. The results define antiviral and pro-inflammatory responses mainly composed of IFNα/β- induced genes, which likely participate in the regulation of the immune response and vascular leakage during acute illness. DENV counteracts the IFNα/β response of the host. The evidence indicates that non-structural proteins of DENV weaken IFNα/β signalling, causing reduced activation of IFNα/β-induced genes. The increased virus uptake, weakened host cell defence, and unrestrained inflammatory response likely predispose patients to develop severe illness. The unveiling of these virus–host interactions leads to a better understanding of dengue pathogenesis, and to innovative diagnostic and therapeutic approaches.

Introduction

Dengue has re-emerged as a global health threat, particularly among children in the tropics of Asia as well as in young adults in the Americas. This mosquito-borne flavivirus causes an estimated 50 million infections annually (World Health Organization, 2004). Most DENV infections result in an illness known as dengue fever (DF). Among these patients, less than 0.1% develop dengue haemorrhagic fever (DHF), a vascular leakage syndrome that can lead to dengue shock syndrome (DSS) and death (Gubler, 2006). Complete understanding of dengue pathogenesis and progress in developing therapeutic treatments against DENV have been hindered by the lack of biological models that mimic disease progression (but see Chapter 6) and by limited information about gene function related to pathogenesis. In recent years, we have gained valuable information from biological assays that have been used for the study of antiviral responses against DENV, and revealed previously unknown mechanisms of host–virus interactions. However validation of experimental data is often obstructed by lack of animal systems or the difficulties inherent in clinical studies.

Type-1 Interferon (IFNα/β) is made and secreted by mammalian cells rapidly after viral infection, stimulating the onset of an immediate antiviral response in the infected and surrounding cells. The recognition of the IFNα/β network as a fundamental mechanism for the establishment of the antiviral response, and the unveiling of a plethora of mechanisms used by different viruses to counteract this response, have made significant impacts in our understanding of viral pathogenesis (Garcia-Sastre, 2002; Garcia-Sastre and Biron, 2006; Pitha and Kunzi, 2007). This elaborate interplay between virus and host, known as interferon modulation, is the subject of increasing interest. The existence of numerous sensing mechanisms with different degrees of specificity for viral components ensures that mammalian cells are prepared to recognize viruses that greatly differ in their structural and functional features. The capacity of this array of sensing mechanisms to recognize pathogens is further enhanced by an assortment of interferon-related pathways activated during viral infection, resulting in the establishment a complex antiviral response (Onomoto *et al.*, 2007; Severa and Fitzgerald, 2007).

DENV appears to be recognized by most of the currently known host cell sensing mechanisms and activates a strong antiviral response in the host led by IFNα/β production. The impairment of IFNα/β signalling in mice (IFNAR$^{-/-}$ IFNGR$^{-/-}$ mice) resulted in high lethality from DENV challenge, demonstrating the importance of this antiviral mechanism in fighting DENV infection (Johnson and Roehrig, 1999). Genomic technology has been increasingly utilized to define genes induced by human host cells in response to infection. Responses of vascular endothelial cells, monocytes, B-cells and dendritic cells have been investigated *in vitro* using gene array technology, and activation of key components of the antiviral response have been monitored in patients (Fink *et al.*, 2007; Simmons *et al.*, 2007; Ubol *et al.*, 2008; Warke *et al.*, 2003). These gene expression arrays define a common response composed of genes involved in the IFNα/β response. Vascular-specific responses to DENV include genes involved in the increased endothelial cell proliferation and angiogenesis response, wound healing, cell adhesion changes, T-cell inhibition and complement activation. These processes probably participate in the DENV-induced vascular leakage or regulation of the immune response during the acute phase of the disease, and studies discussed in this chapter indicate that key players of the host antiviral response are likely to participate in these pathogenic processes during DENV infection. While the intricate IFNα/β response associated

to DENV infection is unveiled, the mechanisms encoded by the virus to counteract this response have also emerged as a significant area of research. Biological assays have shown that although DENV stimulates a strong IFNα/β response, the virus inhibits IFNα/β signalling (Jones *et al.*, 2005; Munoz-Jordan *et al.*, 2003). The understanding of this complex interplay between virus and host may result in better understanding dengue pathogenesis and help advance the development of antiviral approaches.

Activation of the interferon response by DENV

Viral recognition by the host occurs immediately upon entry of the viral particle to the host cells; and these sensing mechanisms are pivotal for the development of the antiviral response. Fig. 7.1 outlines the complex process of DENV sensing by the host cell, salient features of the antiviral response generated during DENV infection, and ways in which the this virus antagonizes this antiviral response. Host cells sense DENV by the two main families of pathogen recognition receptors: the extracellular/endosomal toll-like receptors (TLRs) (Akira and Takeda, 2004; Bowie and Haga, 2005) and the cytoplasmic receptor family of DExD/H box RNA helicases (e.g. RIG-I: retinoic acid inducible gene I, and MDA5: melanoma differentiation-associated gene-5) (Meylan and Tschopp, 2006). TLRs are critically important in the sensing of viral nucleic acid. The endosomal side of TLR3, 7, 8 and 9 is implicated in detecting viral nucleic acids. A structural motif present in intracellular regions of all TLR molecules and interleukin 1 (IL-1) receptor, known as TLRs/IL-1 resistance (TIR) domain, is involved in the recruitment of adaptor molecules. This recognition leads to activation of two major families of transcription factors: the interferon regulatory factors (IRFs), and the nucleic factor-kappa B (NF-kB). This dual signalling cascade activates production of IFNα/β and inflammatory cytokines, leading to maturation of dendritic cells and the establishment of an antiviral response (Severa and Fitzgerald, 2007). Double-stranded RNA (dsRNA) is commonly a by-product of viral replication and transcription generated in host cells infected by many ssRNA viruses including DENV. TLR3, a broad sensor of

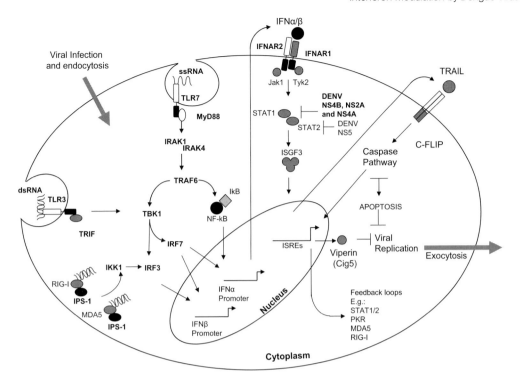

Figure 7.1 Modulation of IFNα/β in the host cell. TLR7 recognition of DENV RNA occurs in endosomal compartments during infection, leading to activation of NF-κB, IFNα/β, IRF7 and IRF3 and activation of IFNα and IFNβ. TLR7/8/9 activation is mediated by MyD88, resulting in downstream activation of IRF7, IKKα/β/γ and MAPK cascades and activation of NF-κB and AP-1. TLR3 activation through TRIF leads to activation of IFNβ. RIG-I and MDA5 activation results in stimulation IRF3, which in turn induces the IFNβ promoter. Secreted IFNα/β activates the IFNα/β receptor and the JAK/STAT pathway, leading to phosphorylation and dimerization of STAT1/2 and formation of the macromolecular factor ISGF3, which translocates to the nucleus and activates ISREs. The nonstructural proteins NS4B, NS2A and NS4A block STAT1 phosphorylation. NS5 binds and facilitates STAT2 degradation. These interactions impair parts of the JAK/STAT pathway and reduce activation of ISREs. Two proteins identified as highly up-regulated by experimental dengue infections and gene expression analysis, TRIAL and Viperin, are represented as part of the IFN response. It is predicted that the IFN antagonistic effect of DENV proteins will result in setbacks of the antiviral response.

dsRNA, resides in multivescicular bodies situated along the endocytic trafficking pathway of the endoplasmic reticulum in dendritic cells (Yoneyama et al., 2005). Flavivirus-derived dsRNA molecules likely activate TLR3. A role for TLR3 in viral pathogenesis of West Nile virus and DENV has been documented. West Nile virus infection leads to a breakdown of the blood-brain barrier, leading to enhanced infection of central nervous tissue in wild-type mice; whereas TLR3-deficient mice have increased viraemia in peripheral nervous tissue and reduced viraemia in the brain (Wang et al., 2004). Therefore, TLR3-mediated response may in fact play a significant role in the activation of the exacerbated inflammatory response of the host against flaviviruses. Up-regulation of TLR3

by DENV probably plays a role in regulation of the inflammatory response developed during infection (Warke et al., 2003). Activation of TLR7 found mainly in endosomal compartments of dendritic cells should then occur during DENV infection, causing the release of high levels of IFNα/β to the serum. DENV ssRNA is also sensed by the host cell. This recognition occurs through direct interactions between TLR7 and the viral RNA. These interactions have been documented in dendritic cells, and are coupled to the viral fusion and uncoating processes that take place in endosomal compartments during early infection of dendritic cells, a process that is enhanced by higher-order structures of the viral RNA, leading to a strong IFNα/β response (Diebold et al., 2004; Lund et al., 2004; Wang et al., 2006). In fact, it is known that DENV activates dendritic cells and induces robust IFNα/β and TNFα production as well as a strong pro-inflammatory response (Libraty et al., 2001). Although TLR8/9 activation has not yet been formally implicated in sensing DENV infection, it is known that yellow fever virus activates these receptors in dendritic cells, leading to IFNα/β and proinflammatory cytokine responses (Querec et al., 2006). In addition, DENV induces IFNα/β expression through the involvement of RIG-I-dependent IRF-3 and PI3K-dependent NF-kB activation, an induction that is likely to occur through TLR7/8/9-mediated recognition of viral RNA (Chang et al., 2006; Loo et al., 2008). Then, dendritic cells, natural targets of DENV infection constitutively expressing TLR7 and IRF7, rapidly mobilize a systemic IFNα/β response.

The involvement of RIG-I and MDA5 in the sensing of DENV infection has been recognized through gene expression analysis (Ramirez-Ortiz et al., 2006). These two proteins share similar signalling features and structural homology. Recognition of the viral RNA occurs through interactions between their caspase activation and recruitment domain (CARD) and the IPS-1/MAVS/CARDIF adaptor protein in association with the outer mitochondrial membrane. While RIG-I recognizes RNA secondary structures and ssRNA 5′-triphosphate ends, MDA5 recognizes annealing inosine- and cytosine-containing RNA strands. These interactions result in activation of a macromolecular signalling complex that stimulates IRF3. In turn, this activation induces the IFNβ promoter. Recent experiments using single and double RIG-I/MDA5 knockout mouse fibroblasts show that DENV triggers both of these responses (Chang et al., 2006; Loo et al., 2008). In differential display reverse transcription polymerase chain reaction (RT-PCR) and oligonucleotide microarrays of human umbilical vein endothelial cells (HUVEC) have shown that MDA5 was defined as a salient, differentially regulated gene during DENV infections (Warke et al., 2003). MDA5 becomes highly expressed during the first 24 h after infection, together with RIG-I and other IFNα/β induced helicases. This up-regulation can also be observed in monocytes, B-cells and dendritic cells in response to DENV

as well as other RNA viruses (Ramirez-Ortiz et al., 2006).

Induction of antiviral response

IFNα/β induction

DENV induces a strong IFNα/β response in natural infections, and IFNα/β is present at high levels for long periods of time in paediatric patients after defervescence (Kurane et al., 1993). The impairment of IFNα/β signalling in mice (IFNAR$^{-/-}$ IFNGR$^{-/-}$ mice), resulting in high lethality and consistent DENV infection in serum and tissue, demonstrated the essential role of IFNα/β in protecting against DENV infections (Johnson and Roehrig, 1999). In addition, the use of STAT1(-/-) mice bearing two different mutant STAT1 alleles demonstrated that such IFNAR-dependent control of DENV infections involved both STAT1-dependent and STAT1-independent mechanisms (Shresta et al., 2005). The STAT1-dependent pathway is likely involved in the control of initial steps of infection (Shresta et al., 2004), whereas the STAT1-independent pathway involves later antiviral mechanisms that control virus spread and prevent disease, including early activation of B and NK cells as well as up-regulation of MHC class I molecules on macrophages and dendritic cells (Shresta et al., 2005). Global gene expression profiling studies have revealed up-regulation of key mediators of the inflammatory cytokine response, the antiviral response through IFNα/β activation, the NF-κB-mediated cytokine/chemokine responses and the ubiquitin proteosome pathway (Fink et al 2007; Sariol et al., 2007; Warke et al., 2003). Ubol et al. (2008) has also shown IFN activation through gene expression analysis of acute phase PMBCs.

By using differential display RT-PCR (DD-RTPCR), quantitative RT-PCR, and Affymetrix oligonucleotide microarrays in HUVECs infected with DENV, eight differentially expressed cDNAs were identified: the inhibitor of apoptosis-1 (AIP-1), 2′–5′ oligoadenylate synthetase (OAS), a 2′–5′ OAS-like (OASL) gene, galectin-9, myxovirus protein A (MxA), regulator of G-protein signalling (RGS), endothelial and smooth muscle cell-derived neuropilin-like protein, and phospholipid scramblase 1 (PLSCR-1). Moreover, this analysis

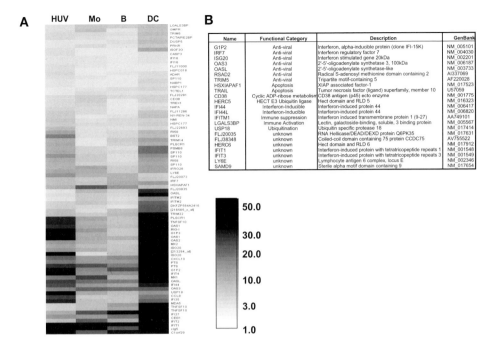

Figure 7.2 Global gene expression in cells infected with DENV. **A** Expression levels of dengue common response signature (79 genes) in endothelial cells (HUVEC), monocytes (Mo), B-cells (B), and dendritic cells (DC) *in vitro*. Gene expression levels were normalized to each of the mock-infected cells. The fold induction on gene expression represents the result of the Affymetrix analysis conducted for each probe set. Normalized expression levels are represented according to the grey-scale key shown. The fold change ranges from 1.0 to 50. Only gene up-regulation is shown. Affymetrix GeneChip U133A was utilized and data obtained were analysed using GeneSpring software (Agilent). Once the fold increased was obtained, Anova analysis was conducted to generate a list of genes that were related to each other at 95% confidence of statistical significance. **B** Gene expression signature derived from the common response in endothelial (HUVEC), B Cells (B), Monocytes (Mo)and dendritic cells (DC) exposed to DENV. The 23 gene IDs, known function and identifications in the Gene Bank are shown. These 23 genes were up-regulated in DENV infections *in vitro*, and confirmed by qRT-PCR.

revealed an additional 269 gene identities that were up-regulated after DENV infection (Warke *et al.*, 2003). Fig. 7.2A shows levels of expression of 79 gene identities (67 genes) found by this Affymetrix microarray in HUVECs, monocytes, B-cells and dendritic cells infected with DENV *in vitro*. From these genes, 23 are shown in Fig. 7.2B. Collectively, the aforementioned microarray analyses have provided a draft of what appears to be some fundamental features of the antiviral response against DENV infection. The conserved features of the IFNα/β and pro-inflammatory responses observed between the four different human cell types and their similarities with the response seen *in vivo*, underscore the importance of the innate immune response during DENV infection.

Gene expression profiles and disease severity

In the absence of animal models, gene expression arrays offer a direct approach to study gene expression during DENV infection, with the potential to become platforms for the discovery of pathogenic responses that would correlate with severity of disease. These studies have revealed an antiviral response and identified differentially regulated genes that could serve as potential markers for occurrence of mild and severe dengue disease (Fink *et al.*, 2007; Simmons *et al.*, 2007; Ubol *et al.*, 2008). This list of potential factors involved in disease outcome shows a DENV response tightly associated with the IFN response (Table 7.1). The chemokines IP-10 and I-TAC (both ligands of the CXCR3 receptor) have also been identi-

Table 7.1 Comparison of gene expression studies using blood cells from acute dengue infections

Gene expression array	Up-regulated genes in DF	Down-regulated genes in DHF	Down-regulated genes in DSS
Sigma-Genosys cDNA array (Fink *et al.*, 2007)	B2M, CBL, **G1P3**, HERC5, Hdm2, IFIH1, **IFI44, IP10 (CXCL10),** IRF9, **ISG15**, I-TAC, MKP-1, **Mx1**, NFKBIA, NFKBIB,**OASL, OAS1**, OAS2, OAS3, PAI1, PSMB9, **RIG-I**, SOCS1, STAT1, TNF, TNFAIP3, UBE1C, UBE2I, USP15, **USP18**, VEGF, **Viperin**		
Amersham GE (Ubol *et al*, 2008)	CD59, CTSL, **CXCL10 (IP10),** CLEC4E, FcGP1A, **GBP1, IFIH1,** IFIT1,IFNα, **ISG12 (IFI27), IFI44, ISG15,** LY6E, **mda5, MX1,** SECTM1, SERPING1 27, **TNFRSF17**, XBP1	CD97, CXCL, 1CXCL7, Factor D complement (DF), IFNγ, **IL1β, IL–8**, IL-10, MAL, MIP–1β, NK4, properdin (PFC), RANTES, TNFα	
cDNA array (Simmons *et al.*, 2007)	CCL5 and TNFSF13B		Annexin A1, **MX1**, IL1RN, IL13, ISG15
Illumina/RT-PCR (C. Simmons, personal communication)	**CCL2, CCL8 (MCP-2), CXCL10 (IP-10), GBP1, G1P3, IDO, IRF-7, LAMP-3, OASL, PLSCR1, STAT1, TRAIL (TNFSF10), USP18**		GBP1, GBP4, IFITM1, IFIT5, IFIT3, IRF7, JAK2, **MX1**, MX2, OASL, **OAS1**, OAS2, OAS3, SOCS1, STAT1, **USP18**
Affymetrix (Bosch *et al*, 2008)	APOBEC1L, BCL2, **CCL2, CCL8 (MCP-2),** CD38, **CXCL10 (IP-10),**G1P3, IFI44, IDO, **IRF-7, ISG12 (IFI27), ISG15, MX1, OAS1, OASL, PLSCR1, RIG-I, STAT1**, TGFBR2, **TNFSF17, TRAIL (TNFSF10),** TS, **USP18, Viperin**	BCL2A1, CCL2, CCL3, CXCL10 (IP-10), CXCCL12, IDO, **IL1β, IL8**, ISG15, LGalS2, LGalS3BP, **OAS1,** PLSCR1, PDL2, RIG-I, TRAIL, **USP18**, WARS	

Up-regulated genes are shown for patients with dengue fever (DF) with respect to patients with another febrile illness. Down-regulated genes are shown for dengue haemorrhagic fever (DHF) or dengue shock syndrome (DSS). The genes that are found across different samples and platforms of analysis are indicated in bold for DF cases or DHF/DSS cases respectively. Genes that were found to be up-regulated in DF and down-regulated in DHF/DSS using the same platform are underlined. The gene codes correspond to full names and gene bank accession numbers located at http://www.ncbi.nlm.nih.gov. Generally, patients show a type I interferon response which was less prominent or inhibited in severe cases. The list of genes may represent new potential markers of disease progression and severity to be further explored in clinical studies.

fied in microarray studies of DENV-infected human cells (Fink *et al.,* 2007) along with CCL8 or MCP-1 (Bosch *et al,* 2008 unpublished data) (Table 7.1). Since these secreted proteins could be detected early after the onset of fever in dengue patients, they may have use in early prognosis.

Among the genes found by analysing differential gene expression profiles is Viperin (Cig5), an IFN-induced gene that reduces viral release for other viruses (Jiang et al., 2008; Wang et al., 2007) and is highly upregulated during DENV infections in rhesus monkeys and humans (Fink et al., 2007; Sariol et al., 2007; Warke et al., 2003). In a recent global gene expression array analysis (Amersham/GE) of PBMCs from 30 patients with DF or DHF, 47% of all altered genes were part of the IFN response, and the majority of the IFN-induced genes were strongly up-regulated in DF compared with the DHF patients, suggesting a significant role of the IFN system during DENV infection (Ubol et al., 2008). Robust production of IFNα/β in PBMCs collected from DF patients was confirmed by ELISA (Ubol et al., 2008). A similar tendency was also found in DF cases compared to DHF cases in a study in Vietnam utilizing Illumina arrays (Cameron Simmons, personal communication). When mRNA levels of paired samples from acutely infected DF and DHF patients were compared using qRT-PCR, a dynamic range in the expression levels of the following genes was observed: CCL8 (MCP-2), CCL2, IP-10 (CXCL10), IDO, IRF-7, LAMP-3, TRAIL (TNFSF10), USP18, PLSCR1, G1P3. Remarkably, a large increase in expression of genes involved in immune response was detected in DF vs. DHF by independent laboratories (Ubol et al, 2008; Cameron Simmons, personal communication). Moreover, Vietnamese children with DSS had substantially lower transcriptional activity of multiple IFN-stimulated genes (ISGs) than did children presenting with DHF without shock (Simmons et al., 2007). We too have observed an overall decrease in the IFN-stimulated genes in DHF compared to DF cases over the acute phase of disease using qRT-PCR arrays (data not shown). Another common finding in gene expression profiles in whole blood or PBMCs is the elevated expression of IFIT2 in DSS patients and CCL5 and TNFSF13B in DF patients (Simmons et al., 2007; Ubol et al., 2008). In the Ubol et al. (2008) study, IFN-inducible genes that were up-regulated in patients with DF included GBP1, IFI27, IFI44, IFIH1/MDA5, IFIT1, ISG15, MX1, and CXCL10/IP10. In contrast to that finding, GBP1 and MX1 were associated with DSS cases (Simmons et al., 2007).

These findings could indicate the exhaustion of immune cells in the more severe condition of the disease or the result of interferon antagonism by viral components during infection (discussed further below).

Among genes that were specific to DENV in HUVECS and not up-regulated by other flaviviruses tested in parallel to DENV, were ST2, a homologous gene of the IL-1 receptor which binds IL-33 and indoleamine 2,3-dioxygenase or IDO. To further evaluate the specific response during DENV infection, a comparative study between DF patients and patients with other febrile illnesses was performed (Becerra et al., 2008). The results indicate that ST2 and IDO were significantly increased in patients with DF with respect to patients with other febrile illnesses (Table 7.2). Dynamic changes in gene expression can occur over the course of illness, which can be observed by qRT-PCR profiles and by Illumina chip hybridization. Therefore, comparisons of gene expression profiles between samples and platforms need to be carefully done. Due to the lack of post-febrile illness gene expression data, our understanding of the antiviral response is solely based on comparisons during acute illness, and we hope to overcome this limitation by conducting prospective clinical studies. As we move forward in this field with larger sample numbers, collected early and across a breadth of clinical disease states, we will learn more about the dynamic interaction between the host and the virus. Future studies should attempt multi-parameter analysis of host gene expression data by bringing in information on the infecting virus via viral genomic sequence analysis or biological characterization. Furthermore, gene expression studies have to date been descriptive in nature. Attempts to identify and prospectively test possible prognostic biomarkers for severe disease should be encouraged.

Despite the differences in the systems used to assess levels of gene expression, results of these studies remarkably converge on several common genes. Strong similarities were detected mainly in the broadly prominent 'immune response' of patients in the febrile stage of a dengue illness. Chief amongst these are: (a) transcripts from IFNα/β inducible genes and signalling molecules of the IFNα/β pathways, (b) transcripts from

Table 7.2 DENV-specific gene expression profile

Genes	CD8 acute	B acute	Mo Acute	CD4 acute	DF (PBMC)	DHF1 (PBMC)	DHF2 (PBMC)
IF127	52.8	32221.5	31494.0	22049.4	0.6	1.3	0.7
IF144	3.2	14.0	90.0	24.7	0.1	0.1	0.2
IF1T1	10.4	93.0	173.0	22.4	0.1	0.1	0.2
IL1RL/ST2	5.9	5.1	1.5	5.0	6.6	4.9	3.8
ISG20	0.9	2.3	82.8	9.4	0.3	0.6	0.3
IDO	637.3	9949.4	12650.0		303.0	932.0	482.0
IRF7	2.4	9.2	8.4	18890.5	0.3	0.4	0.3

Quantitative RT-PCR array shows the levels of expression of RNA obtained from CD8 T-cells, B-cells, monocytes and CD4 T-cells from patients with acute febrile illness. StemCell Technology negative selection kits were used to derive each of these cell types from anticoagulated blood from a febrile patient. The genes shown here were prominently expressed in dengue acute infections normalized to cells from uninfected individuals. PMBC RNA expression levels from one DF and two DHF patients were normalized to RNA expression levels from PBMCs obtained from patients with a febrile illness and confirmed negative for DENV infection or seroconversion. From that normalization, two genes, ILRL2 (ST2) and IDO (highlighted in grey) show specific up-regulation.

select chemokines (IP-10, MCP-1 and MCP-2 in particular), (c) protein ubiquination and (d) cell survival and apoptotic signalling pathways. The possible roles of these gene products in controlling the spread of the disease and the inflammatory response are currently under study. In addition, possible correlates between these up-regulated gene products and disease severity are being investigated. Table 7.2 summarizes the genes that were confirmed to be differentially regulated by qRT-PCR after their detection on the different platforms summarized in Table 7.1 (Bosch *et al.*, unpublished data). Some of these genes are further discussed below.

Tumour necrosis factor-related apoptosis-inducing ligand (TRAIL)

From the dengue common response signature, characterized by the IFNα/β signalling network, one gene, TRAIL, has been identified as a central player in the antiviral response and can be proposed as a potential common linker of the IFNα/β inducible genes. TRAIL is a member of the TNF family that is specifically involved in neuroprotection and growth proliferation in non-cancer cells (Hoffmann *et al.*, 2007; Rimondi *et al.*, 2006; Secchiero *et al.*, 2003; Zauli *et al.*,

2008) and may have a protective physiological role as an antibacterial agent (Hoffmann *et al.*, 2007). TRAIL appears to promote apoptosis in cancer cells by activating the death receptors DR4 and DR5 (Kim and Gupta, 2000) and also to negatively regulate the innate immune response independent of apoptosis (Diehl *et al.*, 2004). *In vitro* and *in vivo* studies have demonstrated tumoricidal and anti-viral activity of TRAIL without significant toxicity towards normal cells or tissues (Sato *et al.*, 2005). IFNα/β enhances expression of TRAIL, while, on the other hand, TRAIL treatment can enhance expression of IFN-inducible genes like IFITM1, IFIT1, STAT1, LGAl3BP, PRKR as well as the IFN-β promoter (Kumar-Sinha *et al.*, 2002). The down-regulation of MCP-2 and IP-10 by TRAIL has been reported in HUVECs, resulting in the reduction of the pro-inflammatory response (Secchiero *et al.*, 2005).

The molecular cross-talk and functional synergy between TRAIL and the IFNα/β pathways may have implications in the physiological role and mechanism of action of TRAIL during infection. Recombinant TRAIL (rTRAIL) treatment strongly inhibited DENV and West Nile Virus replication and DENV antigen levels in DENV-infected dendritic cells by an apoptosis-independent mechanism (Warke *et al.*, 2008). Furthermore, rTRAIL treatment of DENV-infected dendritic cells inhibited the expression of pro-inflammatory cytokines and chemokines produced by infected dendritic cells (IL-6, TNFα, MCP-2, IP-10, and MIP-1β) (unpublished data). These data suggest that TRAIL plays a dual, beneficial role both as an anti-viral and a pro-inflammatory cytokine suppressor during DENV infection. Further investigation of TRAIL as an anti-inflammatory protein in the context of viral infections will help define its role in controlling the pro-inflammatory response triggered after DENV infection.

Interleukin-1 receptor-like 1 precursor (IL1RL1)-ST2 gene

IL-1RL1/ST2 is a member of the interleukin-1 receptor (IL-1R) family of proteins. Alternative splicing of the gene generates three mRNAs, corresponding to a longer membrane-anchored form (ST2L), a shorter released form (sST2) and a membrane bound variant form (ST2V) (Bergers *et al.*, 1994; Tominaga *et al.*, 1999; Yanagisawa *et al.*, 1993). ST2L is selectively expressed on the surface of T-helper 2 (Th2)-polarized T lymphocytes and mast cells and has been described as an activation marker for Th2 cells (Moritz *et al.*, 1998; Xu *et al.*, 1998). It has been shown that sST2 can inhibit IL-1R and TLR4 signalling through the sequestration of MyD88 and Mal proteins (Sweet *et al.*, 2001). Pro-inflammatory stimuli, including LPS and cytokines, induce the expression of sST2 in human and mouse *in vitro* models (Kumar *et al.*, 1997; Oshikawa *et al.*, 2002; Tajima *et al.*, 2003). The administration of sST2 or ST2-Fc fusion protein is able to suppress the production of pro-inflammatory cytokines *in vitro* and *in vivo* (Gadina and Jefferies, 2007; Sweet *et al.*, 2001) and attenuate the inflammatory response *in vivo*. Elevated levels of sST2 have been found in diseases including Th2-associated inflammatory disorders, autoimmune diseases, asthma, sepsis, and myocardial infarction (Brown *et al.*, 2007; Brunner *et al.*, 2004; Kuroiwa *et al.*, 2001; Oshikawa *et al.*, 2001; Sanada *et al.*, 2007; Shimpo *et al.*, 2004; Weinberg *et al.*, 2002). The ST2L molecule is part of a receptor complex for the cytokine IL-33 (Chackerian *et al.*, 2007;

Schmitz *et al.*, 2005). A dual role has been suggested for IL-33: as a nuclear factor with transcriptional regulation activity and as a pro-inflammatory cytokine (Carriere *et al.*, 2007). In a mouse model of cardiac disease, the beneficial anti-hypertrophic effect of IL-33 is blocked by sST2, suggesting that this protein could be acting as a decoy receptor (Sanada *et al.*, 2007). ST2 and IL-33 binding recruits MYD88, IRAK1, IRAK4, and TRAF6, followed by phosphorylation of MAPK kinases (Carriere *et al.*, 2007).

In a small cohort of patients, mostly classified as DF, sST2 levels in serum from DENV infected patients were found to be higher than in patients with other febrile illnesses; levels of sST2 were also higher in secondary infections compared to primary infections (Becerra *et al.*, 2008). The increased levels of sST2 were observed at the late febrile stage and especially at defervescence. Levels of sST2 in dengue patients also correlated with other parameters associated with disease severity. These results prompted us to propose that serum levels of the sST2 protein could indicate when the levels of circulating virus are dropping during a DENV infection. Serum sST2 levels could also be an indicator of the inflammatory response and, as suggested by others (Mathew *et al.*, 1999), low levels may correlate during exacerbated inflammatory response.

Specific endothelial response to DENV: Indoleamine 2,3-dioxygenase (IDO)

IDO was detected as a strongly induced gene after DENV infection of several primary human cells (Warke *et al*, 2003). When IDO expression levels were determined in PMBCs of patients infected with DENV, and compared to the levels of expression of PMBCs from patents with other febrile illnesses, IDO was one of the genes that appear to be more specific to dengue patients. Therefore we decided to study this enzyme in more detail *in vivo*. IDO is an enzyme ubiquitously distributed in mammalian tissues and cells including dendritic cells (Mellor and Munn, 2004) and T-cells (Fallarino *et al.*, 2003). It catalyses the initial and rate-limiting step in the catabolism of L-tryptophan along the kynurenine pathway (Takikawa *et al.*, 2003). *In vivo*, IDO activity in serum increases under pathological conditions such as toxoplasmosis (Daubener *et al.*, 2001), viral, and bacterial infections (Oberdorfer *et al.*, 2003; Obojes *et al.*, 2005), and allograft rejection (Mellor and Munn, 2004). IFNγ is a potent inducer of IDO expression (Musso *et al.*, 1994). IDO-expressing cells can inhibit T-cell proliferation and function by depleting L-tryptophan in the surrounding microenvironment. This process is termed 'immunosuppression by starvation' (Liu *et al.*, 2007). Further studies are needed to investigate whether tryptophan metabolites released by DENV infected endothelial cells are involved in inhibiting CD8 T-cells. We hypothesize that IDO is involved in the establishment of an immunosuppressive condition during DENV infection, as described by others (Mathew *et al.*, 1999).

Gene expression analysis showed that kynureninase in endothelial cells infected with DENV was elevated. This finding suggested that IDO might play a role in DENV induced pathophysiology. Using mass spectrometry methods, reduced levels of L-tryptophan and increased levels of kynurenine were found in DENV-infected patients compared to other febrile illnesses during the acute stage of the disease. This result supports the hypothesis of T and other immune cell inhibition during DENV infection.

Viperin (Cig5)

Several microarrays have shown that viperin is one of the most highly up-regulated genes following DENV infection (Fink *et al.*, 2007; Sariol *et al.*, 2007; Ubol *et al.*, 2008). Viperin is an IFNα/β-inducible protein known to be associated with the endoplasmic reticulum and redistributed to the Golgi apparatus and cytoplasmic vacuoles in response to human cytomegalovirus and hepatitis C infections (Chin and Cresswell, 2001; Helbig *et al.*, 2005). In addition to being up-regulated during DENV infections, overexpressing viperin in A549 cells suppresses DENV replication, indicating that viperin is directly involved in the anti-viral response to DENV. The subcellular localization of viperin during infection suggests a role in trapping the virus in endosomal compartments during viral replication and assembly, but the antiviral mechanisms in which viperin is involved need to be further studied.

Significance of the antiviral response in dengue pathogenesis

The presence of DENV is detected by the human host immediately upon its transmission from the mosquito vector. Langerhans cells and dendritic cells are the initial targets of DENV (Wu *et al.*, 2000). Dendritic cells located in the skin are likely to become activated and migrate to lymph nodes and spleen. The virus travels with serum proteins and immunoglobulins (IgMs and IgGs), forming immune complexes, which are bound by FcR II for uptake. The presence of immune complexes is more prominent in secondary DENV infections, where more severe forms of the disease are observed including more severe plasma leakage. Liver, spleen, and kidney tissues are infected. In the blood, monocytes, macrophages and endothelial cells are the major cell types in which DENV replicates, resulting in further expansion of the virus (Jessie *et al.*, 2004). Ultimately, the infection is systemic.

DF is usually characterized by a fever of 5–7 days, bone and joint ache, retro-orbital pain, nausea and fatigue. While most DF patients recover without intervention, 2–5% develops DHF/DSS, characterized by thrombocytopenia and vascular leakage, causing hypovolaemic shock and death if not promptly treated. Very few determinants of severe dengue have been identified. The theory of 'antibody-dependent enhancement' (ADE) proposes that pre-existing, heterologous, subneutralizing antibodies in complex with the newly infecting virus bind to the Fc receptors in macrophages, resulting in increased uptake and replication of the virus and contributing to the outcome of DHF/DSS associated with secondary infection (Halstead, 2003). The phenomenon of T-cell 'original antigenic sin' (OAS) predicts that the response to a secondary challenge is dominated by the proliferation of cross-reacting memory cells induced by the primary infection, which may be of lower affinity for the virus during secondary infections (Green and Rothman, 2006; Yenchitsomanus *et al.*, 1996). Microarray technology provides an unbiased approach for the study of the host response against DENV infection combined with high throughput quantitative RT-PCR methods in in-vitro models and in dengue patients. Differentially expressed genes have been identified along three major pathways; NF-κB-mediated immune response, the IFNα/β network and the ubiquitin proteosome pathway.

As discussed below, DENV inhibits IFNα/β signalling by suppressing JAK/STAT activation, resulting in reduced host antiviral response (Munoz-Jordan et al., 2005; Munoz-Jordan et al., 2003). Reduced host defence and viral clearance, exacerbated inflammatory response and increased uptake of the virus likely synergize, resulting in increased severity of disease.

Among the many IFN-related response genes identified by microarrays, only a few have been rigorously shown to be specific to DENV. So far, we have attempted to study relevant targets, like endothelium (HUVECs), as a rational design of studying such responses. We give examples of two genes, ST2 and 2,3 indoleamine dioxygenase (IDO) that would be more specific to DENV responses, as these two genes were selected between dengue patients and other febrile illness patients (OFI). The chemokines IP-10 and I-TAC, both ligands of the CXCR3 receptor are highly up-regulated genes in the peripheral blood of acutely ill patients (Fink et al., 2007). In addition, viperin (Cig5) is a highly up-regulated gene in the IFNα/β pathway during DENV infection, and its overexpression in A549 cells results in a significant reduction in viral replication (Fink et al., 2007). Of the 23 genes listed in the dengue specific response (Fig. 7.1), CS1 and MCP-2/CCL8 are known to be up regulated as a direct response to IFNα/β (Indraccolo et al., 2007). CS1 is the first component of the classical pathway of the complement system. Complement is a highly regulated system that plays a crucial role in host defence, but uncontrolled activation can lead to an anaphylactic shock. Several other genes encode for T and NK cell activator IL15 and its receptor (Tsunobuchi et al., 2000), cell adhesion and migration protein (CEACAM1–4L) (Lee et al., 2007), pro-inflammatory IL-1β converting enzyme (CASP1) (Basak et al., 2005) and extracellular matrix remodelling protein (LGMN) (Morita et al., 2007).

The secretion of cytokines and the pro-coagulant activity of platelets are part of immune activation. Sequestration of platelets may give rise to thrombocytopenia and consequently plasma leakage. The secretion of PDL2 by dendritic cells and PD-1 by memory T-cells, as well as the secretion of sST2 and activation of IDO and TRAIL are among the immune controlling events. Cell adhesion genes may be particularly relevant to dengue vascular pathology involving plasma leakage (Pober and Min, 2006). These genes help maintain the integrity of the endothelial cell layer and, at the same time, they play a role in the movement, chemotaxis, and transmigration of leucocytes across the endothelial cell layer (Lush and Kvietys, 2000). Among the genes encoding for chemokines and cytokines were CCL20, CCND1, and IL-6 and adhesion molecules, ICAM1 and VCAM1. Also, the vascular endothelial growth factor (VEGF) and angiopoietin 1 and T-cell regulators such as IDO and complement subunit 1 (CS1) were differentially regulated. IL-6 may be a key factor inducing dengue vascular pathology (Suharti et al., 2002); it helps increase endothelial permeability by altering actin filaments and changing endothelial cell shape (Maruo et al., 1992; Rachman and Rinaldi, 2006), similar to the action of VEGF (Bijuklic et al., 2007). But IL-6 has potent anti-inflammatory and protective properties including the ability to inhibit the production of TNF-α, a cytokine known to induce vascular permeability (Bijuklic et al., 2007; Clark, 2007; Pober, 1998). Although TNF-α was not detected as differentially regulated, other genes among the endothelial-specific DENV-induced response were detected such as CXCL1, Kynureninase (KYNU), Laminin gamma 2 (LAMG2), Tissue Factor protein inhibitor TFPI (Koh and Ng, 2005), and regulator of G protein 2 (RGS2) (Warke et al., 2003), in addition to TNFSF4 encoding OX40-ligand which is an important co-activator of T-cell function (Jenkins et al., 2007; Kadowaki, 2007; Liu et al., 2007; Poulsen and Hummelshoj, 2007).

In future work, biological testing of the properties of the genes comprising the common-response and endothelial cell dengue-specific response could be performed in vitro by utilizing specific siRNA strategies and gene knock out techniques similar to those applied in other studies (Yi et al., 2007). With the increasing success of RNA interference as a tool for repressing gene expression, and the capacity to quantitate DENV genome copies by qRT-PCR as a fast assessment for determining virus loads, we envision that high throughput testing of antiviral cellular proteins can be conducted. Utilizing the techniques described in this chapter have led to interesting discoveries. Among others, the detection of RNA helicases like MDA5 in infected monocytes underscores the importance of DENV sensing mechanisms (Ramirez-Ortiz et al., 2006). Expression of oligo adenylate synthases (OASs), was also found to be differentially regulated in patients. OAS produces 2′,5′-A oligonucleotides, which are required for activation of RNaseL (Silverman, 2007a; Silverman, 2007b), and inhibit West Nile virus infection (Samuel and Diamond, 2005).

Though a variety of different cell types may respond to DENV infection via the common IFNα/β signalling network, we still need to account for the variability in antiviral responses by the different cell types used in these studies. For example, recent work has shown that DENV infection of myeloid dendritic cells results in activation of the early innate immune response, incomplete dendritic cell maturation and production of anti-viral cytokines TNF-α and IFN-α/β (Libraty et al., 2001; Tassaneetrithep et al., 2003). In contrast, vascular endothelial cells have been shown to respond to DENV infection by producing pro-inflammatory cytokines such as IL-6, IL-8, and RANTES, actin cytoskeleton rearrangement and increased permeability to small molecules (Becerra et al., 2008; Peyrefitte et al., 2003; Talavera et al., 2004).

Abrogation of the IFNα/β signalling by DENV

Mammalian cells rapidly produce and secrete IFNα/β after becoming infected by viruses, initiating a refractory state that protects the infected cells and their neighbours against the virus. We have reviewed the main mechanisms employed by the host cells to sense DENV infections and activate IFNα/β production and JAK/STAT signalling. Upon infection, secreted IFNα/β binds to the IFNα/β receptor (IFNAR) on the surface of infected and neighbouring cells, resulting in activation of the JAK/STAT pathway. IFNα/β stimulation of the JAK/STAT pathway results in

activation of hundreds of genes from promoters containing IFN-stimulated regulatory elements (ISRE) and the establishment of an antiviral state (Der et al., 1998; Takaoka and Yanai, 2006) (Fig. 7.1). Experimentally, pretreatment of human hepatoma cells with IFNα/β inhibits DENV replication. This inhibition is retained even when DENV RNA is transfected directly into cells, indicating that IFNα/β affects post-entry steps of viral replication (Diamond and Harris, 2001). In genetically deficient PKR⁻ RNase L⁻ cells, inhibition of DENV replication continued to be observed in the presence of IFNα/β, indicating that this inhibition appeared to be PKR- and RNase L-independent (Diamond and Harris, 2001). Experimental animal models, however limited in their potential to elucidate dengue pathogenesis, have demonstrated that obliteration of the IFNα/β receptor and JAK/STAT pathway results in increased viraemia, underscoring the importance of the early immune response in inhibiting DENV replication (Johnson and Roehrig, 1999; Shresta et al., 2006; Shresta et al., 2005).

Despite the potency of the IFN response, evidence that DENV infection can also counteract this complex and flexible antiviral response is accumulating. Importantly, human studies have shown that the highest amounts of IFNα/β detected by ELISA method are produced in acutely ill dengue patients very early in infection, with IFNα/β levels decreasing with disease progression (Libraty et al., 2002), and dendritic cell activation occurring prior to detection of DENV (Libraty et al., 2001; Pichyangkul et al., 2003). These findings in experimental and natural infections suggested antagonistic effects exerted by DENV to block or circumvent the IFNα/β, allowing the virus to propagate in the presence of what otherwise should be a powerful antiviral response. Experimental confirmation of DENV's ability to block the action of the IFNα/β response was provided by Diamond and Harris (Diamond and Harris, 2001), who showed that DENV infection just prior to IFNα/β treatment resulted in viral replication, whereas a short incubation of cells with IFNα/β was sufficient to completely inhibit viral replication.

More recently, the use of in vitro systems has made it possible to elucidate the interaction of the DENV-encoded proteins with the IFNα/β pathways and the mechanism by which DENV blocks IFNα/β signalling. Of 10 DENV-encoded proteins expressed separately in human alveolar basal epithelial cells (A549), NS2A, NS4A and NS4B enhanced replication of IFN-sensitive viruses in the presence of exogenous IFNα/β, indicating that these proteins may block the action of IFNα/β prior to infection (Munoz-Jordan et al., 2003). Studies in Vero cells showed that DENV NS4B strongly inhibited IFNα/β stimulation of ISRE promoters, whereas NS2A and NS4A produced a milder inhibition of this re-porter construct. However, induction of IFNα/β in response to viral infection in 293T was not inhibited by these or any of the other 10 DENV gene products (Munoz-Jordan et al., 2003). Vero cells are impaired in their ability to produce IFNα/β, and the anti-IFN effect of NS4B in this system was seen upon exogenous treatment with physiological concentrations IFNα/β. Therefore, these results demonstrated that NS4B blocks JAK/STAT signalling if IFNα/β is present. It is also known that NS4B, representing the strong-est IFNα/β antagonist, is capable of inhibiting STAT1 phosphorylation and nuclear transloca-tion upon IFNα/β stimulation (Munoz-Jordan et al., 2003). Co-expression of NS4A and NS4B re-sulted in enhanced inhibition IFNα/β to activate ISRE promoters, and complete obliteration of IFNα/β signalling was obtained by co-expressing NS2A, NS4A and NS4B (Munoz-Jordan et al., 2005; Munoz-Jordan et al., 2003). These findings indicate these three DENV-encoded proteins may synergize to perform their inhibitory func-tion together by interacting with components of the JAK/STAT pathway and completely subvert the IFNα/β response (Fig. 7.1). DENV-infected cells exhibit up-regulation of IRF3 (Warke et al., 2003), which could result in direct activa-tion of ISREs and stimulation of IFN Sensitive Genes (ISGs). Further work will elucidate whether DENV have additional mechanisms to overcome non-JAK/STAT pathways leading to ISRE activation.

The inhibition of JAK/STAT signalling by DENV non-structural region has been further documented. K562 (human chronic myeloid leukaemia) and THP-1 (human monocytic) cell lines stably transfected with DENV replicons that expressed all DENV non-structural proteins

showed reduced levels of STAT2; and replication of an IFN-sensitive virus (encephalomyocarditis virus) was enhanced in these cells (Jones et al., 2005). This impairment of STAT2 may then be an alternate mechanism employed by DENV to block IFNα/β signalling. The NS5 protein of DENV is responsible for STAT2 degradation, contributing then to overall inhibition of IFNα/β signalling during dengue virus infection. This protein binds to STAT2 and targets it to degrada-tion, but only when synthesized as a precursor that is proteolytically cleaved into its mature form (Ashour et al., 2009).

In keeping with the consensus notion that DENV blocks the JAK/STAT pathway, reduced levels of STAT2 and the lack of STAT1 phospho-rylation have been correlated to down-regulation of Tyk2 (Ho et al., 2005; Lin et al., 2004), placing the interaction between dengue non-structural proteins and the IFNα/β system upstream of the JAK/STAT signalling pathway. Transfection experiments also show that cytoplasmic segments between the first and second transmembrane re-gions of NS4B appear to be required for IFNα/β antagonism, possibly implicating this segment in an interaction with the IFNα/β pathway (Lundin et al., 2003; Munoz-Jordan et al., 2003; Qu et al., 2001). This interaction remains to be precisely defined.

Although none of the 10 DENV proteins have yet been found to inhibit IFNα/β induction separately; it is still possible that interactions among them may result in inhibition of IFNα/β expression or IFNα/β signalling. Viral strategies to overcome IFNα/β production by evading or blocking the function of molecular sensors such as TLRs and MDA/RIG-1 have been described for a variety of viruses. The NS3/4A protease from the hepatitis C virus plays an essential role in virus survival by cleaving the cytoplasmic do-main of IPS-1. As a consequence, IPS-1 loses its association to the mitochondria, essential for its binding to RIG-1 and MDA5, and RIG-1/MDA5 signalling is abolished (Lin et al., 2006; Meylan and Tschopp, 2006). The Ebola virus VP-35 also inhibits induction of IFNα/β by sequestering dsRNA and inhibiting RIG-I-mediated IPS-1 and IRF-3 signalling (Cardenas et al., 2006). The lack of evidence for flaviviral mechanisms equivalent to hepatitis C or Ebola that could directly block

IFNα/β expression is somewhat puzzling, as one would imagine that impairment of IFNα/β signalling by the non-structural proteins could only occur in infected cells, and unimpaired IFNα/β production would ultimately alert neighbouring cells of the occurrence of infection. The blocking of IFNα/β signalling, and not IFNα/β activation, is further supported by the demonstration that activation of ISRE promoters insensitive to IRFs is blocked by the DENV NS4B protein (Munoz-Jordan *et al.*, 2005). However, as represented in Fig. 7.1 the IFNα/β system is potentiated by positive feedback loops, and inhibition of the IFNα/β signalling could result in reduced IFNα/β induction. For example, RIG-I and MDA5 are directly activated by IFNα/β, and it is possible that inhibition of the JAK/STAT pathway affects important mechanisms necessary for IFNα/β production, rendering infected and neighbouring cells less responsive to viral infection. Further investigation will more precisely elucidate the extent to which IFNα/β network is antagonized by DENV.

Appropriate folding and post-translational cleavage of the non-structural region of DENV in association with the endoplasmic reticulum are integral, concatenated events leading to completion of DENV replication, and the IFN antagonistic function of the non-structural region of NS4A/B appears to be somewhat synchronized with these processes. Proteolytic processing of the NS4A/B region is an orderly, highly regulated set of events in which the cleavage at the N terminus of the signal peptide (2K) between NS4A and NS4B by the viral peptidase (NS2B/3) is a prerequisite for the cleavage by the host signal peptidase (Lin *et al.*, 1993). In the absence of NS2B/3, the NS4A/B fusion protein remains uncleaved, causing no inhibition of IFNα/β signalling, whereas co-transfection of NS4A/B together with NS2B/3 results in the cleavage of NS4A and NS4B and levels of IFNα/β inhibition comparable to those obtained by co-transfection of NS4A and NS4B (Munoz-Jordan *et al.*, 2003). Correct targeting of NS4B to the endoplasmic reticulum (ER) is also required for anti-interferon function, as deletion of the 2K segment without replacement by another signal peptide resulted in impairment of IFN-antagonistic function (Munoz-Jordan *et al.*, 2005). The ability of NS4B to impair JAK/STAT signalling in Vero cells is conserved in both yellow

fever virus and West Nile virus, possibly indicating a consensus mechanism to block this pathway in mosquito-borne flaviviruses (Guo *et al.*, 2005; Keller *et al.*, 2006; Liu *et al.*, 2005; Munoz-Jordan *et al.*, 2005). In tick-borne flaviviruses, however, the ability to block JAK/STAT signalling is carried by NS5, and not by NS4B (Best *et al.*, 2005). The minimal requirement for this function has been ascribed to residues in two non-contiguous sequences of the RNA-dependent RNA polymerase region of NS5, which appear to come together in the tridimensional structure of this protein. It appears then that mosquito-borne and tick-borne flaviviruses may have evolved separate, yet concurrent, mechanisms of IFN antagonisms. Whereas several lines of evidence support the general mechanisms by which mosquito-borne flaviviruses impair the JAK/STAT pathway, the subtle differences observed by different research groups may be indicative of site-specific and even inter-cellular differences. NS4B, NS5 and other non-structural proteins have been found to antagonized this pathway (Ashour *et al.*, 2009; Jones *et al.*, 2005; Munoz-Jordan *et al.*, 2003); and sequence differences among flavivirus strains may correlate with differences in their ability to block IFNα/β signalling (Ho *et al.*, 2005; Keller *et al.*, 2006). Also, viruses often display multiple and complex mechanisms to antagonize the antiviral response (Haller and Weber, 2007). We may therefore be just beginning to recognize how intricate and redundant these mechanisms may be during DENV infection.

As we gain a more complete understanding of the host response against DENV, the study of the molecular basis of the race between DENV replication and the IFNα/β response will undoubtedly re-focus on understanding the interactions between the most relevant players. How the antagonistic function of DENV non-structural proteins engage in the modulation of this wide array of gene responses in the host will require more research. We have recently learned that different strains of DENV across all serotypes differ in their abilities to activate IFNα/β antiviral genes like EIF2AK2 (PKR), OAS, ADAR and MX, while this response was absent if infection is carried by the DENV-2 strain NGC, a viral strain that strongly suppressed STAT-1 phosphorylation (Umareddy *et al.*, 2008). Understanding these interactions in the context of acute illness is critical, as this will elucidate whether differences in IFNα/β antagonism contribute to the differences in pathogenicity observed among DENV strains. Identifying the molecular mechanisms that DENV utilizes to subvert IFNα/β response may also contribute to the development of strategies for rational attenuation of DENV vaccines.

Conclusions and perspectives

Gene expression profiling studies have revealed key players of the cell response to DENV infection. The data from different research groups converge in three major categories of gene expression patterns: the NF-κB-mediated immune response, the IFNα/β network and the ubiquitin proteosome pathway. The up-regulation of IFNα/β is accompanied by high expression of pro-inflammatory cytokines and complement inhibitors. Studies have recently focused on the differences between DF and DHF. Interestingly, during acute-phase infection, the IFNα/β response represents a substantial subset of differentially regulated genes in PBMCs from acutely ill patients, and this IFNα/β response seems to offer more protection for infection in patients with DF than in patients with DHF. Further studies should be conducted in a time-dependent manner during disease progression and, if possible, isolated subsets of blood cells should be used. The significant up-regulation of IFNα/β inducible gene products found in patients with DF is consistent with the role of IFNα/β in controlling viral replication and the lower viral loads reported in patients with DF compared to patients with DHF. In analysing gene expression profiles, we have revealed key players of IFNα/β modulation which could play a role in disease pathogenesis, including endothelial damage, inflammatory response and capillary leakage. Their differential regulation during the course of illness, as mentioned above, needs to be further analysed. Additionally, potential markers of disease severity have now been identified, some of which are induced as part of the IFNα/β response. Their potential in defining risks for severe illness must be assessed over the course of acute disease in patients.

Finally, the subversion of the antiviral response by DENV needs to be further explored in the context of disease progression and in light of the recently identified components of IFNα/β modulation. The lower IFNα/β response in DHF patients could be both a reflection of the antagonistic function of DENV NS4B and NS5, and the patient's ability to build up an effective IFN response. The specific molecular interactions between DENV protein products and the IFNα/β pathway need to be further studied; especially how individual molecules identified in gene expression profiles are modulated over the course of disease. A more precise definition of these molecular interactions will support discovery of new therapies for dengue disease. Future research will likely be centred in identifying DENV sequences involved in viral antagonism and in studying differences in their ability to block IFNα/β signalling as a systematic approach to better understand virulence and develop attenuated vaccines for efficacy and safety studies.

Acknowledgements

We thank Katherine J. Martin from Bioarray, Inc, for her expert gene expression analysis, Rajas Warke from UMASS Medical School for his ELISA- and qRT-PCR-based gene expression analysis, and the Core Proteomic Facility at UMASS Medical School for data analysis. Cameron Simmons is thanked for intellectual input and sharing of preliminary data. The clinical study site at the Armed Forces Research Institute of Medical Sciences, Bangkok, Thailand (David Vaughn, Timothy Endy, Mammen Mammen, and Ananda Nisalak), the Queen Sirikit National Institute of Child Health, Bangkok, Thailand (Siripen Kalayanarooj), the Banco Municipal de Sangre, Caracas, Venezuela (Norma Bosch and Marion Echenagucia), and the University of Massachusetts Medical School (Francis Ennis, Alan Rothman, Sharone Green, and Daniel Libraty) are thanked for their input. We are in debt to Kris Xhaja for her excellent technical assistance. I. Bosch's work has been funded by NIAID grants P01 AI34533 and U01 AI45440. J. Munoz-Jordan has been funded by the CDC.

References

Akira, S., and Takeda, K. (2004). Toll-like receptor signalling. Nat. Rev. Immunol. 4, 499–511.

Ashour J., Laurent-Rolle, M., Shi, P.Y., and Garcia-Sastre, A. (2009). NS5 of dengue virus mediates STAT2 binding and degradation. J. Virol. 83, 5408–5418.

Basak, C., Pathak, S.K., Bhattacharyya, A., Mandal, D., Pathak, S., and Kundu, M. (2005). NF-kappaB- and C/EBPbeta-driven interleukin-1beta gene expression and PAK1-mediated caspase-1 activation play essential roles in interleukin-1beta release from *Helicobacter pylori* lipopolysaccharide-stimulated macrophages. J. Biol. Chem. *280*, 4279–4288.

Becerra, A., Warke, R.V., de Bosch, N., Rothman, A.L., and Bosch, I. (2008). Elevated levels of soluble ST2 protein in dengue virus infected patients. Cytokine *41*, 114–120.

Bergers, G., Reikerstorfer, A., Braselmann, S., Graninger, P., and Busslinger, M. (1994). Alternative promoter usage of the Fos-responsive gene Fit-1 generates mRNA isoforms coding for either secreted or membrane-bound proteins related to the IL-1 receptor. EMBO J. *13*, 1176–1188.

Best, S.M., Morris, K.L., Shannon, J. G., Robertson, S.J., Mitzel, D.N., Park, G.S., Boer, E., Wolfinbarger, J.B., and Bloom, M.E. (2005). Inhibition of interferon-stimulated JAK-STAT signaling by a tick-borne flavivirus and identification of NS5 as an interferon antagonist. J. Virol. *79*, 12828–12839.

Bijuklic, K., Jennings, P., Kountchev, J., Hasslacher, J., Aydin, S., Sturn, D., Pfaller, W., Patsch, J.R., and Joannidis, M. (2007). Migration of leukocytes across an endothelium-epithelium bilayer as a model of renal interstitial inflammation. Am. J. Physiol. Cell Physio.l *293*, C486–492.

Bowie, A.G., and Haga, I.R. (2005). The role of Toll-like receptors in the host response to viruses. Mol. Immunol. *42*, 859–867.

Brown, A.M., Wu, A.H., Clopton, P., Robey, J.L., and Hollander, J.E. (2007). ST2 in emergency department chest pain patients with potential acute coronary syndromes. Ann. Emerg. Med. *50*, 153–158, 158 e151.

Brunner, M., Krenn, C., Roth, G., Moser, B., Dworschak, M., Jensen-Jarolim, E., Spittler, A., Sautner, T., Bonaros, N., Wolner, E., Boltz-Nitulescu, G., and Ankersmit, H.J. (2004). Increased levels of soluble ST2 protein and IgG1 production in patients with sepsis and trauma. Intensive Care Med. *30*, 1468–1473.

Cardenas, W.B., Loo, Y. M., Gale, M., Jr., Hartman, A.L., Kimberlin, C.R., Martinez-Sobrido, L., Saphire, E.O., and Basler, C. F. (2006). Ebola virus VP35 protein binds double-stranded RNA and inhibits alpha/beta interferon production induced by RIG-I signaling. J. Virol. *80*, 5168–5178.

Carriere, V., Roussel, L., Ortega, N., Lacorre, D. A., Americh, L., Aguilar, L., Bouche, G., and Girard, J.P. (2007). IL-33, the IL-1-like cytokine ligand for ST2 receptor, is a chromatin-associated nuclear factor *in vivo*. Proc. Natl. Acad. Sci. U.S.A. *104*, 282–287.

Chackerian, A.A., Oldham, E.R., Murphy, E. E., Schmitz, J., Pflanz, S., and Kastelein, R.A. (2007). IL-1 receptor accessory protein and ST2 comprise the IL-33 receptor complex. J. Immunol. *179*, 2551–2555.

Chang, T.H., Liao, C.L., and Lin, Y.L. (2006). Flavivirus induces interferon-beta gene expression through a pathway involving RIG-I-dependent IRF-3 and PI3K-dependent NF-kappaB activation. Microbes Infect. *8*, 157–171.

Chin, K.C., and Cresswell, P. (2001). Viperin (cig5), an IFN-inducible antiviral protein directly induced by human cytomegalovirus. Proc. Natl. Acad. Sci. U.S.A. *98*, 15125–15130.

Clark, I.A. (2007). How TNF was recognized as a key mechanism of disease. Cytokine Growth Factor Rev. *18*, 335–343.

Daubener, W., Spors, B., Hucke, C., Adam, R., Stins, M., Kim, K.S., and Schroten, H. (2001). Restriction of Toxoplasma gondii growth in human brain microvascular endothelial cells by activation of indoleamine 2,3-dioxygenase. Infect. Immun. *69*, 6527–6531.

Der, S. D., Zhou, A., Williams, B.R., and Silverman, R.H. (1998). Identification of genes differentially regulated by interferon alpha, beta, or gamma using oligonucleotide arrays. Proc. Natl. Acad. Sci. U.S.A. *95*, 15623–15628.

Diamond, M.S., and Harris, E. (2001). Interferon inhibits dengue virus infection by preventing translation of viral RNA through a PKR-independent mechanism. Virology *289*, 297–311.

Diebold, S.S., Kaisho, T., Hemmi, H., Akira, S., and Reis e Sousa, C. (2004). Innate antiviral responses by means of TLR7-mediated recognition of single-stranded RNA Science *303*, 1529–1531.

Diehl, G.E., Yue, H.H., Hsieh, K., Kuang, A. A., Ho, M., Morici, L.A., Lenz, L.L., Cado, D., Riley, L.W., and Winoto, A. (2004). TRAIL-R as a negative regulator of innate immune cell responses. Immunity *21*, 877–889.

Fallarino, F., Grohmann, U., Vacca, C., Orabona, C., Spreca, A., Fioretti, M.C., and Puccetti, P. (2003). T-cell apoptosis by kynurenines. Adv Exp Med Biol *527*, 183–190.

Fink, J., Gu, F., Ling, L., Tolfvenstam, T., Olfat, F., Chin, K.C., Aw, P., George, J., Kuznetsov, V.A., Schreiber, M., Vasudevan, S.G., and Hibberd, M.L. (2007). Host gene expression profiling of dengue virus infection in cell lines and patients. PLoS Negl. Trop. Dis. *1*, e86.

Gadina, M., and Jefferies, C.A. (2007). IL-33: a sheep in wolf's clothing? Sci STKE *2007*, pe31.

Garcia-Sastre, A. (2002). Mechanisms of inhibition of the host interferon alpha/beta-mediated antiviral responses by viruses. Microbes Infect. *4*, 647–655.

Garcia-Sastre, A., and Biron, C.A. (2006). Type 1 interferons and the virus–host relationship: a lesson in detente. Science *312*, 879–882.

Green, S., and Rothman, A. (2006). Immunopathological mechanisms in dengue and dengue hemorrhagic fever. Curr. Opin. Infect. Dis. *19*, 429–436.

Gubler, D.J. (2006). Dengue/dengue haemorrhagic fever: history and current status. Novartis Found Symp *277*, 3–16; discussion 16–22, 71–13, 251–253.

Guo, J.T., Hayashi, J., and Seeger, C. (2005). West Nile virus inhibits the signal transduction pathway of alpha interferon. J. Virol. *79*, 1343–1350.

Haller, O., and Weber, F. (2007). Pathogenic viruses: smart manipulators of the interferon system. Curr. Top. Microbiol. Immunol. *316*, 315–334.

Halstead, S.B. (2003). Neutralization and antibody-dependent enhancement of dengue viruses. Adv Virus Res. *60*, 421–467.

Helbig, K.J., Lau, D.T., Semendric, L., Harley, H.A., and Beard, M.R. (2005). Analysis of ISG expression in chronic hepatitis C identifies viperin as a potential antiviral effector. Hepatology *42*, 702–710.

Ho, L.J., Hung, L.F., Weng, C.Y., Wu, W.L., Chou, P., Lin, Y. L., Chang, D.M., Tai, T.Y., and Lai, J.H. (2005). Dengue virus type 2 antagonizes IFN-alpha but not IFN-gamma antiviral effect via down-regulating Tyk2-STAT signaling in the human dendritic cell. J. Immunol. *174*, 8163–8172.

Hoffmann, O., Priller, J., Prozorovski, T., Schulze-Topphoff, U., Baeva, N., Lunemann, J. D., Aktas, O., Mahrhofer, C., Stricker, S., Zipp, F., and Weber, J. R. (2007). TRAIL limits excessive host immune responses in bacterial meningitis. J. Clin. Invest, *117*, 2004–2013.

Indraccolo, S., Pfeffer, U., Minuzzo, S., Esposito, G., Roni, V., Mandruzzato, S., Ferrari, N., Anfosso, L., Dell'Eva, R., Noonan, D.M., Chieco-Bianchi, L., Albini, A., and Amadori, A.(2007). Identification of genes selectively regulated by IFNs in endothelial cells. J. Immunol. *178*, 1122–1135.

Jenkins, S.J., Perona-Wright, G., Worsley, A.G., Ishii, N., and MacDonald, A.S. (2007). Dendritic cell expression of OX40 ligand acts as a costimulatory, not polarizing, signal for optimal Th2 priming and memory induction *in vivo*. J. Immunol. *179*, 3515–3523.

Jessie, K., Fong, M.Y., Devi, S., Lam, S. K., and Wong, K.T. (2004). Localization of dengue virus in naturally infected human tissues, by immunohistochemistry and in situ hybridization. J. Infect. Dis. *189*, 1411–1418.

Jiang, D., Guo, H., Xu, C., Chang, J., Gu, B., Wang, L., Block, T.M., and Guo, J. T. (2008). Identification of three interferon-inducible cellular enzymes that inhibit the replication of hepatitis C virus. J. Virol. *82*, 1665–1678.

Johnson, A. J., and Roehrig, J. T. (1999). New mouse model for dengue virus vaccine testing. J. Virol. *73*, 783–786.

Jones, M., Davidson, A., Hibbert, L., Gruenwald, P., Schl aak, J., Ball, S., Foster, G. R., and Jacobs, M. (2005). Dengue virus inhibits alpha interferon signaling by reducing STAT2 expression. J. Virol. *79*, 5414–5420.

Kadowaki, N. (2007). Dendritic cells: a conductor of T-cell differentiation. Allergol. Int. *56*, 193–199.

Keller, B.C., Fredericksen, B. L., Samuel, M. A., Mock, R. E., Mason, P. W., Diamond, M. S., and Gale, M., Jr. (2006). Resistance to alpha/beta interferon is a determinant of West Nile virus replication fitness and virulence. J. Virol. *80*, 9424–9434.

Kim, C.H., and Gupta, S. (2000). Expression of TRAIL (Apo2L), DR4 (TRAIL receptor 1), DR5 (TRAIL receptor 2) and TRID (TRAIL receptor 3) genes in multidrug resistant human acute myeloid leukemia cell lines that overexpress MDR 1 (HL60/Tax) or MRP (HL60/AR). Int. J. Oncol. *16*, 1137–1139.

Koh, W. L., and Ng, M. L. (2005). Molecular mechanisms of West Nile virus pathogenesis in brain cell. Emerg. Infect. Dis. *11*, 629–632.

Kumar-Sinha, C., Varambally, S., Sreekumar, A., and Chinnaiyan, A. M. (2002). Molecular cross-talk between the TRAIL and interferon signaling pathways. J. Biol. Chem. *277*, 575–585.

Kumar, S., Tzimas, M.N., Griswold, D.E., and Young, P.R. (1997). Expression of ST2, an interleukin-1 receptor homologue, is induced by proinflammatory stimuli. Biochem. Biophys. Res. Commun. *235*, 474–478.

Kurane, I., Dai, L.C., Livingston, P.G., Reed, E., and Ennis, F.A. (1993). Definition of an HLA-DPw2-restricted epitope on NS3, recognized by a dengue virus serotype-cross-reactive human CD4+ CD8- cytotoxic T-cell clone. J. Virol. *67*, 6285–6288.

Kuroiwa, K., Arai, T., Okazaki, H., Minota, S., and Tominaga, S. (2001). Identification of human ST2 protein in the sera of patients with autoimmune diseases. Biochem. Biophys. Res. Commun. *284*, 1104–1108.

Lee, H.S., Boulton, I.C., Reddin, K., Wong, H., Halliwell, D., Mandelboim, O., Gorringe, A.R., and Gray-Owen, S.D. (2007). Neisserial outer membrane vesicles bind the coinhibitory receptor carcinoembryonic antigen-related cellular adhesion molecule 1 and suppress CD4+ T lymphocyte function. Infect. Immun. *75*, 4449–4455.

Libraty, D.H., Pichyangkul, S., Ajariyakhajorn, C., Endy, T.P., and Ennis, F.A. (2001). Human dendritic cells are activated by dengue virus infection: enhancement by gamma interferon and implications for disease pathogenesis. J. Virol. *75*, 3501–3508.

Libraty, D.H., Young, P. R., Pickering, D., Endy, T.P., Kalayanarooj, S., Green, S., Vaughn, D. W., Nisalak, A., Ennis, F. A., and Rothman, A. L. (2002). High circulating levels of the dengue virus nonstructural protein NS1 early in dengue illness correlate with the development of dengue hemorrhagic fever. J. Infect. Dis. *186*, 1165–1168.

Lin, C., Amberg, S. M., Chambers, T.J., and Rice, C.M. (1993). Cleavage at a novel site in the NS4A region by the yellow fever virus NS2B-3 proteinase is a prerequisite for processing at the downstream 4A/4B signalase site. J. Virol. *67*, 2327–2335.

Lin, R., Yang, L., Nakhaei, P., Sun, Q., Sharif-Askari, E., Julkunen, I., and Hiscott, J. (2006). Negative regulation of the retinoic acid-inducible gene I-induced antiviral state by the ubiquitin-editing protein A20. J. Biol. Chem. *281*, 2095–2103.

Lin, R.J., Liao, C.L., Lin, E., and Lin, Y.L. (2004). Blocking of the alpha interferon-induced Jak-Stat signaling pathway by Japanese encephalitis virus infection. J. Virol. *78*, 9285–9294.

Liu, W.J., Wang, X. J., Mokhonov, V.V., Shi, P. Y., Randall, R., and Khromykh, A.A. (2005). Inhibition of interferon signaling by the New York 99 strain and Kunjin subtype of West Nile virus involves blockage of STAT1 and STAT2 activation by nonstructural proteins. J. Virol. *79*, 1934–1942.

Liu, Z., Dai, H., Wan, N., Wang, T., Bertera, S., Trucco, M., and Dai, Z. (2007). Suppression of memory CD8 T-cell generation and function by tryptophan catabolism. J. Immunol. *178*, 4260–4266.

Loo, Y.M., Fornek, J., Crochet, N., Bajwa, G., Perwitasari, O., Martinez-Sobrido, L., Akira, S., Gill, M. ., Garcia-

Sastre, A., Katze, M.G., and Gale, M., Jr. (2008). Distinct RIG-I and MDA5 signaling by RNA viruses in innate immunity. J. Virol. *82*, 335–345.

Lund, J.M., Alexopoulou, L., Sato, A., Karow, M., Adams, N. C., Gale, N.W., Iwasaki, A., and Flavell, R.A. (2004). Recognition of single-stranded RNA viruses by Toll-like receptor 7. Proc. Natl. Acad. Sci. U.S.A. *101*, 5598–5603.

Lundin, M., Monne, M., Widell, A., Von Heijne, G., and Persson, M.A. (2003). Topology of the membrane-associated hepatitis C virus protein NS4B. J. Virol. *77*, 5428–5438.

Lush, C.W., and Kvietys, P.R. (2000). Microvascular dysfunction in sepsis. Microcirculation 7, 83–101.

Maruo, N., Morita, I., Shirao, M., and Murota, S. (1992). IL-6 increases endothelial permeability *in vitro*. Endocrinology *131*, 710–714.

Mathew, A., Kurane, I., Green, S., Vaughn, D.W., Kalayanarooj, S., Suntayakorn, S., Ennis, F.A., and Rothman, A.L. (1999). Impaired T-cell proliferation in acute dengue infection. J. Immunol. *162*, 5609–5615.

Mellor, A. L., and Munn, D. H. (2004). IDO expression by dendritic cells: tolerance and tryptophan catabolism. Nat. Rev. Immunol 4, 762–774.

Meylan, E., and Tschopp, J. (2006). Toll-like receptors and RNA helicases: two parallel ways to trigger antiviral responses. Mol. Cell. 22, 561–569.

Morita, Y., Araki, H., Sugimoto, T., Takeuchi, K., Yamane, T., Maeda, T., Yamamoto, Y., Nishi, K., Asano, M., Shirahama-Noda, K., Nishimura, M., Uzu, T., Hara-Nishimura, I., Koya, D., Kashiwagi, A., and Ohkubo, I. (2007). Legumain/asparaginyl endopeptidase controls extracellular matrix remodeling through the degradation of fibronectin in mouse renal proximal tubular cells. FEBS Lett *581*, 1417–1424.

Moritz, D.R., Rodewald, H.R., Gheyselinck, J., and Klemenz, R. (1998). The IL-1 receptor-related T1 antigen is expressed on immature and mature mast cells and on fetal blood mast cell progenitors. J. Immunol. *161*, 4866–4874.

Munoz-Jordan, J.L., Laurent-Rolle, M., Ashour, J., Martinez-Sobrido, L., Ashok, M., Lipkin, W. I., and Garcia-Sastre, A. (2005). Inhibition of alpha/beta interferon signaling by the NS4B protein of flaviviruses. J. Virol. *79*, 8004–8013.

Munoz-Jordan, J.L., Sanchez-Burgos, G.G., Laurent-Rolle, M., and Garcia-Sastre, A. (2003). Inhibition of interferon signaling by dengue virus. Proc. Natl. Acad. Sci. U.S.A. *100*, 14333–14338.

Musso, T., Gusella, G.L., Brooks, A., Longo, D.L., and Varesio, L. (1994). Interleukin-4 inhibits indoleamine 2,3-dioxygenase expression in human monocytes. Blood 83, 1408–1411

Oberdorfer, C., Adams, O., MacKenzie, C. R., De Groot, C.J., and Daubener, W. (2003). Role of IDO activation in anti-microbial defense in human native astrocytes. Adv Exp Med Biol *527*, 15–26.

Obojes, K., Andres, O., Kim, K.S., Daubener, W., and Schneider-Schaulies, J. (2005). Indoleamine 2,3-dioxygenase mediates cell type-specific anti-measles virus activity of gamma interferon. J. Virol. *79*, 7768–7776.

Onomoto, K., Yoneyama, M., and Fujita, T. (2007). Regulation of antiviral innate immune responses by RIG-I family of RNA helicases. Curr. Top. Microbiol. Immunol. *316*, 193–205.

Oshikawa, K., Kuroiwa, K., Tago, K., Iwahana, H., Yanagisawa, K., Ohno, S., Tominaga, S. I., and Sugiyama, Y. (2001). Elevated soluble ST2 protein levels in sera of patients with asthma with an acute exacerbation. Am J Respir Crit Care Med *164*, 277–281.

Oshikawa, K., Yanagisawa, K., Tominaga, S., and Sugiyama, Y. (2002). Expression and function of the ST2 gene in a murine model of allergic airway inflammation. Clin. Exp. Allergy *32*, 1520–1526.

Peyrefitte, C.N., Pastorino, B., Bessaud, M., Tolou, H. J., and Couissinier-Paris, P. (2003). Evidence for *in vitro* falsely-primed cDNAs that prevent specific detection of virus negative strand RNAs in dengue-infected cells: improvement by tagged RT-PCR J. Virol. Methods *113*, 19–28.

Pichyangkul, S., Endy, T.P., Kalayanarooj, S., Nisalak, A., Yongvanitchit, K., Green, S., Rothman, A. L., Ennis, F.A., and Libraty, D.H. (2003). A blunted blood plasmacytoid dendritic cell response to an acute systemic viral infection is associated with increased disease severity. J. Immunol. *171*, 5571–5578.

Pitha, P.M., and Kunzi, M. . (2007). Type I interferon: the ever unfolding story. Curr. Top. Microbiol. Immunol. *316*, 41–70.

Pober, J.S. (1998). Activation and injury of endothelial cells by cytokines. Pathol Biol (Paris) *46*, 159–163.

Pober, J.S., and Min, W. (2006). Endothelial cell dysfunction, injury and death. Handb. Exp. Pharmacol. 135–156.

Poulsen, L. K., and Hummelshoj, L. (2007). Triggers of IgE class switching and allergy development. Ann. Med. *39*, 440–456.

Qu, L., McMullan, L. K., and Rice, C.M. (2001). Isolation and characterization of noncytopathic pestivirus mutants reveals a role for nonstructural protein NS4B in viral cytopathogenicity. J. Virol. *75*, 10651–10662.

Querec, T., Bennouna, S., Alkan, S., Laouar, Y., Gorden, K., Flavell, R., Akira, S., Ahmed, R., and Pulendran, B. (2006). Yellow fever vaccine YF-17D activates multiple dendritic cell subsets via TLR2, 7, 8, and 9 to stimulate polyvalent immunity. J .Exp. Med. *203*, 413–424.

Rachman, A., and Rinaldi, I. (2006). Coagulopathy in dengue infection and the role of interleukin-6. Acta Med Indones *38*, 105–108.

Ramirez-Ortiz, Z. G., Warke, R.V., Pacheco, L., Xhaja, K., Sarkar, D., Fisher, P.B., Shaw, S. K., Martin, K.J., and Bosch, I. (2006). Discovering innate immunity genes using differential display: a story of RNA helicases. J. Cell Physiol. *209*, 636–644.

Rimondi, E., Secchiero, P., Quaroni, A., Zerbinati, C., Capitani, S., and Zauli, G. (2006). Involvement of TRAIL/TRAIL-receptors in human intestinal cell differentiation. J. Cell Physiol. *206*, 647–654.

Samuel, M.A., and Diamond, M.S. (2005). Alpha/beta interferon protects against lethal West Nile virus infection by restricting cellular tropism and enhancing neuronal survival. J. Virol. *79*, 13350–13361.

Sanada, S., Hakuno, D., Higgins, L. J., Schreiter, E.R., McKenzie, A.N., and Lee, R.T. (2007). IL-33 and ST2 comprise a critical biomechanically induced and cardioprotective signaling system. J. Clin. Invest. *117*, 1538–1549.

Sariol, C. A., Munoz-Jordan, J.L., Abel, K., Rosado, L.C., Pantoja, P., Giavedoni, L., Rodriguez, I. V., White, L.J., Martinez, M., Arana, T., and Kraiselburd, E.N. (2007). Transcriptional activation of interferon-stimulated genes but not of cytokine genes after primary infection of rhesus macaques with dengue virus type 1. Clin Vaccine Immunol *14*, 756–766.

Sato, K., Nakaoka, T., Yamashita, N., Yagita, H., Kawasaki, H., Morimoto, C., Baba, M., and Matsuyama, T. (2005). TRAIL-transduced dendritic cells protect mice from acute graft-versus-host disease and leukemia relapse. J. Immunol. *174*, 4025–4033.

Schmitz, J., Owyang, A., Oldham, E., Song, Y., Murphy, E., McClanahan, T.K., Zurawski, G., Moshrefi, M., Qin, J., Li, X., *et al.* (2005). IL-33, an interleukin-1-like cytokine that signals via the IL-1 receptor-related protein ST2 and induces T helper type 2-associated cytokines. Immunity *23*, 479–490.

Secchiero, P., Corallini, F., di Iasio, M. G., Gonelli, A., Barbarotto, E., and Zauli, G. (2005). TRAIL counteracts the proadhesive activity of inflammatory cytokines in endothelial cells by down-modulating CCL8 and CXCL10 chemokine expression and release. Blood *105*, 3413–3419.

Secchiero, P., Gonelli, A., Carnevale, E., Milani, D., Pandolfi, A., Zella, D., and Zauli, G. (2003). TRAIL promotes the survival and proliferation of primary human vascular endothelial cells by activating the Akt and ERK pathways. Circulation *107*, 2250–2256.

Severa, M., and Fitzgerald, K. A. (2007). TLR-mediated activation of type I IFN during antiviral immune responses: fighting the battle to win the war. Curr. Top. Microbiol. Immunol. *316*, 167–192.

Shimpo, M., Morrow, D. A., Weinberg, E.O., Sabatine, M.S., Murphy, S.A., Antman, E.M., and Lee, R.T. (2004). Serum levels of the interleukin-1 receptor family member ST2 predict mortality and clinical outcome in acute myocardial infarction. Circulation *109*, 2186–2190.

Shope, R.E., Kieny, M.P., and Engers, H. (2004). Guidelines for the Evaluation of Dengue Vaccines in Populations Exposed to Natural Infections. (Geneva: World Health Organization).

Shresta, S., Kyle, J.L., Snider, H. M., Basavapatna, M., Beatty, P.R., and Harris, E. (2004). Interferon-dependent immunity is essential for resistance to primary dengue virus infection in mice, whereas T- and B-cell-dependent immunity are less critical. J. Virol. *78*, 2701–2710.

Shresta, S., Sharar, K. L., Prigozhin, D.M., Beatty, P.R., and Harris, E. (2006). Murine model for dengue virus-induced lethal disease with increased vascular permeability. J. Virol. *80*, 10208–10217.

Shresta, S., Sharar, K. L., Prigozhin, D.M., Snider, H.M., Beatty, P.R., and Harris, E. (2005). Critical roles for both STAT1-dependent and STAT1-independent

pathways in the control of primary dengue virus infection in mice. J. Immunol. *175*, 3946–3954.

Silverman, R. H. (2007a). A scientific journey through the 2–5A/RNase L system. Cytokine Growth Factor Rev. *18*, 381–388.

Silverman, R. H. (2007b). Viral encounters with 2′,5′-oligoadenylate synthetase and RNase L during the interferon antiviral response. J. Virol. *81*, 12720–12729.

Simmons, C.P., Popper, S., Dolocek, C., Chau, T.N., Griffiths, M., Dung, N.T., Long, T. H., Hoang, D.M., Chau, N.V., Thao le, T.T., *et al.* (2007). Patterns of host genome-wide gene transcript abundance in the peripheral blood of patients with acute dengue hemorrhagic fever. J. Infect. Dis. *195*, 1097–1107.

Suharti, C., van Gorp, E. C., Setiati, T.E., Dolmans, W.M., Djokomoeljanto, R.J., Hack, C. E., ten, C.H., and van der Meer, J. W. (2002). The role of cytokines in activation of coagulation and fibrinolysis in dengue shock syndrome. Thromb. Haemost. *87*, 42–46.

Sweet, M.J., Leung, B. P., Kang, D., Sog aard, M., Schulz, K., Trajkovic, V., Campbell, C.C., Xu, D., and Liew, F.Y. (2001). A novel pathway regulating lipopolysaccharide-induced shock by ST2/T1 via inhibition of Toll-like receptor 4 expression. J. Immunol. *166*, 6633–6639.

Tajima, S., Oshikawa, K., Tominaga, S., and Sugiyama, Y. (2003). The increase in serum soluble ST2 protein upon acute exacerbation of idiopathic pulmonary fibrosis. Chest *124*, 1206–1214.

Takaoka, A., and Yanai, H. (2006). Interferon signalling network in innate defence. Cell. Microbiol. *8*, 907–922.

Takikawa, O., Truscott, R.J., Fukao, M., and Miwa, S. (2003). Age-related nuclear cataract and indoleamine 2,3-dioxygenase-initiated tryptophan metabolism in the human lens. Adv. Exp. Med. Biol. *527*, 277–285.

Talavera, D., Castillo, A.M., Dominguez, M. C., Gutierrez, A.E., and Meza, I. (2004). IL8 release, tight junction and cytoskeleton dynamic reorganization conducive to permeability increase are induced by dengue virus infection of microvascular endothelial monolayers. J. Gen. Virol. *85*, 1801–1813.

Tassaneetrithep, B., Burgess, T. H., Granelli-Piperno, A., Trumpfheller, C., Finke, J., Sun, W., Eller, M.A., Pattanapanyasat, K., Sarasombath, S., Birx, D.L., Steinman, R.M., Schlesinger, S., and Marovich, M.A. (2003). DC-SIGN (CD209) mediates dengue virus infection of human dendritic cells. J. Exp. Med. *197*, 823–829.

Tominaga, S., Kuroiwa, K., Tago, K., Iwahana, H., Yanagisawa, K., and Komatsu, N. (1999). Presence and expression of a novel variant form of ST2 gene product in human leukemic cell line UT-7/G.M. Biochem. Biophys. Res. Commun. *264*, 14–18.

Tsunobuchi, H., Nishimura, H., Goshima, F., Daikoku, T., Suzuki, H., Nakashima, I., Nishiyama, Y., and Yoshikai, Y. (2000). A protective role of interleukin-15 in a mouse model for systemic infection with herpes simplex virus. Virology *275*, 57–66.

Ubol, S., Masrinoul, P., Chaijaruwanich, J., Kalayanarooj, S., Charoensirisuthikul, T., and Kasisith, J. (2008). Differences in global gene expression in peripheral blood mononuclear cells indicate a significant role of the innate responses in progression of dengue fever but not dengue hemorrhagic fever. J. Infect. Dis. *197*, 1459–1467.

Umareddy, I., Tang, K.F., Vasudevan, S.G., Devi, S., Hibberd, M.L., and Gu, F. (2008). Dengue virus regulates type I interferon signalling in a strain-dependent manner in human cell lines. J. Gen. Virol. *89*, 3052–3062.

Wang, J.P., Liu, P., Latz, E., Golenbock, D.T., Finberg, R.W., and Libraty, D.H. (2006). Flavivirus activation of plasmacytoid dendritic cells delineates key elements of TLR7 signaling beyond endosomal recognition. J. Immunol. *177*, 7114–7121.

Wang, T., Town, T., Alexopoulou, L., Anderson, J. F., Fikrig, E., and Flavell, R. A. (2004). Toll-like receptor 3 mediates West Nile virus entry into the brain causing lethal encephalitis. Nat. Med. *10*, 1366–1373.

Wang, X., Hinson, E.R., and Cresswell, P. (2007). The interferon-inducible protein viperin inhibits influenza virus release by perturbing lipid rafts. Cell Host Microbe *2*, 96–105.

Warke, R.V., Martin, K.J., Giaya, K., Shaw, S. K., Rothman, A.L., and Bosch, I. (2008). TRAIL is a novel antiviral protein against dengue virus. J. Virol. *82*, 555–564.

Warke, R.V., Xhaja, K., Martin, K.J., Fournier, M.F., Shaw, S.K., Brizuela, N., de Bosch, N., Lapointe, D., Ennis, F.A., Rothman, A L., and Bosch, I. (2003). Dengue virus induces novel changes in gene expression of human umbilical vein endothelial cells. J. Virol. *77*, 11822–11832.

Weinberg, E. O., Shimpo, M., De Keulenaer, G. W., MacGillivray, C., Tominaga, S., Solomon, S. D., Rouleau, J. L., and Lee, R. T. (2002). Expression and regulation of ST2, an interleukin-1 receptor family member, in cardiomyocytes and myocardial infarction. Circulation *106*, 2961–2966.

Wu, S. J., Grouard-Vogel, G., Sun, W., Mascola, J.R., Brachtel, E., Putvatana, R., Louder, M. K., Filgueira, L., Marovich, M.A., Wong, H.K., *et al.* (2000). Human skin Langerhans cells are targets of dengue virus infection. Nat. Med. *6*, 816–820.

Xu, D., Chan, W.L., Leung, B. P., Huang, F., Wheeler, R., Piedrafita, D., Robinson, J.H., and Liew, F.Y. (1998). Selective expression of a stable cell surface molecule on type 2 but not type 1 helper T-cells. J. Exp. Med. *187*, 787–794.

Yanagisawa, K., Takagi, T., Tsukamoto, T., Tetsuka, T., and Tominaga, S. (1993). Presence of a novel primary response gene ST2L., encoding a product highly similar to the interleukin 1 receptor type 1. FEBS Lett. *318*, 83–87.

Yenchitsomanus, P.T., Sricharoen, P., Jaruthasana, I., Pattanakitsakul, S.N., Nitayaphan, S., Mongkolsapaya, J., and Malasit, P. (1996). Rapid detection and identification of dengue viruses by polymerase chain reaction (PCR). South-East Asian J. Trop. Med. Public Health *27*, 228–236.

Yi, C.H., Sogah, D.K., Boyce, M., Degterev, A., Christofferson, D.E., and Yuan, J. (2007). A genome-wide RNAi screen reveals multiple regulators of caspase activation. J. Cell. Biol. *179*, 619–626.

Yoneyama, M., Kikuchi, M., Matsumoto, K., Imaizumi, T., Miyagishi, M., Taira, K., Foy, E., Loo, Y.M., Gale, M., Jr., Akira, S., Yonehara, S., Kato, A., and Fujita, T. (2005). Shared and unique functions of the DExD/H-box helicases RIG-I, MDA5, and LGP2 in antiviral innate immunity. J. Immunol. *175*, 2851–2858.

Zauli, G., Rimondi, E., Stea, S., Baruffaldi, F., Stebel, M., Zerbinati, C., Corallini, F., and Secchiero, P. (2008). TRAIL inhibits osteoclastic differentiation by counteracting RANKL-dependent p27Kip1 accumulation in pre-osteoclast precursors. J. Cell. Physiol. *214*, 117–125.

Part IV

Epidemiology and Evolutionary Dynamics

Dengue Virus–Mosquito Interactions

8

Eng-Eong Ooi and Duane J. Gubler

Abstract

Dengue/dengue haemorrhagic fever is the most important vector-borne viral disease globally, with over half of the world's population living at risk of infection. While vaccines for other flaviviruses such as yellow fever, Japanese encephalitis and tick-borne encephalitis have been developed, dengue vaccine development is complicated by the need to incorporate all four virus serotypes into a single formulation. The only way to prevent dengue transmission presently is to reduce the vector population. This chapter focuses on the latest information on mosquito–dengue virus interaction, with the overall goal of identifying areas of research where improved understanding would likely contribute to our ability to predict and prevent cyclical epidemics.

Introduction

Dengue fever/dengue haemorrhagic fever (DF/DHF) is the most important vector-borne viral disease in the world. Frequent and cyclical epidemics throughout the tropical world put an estimated 3 billion people at risk of acute illness each year (Halstead, 2008). This re-emergence coincides with the expanded geographical distribution of its principal vector, *Aedes aegypti*, along with other factors such as neglected public health infrastructure and migration of the human population from rural areas to urban centres (Wilcox *et al* 2007). Many parts of the world have now become hyperendemic for dengue virus (DENV), meaning that all four serotypes of the DENV now be found co-circulating in the same countries. Other *Aedes* species, including *Aedes albopictus* and *Aedes polynesiensis,* also serve as vectors for DENV, albeit less efficiently than *Ae. aegypti* (Gubler, 1987, 1988).

The recent geographical expansion of chikungunya, another virus transmitted by *Ae. aegypti* and *Ae. albopictus,* serves as a bellwether of future DENV and other *Aedes* transmitted arboviruses. The 2007 outbreak of chikungunya fever in Italy (Angelini *et al.*, 2007; Beltrame *et al.*, 2007; Rezza *et al.*, 2007), occurred as a result of the recent expansion in the geographic range of *Ae. albopictus* followed by the importation of chikungunya virus by travellers (Bonilauri *et al.*, 2007). The expansion of *Ae. albopictus* in the US and southern Europe suggests that dengue could once again expand beyond its current distribution in the tropics, especially since the importation of dengue fever cases to non-endemic regions has increased dramatically with increasing epidemic activity in the tropical endemic areas (Gubler 1998, 2006; Gould *et al.*, 2006; Diseases of Environmental and Zoonotic Origin Team European Centre for Disease Prevention and Control, 2007; Hashimoto *et al.*, 2007). However, because *Ae. albopictus* is usually an inefficient epidemic vector, outbreaks in these areas would be expected to be small (Gubler, 2003).

Although efforts to develop anti-dengue drugs are in progress (see Chapters 13 and 14), there is as yet no specific therapeutic treatment for DF or DHF. However, with proper clinical diagnosis and management, DHF mortality rates can be kept to <1% (Nimmannitya, 1997).

Prevention of this disease is thus imperative. While vaccines for other flaviviral diseases such as yellow fever, Japanese and tick-borne encephalitis have been developed, dengue vaccine development is complicated by the need to produce tetravalent immunity to all four virus serotypes in a single formulation. An approved vaccine is thus not likely to be available for 5 to 7 years (see Chapter 12). The only way to prevent DENV transmission, therefore, is to reduce the vector population. This chapter focuses on mosquito–dengue virus interactions (overview of discussion is provided in Fig. 8.1), with the overall goal of identifying areas of research where improved understanding would likely contribute to our ability to predict and prevent the currently observed cyclical epidemics.

Virus–vector relationships

The strong association of *Ae. aegypti* with epidemic dengue compared with transmission by the other mosquito vectors (Gubler, 2004a) serves as a useful starting point for our discussion. This association is perhaps best illustrated by the history of dengue in South-East Asia.

Dengue rapidly emerged as a significant public health burden in South-East Asia following the Second World War, resulting from the significant ecologic disruption and demographic changes. The movement of equipment and people resulted in the transportation of *Ae. aegypti* and the dengue viruses to new geographic areas. The use of containers to store water for domestic use as well as for fire control following destruction of the then existing water systems, along with the presence of discarded war equipment and junk, all served to provide ideal conditions for *Ae. aegypti* to thrive. The movement of Japanese and Allied troops in and out of the region also served to provide susceptible hosts (Gubler, 1997).

While the activities in the Second World War served to expand the geographic distribution of both the DENV and their vectors, post-war urbanization of South-East Asia served to propagate the virus in the region. Millions of people migrated from the rural areas to the cities seeking employment. This resulted in rapid, unplanned growth of urban centres in many South-East Asian countries. Housing, water supply and sewerage systems were inadequate. This mixture, an ideal breeding habitat for the highly domesticated *Ae. aegypti* as well as susceptible human hosts, resulted in increasingly larger and more frequent epidemics of dengue, and the emergence of DHF (Gubler, 2004a).

The importance of large urban centres in South-East Asia in the epidemiology of DF/DHF is underscored by a study that modelled dengue epidemics in Thailand and suggested that epidemic strains of DENV emerged and spread from Bangkok with an approximate 3-year periodicity. These studies underscore the importance of urban centres in the maintenance of DENV during interepidemic periods and the emergence of dengue epidemics.; see also Chapter 10). Following an increase in dengue cases in children residing in Bangkok, epidemic waves appeared outside of the capital progressively, moving away at a speed of 148 km per month (Cummings *et al.*, 2004). Such travelling waves of dengue are also not unique to Thailand (Gubler, 2004b). Dengue virus surveillance in Indonesia in the 1970s showed a similar trend with the 1976–78 epidemics of DHF which began in Jakarta and then spread throughout the country over the next 2 years (Gubler *et al.*, 1979, 2004b

It was the post-Second World War setting of economic and urban growth that epidemic DHF emerged as a leading public health burden in South-East Asia. Manila, Philippines recorded the first outbreak in 1953/54, with a second 2 years later in 1956 (Gubler, 1997). Bangkok had an epidemic in 1958 although sporadic cases of DHF were retrospectively identified in Thailand throughout the 1950s (Hammon, 1973; Halstead, 1980). Subsequently, dengue epidemics have occurred cyclically every 3–5 years, with each epidemic generally being larger in magnitude than those before.

The re-infestation of the American tropics by *Ae. aegypti* following cessation of the Pan American Health Organization *Ae. aegypti* eradication programme in the 1970s was rapidly followed in each country by re-emergence of epidemic dengue, beginning in the Caribbean islands and spreading from Mexico in the north to Argentina in the south (Gubler, 1997; Ramos *et al.*, 2008; Vezzani and Carbajo, 2008). The

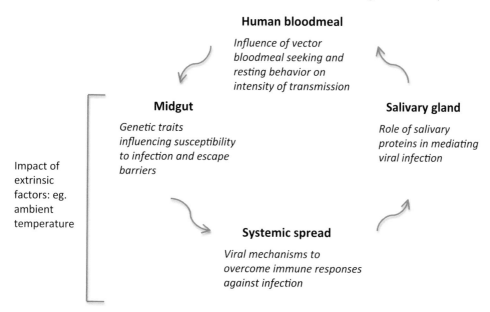

Figure 8.1 Life cycle of dengue virus in the mosquito host: an overview of the factors influencing virus transmission.

primary association of *Ae. aegypti* with transmission of major dengue epidemics gives rise to two important questions.

What is the relationship between vector susceptibility and epidemic transmission?

Epidemiologic observations, along with experimental data, serve to incriminate mosquito species as vectors for a specific pathogen. The direct relationship between *Ae. aegypti* and epidemic dengue thus suggests that this species is a more efficient epidemic vector for DENV than other species. However, experimental infection data are not as clear cut. Significant variation in susceptibility among both *Ae. aegypti* and *Ae. albopictus* strains from different geographic regions has been observed (Gubler and Rosen, 1976a; Gubler *et al.*, 1979; Bennett *et al.*, 2005; Schneider *et al.*, 2007), but in general, *Ae. aegypti* is less susceptible to infection by DENV than *Ae. albopictus* (Gubler *et al.*, 1979; Jumali *et al.*, 1979; Rosen *et al.*, 1985; Gubler *et al.*, 1987). However, susceptibility of the mosquito vector and transmission dynamics are also influenced by the strain of infecting virus (Gubler and Rosen, 1977; Anderson and Rico-Hesse, 2006; Hanley *et al.*, 2008).

The lower susceptibility of *Ae. aegypti* to oral infection could act as a selecting mechanism for more virulent strains of DENV (Gubler, 1987). One of the risk factors for severe dengue is viraemia levels during the febrile phase of illness (Gubler *et al.*, 1978; Gubler *et al.*, 1981; Vaughn *et al.*, 2000; Wang *et al.*, 2003; Tanner *et al.*, 2008). Viraemia titres are influenced by factors intrinsic to both the human host and the virus. The lower susceptibility of *Ae. aegypti* would require dengue viruses capable of growing to high titres in humans in order for the mosquito to be infected through a blood meal, thus selecting for and transmitting viruses that would likely cause more severe human disease (Gubler, 1987; Cologna *et al.*, 2005). Following this logic, *Ae. albopictus* could transmit viral strains that do not replicate to such high titres and thus result in less clinically overt or severe disease, with a resultant reduction in public health burden of dengue compared to transmission by *Ae. aegypti*. This scenario could also explain the role of *Ae. albopictus* as a maintenance vector involved in the silent transmission of DENV during interepidemic periods (Gubler, 1987, 1998).

In addition to strain variation in mosquito vector competence, intrinsic permissiveness for

DENV infection, replication and transmission, has also been observed for both *Ae. aegypti* and *Ae. albopictus* (Gubler and Rosen, 1976a; Gubler and Rosen, 1977; Gubler *et al.*, 1979; Hardy, 1988; Woodring *et al.*, 1996). In sub-Saharan Africa, *Ae. aegypti formosus*, which is a subspecies of *Ae. aegypti*, predominates. Compared to *Ae. aegypti aegypti*, this subspecies has low vector competence for flaviviruses due primarily to a midgut infection barrier (Gubler *et al.*, 1979; Tabachnick *et al.*, 1985).

The mechanism that gives rise to these observed interspecific and inter-strain differences in vector susceptibility to DENV infection has yet to be determined. Once ingested by its mosquito host, DENV must overcome several barriers to be transmitted to a subsequent human host. Firstly, the virus must establish a productive infection in the mosquito midgut, after which it must disseminate from the gut and replicate in other tissues. Ultimately, virus must then infect the salivary glands and be shed in the saliva for transmission to the next vertebrate host. Genetic traits that influence midgut infection and escape barriers have been mapped to several loci on the *Ae. aegypti* chromosomes (Bosio *et al.*, 2000; Gomez-Machorro *et al.*, 2004; Bennett *et al.*, 2005) indicating that vector competence is genetically determined. The use of new genetic tools for studying gene expression, such as siRNA, along with the full genome sequence of *Ae. aegypti* (Nene *et al.*, 2007) and *Ae. albopictus* (Gaines *et al.*, 1996; Olson *et al.*, 2002) could further improve our understanding of the factors that determine vector competence. One such factor is the immune response to DENV infection in the mosquito vector. Waterhouse *et al.* (2007) analysed the immune signalling pathways and response and identified both conservative and rapidly evolving features associated with different functional gene categories and particular aspects of immune reactions in DENV-infected *Ae. aegypti*. Using high-throughput gene expression and RNA interference (RNAi)-based reverse genetic analyses, Xi and colleagues (2008) showed a significant role for the Toll pathway in regulating resistance to DENV in *Ae. aegypti*. Furthermore, their study showed that the mosquito's natural microbiota modulate DENV infection, possibly through basal-level stimulation of the Toll

immune pathway. The availability of such high-throughput genomic and epigenomic approaches will improve our understanding of the critical factors that determine transmissibility of DENV, which may in turn improve our ability to predict and pre-empt dengue epidemics.

The elucidation of the molecular mechanisms underlying vector competence has given rise to the possibility of preventing virus transmission with the use of transgenic mosquitoes that are refractory to virus infection. Indeed, the methodology has already been shown to be effective in the laboratory (Kokoza *et al.*, 2000). Current approaches to developing modified vector populations by introgressing genes that eliminate vector competence include synthetic transposable elements (Adelman *et al.*, 2002), meiotic drive (Mori *et al.*. 2004) and underdominance (Davis *et al.*, 2001). While these approaches are exciting, they will need to be investigated and evaluated intensively in the laboratory and tested in large-cage field trials. Nonetheless, these efforts are likely to generate improved understanding of vector competence at the molecular level.

How do differences in vector behaviour influence transmission dynamics?

Mosquitoes differ in their bloodmeal preferences. *Ae. aegypti* feeds almost exclusively on humans whereas *Ae. albopictus* has a more catholic feeding behaviour. The latter species will feed on a variety of animals, including mammals, birds and reptiles. This lack of host specificity results in dead-ends for DENV, since it does not replicate efficiently in vertebrates other than humans and lower primates (Halstead *et al.*, 1973; Marchette *et al.*, 1973). This difference in feeding pattern could account for the observation that all major epidemics of DHF have been associated with *Ae. aegypti* transmission (Gubler, 1987, 2004a).

The difference in host-seeking behaviour between *Ae. aegypti* and *Ae. albopictus* is likely to result in differences in the basic reproduction number (R_0) of DENV infections in human populations. R_0 is the number of cases that result from the introduction of a single case into a susceptible population and thus provides an indication of the epidemic potential of a pathogen (Anderson and May, 1992). The higher the R_0, the greater the transmission and hence the likelihood of epidemics. There are several different ways of calculating R_0. In its simplest form, R_0 is a product of the infectiousness of a pathogen, the number of contact events that would lead to pathogen transmission, and the duration of transmissibility of the pathogen. Several studies have estimated the R_0 of dengue and a wide range of numbers have been reported, from as low as 1.33 to a high of 11.6. (see review by Halstead, 2008).

Both mosquito feeding and resting behaviour may affect the DENV R_0. *Ae. aegypti* is a cautious feeder. The slightest movement of the host from which it is taking its blood meal is likely to send it fleeing to seek another host. As a result, *Ae. aegypti* is likely to bite or probe 3–5 persons to obtain sufficient nutrients for one gonotrophic cycle; virus can be transmitted to the host with each probe (Gubler and Rosen, 1976b), thus resulting in a proportionately large number of human infections and increasing the R_0. *Ae. albopictus*, on the other hand, lives up to its common name of 'Tiger mosquito'. Unlike *Ae. aegypti*, it is not easily deterred from feeding and will often take a full bloodmeal before leaving the host and seeking a place to rest. Resting behaviour is another factor that can influence survival and thus the R_0. The probability of survival is higher for mosquitoes resting indoors compared to outdoors. *Ae. aegypti* prefers the former whereas *Ae. albopictus* prefers the latter, a difference that may enhance the survival of *Ae. aegypti* relative to *Ae. albopictus*. Also, resting indoors puts *Ae. aegypti* in closer contact with the human host. Thus the differences in the feeding and resting behaviour of these mosquitoes is expected to result in greater survival and contact rates for *Ae. aegypti*, leading to a greater R_0 for dengue transmitted by this mosquito. The R_0 of DENV transmitted by *Ae. aegypti* and *Ae. albopictus* has never been compared, but such a study is difficult to perform since most dengue endemic areas are infested with both species of vectors. However, the outbreak of dengue fever in Hawaii or perhaps the uneven distribution of *Ae. aegypti* and *Ae. albopictus* in Singapore and Taiwan could serve as field sites to gather data useful for such a comparison. Because of the extensive variation in vector competence among mosquito strains, and infectivity or epidemic potential among dengue virus strains, however, estimation of R_0 in one endemic area may not reflect the R_0 in other areas where strains of mosquitoes and viruses are different.

Public health implications

The issues discussed above have relevance to prevention and control of dengue. The differences in the impact of *Ae. aegypti* and *Ae. albopictus* on the evolution and ecology of DENV, could be exploited for disease prevention by directing the focus of preventive measures against the former and not the latter. This, in fact was the practice during the *Ae. aegypti* eradication campaign in the 1950s and 1960s in places like Hawaii and Guam, where *Ae. aegypti* was eliminated and replaced by *Ae. albopictus* (Schliessmann, 1967). With the exception of the small 2001 outbreak of dengue in Hawaii, both islands have remained free of epidemic dengue (Gubler, 2003). These programmes selectively targeted the *Ae. aegypti* populations through larval and adult mosquito surveillance coupled to direct mosquito control efforts. The selective pressure on *Ae. aegypti* allows *Ae. albopictus* to thrive and outcompete *Ae. aegypti* for larval habitats and food. Another strong justification for this strategy is the history of dengue in South-East Asia, where, despite the fact that *Ae. albopictus* is native to the region (Chan *et al.*, 1983; Gubler, 1998), dengue epidemics only emerged after the introduction of *Ae. aegypti* into the region. Furthermore, despite the introduction of *Ae. albopictus* into the United States in 1985 (Hawley *et al.*, 1987), major dengue epidemics have only occurred south of the border with Mexico where *Ae. aegypti* is entrenched (Flisser *et al.*, 2002; Gubler, 2004a; Arredondo-Jiménez and Valdez-Delgado, 2006; Mercado-Hernandez *et al.*, 2006). Vector control campaigns that specifically target *Ae. aegypti* might be an effective approach to roll back dengue in South-East Asia to the low levels of transmission and disease severity that were the status quo prior to the Second World War.

Emerging trend of *Ae. albopictus*?

While the above approach to dengue epidemic prevention warrants consideration, recent events strike a note of caution. The geographic expansion of another virus that is transmitted by both *Ae. aegypti* and *Ae. albopictus*, Chikungunya virus, has raised concern that dengue could soon follow the same track into the Mediterranean regions, including Southern Europe where *Ae. albopictus* has infested many countries (Dalla Pozza and Majori, 1992; Aranda *et al.*, 2006; Klobucar *et al.*, 2006). This concern arises from the observation that *Ae. albopictus* and not *Ae. aegypti* appeared to be the main, and in the case of the outbreak in La Reunion (Reiter *et al.*, 2006) the exclusive, vector in epidemic transmission of the Indian Ocean strain of chikungunya virus (de Lamballerie *et al.*, 2008). Moreover, observations in southern Thailand have suggested that this species of mosquito may have evolved into an anthropophilic blood meal seeker, where all of the 105 *Ae. albopictus* sampled had fed exclusively on humans (Ponlawat and Harrington, 2005). While a mutation in the chikungunya virus genome is likely to have had significant influence on the epidemic transmission (Tsetsarkin *et al.*, 2007), the chikungunya experience raises concerns that *Ae. albopictus* could also transmit DENV with the same capacity as *Ae. aegypti*. However, the natural experiments of Hawaii, Guam and northern Taiwan, where *Ae. aegypti* has been replaced by *Ae. albopictus* for 40 years, suggests that anthropophily in *Ae. albopictus* does not occur often (Gubler, 2004a). Studies on the blood meal preference and biting frequency of multiple strains of *Ae. albopictus* are urgently needed.

In addition to the recent trends in chikungunya virus transmission, the increasing trend of DENV infection among the adult population in Singapore also raises the question whether our understanding of the differences in the vector capacity of *Ae. aegypti* and *Ae. albopictus* remains accurate. Adult DENV infection has been reported in the Americas and in some parts of Asia for many years and can be explained to a large extent by the rural to urban migration that has resulted in changing urban demographics (Wilcox, *et al* 2007). While the trend has not been investigated in depth in most countries, the near 20-year long resurgence of dengue in Singapore has been studied epidemiologically (Ooi *et al.*, 2005). The shift of dengue from an almost exclusively paediatric to a predominantly adult disease has been attributed to a combination of herd immunity (Goh, 1993, 1998) and a change from a domestic-based to a non-domestic-based virus transmission pattern

(Ooi *et al.*, 2001, 2005). The change in transmission pattern, as indicated by a seroprevalence study in Singapore children (Ooi *et al.*, 2001), has implications for the role of the different vectors. Two explanations can be offered for this phenomenon: first, the domestic-focused vector control strategy designed in the late 1960s (Chan, 1983; Ooi *et al.*, 2005) has selected for *Ae. aegypti* strains that thrive extradomestically; alternatively, peridomesticated *Ae. albopictus* may be an important driver of the observed epidemiological trend in Singapore.

Multiple factors could contribute to the emergence of *Ae. albopictus* as the main vector of DENV in Singapore as well as in other places where peri-domestic virus transmission patterns occur. First, the intensity of virus transmission could be affected by the population density of *Ae. albopictus*, which is often not specifically targeted by vector control efforts. Second, high human population density, such as that in Singapore, could mean that the availability of human hosts for blood meals far outnumbers that of the other animal species, thus reducing the likelihood that DENV would be transmitted to dead-end animal hosts. This would influence the third possibility, that *Ae. albopictus* may evolve to seek human blood in preference to blood from other animals in these regions (Ponlawat and Harrington, 2005). These issues need to be addressed with well-structured field studies. Active vector and virological surveillance will need to be carried out alongside periodic serological surveys of the human and animal population to characterize virus transmission. These data will also need to be supplemented with analyses of field-caught blood-fed mosquitoes for the bloodmeal host preferences.

Extrinsic factors

Effect of temperature, rainfall and implications of global warming

Generally, dengue epidemics correlate with fluctuations in ambient temperature and rainfall, although exceptions to this correlation exist (Halstead, 2007). Higher temperature reduces the length of both the mosquito gonotrophic cycle and the viral extrinsic incubation period. Given two adult female mosquitoes with a similar lifespan, the mosquito with a shorter gonotrophic cycle will have more egg-laying episodes and thus take more blood meals. Thus, shortening the gonotrophic cycle will likely increase the mosquito population, which may or may not result in increased virus transmission. In addition, as the extrinsic incubation period is shortened, the probability of transmission is increased.

Adult mosquitoes that emerge from larval stages that experience warmer temperatures are smaller in size. Although not well documented, smaller adults may have to feed more often than larger adults, thus increasing the R_0 of DENV (Focks et al., 1995; Jetten and Focks, 1997). The lowest temperature in the day appears to be a more important determinant of dengue transmission than the average daily temperature, with a correlation coefficient of 0.81 (Yasuno and Tonn, 1970). However, this observation too should be documented.

All the above factors suggest that a rise in temperature may result in increased transmission of DENV. To demonstrate the direct relationship between temperature and dengue incidence, Cheng and colleagues have modelled the effect of temperature on the incidence of DHF and have used sea surface temperatures to predict the onset of dengue epidemics in Indonesia (Cheng et al., 1998). This study suggests that high ambient temperature gives rise to conditions that favour DENV transmission. On the other hand, higher temperature also has a negative impact on the survival of the adult vector. Using field derived data from Bangkok and New Orleans, Focks and colleagues estimated a range from 0.87 to 0.91 attrition per day (Focks et al., 1993). With higher temperature, the attrition rate is likely to increase further since maximum survival rates have been observed to be limited to 20°C to 30°C (Tun-Lin et al., 2000). Such studies however, are heavily influenced by other environmental factors (Halstead, 2007) and a holistic approach is needed to integrate the various effects of temperature on DENV transmission.

Based on the experimental evidence presented above, it seems intuitive that increased temperature equates to increased DENV transmission. However, natural seasonal transmission cycles of dengue suggest otherwise! Dengue viruses occur in all tropical areas of the world, where periodic epidemics occur, the latter being determined by a number of factors related to the environment, the vector, the virus and the human host. With few exceptions, dengue epidemics in endemic countries both above and below the equator, occur during the rainy season, which generally has lower temperature by a few degrees than the hot dry season that precedes the monsoon rains. Thus, epidemiologic observations and natural experiments demonstrate that increased temperatures alone will not result in increased transmission.

Rainfall has also been reported to affect the abundance of Ae. aegypti, although geographical variation in this pattern has been reported. In Puerto Rico, which has a reliable piped water system, a positive correlation between rainfall and vector population density was observed (Moore et al., 1978). Rain-filled containers such as discarded tires were the most productive sources of Ae. aegypti larvae. In contrast, a series of studies in Bangkok showed that rainfall had no influence on the availability of Ae. aegypti larval habitats (Tonn et al., 1969; Pant and Yasuno, 1973). With the exception of ant traps, all containers were filled not by rainfall but manually, such as watering of potted flowering plants. Consequently, there was no observed fluctuation in adult Ae. aegypti population density with the changing rainy and hot dry seasons. Obviously, differences in environmental management could impact the types of larval habitats and hence account for whether the intensity of DENV transmission is affected by seasonal rainfall. These two instances indicate that vector control for preventing dengue virus transmission cannot be generalized but will have to address issues specific for each locality.

Global warming is predicted to result in increased mean temperature and altered rainfall patterns in many parts of the world. Many voices have raised the concern that global warming is likely to increase the geographical expansion of the dengue vectors and the frequency and magnitude of dengue epidemics (Shope, 1992; Patz et al., 1998; Morens and Fauci, 2008). While the relationship between temperature and dengue incidence has been examined in several places, these are mainly in the tropics. Warmer temperatures in regions with temperate climate may allow temperatures suitable for DENV transmission to occur over a longer period of time, but should not influence other conditions such as housing type, piped water systems and living standards in general, all of which influence transmission of dengue viruses. It should be noted that cooler temperatures did not keep dengue epidemics at bay. In the nineteenth century dengue and yellow fever epidemics transmitted by *Ae. aegypti* were well documented in USA as far north as Boston and Philadelphia (Reed *et al.*, 1900). This coincided with the slave trade, as a result of which the movement of infected human hosts from Africa in trans-Atlantic ships brought the viruses to naive populations living in cities that were seasonally infested with *Ae. aegypti*, resulting in explosive epidemics (Gubler, 2004a). At present, there is no good scientific evidence that the recent geographic expansion of DENV virus and its vectors has been due to global warming. Instead, the magnitude of movement of the human population and trade materials, uncontrolled and poorly planned expansion of urban centres, as well as the lack of commitment to disease prevention in dengue endemic regions have all served to produce conditions ideal for DENV transmission (Gubler *et al.*, 2001). It will, however, be important to understand how global warming will alter human behaviour and infrastructure development, as these may impact on the availability of vector breeding sites as well as mosquito–human contact. For instance the widespread use of air-conditioning in Laredo, Texas, USA, serviced by reliable piped water and drainage, resulted in low DENV infection rates even though *Ae. aegypti* was common in Laredo, and dengue epidemics are common in Nuevo Laredo, Mexico, just south of the border (Reiter *et al.*, 2003).

Influence of mosquito salivary antigens on viral infection

Recent reports indicate that mosquito saliva plays a significant role in creating a condition suitable for some viruses to infect and replicate in humans (Schneider and Higgs, 2008). Salivary proteins appear to mediate viral infection through immunomodulation of the human host as well as potentiation of the infectivity of the virus. These studies are still in their infancy but results are interesting and this area could be an important component in the pathogenesis of dengue.

Mosquitoes secrete saliva into the feeding site when taking a blood meal. The salivary proteins prevent vasoconstriction, coagulation, platelet aggregation and inflammation; all of which would reduce the availability of blood to the mosquito. These proteins, many of which remain to be characterized (Valenzuela *et al.*, 2002; Ribeiro *et al.*, 2007), appear to enhance virus infectivity and may affect subsequent clinical outcome. Schneider and colleagues recently demonstrated that mice co-inoculated with a constant titre of West Nile virus (WNV) but varying mosquito saliva showed a dose-response effect on the degree of brain infection (Schneider *et al.*, 2006). This suggests that inoculation of the same virus would result in different clinical outcomes depending on whether the virus was administered by needle or by mosquito. Similar observations have been reported with other arboviruses and these have been reviewed recently by Schneider and Higgs (2008).

One of the ways in which mosquito salivary proteins may enhance virus infectivity is to modulate the early immune response against the virus. The addition of *Ae. aegypti* salivary gland extract into splenocyte cultures *in vitro* resulted in down-regulation of IL-2 and IFNγ production (Cross *et al.*, 1994). In contrast, splenocytes obtained from mice in which *Ae. aegypti* has taken a blood meal produced high levels of IL-4 and IL-10 (Zeidner *et al.*, 1999). This effect could be observed up to 10 days after the bite of the mosquito. More recently, Schneider and colleagues showed that prior exposure to uninfected mosquito saliva increases mortality rate in mice infected with West Nile virus inoculated through the bite of an infective mosquito (Schneider *et al.*, 2007). Collectively, these studies indicate that mosquito saliva has an important effect on the human immune response.

The differences in cytokine profile in splenocytes, particularly the observed increased expression of IL-10, could represent a shift from a cytotoxic T-cell response to a largely humoral response during the early stages following inoculation of DENV into the human host. Since cytotoxic T-cell responses would be more appropriate for an intracellular infection, a suppression of cellular response in favour of B-cell responses could produce conditions that allow DENV virus to survive in the host. Wasserman and colleagues showed that *Ae. aegypti* saliva affected T- and B-cell proliferation in a dose-dependent manner. Lower concentrations of saliva inhibited T-cell proliferation while higher concentrations suppressed pro-inflammatory responses and decreased T-cell viability (Wasserman *et al.*, 2004). In addition, both types 1 and 2 interferons (IFNs) have been observed to be suppressed by mosquito saliva (Schneider *et al.*, 2004). Type 1 IFN appears to have potent antiviral activity for DENV (Diamond *et al.*, 2000; Diamond and Harris, 2001). By suppressing such antiviral cytokine expression following the introduction of DENV into the human host, the mosquito saliva could thus play a significant role in creating an environment suitable for DENV to replicate.

Apart from immunomodulation, mosquito saliva could also support DENV survival by creating a microenvironment composed of susceptible host cells for infection. Mosquito saliva can induce type-I hypersensitivity reactions (Demeure *et al.*, 2005). This could in turn trigger the release of chemokines that result in migration of monocytes into the extravascular space, as has been observed with in mice fed upon by *Anopheles stephensi*. Serum macrophage induced protein-2 (MIP-2) concentration was increased in the skin and lymph nodes of animals and the consequent migration of monocytes to the site of inoculation could serve to benefit the dengue virus. Furthermore, Chen *et al.*, 1998 reported that mice sensitized to mosquito saliva produced high levels of IgE and corresponding differences in IFNγ expression, suggesting that hypersensitivity reaction could create conditions favourable for virus replication.

While studies have been conducted to investigate the effect of *Ae. aegypti* saliva and their immunomodulatory properties in cultured cells and animal models, none has examined how this might affect DENV infection in humans and its clinical outcome. Again, the history of dengue in South-East Asia suggests that this is a study worth doing since DHF epidemics were not observed until after the introduction of *Ae. aegypti* to the region. Besides its ability to transmit the virus to a larger number of people compared to *Ae. albopictus*, there could well be differences in the salivary proteins of these two mosquito species that could affect overall dengue epidemiology and clinical disease. Indeed, the saliva of anthropophilic *Ae. aegypti* has been shown to suppress mammalian splenocyte activation compared to that of *Culex pipiens*, an ornithophilic species (Wanasen, 2004). Investigation into how vector salivary proteins modulate immune response and support viral infection and replication, as well as their impact on disease pathogenesis and outcome should thus be supported and would likely yield information critical to our understanding of dengue.

Conclusions

The DENV–mosquito interaction remains a field for potentially exciting research that may have important implications for understanding transmission dynamics, and infection, pathogenesis and immune response in the human host. Moreover, results from these studies can contribute to improving our understanding of the driving forces behind epidemic transmission of DENV and potentially improve our ability to protect more than half of the world's population from this devastating disease.

References

Adelman, Z.N., Jasinskiene, N., and James, A.A. (2002). Development and applications of transgenesis in the yellow fever mosquito, *Aedes aegypti*. Mol. Biochem. Parasitol. *121*, 1–10.

Anderson, J.R., and Rico-Hesse R. (2006). *Aedes aegypti* vectoral capacity is determined by the infecting genotype of dengue virus. Am. J. Trop. Med. Hyg. *75*, 886–892.

Anderson, R., and May, R. (1992). Infectious Diseases of Humans (Oxford: Oxford University Press).

Angelini, R., Finarelli, A.C., Angelini, P., Po, C., Petropulacos, K., Silvi, G., Macini, P., Fortuna, C., Venturi, G., Magurano, F., Fiorentini, C., Marchi, A., Benedetti, E., Bucci, P., Boros, S., Romi, R., Majori, G., Ciufolini, M.G., Nicoletti, L., Rezza, G., and Cassone, A. (2007). Chikungunya in north-eastern Italy: a summing up of the outbreak. Euro. Surveill. 12, E071122.2.

Aranda, C., Eritja, R., and Roiz, D. (2006). First record and establishment of the mosquito *Aedes albopictus* in Spain. Med. Vet. Entomol. *20*, 150–152.

Arredondo-Jiménez, J.I., and Valdez-Delgado, K.M. (2006). *Aedes aegypti* pupal/demographic surveys in southern Mexico: consistency and practicality. Ann. Trop. Med. Parasitol. 100 Suppl 1,S17-S32.

Beltrame, A., Angheben, A., Bisoffi, Z., Monteiro, G., Marocco, S., Calleri, G., Lipani, F., Gobbi, F., Canta, F., Castelli, F., Gulletta, M., Bigoni, S., Del Punta, V., Iacovazzi, T., Romi, R., Nicoletti, L., Ciufolini, M.G., Rorato, G., Negri, C., and Viale, P. (2007). Imported chikungunya infection, Italy. Emerg. Infect. Dis. *13*, 1264–1266.

Bennett, K. E., Beaty, B. J., and Black, W. C. (2005). Selection of DS3, an *Aedes aegypti* strain with high oral susceptibility to dengue 2 virus and DMEB., a strain with a midgut barrier to dengue escape. J. Med. Entomol. *42*, 110–119.

Bennett, K.E., Flick, D., Fleming, K.H., Jochim, R., Beaty, B.J., and Black, W.C. (2005). Quantitative trait loci that control dengue-2 virus dissemination in the mosquito *Aedes aegypti*. Genetics *170*, 185–194.

Bonilauri, P., Bellini, R., Calzolari, M., Angelini, R., Venturi, L., Fallacara, F., Cordioli, P., Angelini, P., Venturelli, C., Merialdi, G., and Dottori, M. (2008). Chikungunya virus in *Aedes albopictus*, Italy. Emerg. Infect. Dis. *14*, 852–854.

Bosio, C. F., Beaty, B. J., and Black, W. C. (1998). Quantitative genetics of vector competence for dengue-2 virus in *Aedes aegypti*. Am. J. Trop. Med. Hyg. *59*, 965–970.

Bosio, C. F., Fulton, R.E., Salasek, M. L., Beaty, B. J., and Black, W. C. (2000). Quantitative trait loci that control vector competence for dengue-2 virus in the mosquito *Aedes aegypti*. Genetics *156*, 687–698.

Chen, Y.L., Simons, F.E., and Peng,Z. (1998). A mouse model of mosquito allergy for study of antigen-specific IgE and IgG subclass responses, lymphocyte proliferation, and IL-4 and IFN-gamma production. Int. Arch. Allergy Immunol. *116*, 269–277.

Cheng, S., Kalkstein, L.S., Focks, D.A., and Nnaji, A. (1998). New procedures to estimate water temperatures and water depths for application in climate-dengue modeling. J. Med. Entomol. *35*, 646–652.

Cologna, R., Armstrong, P.M., and Rico-Hesse, R. (2005). Selection for virulent dengue viruses occurs in humans and mosquitoes. J. Virol. *79*, 853–859.

Cross, M.L., Cupp, E.W., and Enriquez, F.J. (1994). Differential modulation of murine cellular immune responses by salivary gland extract of *Aedes aegypti*. Am. J. Trop. Med. Hyg. *51*, 690–696.

Dalla Pozza, G., and Majori, G. (1992). First record of *Aedes albopictus* establishment in Italy. J. Am. Mosq. Control Assoc. *8*, 318–320.

Davis, S., Bax, N., and Grewe, P. (2001). Engineered under-dominance allows efficient and economical introgression of traits into pest populations. J. Theor. Biol. *212*, 83–98.

de Lamballerie, X., Leroy, E., Charrel, R.N., Tsetsarkin, K., Higgs, S., and Gould, E.A. (2008). Chikungunya virus adapts to tiger mosquito via evolutionary convergence: a sign of things to come? Virol. J. *5*, 33.

Demeure, C.E., Brahimi, K., Hacini, F., Marchand, F., Péronet, R., Huerre, M., St-Mezard, P., Nicolas, J.F., Brey, P., Delespesse, G., and Mécheri, S. (2005). *Anopheles* mosquito bites activate cutaneous mast cells leading to a local inflammatory response and lymph node hyperplasia. J. Immunol. *174*, 3932–3940.

Diamond, M.S., Roberts, T.G., Edgil, D., Lu, B., Ernst, J., and Harris, E. (2000). Modulation of Dengue virus infection in human cells by alpha, beta, and gamma interferons. J. Virol. *74*, 4957–4966.

Diamond, M.S., and Harris, E. (2001). Interferon inhibits dengue virus infection by preventing translation of viral RNA through a PKR-independent mechanism. Virology *289*, 297–311.

Diseases of Environmental and Zoonotic Origin Team European Centre for Disease Prevention and Control. (2007). Dengue worldwide: an overview of the current situation and the implications for Europe. Euro. Surveill. 12, E070621.1.

Flisser, A., Velasco-Villa, A., Martínez-Campos, C., González-Domínguez, F., Briseño-García, B., García-Suárez, R., Caballero-Servín, A., Hernández-Monroy, I., García-Lozano, H., Gutiérrez-Cogco, L., Rodríguez-Angeles, G., López-Martínez, I., Galindo-Virgen, S., Vázquez-Campuzano, R., Balandrano-Campos, S., Guzmán-Bracho, C., Olivo-Díaz, A., de la Rosa, J., Magos, C., Escobar-Gutiérrez, A., and Correa, D. (2002). Infectious diseases in Mexico. A survey from 1995–2000. Arch. Med. Res. *33*, 343–350.

Focks, D.A., Haile, D.G., Daniels, E., and Mount, G.A. (1993). Dynamic life table model for *Aedes aegypti* (diptera: Culicidae): simulation results and validation. J. Med. Entomol. *30*, 1018–1028.

Focks, D.A., Daniels, E., Haile, D.G., and Keesling, J.E. (1995). A simulation model of the epidemiology of urban dengue fever: literature analysis, model development, preliminary validation, and samples of simulation results. Am. J. Trop. Med. Hyg. *53*, 489–506.

Gaines, P.J., Olson, K.E., Higgs, S., Powers, A.M., Beaty, B.J., and Blair, C.D. (1996). Pathogen-derived resistance to dengue type 2 virus in mosquito cells by expression of the premembrane coding region of the viral genome. J. Virol. *70*, 2132–2137.

Gomez-Machorro, C., Bennett, K. E., Munoz, M. L., and Black, W. C. (2004). Quantitative trait loci affecting dengue midgut infection barriers in an advanced intercross line of *Aedes aegypti*. Insect Mol. Biol. *13*, 637–648.

Gould, E.A., Higgs, S., Buckley, A., and Gritsun, T.S. (2006). Potential arbovirus emergence and implications for the United Kingdom. Emerg. Infect. Dis. *12*, 549–555.

Gubler, D.J. (1987). Current research on dengue. In: Current Topics in Vector Research, K.F. Harris, ed. (New York: Springer Verlag Inc.), pp. 37–56.

Gubler, D.J. (1997). Dengue and dengue hemorrhagic fever: its history and resurgence as a global public health problem. In: Dengue and Dengue Hemorrhagic Fever, D.J. Gubler, and G. Kuno, eds. (Oxford, UK: CAB International), pp. 1–22.

Gubler, D.J. (1998). Dengue and dengue hemorrhagic fever. Clin. Microbiol. Rev. *11*, 480–496.

Gubler, D.J. (2003). *Aedes albopictus* in Africa. Lancet Infect. Dis. *3*, 751–752.

Gubler, D.J. (2004a). The changing epidemiology of yellow fever and dengue. 1900 to 2003: a full circle? Comp. Immunol. Microbiol. Infect. Dis. *27*, 319–330.

Gubler, D.J. (2004b). Cities spawn epidemic dengue viruses. Nat. Med. *10*, 129–130.

Gubler, D.J. (2006). Dengue/dengue haemorrhagic fever: history and current status. In: Novartis Foundation Symposium 277: New Treatment Strategies for Dengue and Other Flaviviral Diseases. p. 3–22

Gubler, D.J., and Rosen, L. (1976a). Variation among geographic strains of *Aedes albopictus* in susceptibility to infection with dengue viruses. Am. J. Trop. Med. Hyg. 25, 318–325.

Gubler, D.J., Rosen L (1976b). A simple technique for demonstrating transmission of dengue virus by mosquitoes without the use of vertebrate hosts. Am J Trop Med Hyg; 25;146–50

Gubler, D.J., and Rosen, L. (1977). Quantitative aspects of replication of dengue viruses in *Aedes albopictus* after oral and parenteral infection. J. Med. Entomol. 13, 469–472.

Gubler, D.J., Reed, D., Rosen, L., and Hitchcock, J.R. Jr. (1978). Epidemiologic, clinical and virologic observations on dengue in the Kingdom of Tonga. Am. J. Trop. Med. Hyg. 27, 581–589.

Gubler, D.J., Nalim, S., Tan, R., Saipan, H., and Sulianti Saroso, J. (1979). Variation in susceptibility to oral infection with dengue viruses among gegraphic strains of *Aedes aegypti*. Am. J. Trop. Med. Hyg. 28, 1045–1052.

Gubler, D.J., Suharyono, W., Sumarmo, Wulur, H., Jahja, E., Sulianti Saroso, J. (1979). Virological surveillance for dengue haemorrhagic fever in Indonesia using the mosquito inoculation technique. Bull. World Health. Organ. 57, 931–936.

Gubler, D.J., Suharyono, W., Gunarso, Lubis, I., and Tan, R. (1981). Epidemic dengue 3 in Central Java associated with low viremia. Am. J. Trop. Med. Hyg. 30, 1094–1099.

Gubler, D.J., Kuno, G., and Waterman, S.H. (1983). Neurologic disorders associated with dengue infection. In Proceedings of the International Conference on Dengue/Dengue Hemorrhagic Fever, Kuala Lumpur, Malaysia, 1983. pp 290–306.

Gubler, D.J., Novak, R.J., Vergne, E., Colon, N.A., Velez, M., and Fowler, J. (1985). *Aedes (Gymnometopa) mediovittatus* (Diptera: Culicidae), a potential maintenance vector of dengue viruses in Puerto Rico. J. Med. Entomol. 22, 469–475.

Gubler, D.J., Reiter, P., Ebi, K.L., Yap, W., Nasci, R., and Patz, J.A. (2001). Climate variability and change in the United States: potential impacts on vector- and rodent-borne diseases. Environ. Health Perspect. 109 Suppl 2, 223–233.

Hammon, WMcD. (1973). Dengue hemorrhagic fever – Do we know its cause? Am. J. Trop. Med. Hyg. 22, 82–91

Halstead, S.B. (1980). Dengue hemorrhagic fever. A public health problem and a field for research. Bull. World Health Organ. 58, 1–21.

Halstead, S.B. (2007). Dengue. Lancet. 370, 1644–1652.

Halstead, S.B. (2008). Dengue virus–mosquito interaction. Annu. Rev. Entomol. 53, 15.1–15.19.

Halstead, S.B., Shotwell, H., and Casals, J. (1973). Studies on the pathogenesis of dengue infection in monkeys. I. Clinical laboratory responses to primary infection. J. Infect. Dis. 128, 7–14.

Hanley, K.A., Nelson, J.T., Schirtzinger, E.E., Whitehead, S.S., and Hanson, C.T. (2008). Superior infectivity for mosquito vectors contributes to competitive displacement among strains of dengue virus. BMC Ecology 8, 1.

Hardy, J.L. (1988). Susceptibility and resistance of vector mosquitoes. In The Arboviruses: Epidemiology and Ecology, T. P. Monath, ed. (Boca Raton, F.L. CRC Press), pp. 87–126

Hashimoto, S., Kawado, M., Murakami, Y., Izumida, M., Ohta, A., Tada, Y., Shigematsu, M., Yasui, Y., Taniguchi, K., and Nagai, M. (2007). Epidemics of vector-borne diseases observed in infectious disease surveillance in Japan, 2000–2005. J. Epidemiol. 17 Suppl, S48–55.

Hawley, W.A., Reiter, P., Copeland, R.S., Pumpuni, C.B., and Craig, G.B. Jr. (1987). *Aedes albopictus* in North America: probable introduction in used tires from northern Asia. Science. 236, 1114–1116.

Jetten, T.H., and Focks, D.A. (1997). Potential changes in the distribution of dengue transmission under climate warming. Am. J. Trop. Med. Hyg. 57, 285–297.

Jumali, Sunarto, Gubler, D.J., Nalim, S., Sutrisno, I., Sulianti Saroso J. (1979). Epidemic Dengue Hemorrhagic Fever in rural Indonesia. III Entomological Studies. Am. J. Trop. Med. Hyg. ; 28:717–24.

Klobucar, A., Merdić, E., Benić, N., Baklaić, Z., and Krcmar, S. (2006). First record of *Aedes albopictus* in Croatia. J. Am. Mosq. Control Assoc. 22, 147–148.

Kokoza, V., Ahmed, A., Cho, W.L., Jasinskiene, N., James, A.A., and Raikhel, A. (2000). Engineering blood meal-activated systemic immunity in the yellow fever mosquito *Aedes aegypti*. Proc. Natl. Acad. Sci. U.S.A. 97, 9144–9149.

Marchette, N.J., Halstead, S.B., Falkler, W.A. Jr., Stenhouse, A., and Nash, D. (1973). Studies on the pathogenesis of dengue infection in monkeys. 3. Sequential distribution of virus in primary and heterologous infections. J. Infect. Dis. 128, 23–30.

Mercado-Hernandez, R., Aguilar-Gueta, Jde. D., Fernandez-Salas, I., and Earl, P.R. (2006). The association of *Aedes aegypti* and *Ae. albopictus* in Allende, Nuevo León, Mexico. J. Am. Mosq. Control Assoc. 22, 5–9.

Moore, C.G., Cline, B.L., and Tiben, E.R. (1978). *Aedes aegypti* in Puerto Rico environmental determinants of larval abundance and relation to dengue virus transmission. Am. J. Trop. Med. Hyg. 27,1225–1231.

Morens, D.M., and Fauci, A.S. (2008). Dengue and hemorrhagic fever: a potential threat to public health in the United States. JAMA 299, 214–216.

Mori, A., Chadee, D.D., Graham, D.H., and Severson, D.W. (2004). Reinvestigation of an endogenous meiotic drive system in the mosquito, *Aedes aegypti* (Diptera: Culicidae). J. Med. Entomol. 41, 1027–1033.

Nene, V., Wortman, J.R., Lawson, D., H aas, B., Kodira, C., Tu, Z.J., Loftus, B., Xi, Z., Megy, K., Grabherr, M., Ren, Q., Zdobnov, E.M., Lobo, N.F., Campbell, K.S., Brown, S.E., Bonaldo, M.F., Zhu, J., Sinkins, S.P., Hogenkamp, D.G., Amedeo, P., Arensburger, P., Atkinson, P.W., Bidwell, S., Biedler, J., Birney, E., Bruggner, R.V., Costas, J., Coy, M.R., Crabtree, J., Crawford, M., Debruyn, B., Decaprio, D., Eiglmeier, K., Eisenstadt, E., El-Dorry, H., Gelbart, W.M., Gomes, S.L., Hammond, M., Hannick, L.I., Hogan, J.R.,, Holmes, M.H., Jaffe, D., Johnston, J.S., Kennedy, R.C., Koo, H., Kravitz, S., Kriventseva, E.V., Kulp, D., Labutti, K., Lee, E., Li, S., Lovin, D.D., Mao, C., Mauceli, E., Menck, C.F., Miller, J.R., Montgomery, P., Mori, A., Nascimento, A.L., Naveira, H.F., Nusbaum, C., O'leary, S., Orvis, J., Pertea, M., Quesneville,

H., Reidenbach, K.R., Rogers, Y.H., Roth, C.W., Schneider, J.R., Schatz, M., Shumway, M., Stanke, M., Stinson, E.O., Tubio, J.M., Vanzee, J.P., Verjovski-Almeida, S., Werner, D., White, O., Wyder, S., Zeng, Q., Zhao, Q., Zhao, Y., Hill, C.A., Raikhel, A.S., Soares, M.B., Knudson, D.L., Lee, N.H., Galagan, J., Salzberg, S.L., Paulsen, I.T., Dimopoulos, G., Collins, F.H., Birren, B., Fraser-Liggett, C.M., and Severson, D.W. (2007). Genome sequence of Aedes aegypti, a major arbovirus vector. Science 316, 1718–1723.

Nimmannitya, S. (1997). Dengue hemorrhagic fever: diagnosis and management. In: Dengue and Dengue Hemorrhagic Fever, D.J. Gubler, and G. Kuno, eds. (Oxford, UK: CAB International), pp133–45.

Olson, K.E., Adelman, Z.N., Travanty, E.A., Sanchez-Vargas, I., Beaty, B.J., and Blair, C.D. (2002). Developing arbovirus resistance in mosquitoes. Insect. Biochem. Mol. Biol. 32, 1333–1343.

Ooi, E.E., Hart, T.J., Tan, H.C., and Chan, S.H. (2001). Dengue seroepidemiology in Singapore. Lancet 357, 685–686.

Ooi, E.E., Goh, K.T., and Gubler, D.J. (2006). Dengue prevention and 35 years of vector control in Singapore. Emerg. Infect. Dis. 12, 887–893.

Pant, C.P., and Yasuno, M. (1973). Field studies on the gonotrophic cycle of Aedes aegypti in Bangkok, Thailand. J. Med. Entomol. 10, 219–223.

Patz, J.A., Martens, W.J., Focks, D.A., and Jetten, T.H. (1998). Dengue fever epidemic potential as projected by general circulation models of global climate change. Environ. Health Perspect. 106, 147–153.

Ponlawat, A., and Harrington, L.C. (2005). Blood feeding patterns of Aedes aegypti and Aedes albopictus in Thailand. J. Med. Entomol. 42, 844–849.

Powell, J.R., Tabachnick, W.J., and Arnold, J. (1980). Genetics and the origin of a vector population: Aedes aegypti, a case study. Science 208, 1385–1387.

Ramos, M.M., Mohammed, H., Zielinski-Gutierrez, E., Hayden, M.H., Lopez, J.L., Fournier, M., Trujillo, A.R., Burton, R., Brunkard, J.M., Anaya-Lopez, L., Banicki, A.A., Morales, P.K., Smith, B., Muñoz, J.L., and Waterman, S.H.; Dengue Serosurvey Working Group. (2008). Epidemic dengue and dengue hemorrhagic fever at the Texas-Mexico border: results of a household-based seroepidemiologic survey, December 2005. Am. J. Trop. Med. Hyg. 78, 364–369.

Reed, W., Carroll, J., Agramonte, A., (1901). The etiology of yellow fever. JAMA., 86:431–40.

Reiter, P., Lathrop, S., Bunning, M., Biggerstaff, B., Singer, D., Tiwari, T., Baber, L., Amador, M., Thirion, J., Hayes, J., Seca, C., Mendez, J., Ramirez, B., Robinson, J., Rawlings, J., Vorndam, V., Waterman, S., Gubler, D., Clark, G., and Hayes, E. (2003). Texas lifestyle limits transmission of dengue virus. Emerg. Infect. Dis. 9, 86–89.

Reiter, P., Fontenille, D., and Paupy, C. (2006). Aedes albopictus as an epidemic vector of chikungunya virus: another emerging problem? Lancet Infect. Dis. 6, 463–464.

Rezza, G., Nicoletti, L., Angelini, R., Romi, R., Finarelli, A.C., Panning, M., Cordioli, P., Fortuna, C., Boros, S., Magurano, F., Silvi, G., Angelini, P., Dottori, M., Ciufolini, M.G., Majori, G.C., and Cassone, A.; CHIKV study group. (2007). Infection with chikungunya virus in Italy: an outbreak in a temperate region. Lancet. 370, 1840–1846.

Ribeiro, J.M., Arcà, B., Lombardo, F., Calvo, E., Phan, V.M., Chandra, P.K., and Wikel, S.K. (2007). An annotated catalogue of salivary gland transcripts in the adult female mosquito, Aedes aegypti. BMC Genomics 8, 6.

Rosen, L., Roseboom, L.E., Gubler, D.J., Lien, C.J., and Chantois, B.N. (1985). Comparative susceptibility of mosquito species and strains to oral and parenteral infection with dengue and Japanese encephalitis viruses. Am. J. Trop. Med. Hyg. 34, 603–615.

Schliessmann, D.J. (1967). Aedes aegypti eradication program of the United States – Progress Report 1965. Am. J. Public Health. 57, 460–465.

Schneider, B.S., Soong, L., Zeidner, N.S., and Higgs, S. (2004). Aedes aegypti salivary gland extracts modulate anti-viral and TH1/TH2 cytokine responses to sindbis virus infection. Viral Immunol. 17, 565–573.

Schneider, B.S., Soong, L., Girard, Y.A., Campbell, G., Mason, P., and Higgs, S. (2006). Potentiation of West Nile encephalitis by mosquito feeding.Viral Immunol. 19, 74–82.

Schneider, B.S., McGee, C.E., Jordan, J.M., Stevenson, H.L., Soong, L., and Higgs, S. (2007). Prior exposure to uninfected mosquitoes enhances mortality in naturally-transmitted West Nile virus infection. PLoS ONE 2, e1171.

Schneider, B.S., and Higgs, S. (2008). The enhancement of arbovirus transmission and disease by mosquito saliva is associated with modulation of the host immune response. Trans. R. Soc. Trop. Med. Hyg. 102, 400–408.

Schneider, J.R., Mori, A., Romero-Severson, J., Chadee, D.D., and Severson, D.W. (2007). Investigations of dengue-2 susceptibility and body size among Aedes aegypti populations. Med. Vet. Entomol. 21, 370–376.

Seet, R.C., Ooi, E.E., Wong, H.B., and Paton, N.I. (2005). An outbreak of primary dengue infection among migrant Chinese workers in Singapore characterized by prominent gastrointestinal symptoms and a high proportion of symptomatic cases. J. Clin. Virol. 33, 336–340.

Shope, R. (1991). Global climate change and infectious diseases. Environ. Health Perspect. 96, 171–174.

Tanner, L., Schreiber, M., Low, J.G., Ong, A., Tolfvenstam, T., Lai, Y.L., Ng, L.C., Leo, Y.S., Thi Puong, L., Vasudevan, S.G., Simmons, C.P., Hibberd, M.L., and Ooi, E.E. (2008). Decision tree algorithms predict the diagnosis and outcome of dengue Fever in the early phase of illness. PLoS Negl. Trop. Dis. 2, e196.

Tonn, R.J., Sheppard, P.M., MacDonald, W.W., and Bang, Y.H. (1969). Replicate surveys of larval habitats of Aedes aegypti in relation to dengue hemorrhagic fever in Bangkok, Thailand. Bull. World Health Organ. 40, 819–29.

Tabachnick, W.J., Wallis, G.P., Aitken, T.H., Miller, B.R., Amato, G.D., Lorenz, L., Powell, J.R., and Beaty, B.J. (1985). Oral infection of Aedes aegypti with yellow fever virus: geographic variation and genetic considerations. Am. J. Trop. Med. Hyg. 34, 1219–1224.

Tsetsarkin, K.A., Vanlandingham, D.L., McGee, C.E., and Higgs, S. (2007). A single mutation in chikungunya virus affects vector specificity and epidemic potential. PLoS Pathog. *3*, e201.

Tun-Lin, W., Burkot, T.R., and Kay, B.H. (2000). Effects of temperature and larval diet on development rates and survival of the dengue vector *Aedes aegypti* in north Queensland, Australia. Med. Vet. Entomol. *14*, 31–37.

Valenzuela, J.G., Pham, V.M., Garfield, M.K., Francischetti, I.M., and Ribeiro, J.M. (2002). Toward a description of the sialome of the adult female mosquito *Aedes aegypti*. Insect Biochem. Mol. Biol. *32*, 1101–1122.

Vaughn, D.W., Green, S., Kalayanarooj, S., Innis, B.L., Nimmannitya, S., Suntayakorn, S., Endy, T.P., Raengsakulrach, B., Rothman, A.L., Ennis, F.A., and Nisalak, A. (2000). Dengue viremia titer, antibody response pattern, and virus serotype correlate with disease severity. J. Infect. Dis. *181*, 2–9.

Vezzani, D., and Carbajo, A.E. (2008). *Aedes aegypti*, *Aedes albopictus*, and dengue in Argentina: current knowledge and future directions. Mem. Inst. Oswaldo Cruz. *103*, 66–74.

Wanasen, N., Nussenzveig, R.H., Champagne, D.E., Soong, L., and Higgs, S. (2004). Differential modulation of murine host immune response by salivary gland extracts from the mosquitoes *Aedes aegypti* and *Culex quinquefasciatus*. Med. Vet. Entomol. *18*, 191–199.

Wang, W.K., Chao, D.Y., Kao, C.L., Wu, H.C., Liu, Y.C., Li, C.M., Lin, S.C., Ho, S.T., Huang, J.H., and King, C.C. (2003). High levels of plasma dengue viral load during defervescence in patients with dengue hemorrhagic fever: implications for pathogenesis. Virology *305*, 330–338.

Wasserman, H.A., Singh, S., and Champagne, D.E. (2004). Saliva of the yellow fever mosquito, *Aedes aegypti*, modulates murine lymphocyte function. Parasite Immunol. *26*, 295–306.

Waterhouse, R.M., Kriventseva, E.V., Meister, S., Xi, Z., Alvarez, K.S., Bartholomay, L.C., Barillas-Mury, C., Bian, G., Blandin, S., Christensen, B.M., Dong, Y., Jiang, H., Kanost, M.R., Koutsos, A.C., Levashina, E.A., Li, J., Ligoxygakis, P., Maccallum, R.M., Mayhew, G.F., Mendes, A., Michel, K., Osta, M.A., Paskewitz, S., Shin, S.W., Vlachou, D., Wang, L., Wei, W., Zheng, L., Zou, Z., Severson, D.W., Raikhel, A.S., Kafatos, F.C., Dimopoulos, G., Zdobnov, E.M., and Christophides, G.K. (2007). Evolutionary dynamics of immune-related genes and pathways in disease-vector mosquitoes. Science *316*, 1738–1743.

Wilcox B.A., Gubler, D.J., and Pizer, H.F. Urbanization and the Social Ecology of Emerging Infectious Diseases. Chapter 4. In: Mayer K., Pizer, HF (Eds.), Social Ecology of Infectious Diseases. Academic Press (Elsevier, Inc), London; 2008

Woodring, J.L., Higgs, S., and Beaty, B.J. (1996). Natural cycles of vector borne pathogens. In Biology of Disease Vectors, W. C. Marquardt and B. J. Beaty eds. Boulder CO: University of Colorado Press), pp. 51–72

World Health Organization (1975). Technical guides for diagnosis, treatment, surveillance, prevention and control of dengue hemorrhagic fever (Geneva: WHO).

Xi, Z., Ramirez, J.L., and Dimopoulos, G. (2008). The *Aedes aegypti* toll pathway controls dengue virus infection. PLoS Pathog. *4*, e1000098.

Yasuno, M., and Tonn, R.J. (1970). A study of the biting habits of *Aedes aegypti* in Bangkok, Thailand. Bull. World Health Organ. *43*, 319–325.

Zeidner, N.S., Higgs, S., Happ, C.M., Beaty, B.J., and Miller, B.R. (1999). Mosquito feeding modulates Th1 and Th2 cytokines in flavivirus susceptible mice: an effect mimicked by injection of sialokinins, but not demonstrated in flavivirus resistant mice. Parasite Immunol. *21*, 35–44.

Evolutionary Dynamics of Dengue Virus

Shannon N. Bennett

Abstract

This chapter addresses the following topics with respect to the evolutionary dynamics of dengue viruses in humans. Beginning with an *introduction* to evolutionary thinking and terminology, particularly with relevance to RNA viruses, the *macroevolutionary* divergence from dengue progenitors into the four serotypes is discussed. The *microevolutionary* dynamics within dengue serotypes and the *forces of evolution* – mutation, migration, recombination, selection and drift – that drive them are outlined. Finally, I summarize frontiers in understanding the *evolutionary ecology* of dengue, epitomized by the study of phylodynamics, in which interactions amongst host, vector and virus population sizes, their dynamic distributions, immune landscapes and evolutionary dynamics are modeled.

Introduction: observing evolution in a virus

Our modern-day understanding of biological evolution was initially based on observations of plants and animals whose forms varied across populations and species according to their environments. Evolution is quite simply change in heritable traits over time (generations) (Darwin, 1859), and its principles have been successfully applied, from animal husbandry to agriculture to vaccines, well before the underlying genetic mechanisms were understood, based on manipulation of these outward forms or *phenotypes*. Evolution is driven by five elements or forces: mutation, the primary source of variation; recombination; gene flow or migration, both of which introduce additional variation into populations; and natural selection and genetic drift, which change the relative proportion of variants in populations over time.

Before the rise of modern genetics, which today allows us to decipher the genetic code of entire viral genomes across space and time, virus evolution was documented based on changes in phenotype, the outward manifestation of a virus's nature that could and did include the kinds of cells or hosts it infected, plaque size and clarity, morphology (following advances in microscopy), and even the disease it caused (its *extended phenotype*, Dawkins, 1982). Vaccine strains of many viruses were commonly derived by passaging them from one cell culture to another until their plaque sizes changed (the viral equivalent of animal husbandry to breed, for example, dairy cows from bison). Whether viruses or cows, both are instances of evolution by *artificial selection* (preferential survival/reproduction by bearers of the targeted trait) for a desirable phenotypic trait, either virulence represented by plaque size (the area of dead cells surrounding a focus of infection) or milk production, respectively. Although the target of selection is a phenotype, selection results in a change in genotype frequencies – the resulting change in phenotypes, the response to selection, is actually a property of the interaction between the organism's underlying genotype and its environment. For instance, artificial selection to minimize plaque size in a given cell line (virus phenotype) will result in the preponderance of a given genotype that may not necessarily produce small plaques in all cell lines or be attenuated *in vivo*.

In fact, evolution can result both from non-random forces such as artificial or *natural selection*, or randomly, by *genetic drift*. Selection acts on a phenotype, not necessarily the one sometimes used to measure evolution, and its strength and direction can vary in space and time with the selective environment. But its effects, the differential survival and reproduction of certain phenotypes, results in changes in the frequencies of the underlying genotypes in a given population over time, *ergo* evolution. Such change may encompass extinction of deleterious genotypes (frequency 0) or *fixation* of beneficial genotypes (frequency 1). These latter genotypes are referred to as being more *fit*. Genetic drift similarly describes change in genotype frequencies over time, but this occurs independent of the encoded phenotype, that is, randomly with respect to the direction of change. For example, a mosquito's blood meal may randomly capture only a subset of genotype variants in a population of circulating dengue viruses within a host, those that happened to be 'in the right place at the right time.' The resultant transmitted population is much smaller than the original, and therefore has less variation possibly to the point of being *fixed* for a single genotype. Incidently, virus strain variation, which influences viraemia (Gubler, 1987), could play an indirect role in genetic drift: low viraemia strains with their lower population size would experience more drift than high-viraemia strains. Thus the rate of change, or evolution, by genetic drift also varies in space and time, but rather than varying directly with the selective environment, it varies with population size (e.g. population bottlenecks accelerate drift), with rather more stochasticity than consistency. Regardless of whether by selection or drift, the rate at which new genotypes become fixed in a population from generation to generation is referred to as the *substitution rate*. Substitution rate is not to be confused with *mutation rate*, the rate at which mutations are generated in a nucleotide sequence, either by copying errors during genome replication, exposure to radiation or chemical mutagens, etc. Substitutions of single nucleotides in the genetic sequence (point mutations), deletions or insertions of nucleotides, or

inversions of a stretch of sequence involving many nucleotides are all types of mutations. Mutations are an important source of variation in a population upon which the forces or mechanisms of evolution act. RNA viruses such as dengue have high mutation rates because they replicate using RNA-dependent RNA polymerases as opposed to DNA polymerases, which can correct replication errors. Based on base-pairing constraints, only 1 in 10^4 replicated bases should be wrong, yet DNA replication has a much lower estimated error rate, at 1 in 10^{10}, due to proofreading (Berg *et al.*, 2002). RNA-dependent RNA polymerases have estimated error rates between 10^{-3} and 10^{-4} (erroneous bases per base, reviewed in Reanney, 1982) similar to predicted rates based on base-pairing constraints, although experimentally some viruses demonstrate lower error rates (e.g. polioviruses, Freistadt *et al.*, 2007; Arias *et al.*, 2008). Dengue intrahost populations before being subjected to strong purifying selection during transmission show mutations rates in the order of 10^{-3} (derived from Holmes, 2003).

Rates of evolution: RNA viruses and dengue

Ultimately, the rate of evolution varies most with generation time, which for dengue viruses is certainly much shorter than for cows. Every lifetime represents differential opportunities for survival, and every reproductive event can be biased in the genotypes passed on. Furthermore, RNA viruses use an RNA-dependent polymerase for genome replication that lacks a proofreading mechanism, contributing by mutation more variants to the population to become the targets of selection or subject to the vagaries of drift. Compound rapid population turnover with large populations sizes and high mutation rates, and you create an enormous potential for evolutionary rapidity and innovation (Domingo and Holland, 1997; Drake and Holland, 1999; Moya *et al.*, 2004).

Dengue viruses do not evolve as fast as many RNA viruses, particularly those that are directly transmitted: estimates are in the order of 10^{-4} substitutions per site per year (Zanotto *et al.*, 1996; Jenkins *et al.*, 2002; Twiddy *et al.*, 2003; Bennett *et al.*, 2006). Limits to the rate of evolution, placed for example by the necessity of transmitting between two very different hosts, are variously referred to as stabilizing selection, negative selection, constraining selection (selective constraints), conserving selection (conservation), and purifying selection. All describe selection against change, the extinction of deleterious genotypes mentioned above. For viruses with a two-host life history, such as arboviruses that transmit between arthropods and humans, we presume that fit genotypes in one host may not necessarily be fit in the other (Beaty and Bishop, 1988; Strauss and Strauss, 1988). This reduction in the rate of adaptive evolution was recently demonstrated experimentally in another arbovirus Venezuelan Equine Encephalitis (Coffey et al., 2008).

Macroevolution

Dengue viruses are members of the family *Flaviviridae*, genus *Flavivirus*, a group whose evolutionary origins are still unclear: discovery of additional flavivirus relatives would help determine whether the ancestor of flaviviruses including dengue was a virus of arthropods or mammals.; see also Chapter 1). The genus consists of 53 species including vectored, non-vectored arthropod and non-vectored vertebrate viruses sorted into monophyletic groups based on mode of transmission and host specificity (Kuno et al., 1998; Gould et al., 2001; de Lambellerie et al., 2002; Holmes and Twiddy, 2003; Cook et al., 2006; Gubler et al., 2006) (see Fig. 11.3). It has been proposed that Flavivirus ancestors were arthropod viruses based on similarities to the mosquito viruses cell fusing agent virus (Cammisa-Parks et al., 1992), Kamiti River virus (Crabtree et al., 2003; Sang et al., 2003) and Culex flavivirus (Hoshino et al., 2007). However, recent characterization of Tamana Bat virus as another distant relative of the genus but well within the family (de Lamballerie et al., 2002) alternatively suggests a non-vectored vertebrate virus ancestor. The fact that the other two genera of the Flaviviridae, Hepacivirus and Pestivirus, consist of non-vectored vertebrate viruses would support the latter (de Lamballerie et al., 2002). Approximately 2000 undiscovered flaviviruses are predicted to exist in the mosquito-borne clade alone (Pybus et al., 2002

Macroevolutionary divergence amongst the flaviviruses in congruence with host type and mode of transmission suggests that it has been shaped in large part by selection for host specificity. Similar forces appear to have been at work in the evolution of dengue viruses, a clade within the mosquito-borne flaviviruses, transmitted amongst primates. Dengue emerged into non-human primates before humans and radiated into four 'species' or serotypes that are antigenically distinct (Wang et al., 2000) approximately one thousand years ago (Twiddy et al., 2003). Their antigenic uniqueness, strong in phenotypic as well as phylogenetic signal, suggests that their diversification was driven by selection pressure to reduce protective cross-reactions in the primate immunological response, and has allowed them to coexist sympatrically (Ferguson et al., 1999). Limits to this diversification may in part explain why there are only 4 serotypes (but see Chapter 11): viraemia-enhancing immunologic cross-reactions (e.g. Halstead, 1988) may help to maintain a degree of similarity amongst serotypes that are at the same time being driven genetically apart to circumvent cross-protective immunity (Adams and Boots, 2006; Cummings et al., 2005). An evolutionary balance may have been struck between positive frequency-dependent selection and diversifying selection. The end result is 62–67% amino acid identity amongst the polyproteins of the 4 serotypes (Westaway and Blok, 1997); dengue viruses are no more similar to each other than they are to other flavivirus species (Kuno et al., 1998).

Although dengue viruses emerged in a sylvatic setting, all four serotypes have independently crossed over into humans, possibly numerous times, but finally establishing their endemic/epidemic forms as human populations grew large enough 125–320 years ago (Zanotto et al., 1996; Twiddy et al., 2003; see Chapter 11 for further review and the ecology). In three of the four serotypes, emergence of sylvatic forms into human populations was accompanied by changes in the envelope protein, clustered non-randomly in the main receptor-binding domain III, that suggest adaptation (Wang et al., 2000). The sylvatic form of dengue virus type 3 (DENV-3) has yet to be isolated, and therefore cannot be subjected to the same genetic comparison.

Microevolution

Dengue serotype emergence into their human endemic/epidemic forms has resulted in dramatic radiations of genetic diversity in 200 years or less (Zanotto *et al.*, 2006; Twiddy *et al.*, 2003). This diversity is structured within the endemic/epidemic forms of each serotype into 'genotypes' (clusters of genetically similar viruses) that are largely characterized by geographical distribution. Before the advent of large-scale genomics, these groupings or 'genotypes' were identified as 'subtypes,' often based on phenotypes such as distribution, epidemic potential and/or clinical severity. Because this chapter often refers to genotype in its original sense, as the unique genetic sequence of an individual, I will use 'subtype' to refer to these genetically similar clusters of viruses that often have distinct geographic distributions as well as similar phenotypes.

The geographic structure of present-day dengue diversity is strongly biased towards Asia, supporting the hypothesis of Asian origins for each serotype's endemic/epidemic form from Asian sylvatic ancestors (discussed in Chapter 11). On the other hand, the geographic structure could as well reflect inherent differences in the ability of strains to spread and establish (Gubler *et al.*, 1978; Gubler *et al.*, 1981b; Rico-Hesse *et al.*, 1997; Rico-Hesse, 2003). The classic example is DENV-2, which consists of Asian, American, American/Asian (representing an Asian colonization of America), and Cosmopolitan subtypes in humans (Lewis *et al.*, 1993; Twiddy *et al.*, 2002a), named for their present-day distributions more than their geographic origins. DENV-1 has five subtypes (including the sylvatic strain): Asia ('I'), Thailand ('II'), Malaysia ('III', the sylvatic strain), South Pacific ('IV'), and Americas/Africa ('V') (Goncalvez *et al.*, 2002; Rico-Hesse, 2003). DENV-3 is structured into four or five clades depending on the authority: subtype I from S. E. Asia/S. Pacific, subtype II from Thailand, subtype III from the Indian subcontinent, subtype IV from the Americas (Lanciotti *et al.*, 1994; Rico-Hesse, 2003) and subtype V from S.E. Asia, sometimes classified with subtype I (Wittke *et al.*, 2002). DENV-4 consists of 2 human subtypes, one largely S.E. Asian ('I'), and one largely American and South Pacific but with an Asian source ('II') (Chungue *et al.*, 1995; Lanciotti *et al.*, 1997) thus referred to as 'Indonesia' by Rico-Hesse (2003) (Fig. 9.1, reviewed in Rico-Hesse, 2003). The diversity of DENV from Asia, across

all serotypes and multiple subtypes, supports an Asian origin although Africa has been under-sampled (Gubler, 1987; Gubler, 1998; Chapter 11). The paucity of dengue from Africa may also be the result of cross-protective immunity wherever Yellow fever virus cocirculates (Holmes, 2004) in addition to a dearth of surveillance and underreporting.

Migration and recombination

Clearly movement or gene flow of virus strains into different geographic regions has been an important source of subsequent diversification, lying at the origin of several subtypes across the four serotypes. In some cases, this has resulted in subtype replacement and local extinction of the previous lineage, as appears to be the case in DENV-2 where the Asian/American subtype is replacing the American subtype in most of the Caribbean and parts of South and Central America (Rico-Hesse *et al.*, 1997; Watts *et al.*, 1999; Loroño-Pino *et al.*, 2004; Bennett *et al.*, 2006). Lineage extinction and replacement over time has also been noted within populations (Wittke *et al.*, 2002; Bennett *et al.*, 2003; Bennett *et al.*, 2006). Finer-scale geographic analyses suggest that the microevolutionary dynamics of dengue viruses generally consist of distinct populations evolving *in situ* and diverging, infused periodically with divergent viral immigrants from different populations. Diversity is then pruned by the occasional lineage extinction. Two populations have been used to demonstrate these microevolutionary patterns. In the Caribbean, Puerto Rico dengue diversifies and evolves *in situ* by both genetic drift and selection into novel lineages (Bennett *et al.*, 2003; Bennett *et al.*, 2006). Yet occasionally Puerto Rico both imports and exports viruses from/to the mainland (Foster *et al.*, 2003; Foster *et al.*, 2004; Carrington *et al.*, 2005), the former sometimes resulting in new local lineages (Fig. 9.2). A recent study in the Kamphaeng Phet province of Thailand also noted *in situ* evolution of dengue within discreet school catchment areas, but although subdividing, it was slow and appeared to be dominated by genetic drift. These dengue populations were frequently infused by viral migration from other schools, leading to the origination of locally new lineages (Jarman *et al.*, 2008).

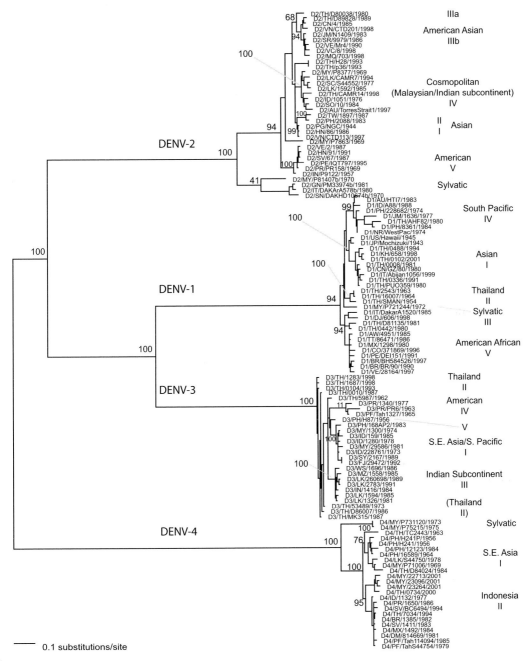

Figure 9.1 Maximum likelihood (ML) phylogeny depicting subtypes (as in 'genotypes') of all four dengue serotypes, DENV-1–4. Labels represent serotype/ISO country code/laboratory strain/year of isolation (as proposed in Schreiber *et al.*, 2007). Bootstrap support values from 100 ML replicates implemented in RAxML Blackbox web-server (Stamatakis *et al.*, 2008) are shown at nodes only for those nodes that distinguish serotypes and subtypes (as in 'genotypes'). Subtype names from: Lewis *et al.*, 1993; Lanciotti *et al.*, 1997; Goncalvez *et al.*, 2002; Twiddy *et al.*, 2002a; Wittke *et al.*, 2002; Rico-Hesse, 2003.

In addition to sporadic viral migration of divergent genotypes amongst discrete independently evolving dengue populations, another potentially important source of genetic diversity and novel lineages is recombination. Recombination and reassortment (the exchange of whole segments of genetic material) are notable drivers of evolutionary innovation in such cases as host

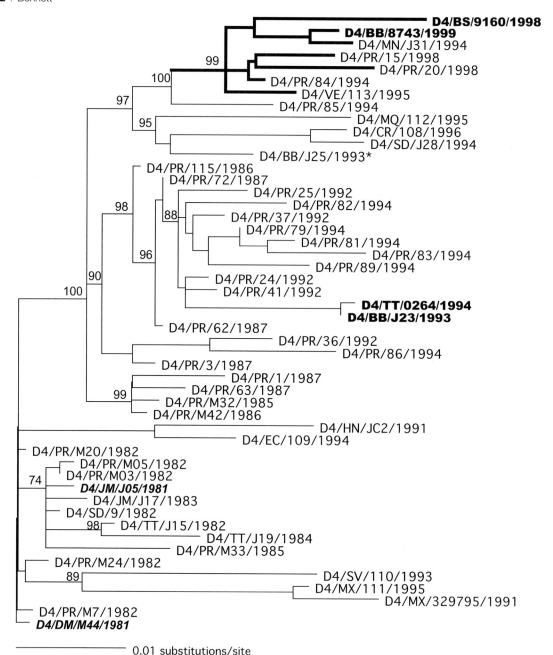

0.01 substitutions/site

Figure 9.2 Maximum likelihood (ML) phylogeny of DENV-4 from Puerto Rico, as well as the Caribbean, Central and South America, showing *in situ* evolution of Puerto Rican dengue after importation of a 1981 strain from Dominica (italics labels), with occasional exports (bold labels) where sufficient data is available. Labels represent serotype/ISO country code/laboratory strain/year of isolation. The 1998 epidemic clade in Puerto Rico (Bennett *et al.*, 2003), shown with darker lines, appeared in Puerto Rico as early as 1994, but since 1993 was not sampled (compare to a divergent Barbados sample from 1993 denoted by an asterix), the origin of this clade remains unclear. Bootstrap support values from 100 ML replicates implemented in RAxML Blackbox web-server (Stamatakis *et al.*, 2008) are shown at nodes.

range expansion by influenza A (Hay *et al.*, 2001) and novel subtype formation in HIV-1 (Steain *et al.*, 2004; Abecasis *et al.*, 2007). Although there is no evidence that recombination has played a role in the evolution of dengue serotypes, it may be of some importance in dengue microevolution (within serotypes). AbuBakar *et al.* (2002) identified a new sub-subtype, IIa, of DENV-4, isolates from patients in Malaysia in 2001 that included a recombinant, MY01-22713, of subtype II (Indonesia 1976) and subtype I (Malaysia 1969) isolated 4 months before the rest. Although the isolates were directly sequenced rather than cloned to ensure a single individual template, the genomic signature was observed with slight variation in four other independent isolates, suggesting that this recombination event is real. Recombination amongst DENV-2 subtypes has been demonstrated in an *Aedes aegypti* mosquito collected in Myanmar (Craig *et al.*, 2003): a clone containing a recombinant genome homologous to both D2 Asian I and Cosmopolitan subtypes was recovered along with numerous clones containing similar sequences to the 'parentals' – a group of sequences closely related to viruses circulating amongst dengue patients in the same locality (Yangon, Myanmar) since 1995, classified as Asian 1 subtype, and another group of the Cosmopolitan subtype which were also circulating in Yangon in 2000, related to DENV-2 viruses sampled from southern China in 1999 (Craig *et al.*, 2003). Recombinants have been identified phylogenetically within all serotypes (e.g. Holmes *et al.*, 1999; Worobey *et al.*, 1999; Tolou *et al.*, 2001; Uzcategui *et al.*, 2001; Domingo *et al.*, 2006; Chen *et al.*, 2008) but do not appear to have established new lineages. Rico-Hesse (2003) cautions that some of these recombinants involve erroneous sequences on NCBI's GenBank and/or should be confirmed by re-isolation in pure cultures and resequenced from clones (Rico-Hesse, 2003). The requirements for recombination, concomitant infection with two different viruses, have been demonstrated in humans (e.g. Gubler *et al.*, 1985) and mosquitoes (Craig *et al.*, 2003), the latter resulting in a recombination event as mentioned above.

Natural selection and genetic drift

Although mutation, gene flow and recombination introduce genetic variation, ultimate diversity, resulting from the differential fixation of substitutions, is shaped by natural selection and genetic drift. Decoding the relative importance of natural selection versus genetic drift in shaping dengue diversity and ongoing evolutionary dynamics

is particularly difficult given epidemiological observations that dengue viruses have continued to expand in range and epidemic potential since their initial emergence into human populations a few hundred years ago. Certain dengue virus strains are associated with the more severe disease phenotype, dengue haemorrhagic fever (DHF, see Chapter 5), others are geographically more widespread than their counterparts, and locally certain strains displace other cocirculating lineages through time. These observations raise the question, are these more fit viruses? For example, the accompaniment of the DENV-2 'American'-to-'American/Asian' subtype replacement by sudden increase in South and Central America and the Caribbean of severe disease (Rico-Hesse *et al.*, 1997; Loroño-Pino *et al.*, 2004) suggests the displacing subtype was more transmissible and/ or virulent. Indeed, the original 'American' subtype has been shown to replicate less efficiently in human target cells (Cologna and Rico-Hesse, 2003) and is less often associated with severe disease even when infecting previously exposed individuals (Watts *et al.*, 1999), although it can cause severe disease in some instances (e.g. Barnes and Rosen, 1974). Overall, the 'Cosmopolitan' subtype of DENV-2 is the most widespread, and shows higher rates of adaptive evolution, represented by elevated non-synonymous substitutions, relative to the other DENV-2 subtypes (Twiddy *et al.*, 2002a). Other mechanisms of selection leading to subtype spread or lineage replacement could include virus–vector interactions, such as increased vectorial capacity for certain virus genotypes (Anderson and Rico-Hesse, 2006; reviewed in Halstead 2008; Hanley *et al.*, 2008). Hanley *et al.* (2008) demonstrate that an invasive and more pathogenic strain of DENV-3, that competitively displaced milder forms circulating in Sri Lanka in the 1980s (Messer *et al.*, 2002; Messer *et al.*, 2003), replicated to greater titre and disseminated more efficiently in the vector. This has also been demonstrated for Tahitian and Asian strains of DENV-3, the first published evidence that virus strains varied in replication efficiency in mosquitoes (Gubler and Rosen, 1977). *In situ* replacement by lineages descended from rare earlier genotypes is suggestive of negative frequency-dependent selection, commonly associated with immune selection, and has been noted in the field (Bennett *et al.*, 2003; Zhang *et al.*, 2005). Immune responses select against recognized antigenic signatures, such that novel or rare genotypes (*sensu strictu*) have a fitness advantage over common ones proportional to how rare they are in the population (that is, inversely proportional to their frequency in a population, hence negative frequency-dependent selection). In dengue, this form of selection could only operate across serotypes, since dengue infection elicits complete serotype-specific immunity that is theoretically life long. Only novelty at cross-reactive antigenic sites, amongst serotypes circulating in a hyperendemic transmission arena, could theoretically confer a fitness advantage in secondarily (or tertiarily) infected hosts.

Measuring natural selection and interpreting its phenotypic targets remains the biggest challenges in determining the ongoing significance of adaptive evolution in dengue evolutionary dynamics and emergence. Two main methods of measuring selection are employed: comparisons of rates of non-synonymous (d_N) to synonymous (d_S) substitutions over an evolutionary history (e.g. in a phylogenetic framework), and measuring changes in genotype frequency leading to nonrandom fixation (i.e. substitution) within a population, usually over shorter timescales. The former, employed in a phylogenetic context, has significant limitations in power, detecting only sites that have changed repeatedly over time (e.g. under fluctuating or diversifying selection pressures, such as that imposed by the immune system), as opposed to substitutions that have remain fixed (e.g. under strong consistent directional selection). Even once sites have been identified bearing statistically significant signals of selection (as in elevated d_N relative to d_S), the challenge remains to interpret their relevance to the virus phenotype and its extended disease phenotype. In dengue, we have a particularly poor understanding of what creates the disease phenotype, which is decoupled from the virus phenotype by the host and vector genotypes/ phenotypes, and the transmission environment itself. *In vitro* models are by nature greatly simplified, focusing on direct viral phenotypes such as replication rate, transmission rates to mosquitoes or vertebrate models, and indirect (i.e. extended) viral phenotypes such as changes in membrane

permeability (reviewed in Chapter 6). Several studies within populations suggest that disease variability can be accounted for by strain or virus genotype variation (e.g. Barnes and Rosen, 1974; Gubler *et al.*, 1978; Gubler *et al.*, 1981b; Leitmeyer *et al.*, 1999; Watts *et al.*, 1999), secondary infection rates or some other epidemiologic/ environmental condition, such as host age or vector species (e.g. Siler *et al.*, 1926; Sabin, 1952; Halstead, 1997; Guzman *et al.*, 2000b, 2002a; Imrie *et al.*, in review), or an interaction amongst these factors (e.g. Rico-Hesse *et al.*, 1998; Guzman *et al.*, 2000a,b; Uzcategui *et al.*, 2001; reviewed in Chapter 5). The nature of these phenotypes being complex and greatly influenced by the environment, they demonstrate extreme variability even for strains with similar genetic signatures, or that have been fixed by positive selection. With the advent of population-level high-throughput sequencing, studies can measure turnover in genotype frequencies either in real time or similarly short time scales as evidence for selection. Bennett *et al.* (2003) document four lineage replacements, three of which were associated with epidemic transmission (1981–82, 1986–87, and 1998): the latest replacement in 1998 was due to rapid fixation of a DENV-4 lineage within 3 years (the new genetic signature made up 10% of the samples in 1994 ($n = 30$), 61% in 1995 ($n = 23$), 65% in 1996 ($n = 17$), 96% in 1997 ($n = 22$) and 100% in 1998 ($n = 14$), Vorndam and Bennett, unpublished data), accompanied by non-synonymous changes in a non-structural gene (NS2A) that occurred eight times more rapidly than predicted random rates based on the coalescent. Experimental studies have documented relevant phenotypes of dengue viruses circulating in nature (e.g. Gubler and Rosen, 1977; Diallo *et al.*, 2005; Anderson and Rico-Hesse, 2006; Hanley *et al.*, 2008; Vasilakis *et al.*, 2008b). To confirm the phenotypic effects of adaptive mutations, one approach is to test individual mutations in a congenic cDNA clone background. Although not yet conducted on dengue, ideal tests for selection include experimental studies of Venzuelan equine encephalitis virus (VEEV), which have documented changes in genotype frequencies of VEEV relative to host selection pressures and the resultant differences in phenotype (Coffey *et al.*, 2008). Many of the field and phylogenetic studies that have documented changes in dengue phenotype accompanying changes in genotype (e.g. in Puerto Rico, Bennett *et al.*, 2003; in Sri Lanka, Messer *et al.*, 2002, 2003; in the South Pacific, Gubler *et al.*, 1978, 1981b) provide ideal systems to begin confirmatory experiments on the phenotypes under selection in dengue. Significant challenges in this regard remain for understanding the role and significance of natural selection in dengue, from better statistical models for selection to rigorous laboratory and field studies to relate virus genotype and its dynamics to changes in virus phenotype and the extended disease phenotype.

In the absence of evidence for selection, substitutions are attributed to genetic drift. Much of the phylogenetic structure within dengue populations consists of lineages that are distinguished by a preponderance of silent or synonymous changes (Twiddy *et al.*, 2002b; Bennett *et al.*, 2003, 2006), augmented by lineage extinction. Both the slow rates of *in situ* evolution noted in the Thai study (Jarman *et al.*, 2008) and the observation that lineage extinction and replacement occurred during interepidemic periods in another Thai study (Wittke *et al.*, 2002), as well as certain lineage replacements in Puerto Rico marked by silent substitutions (Bennett *et al.*, 2003; Bennett *et al.*, 2006) suggest that genetic drift can also be important in dengue microevolutionary processes. Genotypes that are fixed randomly by genetic drift may have phenotypes, and extended phenotypes, significant to disease, but identification of the relevant genetic signatures is even more challenging without the statistical signature of selection to act as a starting place for experimentation.

Dengue's life history makes it equally susceptible to these factors, from strong purifying selection invoked by dual-host tropisms, to seasonal population bottlenecks exacerbating genetic drift. Certain dengue lineages within the serotypes have emerged due to positive selection on both structural and non-structural genes (Twiddy *et al.*, 2002b; Bennett *et al.*, 2003, 2006). Empirical observations on lineage replacement charting the change in genotype frequency over time, where the earlier 1994 year clade dropped from 90% of the total isolates to undetectable, while the 1998/1999 year clade increased from 10% to 100% over the same 5-year period sug-

gests potentially strong positive directional selection (Vorndam, Bennett, unpublished data; Bennett *et al.*, 2003). However, by and large dengue evolution exhibits strong purifying selection, both based on statistical studies of dN/dS ratios (Twiddy *et al.*, 2002b; Holmes, 2003), as well as empirical observations: 'the ratio of non-synonymous to synonymous nucleotide changes within a DEN-3 virus population from a single patient was less than the ratio among the consensus sequences of DEN-3 viruses from different patients, suggesting that many of the non-synonymous nucleotide changes which occurred naturally in the E protein were deleterious and removed by purifying selection' (Wittke *et al.*, 2002). After the emergence of sylvatic dengue serotypes into human populations, there has been very little evolutionary activity in envelope (Holmes, 2004).

The stochastic nature of the dengue virus life cycle exhibited at each transmission event, along with seasonal reductions in vector populations (Gubler, 1987), and annual variation in the abundance of susceptible human hosts, generates innumerable population bottlenecks for the virus, in which genetic drift, the random fixation of genotypes, is more likely to occur. The fact that many lineages are distinguished from others within genotypes by synonymous substitutions (e.g. Bennett *et al.*, 2003), the slow rates of *in situ* evolution noted in the Thai study (Jarman *et al.*, 2008), and the observed bias in turnovers occurring more often during interepidemic periods (Wittke *et al.*, 2002), all suggest an important role for genetic drift in lineage formation and dengue evolutionary dynamics.

The nature of genetic drift predicts that we would expect common genotypes to become fixed more often than rare ones, such that common variants are favoured, all other things being equal. Thus the observed pattern of lineage turnover proceeding from rare cocirculating variants is not well-explained by genetic drift (Bennett *et al.*, 2003). Immune-escape-driven selection (or negative frequency-dependent selection, see above) is at first difficult to reconcile with dengue life history: DENV infection elicits serotype-specific life-long immunity thus eliminating subsequent infections in which rare genotypes within a serotype would be more fit in an exposed individual (e.g. a mechanism of selection in influenza). However, considered in a hyperendemic population of cross-reacting serotypes that may vary in enhancement among heterologous genotypes, selection at cross-reactive sites could pose a plausible mechanism for immune-escape evolution. Zhang et al. (2005) note that lineage replacements in cocirculating DENV-1 and DENV-3 populations in Thailand may have been driven by differential susceptibility to cross-reactive immune responses.

Determining the relative importance of natural selection versus genetic drift in the evolutionary dynamics of DENV has largely relied on above-mentioned phylogeny-based methods of comparing relative rates of non-synonymous to synonymous substitutions (dN/dS ratios). As mentioned, the power of these tests to detect selection typically relies on sites that change frequently over the evolutionary history of a group, and therefore selective sweeps, in which a beneficial substitution becomes rapidly fixed, can be missed (Holmes, 2004). In addition, these tests make the critical but potentially erroneous assumption that synonymous substitutions are phenotypically neutral. Many viruses exhibit codon usage biases, in which the prevalence of nucleotides at presumably neutral sites are skewed from predictions under neutrality. Studies in dengue demonstrate slight codon usage biases, where some constraints against change at synonymous sites is evident but by no means drastic (Jenkins et al., 2001; Jenkins and Holmes, 2003; Auewarakul, 2005). Another consideration is that presumably strong selection for genome packaging may apply considerable pressure to regulatory functions that are often encoded for indirectly through secondary structure.

Within- and between-host evolutionary dynamics

The transmission of dengue occurs between two distinct hosts, in both of which replication occurs, giving rise to alternating opportunities for inter- and intra-host evolution. Replication of virus in both the vertebrate and mosquito is equally error-prone, typical of RNA viruses that use their own polymerase, but differences in the number of replication events in these two very different hosts will be one important factor in the

relative rates of evolution. The vector is infected for its lifetime whereas humans are viraemic for on average 3.5–5 days and up to 12 days (Siler et al., 1926; Simmons et al., 1931; Gubler et al., 1979; Gubler et al., 1981a; reviewed in Kuno, 1997, and Halstead, 2008). After an initial nonfeeding period (less than 2 days, Kuno, 1997) following emergence, Aedes females may become infected after feeding on an infected host: viral replication occurs in most tissues culminating in the potential to transmit after an extrinsic incubation period (necessitating dissemination to the salivary glands) of 10–14 days (Siler et al., 1926; Sabin, 1952). Mosquitoes presumably maintain viraemia for life (Rodhain and Rosen, 1997): field longevity estimates range from 8.5 (Sheppard et al., 1969) to 42 (Trpis and Hausermann, 1986) days. The dynamics of replication, duration of the extrinsic incubation period, and ultimate levels of viraemia depend on environmental factors, mosquito species, strain, size, etc. (e.g. Bosio et al., 1998; Moncayo et al., 2004), virus strain, and infecting dose (reviewed in Rodhain and Rosen, 1997; Halstead, 1997, 2008; Kuno, 1997). Together with the mosquito's lower basal temperature and smaller body size, one expects many fewer replication events in mosquitoes than humans. Estimates of viraemia in mosquitoes in nature are few but in the laboratory under ideal condition can be on the order of 10^7 MID$_{50}$ (50% mosquito infectious dose; Rosen and Gubler, 1974); in humans during peak viraemia estimates range up to $10^{8.3}$ MID$_{50}$ (Gubler et al., 1981a). Regardless, both stages of host infection are liable to generate populations of variant dengue viruses more diverse than the infecting dose. The ultimate diversity of these intrahost populations will depend on the number of virus generations (e.g. duration of infection), virus population size and the forces of evolution such as drift and selection that generally reduce diversity. Clearly the selection pressures within these two hosts could account for some of the greatest differences. Host cell receptors involved in dengue infection, the number and cell types, differ greatly between mosquitoes and humans (Hung et al., 2004; Kuno and Chang, 2005). Antiviral defences also differ greatly between vertebrates and invertebrates, the most obvious difference being the lack of a humoral, or antibody-mediated response in the latter (Kuno and Chang, 2005). Adaptive evolution of dengue viruses during their emergence into human populations has involved adaptations to peridomestic mosquitoes (Moncayo et al., 2004) concentrated in the envelope gene of the virus (Wang et al., 2000). The role of humans in driving adaptive evolution in dengue is less clear – antigenically, sylvatic and endemic strains are similar (Vasilakis et al., 2008a). Lin et al. (2004) showed that the virus populations within mosquitoes compared to humans collected during the same outbreak were generally less diverse based on the envelope gene, confirmed in experimentally infected mosquitoes. Furthermore, this same group conducted whole-genome interrogation of clones of DENV-3 collected from patients and found that envelope was the most heterogeneous within a patient, and NS3 and NS5 the least (Chao et al., 2005). Chao et al. (2005) also report that envelope was under selection in patients, targeted to domain III. The rate of non-synonymous mutations within hosts has been shown to be much higher than at the host population level, but these mutations appear random and 90% are deleterious (Holmes, 2003). Virologists observe that fundamental changes occur in a virus upon isolation in the laboratory: the population of viruses that proliferates from the relatively small subset first put into cell culture (or lab mosquitoes) is probably vastly different in population size, in distribution and diversity than was originally in the patient (C.B. Cropp, personal communication). Viruses represented as a single consensus sequence are actually populations that may vary significantly with respect to these diversity-related characteristics, which may account for interstrain variability observed in experimental infections despite similar consensus sequences (reviewed in Chapter 5).

Nonetheless, most dengue virus genetic studies by default represent a sample with a single consensus sequence even though it may contain a distribution of different virus genomes. Is this representation valid for understanding dengue genetic variation amongst host populations? Lin et al. (2004) compared the distribution of dengue virus envelope sequences within mosquitoes and patients during the same outbreak and found that the major genotype was the same, suggesting that the major variant is transmitted. Aaskov et al. (2006) notes in a comparison of the dengue

1 virus populations collected within patients and mosquitoes in Myanmar that most mutations are deleterious and pruned by purifying selection within hosts, resulting in stable populations of consensus sequences with low d_N/d_S ratios (Holmes, 2003). Observed stop-codon mutants were released from this purifying selection and showed neutral levels of non-synonymous changes in envelope (e.g. arising through mutation and drift) at sites that are normally invariant and therefore presumably significant for viral fitness at the host population level (Aaskov et al., 2006). Thus for host population-level studies, consensus sequencing is probably adequate to represent between-host evolutionary dynamics. However, a dissection of the within-host evolutionary dynamics has clearly shed light on how evolutionary forces are acting on the micro-scale. Transmission of defective lineages of viruses occurred at least over medium timescales (18 months), suggesting that consensus sequencing may mask complex evolutionary dynamics at these timescales (Aaskov et al., 2006). Future directions for research should focus on comparative within and between host evolutionary dynamics, in the same outbreaks and particularly during lineage replacements.

The term 'quasispecies' has been applied to dengue within-host populations (e.g. Wang et al., 2002) without evidence to support the application of this very specific evolutionary model. A quasispecies is, according to the original application (Eigen and Schuster, 1977), a population of viruses at mutation-selection balance, such that genetic drift cannot occur, frequencies of the master genotype(s) are stable through time, and the target of selection is the group, or quasispecies rather than individual genomes, the evidence for any of which in many viruses is controversial (reviewed in Holmes and Moya, 2002). Instead, the term has been generally misapplied to 'any undefined distribution of mutants ... a swarm' (Eigen, 1996), as a surrogate for intrahost genetic variation in viruses, and used as a synonym for high mutation rates (Jenkins et al., 2001; Holmes and Moya, 2002). Standard population genetic models of evolution in which genetic drift and natural selection on individuals maintain dynamic spectra of mutants within hosts are probably sufficient to address the evolutionary dynamics of dengue viruses.

Conclusions: integrating our understanding of the evolutionary ecology of dengue

Dengue evolution occurs against a backdrop of highly variable ecology: incredible population size fluctuations in susceptible hosts, mosquitoes, and resultant cases. Epidemiologic traces of symptomatic cases wherever dengue is endemic show dramatic seasonal and interannual variability. However, human cases are by no means the only place significant populations of dengue viruses are distributed. Since DENV could be serotyped, it has been observed that silent transmission occurs often, particularly between epidemic phases of transmission, when the local strain persists in spite of months without a single case observed. High rates of asymptomatic infection (e.g. Gubler et al., 1978; Burke et al., 1988), transmission by other mosquitoes species besides the main vector Aedes aegypti, as well as vertical transmission amongst mosquitoes (Freier and Rosen, 1987; Freier and Rosen, 1988; Joshi et al., 2002; Angel and Joshi, 2008), create 'reservoirs' of dengue viruses. Dengue virus infection has also been detected in other vertebrates (Thoisy et al., 2008) but their role as traditional reservoirs is highly controversial.

Interactions between mosquito and host populations affect the distribution of dengue viruses in space and time. For example, lower socio-economic status (da Costa et al., 1998) or environmental conditions (Mondini et al., 2008) may bring people into closer contact with mosquitoes to generate new foci of infection. In situations where mosquitoes are relatively rare and/or patchy (e.g. Ooi and Gubler, 2006), and people move around, the latter may act as vectors transporting viruses to the mosquitoes (Adams and Kapan, 2009).

The term phylodynamics captures to some extent the evolutionary integration of these population dynamics by extracting virus effective population size fluctuations from sequence diversity (Drummond et al., 2003; Grenfell et al., 2004). Dengue viruses have thus been shown to increase exponentially (DENV in Puerto Rico/the Caribbean, Carrington et al., 2005) or fluctuate periodically in what appears to be a susceptible-limited state (DENV-4 in Puerto Rico, Bennett et al. unpublished data): the transmission conditions, from immunological to entomological to climatological, and their interactions with virus genotype that together drive these phylodynamics need to be further and systematically explored. Phylodynamic representation of dengue evolutionary ecology and the downstream hypotheses this can generate are important research frontiers that integrate population dynamics with population genetics to better understand how the two mutually influence each other.

From this chapter I hope to have impressed upon the reader that understanding the evolutionary ecology of dengue represents arguably the most important frontier in determining the drivers of ongoing dengue emergence. Interactions amongst host, vector and virus population sizes, their dynamic distributions, immune landscapes and dengue evolutionary dynamics will require innovative laboratory, field, and population genetic techniques to disentangle.

Acknowledgements

Many thanks go to Durrell Kapan, Scott Weaver, Duane Gubler, Argon Steel (IGERT Associate, NSF-DGE 0549515) and Brandi Mueller (IGERT Fellow, NSF-DGE 0549515) for discussions and editorial comments in the writing of this chapter. Salary support was provided by U54AI065359 from the National Institute of Allergy and Infectious Diseases, National Institutes of Health and G12RR003061 from the National Center for Research Resources, National Institutes of Health; bioinformatic and other support was provided by P20RR018727 from the National Center for Research Resources, National Institutes of Health and DOD06187000 from the U.S. Department of Defense.

References

Aaskov, J., Buzacott, K., Thu, H., Lowry, K., and Holmes, E. (2006). Long-term transmission of defective RNA viruses in humans and Aedes mosquitoes. Science 311, 236–238.

Abecasis, A., Lemey, P., Vidal, N., de Oliveira, T., Peeters, M., Camacho, R., Shapiro, B., Rambaut, A., and Vandamme, A. (2007). Recombination confounds the early evolutionary history of human immunodeficiency virus type 1: subtype G is a circulating recombinant form. J. Virol. 81, 8543–8551.

AbuBakar, S., Wong, P., and Chan, Y. (2002). Emergence of dengue virus type 4 genotype IIA in Malaysia. J. Gen. Virol. 83, 2437–2442.

Adams, B., and Boots, M. (2006). Modelling the relationship between antibody-dependent enhancement and immunological distance with application to dengue. J. Theor. Biol. *242*, 337–346.

Adams, B., and Kapan, D.D. (2009). Man Bites Mosquito: understanding the contribution of human movement to vector borne disease dynamics. PLoS 1, in press.

Anderson, J., and Rico-Hesse, R. (2006). *Aedes aegypti* vectorial capacity is determined by the infecting genotype of dengue virus. Am. J. Trop. Med. Hyg. *75*, 886–892.

Angel B., and Joshi V. (2008). Distribution and seasonality of vertically transmitted dengue viruses in Aedes mosquitoes in arid and semi-arid areas of Rajasthan, India. J. Vector Borne Dis. *45*, 56–59.

Arias, A., Arnold, J.J., Sierra, M., Smidansky, E.D., Domingo, E., and Cameron, E. (2008). Determinants of RNA-dependent RNA polymerase (in)fidelity revealed by kinetic analysis of the polymerase encoded by a foot-and-mouth disease virus mutant with reduced sensitivity to ribavirin. J. Virol. *82*, 12346–12355.

Auewarakul, P. (2005). Composition bias and genome polarity of RNA viruses. Virus Res. *109*, 33–37.

Barnes, W.J., and Rosen, L. (1974). Fatal hemorrhagic disease and shock associated with primary dengue infection on a Pacific island. Am. J. Trop. Med. Hyg. *23*, 495–506.

Beaty, B.J., and Bishop, DHL (1988) Bunyavirus–vector interactions. Virus Res. *10*, 289–301.

Bennett, S., Holmes, E., Chirivella, M., Rodriguez, D., Beltran, M., Vorndam, V., Gubler, D., and McMillan, W. (2003). Selection-driven evolution of emergent dengue virus. Mol. Biol. Evol. *20*, 1650–1658.

Bennett, S., Holmes, E., Chirivella, M., Rodriguez, D., Beltran, M., Vorndam, V., Gubler, D., and McMillan, W. (2006). Molecular evolution of dengue 2 virus in Puerto Rico: positive selection in the viral envelope accompanies clade reintroduction. J. Gen. Virol. *87*, 885–893.

Berg J.M., Tymoczko, J.L., Stryer, L., and Clarke N.D. (2002). Chapter 27: DNA Replication, Recombination, and Repair. In Biochemistry, 5th Edition, J.M. Berg, J.L. Tymoczko, and L. Stryer (New York: W.H. Freeman and Company).

Bosio, C.F., Beaty, B.J., and Black, W.C. 4th. (1998). Quantitative genetics of vector competence for dengue-2 virus in *Aedes aegypti*. Am. J. Trop. Med. Hyg. *59*, 965–970.

Burke D.S., Nisalak, A., Johnson, D.E., and Scott R.M. (1988). A prospective study of dengue infections in Bangkok. Am. J. Trop. Med. Hyg. *38*, 172–180.

Cammisa-Parks, H., Cisar, L., Kane, A., and Stollar, V. (1992). The complete nucleotide sequence of cell fusing agent (CFA): homology between the nonstructural proteins encoded by CFA and the nonstructural proteins encoded by arthropod-borne flaviviruses. Virology *189*, 511–524.

Carrington, C. V., Foster, J. E., Pybus, O. G., Bennett, S. N., and Holmes, E. C. (2005). Invasion and maintenance of dengue virus type 2 and type 4 in the Americas. J. Virol. *79*, 14680–14687.

Chen, S., Yu, M., Jiang, T., Deng, Y., Qin, C., Han, J., and Qin, E. (2008). Identification of a recombinant dengue virus type 1 with 3 recombination regions in natural populations in Guangdong province, China. Arch. Virol. *153*, 1175–1179.

Chungue, E., Cassar, O., Drouet, M., Guzman, M., Laille, M., Rosen, L., and Deubel, V. (1995). Molecular epidemiology of dengue-1 and dengue-4 viruses. J. Gen. Virol. *76*, 1877–1884.

Coffey, L., Vasilakis, N., Brault, A., Powers, A., Tripet, F., and Weaver, S. (2008). Arbovirus evolution *in vivo* is constrained by host alternation. Proc. Natl. Acad. Sci. U.S.A. *105*, 6970–6975.

Cologna, R., and Rico-Hesse, R. (2003). American genotype structures decrease dengue virus output from human monocytes and dendritic cells. J. Virol. *77*, 3929–3938.

Cook, S., Bennett, S., Holmes, E., De Chesse, R., Moureau, G., and Lamballerie, X. (2006). Isolation of a new strain of the flavivirus cell fusing agent virus in a natural mosquito population from Puerto Rico. J. Gen. Virol. *87*, 735–748.

Chao, D., King, C., Wang, W., Chen, W., Wu, H., and Chang, G. (2005). Strategically examining the full-genome of dengue virus type 3 in clinical isolates reveals its mutation spectra. Virol J. *2*, 72.

da Cost,a A.I., and Natal ,D. (1998) [Geographical distribution of dengue and socioeconomic factors in an urban locality in southeastern Brazil]. Rev Saude Publica *32*, 232–236.

Crabtree, M., Sang, R., Stollar, V., Dunster, L., and Miller, B. (2003). Genetic and phenotypic characterization of the newly described insect flavivirus, Kamiti River virus. Arch. Virol. *148*, 1095–1118.

Craig, S., Thu, H., Lowry, K., Wang, X., Holmes, E., and Aaskov, J. (2003). Diverse dengue type 2 virus populations contain recombinant and both parental viruses in a single mosquito host. J. Virol. *77*, 4463–4467.

Cummings, D., Schwartz, I., Billings, L., Shaw, L., and Burke, D. (2005). Dynamic effects of antibody-dependent enhancement on the fitness of viruses. Proc. Natl. Acad. Sci. U.S.A. *102*, 15259–15264.

Darwin, C.R. (1859). On the Origin of Species by Means of Natural Selection. (London: John Murray).

Dawkins, R. (1982). The Extended Phenotype. (Oxford: Oxford University Press).

Diallo M., Sall, A.A., Moncayo, A.C., Ba, Y., Fernandez, Z., Ortiz, D., Coffey, L.L., Mathiot, C., Tesh, R.B., and Weaver S.C. (2005). Potential role of sylvatic and domestic African mosquito species in dengue emergence. Am. J. Trop. Med. Hyg. *73*, 445–449.

Domingo, E., and Holland, J. (1997). RNA virus mutations and fitness for survival. Annu. Rev. Microbiol. *51*, 151–178.

Domingo C., Palacios, G., Jabado, O., Reyes, N., Niedrig, M., Gascon, J., Cabrerizo, M., Lipkin, W.I., and Tenorio A. (2006). Use of a short fragment of the C-terminal E gene for detection and characterization of two new lineages of dengue virus 1 in India. J. Clin. Microbiol. *44*, 1519–1529.

Drake, J., and Holland, J. (1999). Mutation rates among RNA viruses. Proc. Natl. Acad. Sci. U.S.A. *96*, 13910–13913.

Drummond, A.J., Pybus, O.G., Rambaut, A., Forsberg, R., and Rodrigo, A.G. (2003) Measurably evolving populations. TREE 18, 481–488.

Eigen, M. (1996). On the nature of virus quasispecies. Trends Microbiol. 4, 216–218.

Eigen, M., and Schuster, P. (1977). A principle of natural self-organization. Naturwissenschaften. Available at: http://www.springerlink.com/index/R133207N06736808.pdf.

Ferguson, N., Anderson, R., and Gupta, S. (1999). The effect of antibody-dependent enhancement on the transmission dynamics and persistence of multiple-strain pathogens. Proc. Natl. Acad. Sci. U.S.A. 96, 790–794.

Foster, J., Bennett, S., Carrington, C., Vaughan, H., and McMillan, W. (2004). Phylogeography and molecular evolution of dengue 2 in the Caribbean basin, 1981–2000. Virology 324, 48–59.

Foster, J., Bennett, S., Vaughan, H., Vorndam, V., McMillan, W., and Carrington, C. (2003). Molecular evolution and phylogeny of dengue type 4 virus in the Caribbean. Virology 306, 126–134.

Freier J.E., and Rosen L. (1987). Vertical transmission of dengue viruses by mosquitoes of the Aedes scutellaris group. Am. J. Trop. Med. Hyg. 37, 640–647.

Freier J.E., and Rosen L. (1988). Vertical transmission of dengue viruses by Aedes mediovittatus. Am. J. Trop. Med. Hyg. 39, 218–222.

Freistadt M.S., Vaccaro, J.A., and Eberle, K.E. (2007). Biochemical characterization of the fidelity of poliovirus RNA-dependent RNA polymerase. Virology J. 4, 44.

Goncalvez, A., Escalante, A., Pujol, F., Ludert, J., Tovar, D., Salas, R., and Liprandi, F. (2002). Diversity and evolution of the envelope gene of dengue virus type 1. Virology 303, 110–119.

Gould, E.A., de Lamballerie, X., Zanotto, P.M.D., and Holmes, E.C. (2001). Evolution, epidemiology, and dispersal of flaviviruses revealed by molecular phylogenies. Adv. Virus Res. 57, 71–103.

Grenfell, B., Pybus, O., Gog, J., Wood, J., Daly, J., Mumford, J., and Holmes, E. (2004). Unifying the epidemiological and evolutionary dynamics of pathogens. Science 303, 327–332.

Gubler, D.J. (1987). Dengue and dengue hemorrhagic fever in the Americas. P. R. Health Sci. J. 6, 107–111.

Gubler, D.J. (1998) Dengue and dengue hemorrhagic fever. Clin. Microbiol. Rev. 11, 480–496.

Gubler, D.J. (2006). Dengue/dengue haemorrhagic fever: history and current status. In New Treatment Strategies for Dengue and Other Flaviviral Diseases, G. Bock and J. Goode, ed. (Novartis Foundation Symposia 277: John Wiley & Sons, Ltd.), pp. 3–16.

Gubler, D.J., Kuno, G., and Markoff L. (2006). Flaviviruses. In Fields Virology 5th Ed., Knipe, D.M., Howley, P.M., Griffin, D.E., Lamb, R.A., Martin, M.A., Roizman, B., and Strauss, S.L., ed. (Lippincott Williams & Wilkins).

Gubler, D.J., Kuno, G., Sather, G.E., and Waterman, S.H. (1985). A case of natural concurrent human infection with two dengue viruses. Am. J. Trop. Med. Hyg. 34, 170–173.

Gubler, D., Reed, D., Rosen, L., and Hitchcock, J. (1978). Epidemiologic, clinical, and virologic observations on dengue in the Kingdom of Tonga. Am. J. Trop. Med. Hyg 27, 581–589.

Gubler, D.J., and Rosen, L. (1977). Quantitative aspects of replication of dengue viruses in Aedes albopictus (Diptera: Culicidae) after oral and parenteral infection. J Med Entomol. 13, 469–472.

Gubler, D.J., Suharyono, W., Tan, R., Abidin, M., and Sie, A. (1981a). Viraemia in patients with naturally acquired dengue infection. Bull. World Health Org. 59, 623–630.

Gubler, D.J., Suharyono, W., Lubis, I., Eram, S., and Gunarso, S. (1981b). Epidemic dengue 3 in Central Java, associated with low viremia in man. Am. J. Trop. Med. Hyg. 30, 1094–1099.

Gubler, D.J., Suharyono, W., Lubis, I., Eram, S., Sulianti Saroso, J. (1979). Epidemic dengue hemorrhagic fever in rural Indonesia. I. Virological and epidemiological studies. Am. J. Trop. Med. Hyg. 28, 701–710.

Guzman, M., Kouri, G., and Halstead, S. (2000a). Do escape mutants explain rapid increases in dengue case-fatality rates within epidemics? Lancet 355, 1902–1903.

Guzman, M., Kouri, G., Valdes, L., Bravo, J., Alvarez, M., Vazques, S., Delgado, I., and Halstead, S. (2000b). Epidemiologic studies on Dengue in Santiago de Cuba, 1997. Am. J. Epidemiol. 152, 793–799.

Guzman, M., Kouri, G., Bravo, J., Valdes, L., Vazquez, S., and Halstead, S. (2002a). Effect of age on outcome of secondary dengue 2 infections. Int. J. Infect. Dis. 6, 118–124.

Guzmán, M.G., Kourí, G., Valdés, L., Bravo, J., Vázquez, S., and Halstead, S.B. (2002b). Enhanced severity of secondary dengue-2 infections: death rates in 1981 and 1997 Cuban outbreaks. Revista Panamericana de Salud Pública/Pan American Journal of Public Health. 11, 223–227.

Halstead, S. (1988). Pathogenesis of dengue: challenges to molecular biology. Science 239, 476–481.

Halstead, S. (1997). Epidemiology of dengue and dengue hemorrhagic fever. In Dengue and Dengue Hemorrhagic Fever, D.J. Gubler and G. Kuno, eds. (New York, USA: CABI Publishing), pp. 23–44.

Halstead, S. (2008). Dengue virus–mosquito interactions. Annu. Rev. Entomol. 53, 273–291.

Hanley, K., Nelson, J., Schirtzinger, E., Whitehead, S., and Hanson, C. (2008). Superior infectivity for mosquito vectors contributes to competitive displacement among strains of dengue virus. BMC Ecol. 8, 1.

Hay, A., Gregory, V., Douglas, A., and Lin, Y. (2001). The evolution of human influenza viruses. Philos. Trans. R. Soc. Lond., B., Biol. Sci. 356, 1861–1870.

Holmes, E. (2003). Patterns of intra- and interhost nonsynonymous variation reveal strong purifying selection in dengue virus. J. Virol. 77, 11296–11298.

Holmes, E. (2004). The phylogeography of human viruses. Mol. Ecol. 13, 745–756.

Holmes, E., and Moya, A. (2002). Is the quasispecies concept relevant to RNA viruses? J. Virol. 76, 460–465.

Holmes, E., and Twiddy, S. (2003). The origin, emergence and evolutionary genetics of dengue virus. Infect. Genet. Evol. 3, 19–28.

Holmes, E., Worobey, M., and Rambaut, A. (1999). Phylogenetic evidence for recombination in dengue virus. Mol. Biol. Evol. *16*, 405–409.

Hoshino, K., Isawa, H., Tsuda, Y., Yano, K., Sasaki, T., Yuda, M., Takasaki, T., Kobayashi, M., and Sawabe, K. (2007). Genetic characterization of a new insect flavivirus isolated from *Culex pipiens* mosquito in Japan. Virology *359*, 405–414.

Hung, J., Hsieh, M., Young, M., Kao, C., King, C., and Chang, W. (2004). An external loop region of domain III of dengue virus type 2 envelope protein is involved in serotype-specific binding to mosquito but not mammalian cells. J. Virol. *78*, 378–388.

Imrie, A., Roche, C., Zhao, Z., Bennett, S.N., Laille, M., Effler, P., and Cao-Lormeau, V.M. (In review). Homology of dengue virus type 1 complete genome sequences from dengue fever- and dengue hemorrhagic fever-associated epidemics in Hawaii and French Polynesia. Virus Res.

Jarman R.G., Holmes, E.C., Rodpradit, P., Klungthong, C., Gibbons, R.V., Nisalak, A., Rothman, A.L., Libraty, D.H., Ennis, F.A., Mammen, M.P. Jr, and Endy T.P. (2008). Microevolution of Dengue Viruses Circulating among Primary School Children in Kamphaeng Phet, Thailand. J. Virol. *82*, 5494–5500.

Jenkins, G., and Holmes, E. (2003). The extent of codon usage bias in human RNA viruses and its evolutionary origin. Virus Res. *92*, 1–7.

Jenkins, G., Pagel, M., Gould, E., de A Zanotto, P., and Holmes, E. (2001). Evolution of base composition and codon usage bias in the genus Flavivirus. J. Mol. Evol. *52*, 383–390.

Jenkins, G., Rambaut, A., Pybus, O., and Holmes, E. (2002). Rates of molecular evolution in RNA viruses: a quantitative phylogenetic analysis. J. Mol. Evol. *54*, 156–165.

Joshi, V., Mourya, D., and Sharma, R. (2002). Persistence of dengue-3 virus through transovarial transmission passage in successive generations of *Aedes aegypti* mosquitoes. Am. J. Trop. Med. Hyg. *67*, 158–161.

Kuno, G. (1997). Factors influencing the transmission of dengue viruses. In Dengue and Dengue Hemorrhagic Fever, D.J. Gubler and G. Kuno, eds. (New York, USA: CABI Publishing), pp. 61–88.

Kuno, G., and Chang, G. (2005). Biological transmission of arboviruses: reexamination of and new insights into components, mechanisms, and unique traits as well as their evolutionary trends. Clin. Microbiol. Rev. *18*, 608–637.

Kuno, G., Chang, G., Tsuchiya, K., Karabatsos, N., and Cropp, C. (1998). Phylogeny of the genus Flavivirus. J. Virol. *72*, 73–83.

de Lamballerie, X., Crochu, S., Billoir, F., Neyts, J., de Micco, P., Holmes, E., and Gould, E. (2002). Genome sequence analysis of Tamana bat virus and its relationship with the genus Flavivirus. J. Gen. Virol. *83*, 2443–2454.

Lanciotti, R., Gubler, D., and Trent, D. (1997). Molecular evolution and phylogeny of dengue-4 viruses. J. Gen. Virol. 78 (Pt 9), 2279–2284.

Lanciotti, R., Lewis, J., Gubler, D., and Trent, D. (1994). Molecular evolution and epidemiology of dengue-3 viruses. J. Gen. Virol. 75 (Pt 1), 65–75.

Leitmeyer, K., Vaughn, D., Watts, D., Salas, R., Villalobos, I., de Chacon, Ramos, C., and Rico-Hesse, R. (1999). Dengue virus structural differences that correlate with pathogenesis. J. Virol. *73*, 4738–4747.

Lewis, J., Chang, G., Lanciotti, R., Kinney, R., Mayer, L., and Trent, D. (1993). Phylogenetic relationships of dengue-2 viruses. Virology *197*, 216–224.

Lin, S., Hsieh, S., Yueh, Y., Lin, T., Chao, D., Chen, W., King, C., and Wang, W. (2004). Study of sequence variation of dengue type 3 virus in naturally infected mosquitoes and human hosts: implications for transmission and evolution. J. Virol. *78*, 12717–12721.

Loroño-Pino, M.A., Farfán-Ale, J.A., Zapata-Peraza, A.L., Rosado-Paredes, E.P., Flores-Flores, L.F., García-Rejón, J.E., Díaz, F.J., Blitvich, B.J., Andrade-Narváez, M., Jiménez-Ríos, E., Blair, C.D., Olson, K.E., Black, W. 4th, and Beaty, B.J. (2004). Introduction of the American/Asian genotype of dengue 2 virus into the Yucatan State of Mexico. Am. J. Trop. Med. Hyg. *71*, 485–492.

Messer, W., Gubler, D., Harris, E., Sivananthan, K., and De Silva, A. (2003). Emergence and global spread of a dengue serotype 3, subtype III virus. Emerg. Infect. Dis. 9, 800–809.

Messer, W., Vitarana, U., Sivananthan, K., Elvtigala, J., Preethimala, L., Ramesh, R., Withana, N., Gubler, D., and De Silva, A. (2002). Epidemiology of dengue in Sri Lanka before and after the emergence of epidemic dengue hemorrhagic fever. Am. J. Trop. Med. Hyg. *66*, 765–773.

Moncayo, A.C., Fernandez, Z., Ortiz, D., Diallo, M., Sall, A., Hartman, S., Davis, C.T., Coffey, L., Mathiot, C.C., Tesh, R.B., and Weaver, S.C. (2004) Dengue emergence and adaptation to peridomestic mosquitoes. Emerg. Infect. Dis. *10*, 1790–1796.

Mondini A., and Chiaravalloti-Neto F. (2008) Spatial correlation of incidence of dengue with socioeconomic, demographic and environmental variables in a Brazilian city. Sci. Total Environ. *393*, 241–248.

Moya, A., Holmes, E., and González-Candelas, F. (2004). The population genetics and evolutionary epidemiology of RNA viruses. Nat. Rev. Microbiol. *2*, 279–288.

Ooi, E., Goh, K., and Gubler, D. (2006). Dengue prevention and 35 years of vector control in Singapore. Emerg. Infect. Dis. *12*, 887–893.

Pybus, O., Rambaut, A., Holmes, E., and Harvey, P. (2002). New inferences from tree shape: numbers of missing taxa and population growth rates. Syst. Biol. *51*, 881–888.

Rico-Hesse, R. (2003). Microevolution and virulence of dengue viruses. Adv. Virus. Res. *59*, 315–341.

Rico-Hesse, R., Harrison, L.M., Nisalak, A., Vaughn, D.W., Kalayanarooj, S., Greene, S., Rothman, A. L., and Ennis, F.A. (1998). Molecular evolution of Dengue type 2 virus in Thailand. Am. J. Trop. Med. Hyg. *58*, 96–101.

Rico-Hesse R., Harrison, L.M., Salas, R.A., Tovar, D., Nisalak, A., Ramos, C., Boshell, J., de Mesa M.T., Nogueira, R.M., and da Rosa A.T. (1997). Origins of dengue type 2 viruses associated with increased pathogenicity in the Americas. Virology *230*, 244–251.

Reanney, D.C. (1982) The evolution of RNA viruses. Annu. Rev. Microbiol. *36*, 47–73.

Rodhain, F., and Rosen, L. (1997). Mosquito vectors and dengue virus–vector relationships. In Dengue and Dengue Hemorrhagic Fever, D.J. Gubler and G. Kuno, eds. (New York, USA: CABI Publishing), pp. 45–60.

Rosen, L., and Gulber, D.J. (1974). The use of mosquitoes to detect and propagate dengue viruses. Am. J. Trop. Med. Hyg. 23, 1153–1160.

Sabin, A.B. (1952). Research on dengue during World War I.I. Am. J. Trop. Med. Hyg. 1, 30–50.

Sang, R., Gichogo, A., Gachoya, J., Dunster, M., Ofula, V., Hunt, A., Crabtree, M., Miller, B., and Dunster, L. (2003). Isolation of a new flavivirus related to cell fusing agent virus (CFAV) from field-collected flood-water Aedes mosquitoes sampled from a dambo in central Kenya. Arch Virol 148, 1085–1093.

Schreiber, M.J., Ong, S.H., Holland, R.C.G., Hibberd, M.L., Vasudevan, S.G., Mitchell, W.P., and Holmes, E.C. (2007). DengueInfo: A web portal to dengue information resources. Infection, Genetics and Evolution 7, 540–541.

Sheppard, P.M., MacDonald, W.W., Tonn, R.J., and Grab, B. (1969). The dynamics of an adult population of Aedes aegypti in relation to dengue hemorrhagic fever in Bangkok. J. Anim. Ecol. 38, 661–702.

Siler, J.F., Hall, M.W., and Hitchens, A.P. (1926). Dengue: its history, epidemiology, mechanism of transmission, etiology, clinical manifestations, immunity, and prevention. Philipp. J. Sci. 29, 1–304.

Simmons, J.S., St. John, J.H., and Reynolds, F.H.K. (1931). Experimental studies of dengue. Philipp. J. Sci. 44, 1–252.

Stamatakis, A., Hoover, P., and Rougemont, J. (2008). A Rapid Bootstrap Algorithm for the RAxML Web-Servers. Systematic Biology 75, 758–771.

Steain, M., Wang, B., Dwyer, D., and Saksena, N. (2004). HIV-1 co-infection, superinfection and recombination. Sexual health 1, 239–250.

Strauss, J., and Strauss, E. (1988). Evolution of RNA viruses. Annu. Rev. Microbiol. 42, 657–683.

Thoisy B.D., Lacoste, V., Germain, A., Muñoz-Jordán J., Colón, C., Mauffrey, J.F., Delaval, M., Catzeflis, F., Kazanji, M., Matheus, S., Dussart, P., Morvan, J., Setién, A.A., Deparis, X., and Lavergne A. (2008). Dengue Infection in Neotropical Forest Mammals. Vector Borne Zoonotic Dis. Available at: http://www.liebertonline.com/doi/abs/10.1089/vbz.2007.0280.

Tolou, H., Couissinier-Paris, P., Durand, J., Mercier, V., de Pina, J., de Micco, P., Billoir, F., Charrel, R., and de Lamballerie, X. (2001). Evidence for recombination in natural populations of dengue virus type 1 based on the analysis of complete genome sequences. J. Gen. Virol. 82, 1283–1290.

Trpis, M., and Haussermann, W. (1986). Dispersal and other population parameters on Aedes aegypti in an African village and their possible significance in epidemiology of vector-borne diseases. Am. J. Trop. Med. Hyg. 35, 1263–1279.

Twiddy, S., Farrar, J.J., Chau, N.V., Wills, B., Gould, E.A., Gritsun, T., Lloyd, G., and Holmes, E.C. (2002a).

Phylogenetic relationships and differential selection pressures among genotypes of dengue-2 virus. Virology 298, 63–72.

Twiddy, S., Holmes, E., and Rambaut, A. (2003). Inferring the rate and time-scale of dengue virus evolution. Mol. Biol. Evol. 20, 122–129.

Twiddy, S., Woelk, C., and Holmes, E. (2002b). Phylogenetic evidence for adaptive evolution of dengue viruses in nature. J. Gen. Virol. 83, 1679–1689.

Uzcategui, N., Camacho, D., Comach, G., Cuello de Uzcategui, R., Holmes, E., and Gould, E. (2001). Molecular epidemiology of dengue type 2 virus in Venezuela: evidence for in situ virus evolution and recombination. J. Gen. Virol. 82, 2945–2953.

Vasilakis, N., Durbin, A., da Rosa, A., Munoz-Jordan, J., Tesh, R., and Weaver, S. (2008a). Antigenic relationships between sylvatic and endemic dengue viruses. Am. J. Trop. Med. Hyg. 79, 128–132.

Vasilakis N., Fokam, E.B., Hanson, C.T., Weinberg, E., Sall, A.A., Whitehead, S.S., Hanley, K.A., and Weaver S.C. (2008b). Genetic and phenotypic characterization of sylvatic dengue virus type 2 strains.Virology 377, 296–307.

Wang, E., Ni, H., Xu, R., Barrett, A., Watowich, S., Gubler, D., and Weaver, S. (2000). Evolutionary relationships of endemic/epidemic and sylvatic dengue viruses. J. Virol. 74, 3227–3234.

Wang, W., Lin, S., Lee, C., King, C., and Chang, S. (2002). Dengue type 3 virus in plasma is a population of closely related genomes: quasispecies. J. Virol. 76, 4662–4665.

Watts, D., Porter, K., Putvatana, P., Vasquez, B., Calampa, C., Hayes, C., and Halstead, S. (1999). Failure of secondary infection with American genotype dengue 2 to cause dengue haemorrhagic fever. Lancet 354, 1431–1434.

Westaway, E.G., and Blok, J. (1997). Taxonomy and evolutionary relationships of flaviviruses. In Dengue and Dengue Hemorrhagic Fever, D.J. Gubler and G. Kuno, eds. (New York, USA: CABI Publishing), pp. 147–174.

Wittke V., Robb, T.E., Thu, H.M., Nisalak, A., Nimmannitya, S., Kalayanrooj, S., Vaughn, D.W., Endy, T.P., Holmes, E.C., and Aaskov J.G. (2002). Extinction and rapid emergence of strains of dengue 3 virus during an interepidemic period. Virology 301, 148–156.

Worobey, M., Rambaut, A., and Holmes, E. (1999). Widespread intra-serotype recombination in natural populations of dengue virus. Proc. Natl. Acad. Sci. U.S.A. 96, 7352–7.

Zanotto, P., Gould, E., Gao, G., Harvey, P., and Holmes, E. (1996). Population dynamics of flaviviruses revealed by molecular phylogenies. Proc. Natl. Acad. Sci. U.S.A. 93, 548–553.

Zhang, C., Mammen, M., Chinnawirotpisan, P., Klungthong, C., Rodpradit, P., Monkongdee, P., Nimmannitya, S., Kalayanarooj, S., and Holmes, E. (2005). Clade replacements in dengue virus serotypes 1 and 3 are associated with changing serotype prevalence. J. Virol. 79, 15123–15130.

Temporal and Spatial Dynamics of Dengue Virus Transmission

10

Derek Cummings

Abstract

The geographical range in which dengue virus (DENV) is transmitted expanded rapidly in the latter decades of the twentieth century. Within locations where dengue (DEN) is endemic, incidence varies widely from year to year, often exhibiting multi-annual cycles. At subnational and even subcommunity spatial scales, incidence varies from place to place within seasons. In this chapter, I describe the variability in DEN incidence that has been observed at multiple spatial and temporal scales and some of the mechanisms that are thought to drive this variability. The chapter is divided into three parts: statics, emergence and dynamics. In the first part of this chapter, I consider geographical patterns of DEN that have remained fairly static for decades. I describe the spatial distribution of DEN globally and mechanisms that dictate the presence or absence of endemic DEN over long time scales. In the second, emergence, I describe epidemics in which DEN has recently emerged or re-emerged, and I consider mechanisms that dictate the speed and success with which this occurs. In the third part, dynamics, I describe mechanisms that are thought to drive temporal variation in incidence within endemic settings. A key focus is on mechanisms of cycles in DEN incidence and, extending this to include space, on mechanisms that create spatial–temporal patterns in incidence, including synchrony of incidence between regions and traveling waves in incidence. Finally, I end with a discussion of the utility of determining the mechanisms that drive spatial–temporal dynamics of
DEN, specifically, the improvements in public health that might be created by an understanding of these mechanisms.

Introduction

> ... the increase in the number of cases can be attributed to the cold weather marked by sporadic rains that provide the dengue mosquito with ideal breeding conditions.

Manila Times, 1 February 2008 (Jalbuena, 2008)

> The surge in dengue fever infections was mainly because of climate change which accelerated the life cycle of the mosquito, enabling them to produce a minimum infective dose of the dengue virus faster than before.

Bangkok Post, 14 March 2008 (Treeratkuarkul, 2008)

> ... outbreaks run in cycles, occurring roughly every four years.

Associated Press, 30 June 2007

Dramatic upsurges in DEN cases are frequent subjects of articles in the press. Each time a large increase occurs, public health officials, members of the press and others look for causes. The quotes above show a number of the candidate causes put forth. Changes in climate, failures of health authorities, and natural cycles in incidence have been cited as reasons for large increases in cases in particular settings. Few other diseases exhibit the amount of inter-annual variability that DEN exhibits. Annual incidence of DEN illness in Thailand over the last twenty years has varied by a factor of ten from small years to large years (Bureau of Epidemiology, 2000). Prediction of increases in incidence could help prepare public health systems to respond to the large number of people that require hospitalization during outbreaks. Understanding the causes of increases and cycles in incidence may also reveal fundamental drivers of the population dynamics of DEN. In addition to year-to-year variation, there appears to be a longer-term trend towards more cases of DEN globally. Discovery of the mechanisms that drive inter-annual variability may help us determine whether this increase is likely to continue and determine the best ways to stop it.

A place to start to understand the conditions that favour high DEN incidence is to differentiate places in the globe where endemic transmission is absent or present based on environmental, climate or other characteristics. The global distribution of DEN disease is limited to places with high ambient temperatures and abundant rainfall. However, not all environments with high temperature and rainfall exhibit endemic DENV transmission. Similarly, within endemic settings, there is quite a bit of spatial variability in incidence rates.

As recently as 1970, endemic DENV transmission occurred in only a small number of locations within the Americas. Vector control campaigns targeting yellow fever, which shares *Aedes aegypti* with DENV as a primary vector, are credited with reducing vector populations to levels below which transmission could be maintained (Halstead, 2006). However, with the reduction of resources to these control programmes and reintroduction of the vector, DEN incidence has risen steadily over the last three decades (see Chapters 5 and 8). The emergence of DEN in these New World settings where transmission had been absent for many years provides a very different view of DEN than is available from endemic settings where DENV has been transmitted for many decades. Detailed observations of the spatial and temporal evolution of DEN and newly introduced DENV serotypes in settings in which little immunity exists has allowed researchers to characterize the temporal

speed of outbreaks moving across a completely susceptible landscape. Heterogeneity of epidemic spread exhibited within communities suggests fundamental heterogeneity in the transmissibility of DENV at large and small spatial scales.

Mathematical models predict that the size of the immune population, and, conversely, the susceptible population has a strong impact on incidence. Immunity waxes and wanes in cycles similar to predator–prey interactions in animal systems, as increases in incidence reduce susceptible populations and slow transmission, and births increase the susceptible fraction over time. Estimates of the impact of extrinsic factors such as climate, urbanization, land use and vector density must adjust for the impact of cycling of immunity on transmission dynamics. Several models have been put forth to describe the cycling of DENV immunity in populations. These models are critical tools to predict the impact of mass vaccination when a vaccine becomes available. These models have also been used as inferential frameworks to estimate the impact of climate variation on incidence.

At the expense of detail, this review aims to provide a broad perspective on the spatial and temporal dynamics of DEN, addressing both small and large spatial scales as well as short and long time scales. The review covers an array of empirical methods to characterize the spatial and temporal distribution of DEN as well as methods that have been used to analyse this distribution.

I discuss multiple mechanisms that might drive spatial and temporal variability. These mechanisms can broadly be categorized as environmental, climatic, demographic and virologic. This review does not present an exhaustive list of the factors that might alter transmission. However, a number of these mechanisms are discussed and recur throughout the chapter as frequent hypothesized drivers of spatial and temporal variability.

Statics: the spatial distribution of endemic transmission

Global distribution of dengue
Estimates of 50 million DENV infections and 200,000 to 500,000 cases of dengue haemorrhagic fever (DHF) occurring each year are often cited,

even in the current volume. The methods used to generate these estimates appear to be lost to the literature and so this should be interpreted as a rough guess of global annual incidence (Gubler, 1998, 2002; World Health Organization, 2002). The average number of cases reported to WHO's DengueNet from 1980–1985 was ~200,000 and has grown to an average of over 800,000 for the period 2000–2005. These numbers are almost certainly a significant underestimate of the actual incidence of symptomatic disease. Such under-reporting must also affect current estimates of the geographic range of DENV transmission. The spatial distribution of transmission overlaps to a large degree with the distribution of the primary vector, *Ae. aegypti*. However, any map based solely on reported cases must underestimate the spatial extent of dengue cases given the large amount of under-reporting.

Several efforts have been made to use spatial information on climate and demography to predict the presence or absence of DEN globally, based on the spatial characteristics of those places that do report DEN. Hopp and Foley used a database of historical DEN cases reported from 32 ministries of health across the globe, reports of outbreaks in the medical literature, and historical climate data to build a model to predict the absence or presence of DEN for a given $1° \times 1°$ grid of the globe (Hopp and Foley, 2003). Their approach modeled the population dynamics of *Ae. aegypti* globally using water availability and daily climate variation. These modeled mosquito densities were correlated with case data. Globally, the correlation of modeled mosquito densities and case numbers varied widely. The authors attributed this to of the poor quality of data on DEN incidence from many areas of the globe.

A more recent effort by Rogers *et al.* (2006) conducted a full literature review for the geographical location of serologically confirmed DEN illness from 1960–2005 (Rogers *et al.*, 2006). Rather than using environmental and climate information as measured by satellite imagery to fit a mosquito model, they used discriminant analytic methods to predict point locations of serologically confirmed dengue directly. Model fit was assessed using an information theoretic approach. The most important variables used to model the distribution of DEN were satellite-derived measures of ambient temperature. Less important, but still statistically significant, were measures of moisture and rainfall. The spatial resolution of their model is much higher than previous efforts. The fit of their model suggests that environmental and climatic information can explain much of the presence and absence of DEN worldwide.

One application for which models of the global distribution of DEN prevalence have been used is to assess the impact of climate change on this distribution. It has been suggested that the warming of the globe will lead to an expansion of the area that supports endemic dengue transmission (Hales *et al.*, 2002; Jetten and Focks, 1997). However, the impact of climate change on the global DEN distribution is difficult to predict. It may be that increases in regions that support transmission because of increases in temperature may be offset by losses in areas that become unsuitable for transmission, as is predicted for malaria (Rogers and Randolph, 2000).

The areas of the globe that support endemic DENV transmission have been successfully modeled using satellite derived measures of climate and land use. Continued refinement and validation of these models will require enhanced surveillance of dengue infections and disease. Though several authors have suggested that global climate change may lead to increases in dengue cases, further research is needed to provide detailed and methodologically validated predictions of the change in dengue incidence with climate change.

Subnational level variability

The work of Rogers *et al.* (2006) indicates that substantial variability in the probability of DENV transmission exists, even within countries. Micro-scale variation in DEN risk has been demonstrated through survey data at community scales. Multiple methods have been used to characterize spatially explicit risk of DEN disease and infection. Geo-referenced case reports, serological data showing evidence of recent exposure, and serological data indicating recent or past exposure have been measured among many endemic populations along with environmental, demographic and socioeconomic covariates. These surveys have identified substantial variation in mosquito densities and hazards of DEN at distances on the order of 10 meters (Mammen *et al.*, 2008). In contrast to the macro-scale studies described in the statics section above, micro-scale variation shows no clear association with particular environmental, climatic or demographic characteristics across settings.

Geo-referenced case reports are available from a number of settings, and have been correlated with census data within cities and wider spatial units (Galli and Chiaravalloti Neto, 2008; Mondini *et al.*, 2005; Teixeira and Medronho Rde, 2008). A number of covariates have been associated with incidence of dengue cases, including socioeconomic status (Mondini and Chiaravalloti-Neto, 2008), water and sanitation infrastructure, population density and rates of literacy (Mondini and Chiaravalloti-Neto, 2008). Case incidence, even when confirmed by serology, can represent a biased sample of total incidence because individuals may have different access to care and receive care at health care facilities that have different rates of reporting (e.g. differences in reporting of public versus private health care facilities). Inferences drawn from case data of the association of environmental, socioeconomic and climatic variation with incidence may be impacted by the presence of these biases.

The most easily interpretable studies of the presence or absence of evidence of exposure to DEN come from serological surveys that detect long-lived antibody indicative of exposure at any time in the past (Siqueira *et al.*, 2004; Van Benthem *et al.*, 2005). Studies that measure recent exposure through the detection of IgM are more difficult to interpret because risk of recent exposure may be inversely related to lifetime risk of exposure (Vanwambeke *et al.*, 2006). Those individuals who live in the riskiest environments are likely to obtain infection early in life, and be at low risk of infection at the time of study due to acquired immunity.

A large number of studies have sought to determine the impact of socioeconomic status on the prevalence of serological evidence of immunity to DENV. In a number of settings, areas with low socioeconomic status have been associated with increased incidence of DENV infection. Studies using serological surveys have found that socioeconomic status is inversely related to the

hazard of DENV infection in Brazil (Caiaffa *et al.*, 2005; Siqueira *et al.*, 2004), Argentina (Carbajo *et al.*, 2004) and the United States (Reiter *et al.*, 2003). However, in another set of studies conducted in Brazil, (Teixeira and Medronho Rde, 2008), Vietnam (Bartley *et al.*, 2002a) and Mexico (Espinoza-Gomez *et al.*, 2003), low socioeconomic has been found to be unrelated to infection. In one additional setting, Ceara State, Brazil, low socioeconomic status has been found to be associated with reduced risk of acquisition of DEN, perhaps mediated by water storage practices around households of high socioeconomic status (Vasconcelos *et al.*, 1998).

Land use and characteristics of the peridomestic environment have also been studied extensively as risk factors for serological evidence of past DENV infection. Van Benthem (2005) found that individuals in Thailand living in structures surrounded by agricultural or forested land were at lower risk of seropositivity than those surrounded by urban land (Van Benthem *et al.*, 2005). Low rise structures, as opposed to high rise buildings, have been found to be associated with increased risk of acquisition of DEN (Goh *et al.*, 1987).

Several authors have attempted to associate spatial variation in vector densities, rather than DEN, with environmental conditions including urbanization and land use. Rios-Velasquez (2007) found little variation in household abundances of *Ae. aegypti* adults among four settings in Manaus, Brazil that varied by socioeconomic status, infrastructure and vegetation coverage (Rios-Velasquez *et al.*, 2007).

Though spatial variation in DEN incidence at large spatial scales appears to be fairly predictable using environmental and climatic information, micro-scale variation appears to be harder to explain and more setting dependent. Dengue seroprevalence studies have revealed a significant amount of spatial variation, but the impact of particular environmental or socioeconomic covariates appears to vary across settings. To obtain predictions of the risk of DENV infection at small spatial scales required to target local prevention efforts, studies from similar ecological and climatic settings must be consulted, or seroprevalence studies repeated. Otherwise, information from other settings on the factors that place particular populations at risk may mislead public health practitioners to target populations that may not have the highest risk.

Emergence

A primary cause of the increase in cases over the last few decades has been the geographic expansion of DENV. The number of countries that experience endemic transmission of DENV has grown dramatically over the last 40 years. In 1970, 10 countries reported endemic DENV transmission (World Health Organization, 2008). By 2000, this number had grown to 70 (World Health Organization, 2008). At the same time, the number of countries reporting DHF has increased from fewer than 5 to over 25 (World Health Organization, 2008).

The observation of an emergent DEN outbreak offers a rare opportunity to measure the spatial and temporal spread of disease in the absence of immunity. The spatial distribution of immunity is difficult to characterize, and thus knowing that it is uniformly absent at the beginning of an emergent outbreak makes interpretation of dynamics much easier.

Two outbreaks in immunologically naïve populations yield clues to specific aspects of the spatial temporal transmission of DEN. The first observed outbreak of DEN in San Salvador, Brazil, in the 1950s offered the opportunity to estimate the impact of land use and socioeconomic status on the speed of temporal spread through the city. Barreto and colleagues used reported cases among the immunologically naive population to estimate the speed of waves moving through the populations of different socioeconomic status (Barreto *et al.*, 2008). They found that an area of the city with low access to water and sanitation and low socioeconomic indices served as an epicentre and appeared to have increased speed of geographic spread, compared with areas of the city with higher socioeconomic indices and higher access to water and sanitation.

An outbreak on the Chilean Easter Island allowed Favier and colleagues to investigate heterogeneity in transmissibility similar to that investigated by Barreto (Favier *et al.*, 2005). Favier *et al.* fit mathematical models of transmission assuming homogeneity in contact rates between individuals as mediated by the mosquito vector, and models that explicitly included heterogeneity in contact rates between households and between communities. They found that models that included heterogeneity fit the data better than models that did not. Their finding lends additional support to the importance of micro-scale variation in characteristics that impact transmission, though they did not characterize the mechanisms that might lead to heterogeneity.

In these two detailed studies of the spread of an emergent dengue virus, the authors have found that the speed at which dengue spreads spatially depends upon local characteristics, with places of low socioeconomic status and sanitation appearing to support faster dispersal of dengue.

Dynamics

Temporal dynamics: seasonality

Within those settings where DEN is present, incidence varies widely from year-to-year and from month-to-month. In almost all locations, DEN is limited to a particular time of year, usually the months with the highest temperatures and precipitation. Several authors have sought to identify the cause of seasonality in DEN incidence. Temperature and rainfall have been the climate covariates most often associated with seasonal DEN incidence in correlation studies. In several locations of SE Asia, the timing of the rainy season has been found to be associated with the timing of the DEN season (Arcari *et al.*, 2007; Nakhapakorn and Tripathi, 2005; Rab aa and Cummings, 2006). Increases in temperature have also been associated with seasonal increases (Arcari *et al.*, 2007; Hurtado-Diaz *et al.*, 2007).

One of the most robust and innovative studies of the association of particular climate variation on the seasonality of DEN was conducted by Bartley and colleagues using data from Thailand (Bartley *et al.*, 2002b). The authors used a transmission model that incorporated specific aspects of the vector life cycle and interaction with human hosts. Using this model, the authors found that the period of mosquito infection required for transmission (most often referred to as the extrinsic incubation period), vector biting rate and vector mortality, which vary with ambient temperature, dictate the timing of the DEN season. Support for these three factors, particularly the impact of temperature on the extrinsic incubation period is echoed by Focks (Focks *et al.*, 1993) and Pant (Pant *et al.*, 1973).

Cycling of immunity

Many infectious diseases show cyclic variation in incidence. Cycles of measles, pertussis, and chickenpox, have been shown to exhibit stable multiannual cycles. The cause of these cycles was a focus of debate for much of the first half of the 20th century. For many childhood diseases, cycles have been shown to be consistent with a predator–prey interaction between the pathogen and susceptible humans. Mathematical models that incorporate a non-linear transmission term, in which new cases arise in time as a function of the product of the number of individuals susceptible to infection and the number currently infectious, have been used to reproduce patterns of incidence for each of the diseases named above. These models predict that large booms in incidence bring susceptible individuals to very low numbers. Then, a period of time must elapse before the susceptible population grows through the birth of susceptible hosts or the waning of herd immunity (for pathogens for which immunity is not life-long). The time scale at which susceptible individuals are re-introduced to the population

and the natural history of the pathogen (i.e. generation time, recovery time) dictate the time scale of oscillations, which are often multi-annual (periods greater than on year).

Dengue incidence in almost all endemic settings shows multi-annual cycling. Fig. 10.1 shows the aggregate annual incidence of DEN disease in 23 countries that had full reporting to Denguenet from 1980–2004 (http://www.who.int/globalatlas/DataQuery/default.asp). Clear multiannual cycling is shown, even in these aggregated data. In the right-hand side of Fig. 10.1, the results of a Fourier transform of each of the 23 time series of annual incidence in each country is shown. The Fourier transform was used to identify the dominant multiannual period. Fifteen of 23 countries exhibit a two or three year periodicity. Examples of two of these time series, Thailand and Vietnam, are shown in Fig. 10.2. Simple models of cycling in immunity predict cycles with a period of multiannual cycles of between 3 and 4 years (Hay *et al.*, 2000). This period changes with model assumptions, but is fairly robust to the inclusion of cross-immunity (Adams *et al.*, 2006; Wearing and Rohani, 2006), short-term cross-immunity (Nagao and Koelle, 2008) and enhancement of transmissibility among those experiencing their second infection (Cummings

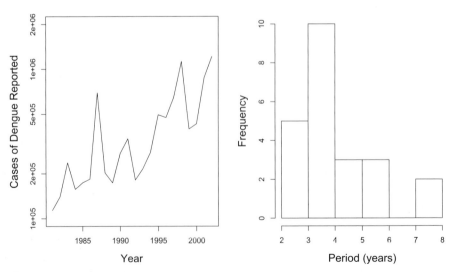

Figure 10.1 Annual cases of dengue identified in 23 countries and the period of multiannual oscillations. **A** Annual cases of dengue identified in 23 countries that have reported to DengueNet for each year 1980–2002. **B** Histogram of multiannual period of dengue cases reported by those 23 countries.

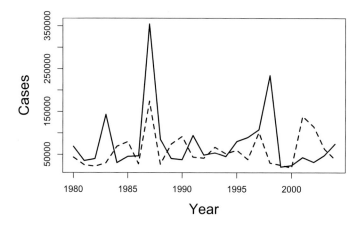

Figure 10.2 Annual cases of dengue disease in Thailand (dotted line) and Vietnam (solid line).

et al., 2005). Recently, several authors have suggested that the inclusion of short-term cross immunity among the four DENV serotypes is critical to producing simulation results that show similar phase decorrelation among serotypes, as is shown in data from Thailand (Adams et al., 2006; Earn et al., 2000). More work is needed to determine the interplay of multiple types of immune responses.

Spatial–temporal patterns

In a small number of settings, sufficient data are available to characterize the synchronization of epidemics across a country or region. Traveling waves in multiannual oscillations of DEN incidence in Thailand were first observed by Cummings et al. (Cummings et al., 2004) and have been confirmed by several other authors (Cazelles et al., 2005; Vecchio et al., 2006). The cause of these waves is unknown. One hypothesis is that they are caused by the reintroduction of particular serotypes to small rural populations that experience frequent stochastic fadeouts of particular serotypes. Another hypothesis is that the waves are caused by heterogeneity in the force of infection across Thailand. Further research is needed to determine the mechanism driving these waves.

Regional level synchrony has not been assessed for any region of the world. As surveillance systems develop and more data become available, a clearer picture of whether countries within a region experience epidemics at the same time may merge.

Public health impact

A reasonable question to ask of studies aimed at characterizing the spatial and temporal dynamics of DEN is 'what is the benefit to public health?' A quantitative description of the variability of DEN incidence would allow public health authorities to target resources to those places and periods of time that have the highest risk of large numbers of cases. Prevention efforts may still be challenging, even given full warning of epidemics. In some endemic areas, little may be done to control epidemics even if they are anticipated well in advance. However, case management and supportive therapy may be improved if large increases in incidence can be anticipated and planned for. In addition, as DEN vaccines become available (see Chapter 12), information on the spatial variation in the amount of transmission may be used to target immunization programmes optimally. The design of studies aimed at evaluating the impact of intervention campaigns may be improved by a quantitative description of incidence in the study setting over time and space.

Much of the surveillance data to assess regional synchrony is collected by national surveillance, but has not been analysed with an aim to compare the timing of dengue epidemics in different settings. This should be a priority for future research.

Conclusions

Although the causes of spatial and temporal variation in DEN incidence in many settings are not fully characterized, many potential mechanisms have been identified, and found to be associated with increased transmission. The data available to date have allowed researchers to identify locations that are at increased risk of transmission and at increased risk of successful emergence, as well as times that are at increased risk of incidence. At smaller spatial scales, multiple factors have been found to dictate the spatial distribution of dengue. Targeted intervention campaigns should not depend upon research from other settings to predict areas with high incidence. Instead, where possible, local information on the characteristics that predict high incidence should be conducted. Mechanistic models that can incorporate the waxing and waning of immunity provide inference frameworks to more effectively estimate the impact of particular factors on transmission. As surveillance systems continue to improve, an increased understanding of the interplay between multiple factors in each setting might lead to successful prediction of spatial and temporal patterns of DENV transmission.

Acknowledgement

Derek A.T. Cummings is supported by a Career Award at the Scientific Interface from the Burroughs Wellcome Fund.

References

Adams, B., Holmes, E.C., Zhang, C., Mammen, M.P., Nimmannitya, S., Kalayanarooj, S., and Boots, M. (2006). Cross-protective immunity can account for the alternating epidemic pattern of dengue virus serotypes circulating in Bangkok. Proc. Natl. Acad. Sci. U.S.A. 103, 14234–14239.

Arcari, P., Tapper, N., and Pfueller, S. (2007). Regional variability in relationships between climate and dengue/DHF in Indonesia. Singapore J of Trop Geog 28, 251–272.

Barreto, F.R., Teixeira, M.G., Costa Mda, C., Carvalho, M.S., and Barreto, M.L. (2008). Spread pattern of the first dengue epidemic in the city of Salvador, Brazil. BMC Public Health 8, 51.

Bartley, L.M., Carabin, H., Vinh Chau, N., Ho, V., Luxemburger, C., Hien, T.T., Garnett, G.P., and Farrar, J. (2002a). Assessment of the factors associated with flavivirus seroprevalence in a population in Southern Vietnam. Epidemiol Infect 128, 213–220.

Bartley, L.M., Donnelly, C.A., and Garnett, G.P. (2002b). The seasonal pattern of dengue in endemic areas: mathematical models of mechanisms. Trans Roy Soc Trop Med Hyg 96, 387–397.

Bureau of Epidemiology (2000). Annual epidemiological surveillance report (Nonthaburi: Ministry of Public Health).

Caiaffa, W.T., Almeida, M.C., Oliveira, C.D., Friche, A.A., Matos, S.G., Dias, M.A., Cunha Mda, C., Pessanha, E., and Proietti, F.A. (2005). The urban environment from the health perspective: the case of Belo Horizonte, Minas Gerais, Brazil. Cad Saude Publica 21, 958–967.

Carbajo, A.E., Gomez, S.M., Curto, S.I., and Schweigmann, N.J. (2004). [Spatio-temporal variability in the transmission of dengue in Buenos Aires City]. Medicina (B Aires) 64, 231–234.

Cazelles, B., Chavez, M., McMichael, A.J., and Hales, S. (2005). Nonstationary influence of El Nino on the synchronous dengue epidemics in Thailand. Plos Medicine 2, 313–318.

Cummings, D.A.T., Irizarry, R.A., Huang, N.E., Endy, T.P., Nisalak, A., Ungchusak, K., and Burke, D.S. (2004). Travelling waves in the occurrence of dengue haemorrhagic fever in Thailand. Nature 427, 344–347.

Cummings, D.A.T., Schwartz, I.B., Billings, L., Shaw, L.B., and Burke, D.S. (2005). Dynamic effects of anti body-dependent enhancement on the fitness of viruses. Proc Natl Acad of Sci USA 102, 15259–15264.

Earn, D.J.D., Rohani, P., Bolker, B.M., and Grenfell, B.T. (2000). A simple model for complex dynamical transitions in epidemics. Science 287, 667–670.

Espinoza-Gomez, F., Hernandez-Suarez, C.M., Rendon-Ramirez, R., Carrillo-Alvarez, M.L., and Flores-Gonzalez, J.C. (2003). [Interepidemic transmission of dengue in the city of Colima, Mexico]. Salud Publica Mex 45, 365–370.

Favier, C., Schmit, D., Muller-Graf, C.D., Cazelles, B., Degallier, N., Mondet, B., and Dubois, M.A. (2005). Influence of spatial heterogeneity on an emerging infectious disease: the case of dengue epidemics. Proc Biol Sci 272, 1171–1177.

Focks, D.A., Haile, D.G., Daniels, E., and Mount, G.A. (1993). Dynamic life table model for Aedes aegypti (diptera: Culicidae): simulation results and validation. J Med Entomol 30, 1018–1028.

Galli, B., and Chiaravalloti Neto, F. (2008). [Temporal-spatial risk model to identify areas at high-risk for occurrence of dengue fever]. Rev Saude Publica 42, 656–663.

Goh, K.T., Ng, S.K., Chan, Y.C., Lim, S.J., and Chua, E.C. (1987). Epidemiological aspects of an outbreak of dengue fever/dengue haemorrhagic fever in Singapore. South-East Asian J Trop Med Pub Health 18, 295–302.

Gubler, D.J. (1998). The global pandemic of dengue/dengue haemorrhagic fever: current status and prospects for the future. Ann Acad Med Singapore 27, 227–234.

Gubler, D.J. (2002). Epidemic dengue/dengue hemorrhagic fever as a public health, social and economic problem in the 21st century. Trends Microbiol 10, 100–103.

Hales, S., de Wet, N., Maindonald, J., and Woodward, A. (2002). Potential effect of population and climate changes on global distribution of dengue fever: an empirical model. Lancet *360*, 830–834.

Halstead, S.B. (2006). Dengue in the Americas and South-East Asia: Do they differ? Revista Panamericana de Salud Publica-Pan Am J of Pub Health *20*, 407–415.

Hay, S.I., Myers, M.F., Burke, D.S., Vaughn, D.W., Endy, T., Ananda, N., Shanks, G.D., Snow, R.W., and Rogers, D.J. (2000). Etiology of interepidemic periods of mosquito-borne disease. Proc. Natl. Acad. Sci. U.S.A. *97*, 9335–9339.

Hopp, M.J., and Foley, J.A. (2003). Worldwide fluctuations in dengue cases related to climate variability. Climate Rese. *25*, 85–94.

Hurtado-Diaz, M., Riojas-Rodriguez, H., Rothenberg, S.J., Gomez-Dantes, H., and Cifuentes, E. (2007). Short communication: Impact of climate variability on the incidence of dengue in Mexico. Trop Med & Int Health *12*, 1327–1337.

Jalbuena, K.R. (2008). Dengue cases double – DOH. In Manila Times (Manila, The Manila Times Publishing Corp.) http://www.manilatimes.net/national/2008/feb/01/yehey/top_stories/20080201top6.html (accessed 9 February 2009).

Jetten, T.H., and Focks, D.A. (1997). Potential changes in the distribution of dengue transmission under climate warming. Am. J. Trop. Med. Hyg. *57*, 285–297.

Mammen, M.P., Pimgate, C., Koenraadt, C.J., Rothman, A.L., Aldstadt, J., Nisalak, A., Jarman, R.G., Jones, J.W., Srikiatkhachorn, A., Ypil-Butac, C.A., Getis A., Thammapalo S., Morrison A.C., Libraty D.H., Green S., and Scott D.W. (2008). Spatial and temporal clustering of dengue virus transmission in Thai villages. PLoS Med. *5*, e205.

Mondini, A., Chiaravalloti Neto, F., Gallo y Sanches, M., and Lopes, J.C. (2005). [Spatial analysis of dengue transmission in a medium-sized city in Brazil]. Rev Saude Publica *39*, 444–451.

Mondini, A., and Chiaravalloti-Neto, F. (2008). Spatial correlation of incidence of dengue with socioeconomic, demographic and environmental variables in a Brazilian city. Sci Total Environ *393*, 241–248.

Nagao, Y., and Koelle, K. (2008). Decreases in dengue transmission may act to increase the incidence of dengue hemorrhagic fever. Proc. Natl. Acad. Sci. U.S.A. *105*, 2238–2243.

Nakhapakorn, K., and Tripathi, N.K. (2005). An information value based analysis of physical and climatic factors affecting dengue fever and dengue haemorrhagic fever incidence. Int J Health Geog *4*, 13.

Pant, C.P., Jatanasen, S., and Yasuno, M. (1973). Prevalence of *Aedes aegypti* and Aedes albopictus and observations on the ecology of dengue haemorrhagic fever in several areas of Thailand. South-East Asian J. Trop. Med. Pub. Health *4*, 113–121.

Rab aa, M.A., and Cummings, D.A. (2006). The relative timing of seasonal weather patterns and dengue incidence across the South-East Asian region. Am. J. Trop. Med. Hyg. *75*, 137–138.

Reiter, P., Lathrop, S., Bunning, M., Biggerstaff, B., Singer, D., Tiwari, T., Baber, L., Amador, M., Thirion, J.,

Hayes, J., Seca C., Mendez J., Ramirez, B., Robinson, J., Rawlings, J., Vorndam, V., Waterman, S., Gubler,D., Clark, G., and Hayes, E. (2003). Texas lifestyle limits transmission of dengue virus. Emerg. Infect. Dis. *9*, 86–89.

Rios-Velasquez, C.M., Codeco, C.T., Honorio, N.A., Sabroza, P.S., Moresco, M., Cunha, I.C., Levino, A., Toledo, L.M., and Luz, S.L. (2007). Distribution of dengue vectors in neighborhoods with different urbanization types of Manaus, state of Amazonas, Brazil. Mem. Inst. Oswaldo Cruz *102*, 617–623.

Rogers, D.J., and Randolph, S.E. (2000). The global spread of malaria in a future, warmer world. Science *289*, 1763–1766.

Rogers, D.J., Wilson, A.J., Hay, S.I., and Graham, A.J. (2006). The global distribution of yellow Fever and dengue. Adv. Parasitol. *62*, 181–220.

Siqueira, J.B., Martelli, C.M., Maciel, I.J., Oliveira, R.M., Ribeiro, M.G., Amorim, F.P., Moreira, B.C., Cardoso, D.D., Souza, W.V., and Andrade, A.L. (2004). Household survey of dengue infection in central Brazil: spatial point pattern analysis and risk factors assessment. Am. J. Trop. Med. Hyg. *71*, 646–651.

Teixeira, T.R., and Medronho Rde, A. (2008). [Sociodemographic factors and the dengue fever epidemic in 2002 in the State of Rio de Janeiro, Brazil]. Cad Saude Publica *24*, 2160–2170.

Treeratkuarkul, A. (2008). Dengue hits hard. In Bangkok Post (Bangkok, Thailand, The Post Publishing Public Company Limited), 14 March 2008.

Van Benthem, B.H., Vanwambeke, S.O., Khantikul, N., Burghoorn-M aas, C., Panart, K., Oskam, L., Lambin, E.F., and Somboon, P. (2005). Spatial patterns of and risk factors for seropositivity for dengue infection. Am. J. Trop. Med. Hyg. *72*, 201–208.

Vanwambeke, S.O., van Benthem, B.H., Khantikul, N., Burghoorn-M aas, C., Panart, K., Oskam, L., Lambin, E.F., and Somboon, P. (2006). Multi-level analyses of spatial and temporal determinants for dengue infection. Int J Health Geog *5*, 5.

Vasconcelos, P.F., Lima, J.W., da Rosa, A.P., Timbo, M.J., da Rosa, E.S., Lima, H.R., Rodrigues, S.G., and da Rosa, J.F. (1998). [Dengue epidemic in Fortaleza, Ceara: randomized seroepidemiologic survey]. Rev. Saude Publica *32*, 447–454.

Vecchio, A., Primavera, L., and Carbone, V. (2006). Periodic and aperiodic traveling pulses in population dynamics: an example from the occurrence of epidemic infections. Phys. Rev. E Stat. Nonlin. Soft Matter Phys. *73*, 031913.

Wearing, H.J., and Rohani, P. (2006). Ecological and immunological determinants of dengue epidemics. Proc. Natl. Acad. Sci. U.S.A. *103*, 11802–11807.

World Health Organization (2002). Dengue and dengue haemorrhagic fever. http://www.who.int/mediacentre/factsheets/fs117/en/

World Health Organization (2008). DengueNet (Geneva, Switzerland, World Health Organization). http://www.who.int/globalatlas/DataQuery/default.asp

Dengue Virus Emergence from its Sylvatic Cycle

Nikos Vasilakis, Kathryn A. Hanley and Scott C. Weaver

Abstract

Dengue viruses (DENV) are members of the genus *Flavivirus* in the family *Flaviviridae* and include four antigenically distinct serotypes (DENV-1–4). In the last half-century, DENV have emerged as the most important arboviral pathogens in tropical and subtropical regions throughout the world, putting a third of the human population worldwide at risk of infection. The transmission of DENV includes a sylvatic, enzootic cycle, most likely between non-human primates and arboreal *Aedes* mosquitoes, and an urban, endemic/epidemic cycle (henceforth referred to as endemic) between peridomestic *Aedes* mosquitoes and human reservoir hosts. Phylogenetic analyses suggest that the four currently circulating urban DENV serotypes emerged independently from ancestral sylvatic progenitors in the forests of South-East Asia after the establishment of urban populations large enough to support continuous inter-human transmission. In this chapter we examine the sylvatic origins of DENV, including ecology, adaptation for urban transmission, and molecular epidemiology, as well as the forces that have shaped the molecular evolution of extant sylvatic DENV strains. The potential for sylvatic DENV to re-emerge into the human transmission cycle in the face of immunity to current urban strains or vaccine candidates currently under development is also discussed. The lines of information addressed in this chapter will provide an overview of how sylvatic DENV population dynamics and transmission influence pathogenesis and how study of sylvatic DENV will improve our ability to understand and predict DENV emergence.

Introduction

Dengue viruses (DENV) are arthropod-borne viruses (arboviruses) in the genus *Flavivirus* (family *Flaviviridae*) that utilize *Aedes* (*Stegomyia*) spp., primarily *Ae. aegypti* and *Ae. albopictus*, as vectors for endemic and epidemic transmission (Fig. 11.1). There are four antigenically distinct but genetically related serotypes (DENV-1, -2, -3 and -4) within the dengue (DEN) antigenic complex (Calisher *et al.*, 1989). Relative to other arboviruses, DENV are extremely restricted in their natural vertebrate host range, which according to most evidence includes only primates. Currently, all four DENV serotypes can be found in nearly all urban and periurban environments throughout the tropics and subtropics where *Aedes* (Stegomyia) *aegypti aegypti* is present. By current estimates this distribution of the principal vector puts nearly a third of the global human population at risk of infection. The impact of DENV infections on human health is enormous – DENV are responsible for ca. 100 million infections per year, resulting in approximately 500,000 cases of dengue haemorrhagic fever/dengue shock syndrome (DHF/DSS) with a case fatality rate of about 5% (Halstead, 1997). This is the highest incidence of human morbidity and mortality among all arboviruses. Although the true economic burden due to DEN infections is difficult to estimate with accuracy due to underreporting of disease and underutilization of health

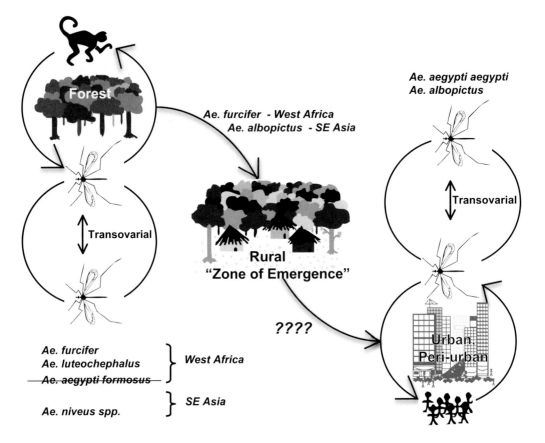

Figure 11.1 The transmission cycles of DENV, showing the sylvatic origins and the 'zone of emergence' where these cycles contact human populations in rural areas of West Africa (DENV-2) and Asia (probably all 4 DENV serotypes).

services, it is estimated that costs associated with a dengue infection significantly exceed the average monthly income of an affected patient (Clark *et al.*, 2005; Garg *et al.*, 2008).; see also Chapter 5

Human DEN results in a range of clinical manifestations. The great majority of infections are manifested as asymptomatic or subclinical infections, while symptomatic infections commonly manifest as flu-like disease [classical dengue fever (DF)] characterized by sudden onset of fever, arthralgia, myalgia, retro-orbital headaches, maculopapular rash, and leucopenia. The most severe form of the disease (DHF/DSS) is characterized by vascular leakage and/or haemorrhage (Burke *et al.*, 1988; WHO, 1997). DHF/DSS is a disease associated with children under the age of 15 years in hyperendemic regions of DENV activity (circulation of all 4 DENV serotypes) (Kalayanarooj and Nimmannitya, 2003; Kongsomboon *et al.*, 2004; Witayathawornwong,

2005), although recent evidence from South-East Asia and Latin America suggests that adults may also be at risk of developing the disease (Guilarde *et al.*, 2008; Hanafusa *et al.*, 2008; Koh *et al.*, 2008; Lee *et al.*, 2008; Siqueira *et al.*, 2005; Wichmann *et al.*, 2004).; see also Chapter 5

The origins of DENV in Asia

The four DENV serotypes that currently wreak such havoc with human health apparently emerged from sylvatic strains in the forests of South-East Asia. The persistence of these ancestral strains, which are still extant in both South-East Asia and West Africa, enables the study of those initial emergence events. However, continued circulation of sylvatic strains also poses a risk for the contemporary emergence of sylvatic strains into the human population.

An enzootic, sylvatic DENV cycle was first proposed fifty years ago, based on pioneering work by Gordon Smith. Smith, working in Penang, Malaya, first correlated earlier reports (More, 1904; Skae, 1902) describing a series of dengue epidemics, affecting most port cities of the Pacific coast and victimizing primarily recent immigrants, with the distribution of *Ae. aegypti aegypti* (Smith, 1956b). In the early part of the twentieth century, the distribution of this vector was confined solely in the seaports and along the coastline, but did not include inland locations. However, by the mid-1950s, Smith demonstrated serological evidence (neutralizing antibodies) of DENV infection in humans inhabiting a variety

of Malayan habitats and locations (Table 11.1) (I.M.R., 1956; Smith, 1956b, 1958), including rural areas where *Ae. aegypti aegypti* was absent. At the time Smith was conducting his urban eco-logical studies, the endemic transmission cycle was known only to involve *Ae. aegypti aegypti*. However, the work of Simmons (Simmons *et al.*, 1931) 25 years earlier had demonstrated experi-mentally that *Ae. albopictus* was also an efficient DENV vector. Incrimination of *Ae. albopictus* as an important, natural DENV vector in nature came later, when it was implicated in a series of large epidemics in port cities of Japan during

Table 11.1 Rural communities studied by Smith[a] for dengue seroprevalence

Location	Ecotope	Vectors (subgenus)[b]	Inhabitants (tribe)	DENV-1[c]	DENV-2[d]
Bukit Lanong Forest Preserve	Primary hill forest with variable amounts of clearing	*Aedes* (Finlaya) spp. *Heizmannia* spp. *Armigeres* spp.	Aborigines (Jakun)	Yes	Yes
Ulu Langat Forest Preserve	Primary hill forest with variable amounts of clearing	*Aedes* (Finlaya) spp. *Heizmannia* spp. *Armigeres* spp.	Ethnic Malays	Yes	Yes
Rantau Panjang	Coastal swamp and mangrove	*Aedes* (Aedes) spp. *Aedes* (Stegomyia) spp. *Aedes* (Skuse) spp. *Culex* (Culex) spp.	Ethnic Malays	Yes	Yes
Kampong Sireh	Coastal swamp and mangrove	*Aedes* (Aedes) spp. *Aedes* (Stegomyia) spp. *Aedes* (Skuse) spp. *Culex* (Culex) spp.	Ethnic Malays	Yes	Yes
Kampong Terachi	Narrow valley w/ rice paddies and surrounded by forest	*Anopheles* spp. *Culex* (Culex) spp. *Ae. albopictus*	Ethnic Malays	Yes	Yes
Kota Kuala Muda	Coastal rice plain surrounded by wet padi and coconut plantations	*Culex* spp. *Mansonia* spp. *Anopheles* spp. *Aedes* spp.	Ethnic Malays	Yes	Yes
Cameron Highlands	Tea plantations surrounded by mountain forests. Elevation – 5,000 ft	N/A	Aborigines (Senoi)	Yes	Yes

a   Information for this table assembled from IMR (1956), Smith (1956b).

b   Indicates predominant human-biting mosquito vectors.

c   Refers to presence of DENV-1 neutralizing antibodies.

d   Refers to presence of DENV-2 neutralizing antibodies, but at the time was unclear whether they represent serological cross-reactions with other flaviviruses or represent true infections with DENV-2.

the Second World War (Hota, 1952; Kimura and Hotta, 1944). In 1956, Smith demonstrated the presence of DENV antibodies in arboreal mammals such as non-human primates, civets, squirrels and slow lorises (*Nycticebus coucang*, a lower non-human primate), whereas few ground-dwelling vertebrates were seropositive (Smith, 1956b). In a subsequent study, he confirmed that only non-human primates among the arboreal vertebrates were positive for DENV-antibodies (Smith, 1958). Collectively, Smith's pioneering studies suggested for the first time that DENV neutralizing antibodies were prevalent in humans inhabiting forest, rural and urban ecotopes, and that an arboreal canopy-dwelling vector might be responsible for DENV transmission among non-human primates in a sylvan environment. Although these data suggested the existence of a DENV transmission cycle other than the recognized urban endemic cycle, the evidence was inconclusive because viruses were not isolated from putative arboreal vectors.

A decade later, in a preliminary study of DENV ecology, Rudnick and colleagues corroborated Smith's earlier conclusions that DEN was a zoonosis. Working in various ecotopes (primary dipterocarp, freshwater peat swamp and mangrove swamp forests) in Malaysia where both regular human activity and the urban vector *Ae. aegypti aegypti* were absent, he demonstrated the presence of DENV neutralizing antibodies in *Macaca fascicularis* (68.7%), *M. nemestrina* (50.0%) and *Presbytis cristata* (47.8%) monkeys (Rudnick, 1965). However, attempts to isolate virus from these monkeys, as well as from over 25,000 mosquitoes (eight genera and more than 69 species) were unsuccessful (Rudnick, 1965; Rudnick, 1986). Follow-up studies concentrated in the isolated the Jugra Forest Reserve (a sprawling mangrove swamp forest on Carey Island), the Tanjong Rabok and Telok Forest Reserves (freshwater peat-swamp forests), and Gunong Besut North (a primary dipterocarp forest) utilizing wild-captured *M. fascicularis* and/or *P. obscura* monkeys as sentinels in the forest canopy. From these sentinels were isolated DENV-1 (P72–1244), DENV-2 (P8–1407, P72–1273 and P72–1274 strains), and DENV-4 (P75–481, P73–1120 and P75–514 strains). DENV-3 was not isolated, but its existence was suggested

by the seroconversion of sentinel non-human primates. Surprisingly, no virus isolation was reported from 19 *M. nemestrina* sentinel monkeys exposed on the ground level (Rudnick and Lim, 1986), suggesting that transmission was focused in the canopy.

Rudnick's mosquito collections in Malaysia also eventually led to sylvatic DENV isolations. In 1969, the P8–377 strain of DENV-2 was isolated from *Ae. albopictus*, a vector found only at the ground level of forests, in the Malaysian study site at Carey Island. Subsequent phylogenetic analyses classified this strain as an endemic isolate (Wang *et al.*, 2000) (Fig. 11.2). In 1975, the DENV-4 P75–215 strain was isolated from *Aedes* (Finlaya) *niveus s.l.*, a primatophilic canopy-dwelling vector (Rudnick and Lim, 1986). This mosquito is abundant in the forest canopy and is attracted to humans. In fact, Rudnick significantly increased the capture numbers of this vector in the forest canopy with the use of human bait. Furthermore, all member species of this vector group [*Ae.* (Theobald) *pseudoniveus*, *Ae.* (Edwards) *subniveus*, *Ae.* (Colless) *vanus*, *Ae.* (Theobald) *albolateralis*, *Ae.* (Barraud) *niveoides* and *Ae.* (Barraud) *novoniveus*; *Ae.* (Theobald) *pseudoniveus*, *Ae. subniveus*] are primatophilic and known to descend to the ground level to feed on humans, a behaviour that could facilitate transfer of sylvatic DENV from the forest to peridomestic environments (Rudnick and Lim, 1986).

To further support the hypothesis that DENV are maintained in a sylvatic transmission cycle, a serosurvey of the aboriginal Orang Asli tribe (forest-dwelling people who had little contact with the general population) was conducted. The 175 surveyed participants, who represented various small communities from fringe- to deep- forest, included 20 residents of highly isolated, deep forest inhabitants. Of these populations, 80% (16/20) were positive for DENV neutralizing antibodies. Although no clinical DEN was reported in this group overall, 82% of the participants (144/175) were seropositive (DENV neutralizing antibodies) with increasing seroprevalence with age (Rudnick and Lim, 1986). This suggested the regular exposure of these isolated populations to DENV. Similarly high rates of DENV neutralizing antibodies were reported among isolated aborigines in the

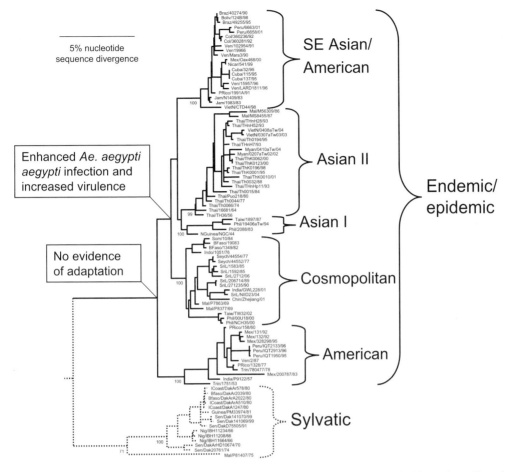

Figure 11.2 Phylogenetic relationships of DENV-2. The phylogeny was inferred based on E nucleotide sequences in the GenBank library, using Bayesian analysis (one million reiterations) and all horizontal branches are scaled according to the number of substitutions per site. Bayesian probability values are shown for key nodes. Strains are abbreviated as follows: Country abbreviation/strain/year.

Philippines who inhabited areas where *Ae. aegypti* was absent at that time (Hammon *et al.*, 1958), suggesting sylvatic transmission.

More recently, serologic surveys of Bornean orangutans demonstrated the presence of DENV-2 neutralizing antibodies in 28% (11/40) and 32% (10/31) of wild and semi-captive orangutans, respectively (Wolfe *et al.*, 2001). The wild orangutans were captured deep within the Tabin Wildlife Reserve, a vast tract of protected primary and secondary lowland tropical rain forest. Although it is possible that the wild orangutans were infected by exposure to mosquitoes that fed on domestic animals or viraemic agricultural or timber workers, the low human population density and absence of any human settlements within a 200 km radius of the capturing locations makes

this scenario unlikely. Thus this finding suggests the presence of a sylvatic transmission cycle in the forests of Borneo. However, the role of wild orangutans in the maintenance of a potential zoonotic DENV cycle remains unknown. Their low population density of ca. two individuals per square kilometre (km^2) makes it rather unlikely that orangutans alone could serve as reservoir hosts for such a cycle, and several species of non-human primates that are more abundant in the area may also play a role in the maintenance of sylvatic DENV (Wolfe *et al.*, 2001).

The seminal work of Smith laid the groundwork for Rudnick and his colleagues to demonstrate that enzootic, sylvatic DENV transmission cycles occur in the forest canopies of Malaysia and probably other locations tropical Asia where the zoonotic primate reservoir hosts and arboreal vectors occur. These cycles are maintained by various members of *Ae. niveus* spp., most likely utilizing *Macaca* and *Presbytis* spp. primates as reservoir and amplification hosts. Interestingly, although serologic evidence suggests sylvatic transmission of all four DENV serotypes, recent phylogenetic analyses using complete DENV genomes (Vasilakis and Weaver, 2008) indicate that the only isolate of DENV-1 that was originally classified as sylvatic may actually be an endemic isolate (although some methods produce trees consistent with an ancestral position in the DENV-1 clade; Fig. 11.3). Sylvatic DENV-3 strains have not been isolated but are believed to exist in Malaysia based on the seroconversion of sentinel monkeys (Rudnick, 1978).

Discovery of sylvatic transmission cycles in Africa

Around the same time that Rudnick and colleagues confirmed the existence of sylvatic DENV in Asia, evidence of similar zoonotic cycles in West Africa was uncovered. Until 1974, sylvatic DENV transmission cycles were only suspected to occur in this region based on DENV-2 seroconversions

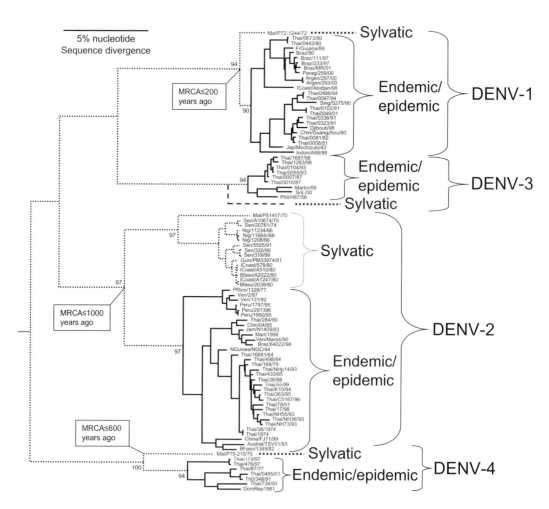

Figure 11.3 Phylogenetic tree derived from complete open reading frame sequences of dengue virus strains from the GenBank library. The tree was drawn using the neighbour joining method. Numbers indicate bootstrap values for major clades to the right. Virus strains and internal branches representing hypothetical ancestors are in dotted lines for sylvatic and continuous lines for endemic strains. The hypothetical sylvatic dengue virus-3 is believed to exist based on seroconversions in sentinel monkeys exposed in Malaysia (Rudnick, 1984a).

of wild monkeys inhabiting both gallery and lowland forests of Nigeria (Fagbami *et al.*, 1977). However, human DEN cases in eastern Senegal (Robin *et al.*, 1980a) prompted the establishment of a longitudinal surveillance programme near Kedougou that continues to this day. This programme led to the isolation of many arboviruses from mosquitoes that inhabit gallery forests near Kedougou, including DENV-2 (Cornet *et al.*, 1984a). Isolations from the sylvatic mosquitoes *Aedes furcifer, Ae. taylori,* and *Ae. luteocephalus* during a DENV-2 epidemic suggested a possible link between a forest cycle and human disease (Traore-Lamizana *et al.*, 1994). These findings also prompted the establishment of a comprehensive, decade-long (1974–1985) serosurvey of both human and non-human primate populations inhabiting the same region of Senegal. This study (i) corroborated the hypothesis that non-human primates serve as reservoir and amplification hosts, (ii) demonstrated the existence of successive epizootics among non-human primates (1974–1975 and 1981–1982) and (iii) failed to provide evidence of symptomatic, spillover (the transmission of an enzootic virus into people due to amplification of the enzootic cycle) epidemics (Saluzzo *et al.*, 1986a). These observations suggested that sylvatic strains are either confined to forest habitats and/or produce subclinical and/or relatively mild human illness without haemorrhagic manifestations (see 'Spillover epidemics and human contact' section below)

At the same time, additional lines of evidence provided strong support for DENV-2 enzootic transmission cycles in other parts of West Africa. During the 1980 rainy season (May–October), 28 strains of DENV-2 were isolated from primatophilic forest mosquitoes [*Ae. (Diceromyia) taylori, Ae. (Diceromyia) furcifer, Ae. (Stegomyia) opok, Ae. luteocephalus* and *Ae. (Stegomyia) africanus*] in Sokala-Sobara, a sub-Sudanese savannah ecotope. Mosquito DENV-2 isolates from Côte d'Ivoire (Cordellier *et al.*, 1983), 68 strains, mainly from *Ae. luteocephalus* (65 strains), in Upper Volta (present-day Burkina Faso) (Hervy *et al.*, 1984; Roche *et al.*, 1983), and several strains in the Republic of Guinea (Rodhain, 1991) extended the known distribution of sylvatic DENV cycles. Furthermore during the epizootic/epidemic of 1981–1982 in eastern Senegal, a DENV-2 strain

was isolated from *Erythrocebus patas* monkeys (Cornet *et al.*, 1984b). These surveillance studies strongly suggested the zoonotic nature of DENV-2 circulation, reinforced by the absence of any clinical DEN cases and negative seroprevalence among nearby human inhabitants.

Spillover epidemics and disease

The few epidemiological studies of the spillover of sylvatic DENV into humans that have been conducted indicate that such infections may be missed altogether due to the absence of clinical disease. Moreover, disease associated with sylvatic infections may be misclassified as endemic DENV due to erroneous assumptions about the absence of spillover. As an example of the former, during the 1981–1982 epizootic in eastern Senegal, approximately 11% of sera from children in the region tested positive by compliment fixation (CF), indicating probable DENV-2 infection in the absence of clinical disease (Saluzzo *et al.*, 1986a). As an example of the latter, an outbreak of febrile illness in Ibadan, Nigeria during the mid-1960s, discussed in detail below, was originally attributed to endemic DENV but recent phylogenetic analyses of DENV strains obtained from human sera classified these strains as sylvatic DENV (Vasilakis *et al.*, 2008c). These observations provide epidemiological support that limited spillover epidemics from enzoonotic cycles are possible in rural peridomestic and urban settings, but also caution that such spillover events may not be recognized due to lack of adequate surveillance and too-hasty attribution of disease to the endemic ecotype.

Sylvatic DENV infection can cause disease, although our understanding of human illness due to sylvatic DENV derives mainly from a handful of serendipitous investigations of febrile illness in Senegal that led to virus isolation and characterization. Almost all of these cases were concurrent with amplification of the prevalence of sylvatic DENV in arboreal *Aedes* (see below). Cases included (i) a 6 year old Senegalese girl living in the prefecture of Bandia whose blood sample in 1970 yielded the DakAr HD10674 strain of DENV-2 (Robin *et al.*, 1980b); (ii) a Caucasian scientist whose infection occurred while investigating the 1983 DENV amplification cycle in south-eastern

Senegal. His clinical illness was characterized as severe, with persistent rash and arthralgia lasting for approximately a year; (iii) a Caucasian male who in 1983 upon return from the south-western Senegalese province of Casamance, developed a febrile illness characterized by classical DF signs and symptoms including a maculopapular rash on the fifth day after onset (Saluzzo *et al.*, 1986b); (iv) a 31 year-old Caucasian man who 2 days after his return to Dakar from military manoeuvres in eastern Senegal developed a flu-like illness, characterized by sudden onset of classic DF signs and symptoms. Serological analyses performed 77 days after the onset of symptoms indicated past exposure to WNV, ZIKV and WESSV viruses (Monlun *et al.*, 1992); (v) a 15-year-old Senegalese boy living in the vicinity of Kedougou, eastern Senegal, who presented with a malaria-like disease and developed mild arthralgia that lasted nearly a month without any other symptoms. The last two cases occurred during the 1990 amplification detected by mosquito infections in the south-eastern Senegalese province of Kedougou, and led to DENV isolation from their sera. Importantly, virus isolation from humans was concomitant with virus isolation from sylvatic *Aedes* mosquitoes, reinforcing the notion of a sylvatic transmission cycle (Monlun *et al.*, 1992), but also the possibility of limited spill-over into adjacent humans. Additionally, these cases documented that clinical illness due to sylvatic DENV infection can be indistinguishable from classic DF due to infection with the ecologically and genetically distinct DENV from the endemic transmission cycle. However in 2008, a 20 year-old university student in Serawak (a Malaysian state in the island of Borneo), developed DHF grade II after returning from a month-long vacation in peninsular Malaysia, in a location near to Rudnick's study sites. Surprisingly, phylogenetic analysis of the virus isolate (DKD811) revealed a close relationship to the 1970 sylvatic P8-1407 isolate (Cardosa *et al.*, 2009). The implications of this case are significant for the following reasons: (i) the virus was probably maintained in non-human primates without detection for nearly 4 decades; (ii) it is the first indication that sylvatic DENV human infections can present with severe manifestations of DEN disease; (iii) it confirms our previous evidence from West Africa that re-

introduction of sylvatic DENV into the endemic transmission cycle is possible (Vasilakis *et al.*, 2008c). In light of these findings it is clear that comprehensive ecological and epidemiological studies are urgently needed to assess the role of non-human primates and possibly other vertebrate hosts in the maintenance of sylvatic DENV, as well as the degree and routes of ecological contact between humans and sylvatic DENV.

When conducting such surveillance, it will be important to recognize the possibility for sylvatic DENV spillover to occur in urban as well as rural settings, as illustrated by the following example. The first isolation of DENV in Africa from human sera occurred in Nigeria in the mid-1960s (Carey *et al.*, 1971), although the first reports of human DEN in Africa date back to the nineteenth and early twentieth century during outbreaks of febrile illness at the East and South Coast of the continent (Christie, 1881; Edington, 1927; Hirsch, 1883). Serological surveillance studies, carried out under the auspices of the Rockefeller Foundation, of febrile patients attending the Outpatient Clinic of the University College Hospital in the city of Ibadan (meaning, 'at the edge of the forest' in the local dialect), had demonstrated that DENV were endemic, as demonstrated by the isolation of 57 DENV-1 and -2 strains from both humans and mosquitoes within a span of 6 years (1964–1970) (Anonymous, 1969, 1971–1972). Specifically, from 1964–1968, 14 of the 32 strains of DENV isolated from febrile patients were classified as DENV-2, of which 10 strains were isolated in 1966 (Carey *et al.*, 1971). Phylogenetic analysis that included the complete genomic sequences of 3 of the 1966 DENV-2 isolates (the only ones known to exist in reference collections) indicated that they were genetically distinct from endemic DENV-2 isolates, and fell within the sylvatic DENV-2 clade (Vasilakis *et al.*, 2008c). Although there are no written records on the above patients' locations of residence or places of exposure, all had resided within the Ibadan city limits. Thus this study documents for the first time that limited spillover epidemics can occur in urban settings.

Furthermore, a retrospective serological survey provided evidence that non-human primates were involved in the sylvatic DENV transmission cycle in Nigeria (Fagbami *et al.*, 1977). In that study, 38% of sera collected from non-human primates within the rainforest were positive for DENV antibodies, which was similar to levels observed (43%) for humans living in communities within the forest. The prevalence of DENV antibodies in non-human primates living within the forest gallery in the Nupeko forest, a rainforest preserve located along the upper middle Niger river, was the highest at 74% (Fagbami *et al.*, 1977). Like the studies of Smith and Rudnick in South-East Asia, this study also examined the prevalence of DEN neutralizing antibodies in the human population within distinct ecotopes (Table 11.2), demonstrating that: (i) DENV-specific immunity was found throughout different ecotopes within Nigeria. Although the prevalence of neutralizing antibodies was higher in urban than in rural communities, the percentage of DENV-positive sera in rural settings were also high; (ii) the prevalence of DENV-specific antibodies increased with age, suggesting endemicity; (iii) the highest prevalence of DENV neutralizing

antibodies was observed in the derived savannah (63%) and rainforest (42%) ecotopes, respectively. However, an inherent limitation of this study lies in defining what constitutes an 'urban' setting. The authors' classification of urban as any community whose inhabitants numbered more than 20,000 might not have accurately reflected the ecological or entomological realities within each ecotope (Fagbami *et al.*, 1977). From an entomological perspective, the major DENV vector, *Ae. aegypti* was present in both rural and urban settings throughout Nigeria, albeit at lower densities in rural communities adjacent to forests (Lee, 1969; Lee *et al.*, 1974). Since the derived savannah ecotope in Nigeria shares many similarities with the moist savannah ecotopes that surround sylvan environments in rural areas of Senegal and South-East Asia [which Germain *et al.* (1976) named 'zones of emergence'] one can speculate that some of these human infections may be due to sylvatic DENV infections. Excepting the Ibadan outbreak described above,

Table 11.2 Prevalence of dengue virus type 2 neutralizing antibodies in human sera from four ecotopes in Nigeria and rural and urban communities within these ecotopes

Ecotope	Age (years)	DENV-2 prevalence[b] (% positive)	Urban[c] (% positive)	Rural (% positive)
Rainforest	Children 0–19	116/417 (28%)	73/243 (30%)	43/164 (26%)
	Adults 20 +	317/604 (52%)	211/412 (51%)	106/202 (52%)
	Total	433/1021 (42%)	284/655 (43%)	149/366 (41%)
Derived savannah	Children 0 – 19	175/297 (59%)	175/264 (66%)	10/33 (29%)
	Adults 20 +	76/101 (75%)	61/88 (69%)	5/13 (38%)
	Total	257/398 (65%)	236/252 (94%)	15/46 (33%)
Southern Guinea savannah	Children 0–19	16/85 (19%)	4/8 (50%)	12/77 (16%)
	Adults 20 +	57/130 (44%)	13/30 (43%)	44/100 (44%)
	Total	73/215 (34%)	17/38 (45%)	56/177 (32%)
Plateau	Children 0 – 19	3/25 (12%)	3/25 (12%)	NT
	Adults 20 +	51/153 (33%)	51/153 (33%)	NT
	Total	54/178 (30%)	54/178 (30%)	NT

a Information for this table was compiled with data obtained from Fagbami *et al.* (1977).

b Prevalence was based on PRNT test.

c Denotes communities within each ecotope with population >20,000 according to 1963 census.

NT, not tested.

the limited epidemiological information available and the paucity of serosurvey data prevent an accurate assessment of the overall human exposure to sylvatic DENV-2 in Nigeria, Senegal and other parts of West Africa.

The ecology of sylvatic DENV transmission in Africa

By current estimates there are approximately 500 strains of sylvatic DENV deposited in various arbovirus collections around the world. The overwhelming majority of these are DENV-2 collected in West Africa as a result of the collection efforts of Institut Pasteur in Senegal (Diallo *et al.*, 2003). This continuous collection in combination with historical record has revealed sylvatic amplification cycles of DENV-2 in West Africa, in which silent intervals (lack of virus isolates from mosquitoes) of about 8 years in length terminate in abrupt spikes in DENV circulation (as detected in 1974, 1980–1982, 1989–1990, 1999–2000). A recurring observation in these cycles is that, despite isolation of several DENV-2 strains from mosquito pools, few human clinical cases were recorded in the region (Diallo *et al.*, 2003). Domestic *Ae. aegypti*, the principle vector of epidemic DENV worldwide, is scarce or possibly absent from this area of eastern Senegal, whereas the sylvatic *Ae. aegypti formosus* form is abundant. Recent findings suggest that the zoophilic *Ae. aegypti formosus* plays little or no role in sylvatic DENV-2 transmission because it is relatively refractory to infection (Bosio *et al.*, 1998; Diallo *et al.*, 2008; Diallo *et al.*, 2005). However, during the last recorded amplification cycle of 1999–2000 in eastern Senegal, six sylvatic DENV-2 strains were isolated within four villages (Ngari, Silling, Bandafassi, and Kenioto) from collections of the arboreal *Ae. furcifer* vector. This suggests that *Ae. furcifer* could act as a bridge vector for sylvatic DENV dissemination into human habitats (Diallo *et al.*, 2003).

Collectively, the entomological and ecological data from various ecotopes (isolated patches of forest, forest-savannah and gallery forests) in Senegal, Côte d'Ivoire and surrounding countries suggest that non-human primates, mainly *Erythrocebus patas,* are the principal and possibly the only amplification and reservoir vertebrate hosts for sylvatic DENV-2. Various sylvatic *Aedes*

spp. mosquitoes, including *Ae. taylori, Ae. furcifer,* and *Ae. luteocephalus* (especially the last two) appear to be the principal vectors in an active sylvatic focus in Kedougou, Eastern Senegal. Sylvatic DENV-2 activity has also been documented in Ibadan, Nigeria, although this probable focus has not been systematically sampled during the last 50 years (Vasilakis *et al.*, 2008c). The assumption that *Erythrocebus patas* is the most important reservoir host is based on its widespread range throughout West Africa in various ecotopes, and its habit of returning to its sleeping sites (tree canopies) at dusk, when primatophilic *Aedes* spp. feed. However other species have also been implicated (Cordellier *et al.*, 1983; Fagbami *et al.*, 1977; Saluzzo *et al.*, 1986a). Although the conclusion that non-primate vertebrates are not important as DENV hosts is based mainly in the absence of virus isolation from non-primatophilic mosquitoes, the involvement of other vertebrate species in the sylvatic transmission cycles cannot be ruled out at present time.

Does sylvatic DENV occur in the Americas?

It must be emphasized that sylvatic DENV cycles have received very little study and are probably more widespread than Malaysia and West Africa. These cycles are difficult to detect in many tropical locations due to the presence of endemic transmission of human strains that may spill over into primates, and it is impossible to distinguish between the endemic/epidemic and sylvatic DENV genotypes in serological studies (Vasilakis *et al.*, 2008a). Thus, a sylvatic cycle can only be definitively identified by virus isolation and genetic characterization, although suggestive evidence can be obtained from seroprevalence in non-human primates without proximity to humans. Additionally the possibility of additional DENV serotypes in sylvatic cycles cannot be ruled out. If such serotypes have not emerged into an endemic cycle, it is quite possible that they have gone undetected but have the potential to impact human health in the future.

Interest in potential sylvatic DENV cycles in the Americas was prompted by the discovery of zoonotic, sylvatic, non-human primate cycles of yellow fever virus (YFV) in the Americas many decades ago (Monath, 2001). Yellow fever virus was apparently introduced into the Americas on sailing ships from Africa engaged in the slave trade (Bryant *et al.*, 2007), a mechanism that probably also brought DENV. Several attempts to document sylvatic transmission cycles in the Americas have been unsuccessful. Several species of New World non-human primates, including *Cebus capucinus, Ateles geoffroyi, Ateles fusciceps, Alouatta palliata, Marikina geoffroyi, Saimiri orstedii and Aotus trivirgatus* are susceptible to DENV-1 and DENV-2 infection. Like many old world species, they develop viraemia in the absence of clinical illness, as well as neutralizing antibodies. However, the viraemia profiles are believed to be insufficient to initiate oral mosquito infection (Rosen, 1958) (see Chapter 6).

Contrary to these lines of evidence, circumstantial and unconfirmed evidence suggest that a yet-to-be-sampled DENV enzootic cycle may in the tropical forests of the New World. Working in the isolated forest region of Rincon del Tigre in Bolivia, where *Ae. aegypti* is not present, Roberts *et al.* (1984) demonstrated, by plaque reduction neutralization titre, seroconversion to DENV-2. Furthermore, a serosurvey of 27 wild forest mammal species in French Guyana demonstrated DENV-2 neutralizing antibodies in various Xenathra (armandillo), Marsupialia (brown, four-eyed opossum), Rodentia (porcupine) and Artiodactyla (brocket deer) (de Thoisy *et al.*, 2004; de Thoisy *et al.*, 2008). In theory, *Ae. (Gymnometopa) mediovittatus*, a forest mosquito that is also adapted to peridomestic habitats and shares larval sites with *Ae. aegypti*, could support such cycles (Gubler *et al.*, 1985).

The French Guyana studies raise the question of whether infections in wild animals represent spillover from endemic DENV cycles. Although the DENV sequences derived from wild animals suggest otherwise, the transfer of DENV from urban areas to smaller outlying communities has been suggested previously (Hayes *et al.*, 1996; Rudnick and Lim, 1986), but is always associated with the presence *Ae. aegypti*. In Asia, although Smith's initial studies had demonstrated the presence of DENV antibodies in tree-dwelling vertebrates including non-human primates, civets, squirrels and slow lorises, and in few ground-dwelling vertebrates (Smith, 1956b); later he was able to serologically confirm only that

non-human primates among the tree-dwelling vertebrates were seropositive (Smith, 1958). It is also conceivable that domestic animals could play a role in such 'reverse spillover'. Rudnick demonstrated by HI the presence of DENV antibodies in 58% of pigs examined during a DENV epidemic in Penang, Malaysia. Only 20.6% of these were positive by the more specific PRNT, and only at low titres (Rudnick and Lim, 1986; Rudnick *et al.*, 1967).

One hypothesis for the periodicity of sylvatic DENV amplification cycles is non-human primate population dynamics. These primates have a relatively long lifespan of ca. 20 years and females usually produce only a single offspring per year after a gestation of ca. 8 months. Since DENV infection produces lifelong homotypic immunity, the number and turnover of susceptible non-human primates may be too small to maintain DENV-2 transmission while a critical threshold of herd immunity is exceeded. These factors, along with deforestation and other pressures such as hunting on non-human primate populations, raise doubts about the sustainability of an enzootic DENV transmission cycle that relies solely on non-human primates as amplification and reservoir hosts. Although vertical transmission may contribute to DENV maintenance in the forest, it is also possible that other mammals could serve occasionally or regularly as reservoir hosts or that sylvatic DENV is maintained as a 'wandering epizootic', as has been proposed for yellow fever virus in South America (Bryant *et al.*, 2003; Chippaux *et al.*, 1993; Vasconcelos *et al.*, 1997).

It has become increasingly clear that comprehensive ecological and epidemiological studies of the potential for the establishment of sylvatic DENV cycles in the Americas are needed. Considering these many similarities with YFV and the lines of evidence described above, it is certainly plausible that, with increased levels of endemic DENV transmission in the Americas, opportunities for the establishment of an enzootic, sylvatic cycle will increase, if one does not exist already, and if DENV cannot establish a sylvatic cycle among non-human primates in the neotropics, discovery of the reasons would greatly enhance our understanding of arbovirus emergence mechanisms and limitations.

The sylvatic DENV 'zone of emergence'

Germain *et al.* (1976) coined the term 'zone of emergence' to describe the ecotope of moist savannahs surrounding sylvan environments in rural areas of Africa and Asia where enzootic vectors often reach high densities, and DENV can transfer between non-human primates and humans (Fig. 11.1). In Asia, the findings of Rudnick suggest that zoonotic *Ae.* (*Finlaya*) *niveus* vectors descend to the ground, where *Ae. albopictus* mosquitoes are abundant, to feed on humans. This behaviour may facilitate the transfer of sylvatic DENV from the forest into human habitats. In a 1977–1978 companion study, Rudnick and colleagues measured the incidence of DENV infection in rural areas (in the Ulu Langat subdistrict) with large plantations of rubber trees, coconut and oil palms adjacent to forests. These study areas were characterized by small, relatively immobile human communities and a lack of *Ae. aegypti aegypti.* The results established that: (i) the highest rates of DENV infection were among the rural populations living adjacent to the forest; (ii) mild fevers of short duration, presumably due to DENV infection were occasionally described; and (iii) *Ae. albopictus* was the principal vector at ground level (Rudnick, 1986). In contrast, the lowest rates of DENV infection were observed in villages with high population densities of relatively mobile humans who had no close contact or association with the plantations or the forest, and who resided where only low densities of *Ae. albopictus* were present. This situation parallels that in rural areas adjacent to forests of West Africa, where the principal bridge vector between the forest and village appears to be *Ae. furcifer* (Diallo *et al.*, 2003; Diallo *et al.*, 2005). Similarly, in West Africa (albeit at lower incidence rates), DENV-2 may circulate among rural populations in the absence of detected clinical illness, probably with the presentation of mild signs and symptoms (Monlun *et al.*, 1992; Saluzzo *et al.*, 1986a).

Other species of *Aedes* mosquitoes, such as *Ae.* (*Stegomyia*) *polynensiensis, Ae. mediovittatus,* and *Ae.* (*Stegomyia*) *scutellaris* have been implicated in DENV transmission in some localities of the Pacific based on epidemiological observations (Daggy, 1944; Mackerras, 1946; Rosen *et al.*, 1954) and experimental transmission (Gubler *et* *al.*, 1985; Rosen *et al.*, 1985). *Ae.* (*Protomacleaya*) *triseriatus* could also be considered a potential DENV vector based on experimental transmission studies (Freier and Grimstad, 1983), but its distribution is temperate and does not overlap with DENV-endemic locations. In the latter studies, the mosquitoes tested exhibited a higher susceptibility to oral DENV infection than did *Ae. aegypti aegypti.* This difference has been also reported with respect to *Ae. albopictus* (Jumali *et al.*, 1979; Moncayo *et al.*, 2004; Rosen *et al.*, 1985) (see Chapter 8).

The origin of dengue viruses

The geographic origins of DENV have been discussed for many decades. The first historical record of dengue-like illness in China dates to over 1000 years ago (Gubler, 1997). However, even today DEN is difficult to diagnose clinically because its signs and symptoms overlap with those of many other tropical diseases. Therefore, it is difficult to know when DEN was first described. The first detailed clinical descriptions of what was almost certainly DEN were made by Benjamin Rush, who described an outbreak of 'bilious remitting fever,' and reported the vernacular name of 'break-bone fever' to describe a 1780 epidemic in Philadelphia (Rush, 1789). The use of the latter term in San Juan, Puerto Rico as early as 1763 suggests that DEN was well known in the Caribbean even earlier (Rigau-Perez, 1998). During the same era, David Bylon described a 1779 epidemic in Jakarta, Indonesia, which he called '*knokkel-koorts*' or knuckle fever (Bylon, 1780). Both physicians emphasized the severity of pains, the rash, and the tendency to relapse. Bylon's report suggests that DEN was known in Jakarta before 1779, but had never before reached epidemic proportions (Bylon, 1780; Pepper, 1941).

Early hypotheses for the ultimate origins of DENV focused on Africa, the location where the principal DENV vector, *Ae. aegypti*, originated (Smith, 1956b). The African origin of yellow fever virus, which is also vectored in its epidemic cycle by this mosquito species, also pointed towards an Africa DENV origin. Yellow fever virus is believed to have arisen in Africa and was introduced into the Americas during the slave trade ca. 300–400 years ago, where it established sylvatic, enzootic cycles that persist today (Bryant *et al.*, 2007).

As a member of the *Flavivirus* genus of the family *Flaviviridae*, the geographic distribution of close relatives might provide insights into the location of the DENV progenitor. A complete phylogeny of all available, partial flavivirus non-structural protein 5 sequences is shown in Fig. 1.1. Unfortunately, the closest relative of DENV is not robustly identified phylogenetically (bootstrap value for Kedougou virus-DENV grouping is <50%) and a wide variety of mosquito-borne viruses, mainly transmitted by *Culex* spp. vectors, fall into the robust sister group that is well supported (bootstrap 92%). These viruses occur in diverse locations in most of the continents, shedding no light on the possible geographic origins of DENV.

Evidence for sylvatic, Asian DENV ancestors of endemic strains

The advent of modern molecular genetics and phylogenetics provided the first opportunity to trace the evolution of DENV. The first phylogenetic studies that utilized DENV nucleotide sequences focused on DENV-1 and -2, and utilized E/NS1 partial gene sequences of 240 nucleotides (Rico-Hesse, 1990). The DENV-2 analysis revealed a genetically distinct lineage restricted to West Africa, and the origins of these strains suggested an independently evolving sylvatic cycle there. Support for independent sylvatic DENV lineages in Malaysia came from the more comprehensive phylogenies of Wang *et al.* produced from complete envelope protein gene sequences of all 4 DENV serotypes, including both endemic/epidemic and sylvatic strains (Wang *et al.*, 2000). The basal position of sylvatic lineages of DENV-1, -2, and -4 suggested that the endemic/epidemic lineages of these three DEN serotypes diverged from sylvatic progenitors from 100 to 1500 years ago. An updated version of this tree is presented in Fig. 11.3. These conclusions were supported by later studies by Holmes and Twiddy, who obtained slightly more recent estimates for most recent common ancestors of sylvatic and endemic/epidemic strains (Holmes and Twiddy, 2003). These very recent estimates for divergence of the extant, endemic/epidemic lineages suggest that epidemics that occurred during the 18th century or earlier could have

been caused by distinct, earlier emergences from sylvatic cycles, and that these lineages may have gone extinct and been replaced by more recently emerged strains. Moreover, in contrast to the 'out of Africa' hypothesis suggested by Smith, Wang *et al.* (2000) postulated an Asian origin of DENV based on the occurrence of all four sylvatic cycles there, but only sylvatic DENV-2 in Africa.

The DENV-2 sylvatic lineages have received the most attention because they occur both in Africa and in Asia, and multiple African strains are available for study. The Asian and African sylvatic lineages share a common ancestor that occurred on the order of hundreds of years ago, suggesting that the African strains originated from an introduction from Asia. This temporal timeframe is consistent with movement of DENV on sailing ships, where human–mosquito–human transmission could occur onboard while *Ae. aegypti* took advantage of stored water as its larval habitat and humans for blood feeding. This hypothesis is also supported by the vector competence of sylvatic, African vectors for endemic/epidemic DENV-2 strains that originated in Asia (Diallo *et al.*, 2005).

The topology of the original DENV-2 trees (Rico-Hesse, 1990; Wang *et al.*, 2000) has recently been supported by phylogenies derived from complete DENV genomic sequences (Vasilakis *et al.*, 2007a). An updated version of this tree is presented in Fig. 11.2. Interestingly, these new analyses indicate no major differences in the rate of sequence evolution between the sylvatic and endemic/epidemic DENV-2 lineages, or in the patterns of selection revealed by maximum likelihood dN/dS analyses. Both findings are somewhat surprising for the following reasons: (1) rates of evolution might be expected to differ between the sylvatic and endemic/epidemic lineages if rates of transmission, and thus DENV replication, differ in the two cycles; and (2) the sylvatic lineages, which have probably circulated in their current niches for centuries or longer, would be expected to have reached a higher fitness plateau than the more recently emerged endemic lineages. The lack of support for the latter hypothesis must be tempered by limitations on the ability to detect positive selection using phylogenetic methods, which recognize repeated non-synonymous mutations of the same codon but not unique amino acid changes that could nevertheless be under strong selection.

Emergence and spread of endemic strains

Although none of the phylogenetic studies cited above could provide direct evidence that the sylvatic cycle of DENV is ancestral to endemic strains, models for endemic DENV transmission suggest that the minimum population size required to support continuous circulation among humans range from 10,000 to 1 million (Kuno, 1997). Human populations in Asia are believed to have approached these sizes on the order of 4000 years ago when urban civilizations first arose (Burns *et al.*, 1986). The divergence estimates described above are consistent with these estimates of adequate human population sizes for endemicity, as well as the first historical record of dengue-like illness in China dating to over 1000 years ago (Gubler, 1997). Therefore, the sylvatic DENV lineages are believed to be ancestral and the endemic/epidemic lineages are apparently derived and evolved independently after the evolution of the four sylvatic DENV cycles (Gubler, 1997; Wang *et al.*, 2000).

Significant amplification and geographic spread of endemic DENV occurred followed the acquisition of *Ae. aegypti aegypti* as the principal DENV vector. Having originated in Africa, where its ancestral form, *Ae. aegypti formosus*, uses tree-holes as larval habitats (Tabachnick and Powell, 1979), the *aegypti* subspecies adapted in Africa to live in close association with people by using artificial water containers as its larval habitat. Following this peridomestication and subspeciation, *Ae. aegypti aegypti* was transported via trade routes to nearly all tropical and subtropical locations. The derived form *Ae. aegypti aegypti* now lives in close contact with people in urban settings by relying on artificial water containers for its larval habitats. Furthermore, its reliance on blood (instead of plant carbohydrates) for its energy needs, as well as its endophilicity, have resulted in a high frequency of multiple host contacts during a single gonotrophic cycle (Harrington *et al.*, 2001). The potent combination of these behavioural and ecological traits has contributed to *Ae. aegypti's* success as an endemic and epidemic vector and enabled the globalization of DENV.

Evolution of DENV antigenic diversity

An interesting question raised by DENV phylogenies and antigenic relationships is why four serotypes have evolved (although, as discussed above additional sylvatic DENV serotypes could exist without detection). The initial divergence of the four serotypes is believed to have occurred in the sylvatic cycle prior to the divergence of the extant endemic/epidemic DENV lineages (Wang *et al.*, 2000). Therefore, assuming that primate host availability is limiting for sylvatic DENV circulation, cross-reactive immunity should have had a strong impact on the diversification of DENV strains (Holmes, 2004). Early divergence of a single DENV ancestor would have been limited by cross-protective immunity, which presumably was complete when little antigenic diversity was present (Fig. 11.4). Thus, if all DENV variants were essentially competing for the same host population, competitive exclusion would have periodically reduced genetic diversity through selective sweeps. However if the four serotypes of DENV evolved in ecological or geographical isolation, they would have avoided this periodic 'pruning' of diversity. Ecological isolation would require either geographical isolation, specialization of different DENV lineages in distinct primate hosts or specialization for different vector species with distinct primate host contacts. Neither of the latter two scenarios is supported by the broad host ranges of sylvatic strains for both humans and non-human primates (Vasilakis *et al.*, 2007b), and for at least two sylvatic West African vectors (Diallo *et al.*, 2005). This information suggests that the four extant DENV serotypes may have evolved allopatrically in different regions of Asia. The limitation of DENV to four main serotypes may reflect, at least in part, limitations on geographic distribution during this ancient period of diversification in the forest cycle.

Following allopatric establishment of the four DENV serotypes, the endemic/epidemic lineages emerged independently as described above. Initially, these lineages may have been maintained in their ancestral distributions and endemic strains may have emerged repeatedly before the divergence of the extant endemic strain. The advent of transoceanic sailing for trade and transportation provided for the human dispersal of the

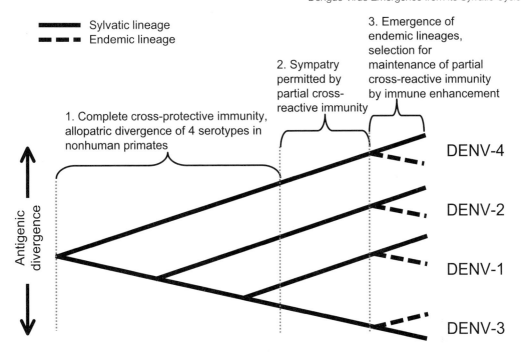

Figure 11.4 Hypothetical phases during the evolution of the 4 DENV serotypes and the development of partial cross-protective immunity. Solid line branches represent sylvatic transmission among non-human primates by arboreal mosquitoes and dotted line branches represent endemic/epidemic transmission among humans by peridomestic vectors.

four DENV serotypes and the elimination of their allopatric distributions. However, presumably the four serotypes had diverged antigenically to the point of limited, cross-reactive immunity observed today, which lasts only for months. This condition released the DENV serotypes from direct competition for hosts, allowing for sympatry of the four serotypes, a configuration that is becoming increasingly common during the contemporary DENV pandemic (see Chapter 8, this volume).

While the absence of long-term cross-neutralization allows sympatry among multiple DENV serotypes, enhancement of virus replication following heterologous infection may actually favour coexistence of multiple serotypes. If such enhancement results in increased transmission, DENV serotypes would benefit from prior and concurrent circulation of different serotypes in the same location (Ferguson et al., 1999). As described in Chapter 5, there is strong evidence that secondary DENV infections are associated with increased pathogenicity, often leading to dengue haemorrhagic fever and shock syndrome

(Rothman, 2003), and that severe disease is associated with increased viraemia (Vaughn et al., 2000). Furthermore, Ae. aegypti is only moderately susceptible to DENV infection (see Chapter 9), so selection for high viraemia facilitated by immune enhancement or other mechanisms is certainly plausible.

Evolution of DENV virulence

Monath speculated that the genetically and ecologically distinct sylvatic DENV exhibit a lower virulence potential for humans than endemic strains and thus limited spillover potential to the endemic transmission cycle (Monath, 1994). However several lines of evidence that have emerged over the last 20 years do not support these notions. First, experimental evidence suggests that sylvatic DENV do not require adaptation for optimal replication in either models for human hosts (Vasilakis et al., 2007b) or authentic mosquito vectors (K. Hanley, N. Vasilakis, unpublished). However, experimental studies are urgently needed to evaluate the replication kinetics and disease manifestation of sylvatic

DENV strains in non-human primates. Second, the historical and medical record of documented sylvatic DENV human infections in West Africa (Monlun *et al.*, 1992; Robin *et al.*, 1980b; Saluzzo *et al.*, 1986b), as well as the recent evidence from Malaysia (Cardosa *et al.*, 2009) demonstrates that sylvatic DENV infections share the virulence potential of DENV from the endemic transmission cycle. Lastly, several studies have suggested that spillover of sylvatic DENV is a fairly common occurrence. Although observations dating back in the 1920s by Siler (Siler *et al.*, 1926) and Simmons (Simmons *et al.*, 1931) in the Philippines suggested (without being confirmed to this day) the presence of sylvatic DENV-caused disease in isolated human populations deep in forested areas, the serologic studies of Smith (Smith, 1956b) and Rudnick (Rudnick, 1986; Rudnick and Lim, 1986) in southeast Asia and Fagbami (Fagbami *et al.*, 1977) in West Africa were the first to suggest that spillover DENV infections from the sylvatic transmission cycle may take place in human populations (forest and plantation workers) in ecotopes within the forest or at the edge of forests (zones of emergence) (Tables 11.1 and 11.2). Conclusive evidence of sylvatic DENV spillover into endemic transmission cycle was only recently obtained (Vasilakis *et al.*, 2008c; Cardosa *et al.*, 2009) (see Potential for Sylvatic DENV Re-emergence: Epidemics and Human Contact).

The most recent chapter in the evolution of DEN appears to also include selection for enhanced transmission and incidentally, for more severe disease. The endemic/epidemic DENV-2 strains can be subdivided into a variety of genotypes, minimally Asian and American (Fig. 11.2). There is strong evidence that the Asian genotype is more virulent and more likely to result in DHF/DSS than the American genotype even after a secondary infection (reviewed in Chapter 5). Experimental data also suggest that the Asian genotype has been selected for enhanced ability to replicate in key human targets such as macrophages and dendritic cells (Cologna *et al.*, 2005; Cologna and Rico-Hesse, 2003; Vasilakis *et al.*, 2007b). In addition, the Asian DENV-2 lineage appears to have evolved enhanced transmission by *Ae. aegypti*, facilitating its spread in the Americas (Armstrong and Rico-Hesse, 2001, 2003). These recent evolutionary events have dramati-

cally increased the incidence of DHF/DSS due to DENV-2, especially in the Americas. A similar pattern has been documented for the evolution of DENV-3. Group A DENV-3, a strain associated with mild disease, was endemic to Sri Lanka until it was displaced by Group B DENV-3, an event that resulted in an outbreak of DHF (Messer *et al.*, 2003). Not only was Group B associated with more disease that Group A, but it also proved to be more infectious for mosquitoes (Hanley *et al.*, 2008), and both traits likely enhanced its capacity to spread and displace endemic strains.

Comparison of the ecology of DENV and yellow fever virus

The lines of evidence presented in the previous sections clearly indicate several similarities between sylvatic DENV and sylvatic yellow fever viruses (YFV). YFV also has a zoonotic presence in the sylvan and transitional vegetation ecotopes of Africa and (unlike DENV) South America. Like DENV it is maintained by transmission between arboreal, primatophilic and tree-hole inhabiting mosquitoes [the main vectors are *Haemagogus* (*Haemagogus*) *janthinomys* and *Ae.* (*Stegomyia*) *africanus* in South America and Africa respectively] and non-human primates (Balfour, 1914; Germain *et al.*, 1976; Haddow, 1969; Monath, 1994; Smithburn and Haddow, 1949; Soper *et al.*, 1933). However, unlike sylvatic DENV, sylvatic YFV is responsible for explosive human epidemics. For example, emergence of sylvatic YFV in the rural savannas spanning the width of Africa, from Ethiopia to Senegal to Nigeria, between 1962 and 1987 was responsible for 8 major human epidemics causing enormous morbidity and mortality (Baudon *et al.*, 1986; Carey *et al.*, 1972; Cornet *et al.*, 1968; Germain *et al.*, 1980; Meunier *et al.*, 1988; Moore *et al.*, 1975; Olaleye *et al.*, 1988; Rodhain, 1991; Serie *et al.*, 1968). In South America, mostly Brazil, two-thirds of the geographic area of which is characterized as an enzootic focus, enzootic-fuelled YFV cases and/or epidemics have been diagnosed yearly since 1930, with over 1000 human confirmed cases (Vasconcelos *et al.*, 2001; Vasconcelos *et al.*, 1997). In the last 20 years alone, two major sylvatic YFV epidemics have occurred, with 163 confirmed human cases and 39 fatalities (all during the 2000 epidemic) (Vasconcelos *et al.*, 1997,

2001). The severity of sylvatic YFV infection to non-native and immunologically naïve humans is illustrated by the fatal outcome of infection in an American and Swiss tourist visiting the Amazon in 1996 (WHO, 1996a,b). Overall the real morbidity and mortality impact of sylvatic YFV in humans is not fully appreciated; it is widely believed that official reports underestimate the actual number by a factor of 10–20 (Rodhain, 1991). Three major reasons are responsible for this discrepancy: (i) difficulty of clinical diagnosis from other infections (ii) sociological factors related to affected patients (most are migratory agricultural workers) and (iii) occurrence in remote areas with limited access to health care.

Although we do not know the clinical manifestations of sylvatic DENV infections in non-human primates in either Africa or South-East Asia, in the Americas we do know that clinical outcome of sylvatic YFV is dependent on the species susceptibility. Several non-human primate species including marmosets, howler (*Alouatta* spp.), night (*Aotus* spp.), spider (*Ateles* spp.), capuchin (*Cebus* spp.), woolly (*Logothrix* spp.) and squirrel (*Saimiri* spp.) monkeys are involved in the maintenance of the enzootic YFV cycle. With the exception of woolly and capuchin monkeys, infection with YFV leads to severe disease with fatal outcome (Rodhain, 1991). However, in Africa sylvatic YFV infection results in subclinical or mild manifestation of the disease in most non-human primates with the exception of grivet monkeys (*Cercopithecus aethiops centralis*) where severe disease manifestations have been reported (Hugues, 1943; Rodhain, 1991; Ross and Gillett, 1950). In a similar fashion to sylvatic DENV recrudescence (Diallo et al., 2003), sylvatic YFV amplification occurs with a periodicity of about 5 years in West Africa (Germain, 1981, 1984), and 8–10 years in South America (Rodhain, 1991). It may therefore be possible to identify common ecological drivers of the population dynamics of both YFV and DENV.

Adaptation and emergence for urban transmission

Several phylogenetic analyses (Rico-Hesse, 1990; Twiddy et al., 2003; Vasilakis et al., 2008b; Wang et al., 2000) and ecological studies (Cordellier et al., 1983; Hervy et al., 1984; Monlun et al., 1992;

Roche et al., 1983; Rudnick, 1965, 1978, 1984b, 1986; Smith, 1956b, 1958) have demonstrated that the ancestral sylvatic DENV viruses are both ecologically and evolutionarily independent from the current endemic DENV circulating within endemic transmission cycles. As described above (see 'Evidence for a sylvatic DENV ancestor of endemic strains' and 'Evolution of DENV antigenic diversity' sections), four independent evolutionary events that most likely occurred repeatedly in the forests of South-East Asia resulted in the emergence of the four endemic DENV serotypes from sylvatic progenitors. Subsequently, the studies of Wang and colleagues suggested that DENV emergence was also facilitated by vector switching – from arboreal, primatophilic *Aedes* mosquito species to peridomestic, anthropophilic *Aedes* species – and probably accompanied by host expansion, from non-human primates to humans (Wang et al., 2000). The explosive increase of urban populations, especially during the last 200 years, and the colonization of most tropical regions by the anthropophilic peridomestic *Ae. aegypti aegypti* and *Ae. albopictus* mosquito vectors, have provided a unique opportunity for adaptation and emergence into the major urban and periurban centres of the tropics worldwide.

As the historical record indicates, *Ae. albopictus* was, until the late nineteenth and early twentieth centuries, the principal anthropophilic mosquito vector responsible for the establishment of sustained human transmission cycles in the tropics (Smith, 1956b; Stanton, 1919; Theobald, 1901). *Ae. aegypti aegypti* mosquitoes are of African origin (Christophers, 1960; Edwards, 1932), and their widespread colonization of the tropics nearly worldwide did not occur until the establishment of extensive trade routes in the seventeenth and eighteenth centuries, facilitated by the movement of people and the their water storage practices. As initially recognized by Leichtenstern and subsequently others, DEN was a disease of ports and towns in coastal regions, where the disease would travel inland along rivers (Barraud, 1928; Bylon, 1780; Leichtenstern, 1896; More, 1904; Skae, 1902; Smith, 1956b; Steadman, 1828).

Support for *Ae. albopictus* as the original peridomestic vectors comes from their longer historical contact with DENV in Asia, as well as from the notion that sylvatic DENV strains required adaptation to these vectors to establish endemic cycles (Moncayo *et al.*, 2004). In that study, Moncayo and colleagues compared the oral infectivity and dissemination profiles of endemic (Asian genotypes 1349 and NGC) versus sylvatic DENV-2 strains from West Africa (PM33974 and DakAr 2022) and Asia (P8–1407) in both endemic vectors spp. (*Ae. aegypti aegypti* and *Ae. albopictus*) collected from diverse geographic locations in southeast Asia and the Americas. As expected, *Ae. albopictus* mosquitoes, regardless of their geographic origin, were more susceptible to endemic DENV-2 strains than *Ae. aegypti aegypti* (94% and 69% respectively), suggesting a higher degree of adaptation. However, when the infection and dissemination rates from each group were pooled, endemic DENV-2 demonstrated higher infection rates than sylvatic DENV-2 strains in most populations of both mosquito species (Moncayo *et al.*, 2004). However, a limitation of that study was that American genotype DENV-2 strains, which are now known to exhibit reduced infectivity for *Ae. aegypti aegypti* compared with Asian genotypes (Armstrong and Rico-Hesse, 2001; Armstrong and Rico-Hesse, 2003), were not included. More recent studies that included an expanded repertoire of sylvatic and endemic DENV strains suggest that the infection profiles of sylvatic and American DENV-2 genotypes for *Ae. aegypti aegypti* are similar, and that adaptation for enhanced transmission by this vector occurred after the emergence of the Asian genotype (K. Hanley and N. Vasilakis, unpublished).

Adaptation and emergence of DENV in West Africa

Another interesting question regarding DENV emergence is whether the endemic strains have lost fitness for transmission by the sylvatic vectors. This question has epidemiologic relevance because, if the endemic strains could re-invade the forest cycle, a reservoir for efficient reestablishment of urban transmission could exist. This hypothesis was addressed by Diallo *et al.*, who compared the DENV-2 genotypes with experimental infections of sylvatic and various peridomestic populations of Senegalese mosquitoes, using both sylvatic (PM33974 and DakAr 2022) and endemic (1349 and NGC) DENV-2 strains

(Diallo *et al.* 2005). The sylvatic mosquitoes were collected in the gallery forest in an enzootic DENV-2 focus in south-eastern Senegal near Kedougou (Cornet 1993; Cornet *et al.* 1984b; Diallo *et al.* 2003; Rodhain 1991; Saluzzo *et al.* 1986a), whereas the peridomestic mosquitoes were collected in the central Senegalese area of Koung Koung Sérère. Several unexpected observations demonstrated that: (i) the sylvatic mosquito vectors *Ae. furcifer* and *Ae. luteocephalus* were highly susceptible to both sylvatic and endemic DENV-2 strains, whereas sylvatic *Aedes* (Stegomyia) *vittatus* and both sylvatic and peridomestic populations of *Ae. aegypti* (*Ae. aegypti formosus* and *Ae. aegypti aegypti* respectively) were relatively refractory to all DENV-2 strains tested; (ii) there was a lack of correlation between infection and dissemination rates; for example, *Ae. furcifer* and *Ae. luteocephalus* exhibited high infection rates but low dissemination rates, whereas *Ae. aegypti* and *Ae. vittatus* exhibited low infection but high dissemination rates, which suggests a trade-off between infection and dissemination rates that influences the transmission potential of these vectors; and (iii) the two Senegalese *Ae. aegypti* populations (peridomestic *Ae. aegypti aegypti* and sylvatic *Ae. aegypti formosus*) were not significantly different in their low susceptibility to DENV-2, which suggests limitations on the local emergence potential of enzootic strains. Overall, this study refuted the hypothesis that any adaptation of endemic strains to urban vectors was species-specific, and supported the hypothesis that endemic strains have the potential to become established in a forest cycle.

Genetic variation among *Ae. aegypti* populations from different geographic areas may explain differences between these results versus others reporting greater susceptibility of *Ae. aegypti aegypti* relative to *Ae. aegypti formosus* to DENV-2 (Bosio *et al.*, 1998; Vazeille-Falcoz *et al.*, 1999). Diallo and colleagues used *Ae. aegypti* populations collected from 2 distinct environments; mosquitoes from Kedougou are zoophilic and colonize tree holes and other sylvatic larval habitats, whereas those from Koung Koung Sérère are anthropophilic, are usually found within human dwellings, and use artificial water containers for their larvae (Diallo *et al.*, 2005). Recently, Diallo and colleagues expanded on the vector competence of Senegalese *Ae. aegypti* populations by sampling vectors (including larval, nymphal and adults) from locations representing almost all bioclimatic zones and habitats of Senegal, including (i) Kedougou in the Sudano-Guinean zone; (ii) Koung Koung Sérère and Ndougoubène in the Sudanian zone; (iii) Ngoye and Dakar in the Sahelo-Sudanian zone; and (iv) Barkédji in the Sahelian zone (there were no collections from the Guinean zone) (Diallo *et al.*, 2008). However, the lack of reliable methods to distinguish Ae. aegypti aegypti from Ae. aegypti formosus underscores the need of systematic genetic investigations to better characterize these mosquito populations and assess their potential as DENV vectors in west Africa.; see also The ecology of sylvatic DENV transmission in Africa section above). Although only one sylvatic (strain ArD 140875) and one endemic (strain ArD 6894) DENV-2 strain were evaluated, the results confirmed the low susceptibility of Senegalese mosquitoes to DENV infection, and demonstrated that (i) Senegalese Ae. aegypti populations show low infection rates (0–26%) for both sylvatic and endemic DENV-2; (ii) there was no significant variation in the susceptibility of these mosquito populations regardless of their geographic origin, and; (iii) all mosquito populations studied developed disseminated infections (10–75%) except those collected from Kedougou (0%) (Diallo *et al.*, 2008). Previous studies identified the peridomestic Ae. aegypti aegypti only in Asia and the New World, whereas in West Africa, including Senegal, only the sylvatic form Ae. aegypti formosus was reported (Powell *et al.*, 1980; Tabachnick and Powell, 1979

Collectively, the data from both Senegalese studies confirm that *Ae. aegypti* is a recent DENV vector and that west African populations were not involved in the initial urban emergence that has resulted in the ongoing DEN pandemic.

Phenotypic characterization of sylvatic DENV *in vivo, ex vivo* and *in vitro*

Because the vector competence data summarized above do not support the hypothesis that sylvatic DENV strains required adaptation to peridomestic vectors to establish endemic transmission cycles, the complementary question

arises of whether emergence of endemic DENV was facilitated by adaptation of sylvatic strains to humans as reservoir and amplification hosts. This hypothesis was evaluated in two surrogate model hosts of human infection: (i) severe combined immune deficient (SCID) mice xenografted with human hepatoma cells (Huh-7), and; (ii) monocyte-derived dendritic cells (moDCs) (Vasilakis *et al.*, 2007b). The replication profiles of 6 DENV-2 strains that represented all major genotypes, including sylvatic strains isolated from Africa and Asia, and Asian, African and American endemic strains, were evaluated. Inclusion of both Asian and American endemic strains was based upon their reported differences in human virulence (Cologna *et al.*, 2005; Rico-Hesse *et al.*, 1997). As expected, the *in vivo* (xenografted mice) replication profiles of the DENV-2 strains examined exhibited significant strain-specific differences among mean replication titres. However, the complex pattern of differences did not reveal a consistent or overall difference between sylvatic and endemic strains, or between Asian endemic and American endemic strains. For example, the titre of Asian endemic strain 16681 was significantly higher than any of the other 3 endemic strains (1349, 1328 and IQT-1950) examined, but not significantly different from sylvatic strain P8-1407 (Table 11.3). Furthermore, the sylvatic strains (P8-1407 and PM33974) replicated to lower titres than the endemic Asian strains (16681 and 1349), but did not differ consistently from the endemic American strains (1328 and IQT-1950) (Vasilakis *et al.*, 2007b) (Table 11.3).

In the *ex vivo* moDC analyses, significant inter-strain variation in mean replication titres among the various DENV-2 strains was also observed, but no overall difference between sylvatic and endemic strains was detected. Importantly, however, Asian and American endemic strains did differ significantly in overall replication, with Asian endemic strains replicating to significantly higher titres than either American or sylvatic strains, which were similar to each other. This finding suggests that when the sylvatic strains first emerged in to humans they may have initially caused relatively mild disease similar to that associated with the American DENV-2 strains, and only later evolved to cause the severe disease that is typical of the Asian DENV-2 strains.

Table 11.3 Replication profile of select endemic and sylvatic DENV-2 *in vivo* (SCID – Huh-7 mice)

Virus[a]	Location/year[b]	Epidemiological type[c]	Mean peak virus titre[d,e] ($\log_{10}$ ffu/ml ± SE)	GenBank accession no.
16681	Thailand/1964	Endemic	5.9 ± 0.2	U87411
1349	Burkina Faso/1982	Endemic	4.9 ± 0.3	EU056810
1328	Puerto Rico/1977	Endemic	2.9 ± 0.3	EU056812
IQT-1950	Peru/1995	Endemic	4.0 ± 0.3	EU056811
P8-1407	Malaysia/1970	Sylvatic	5.4 ± 0.2	EF105379
PM33974	Republic of Guinea/1981	Sylvatic	3.6 ± 0.2	EF105378

a   Groups of SCID-Huh-7 mice were inoculated into the tumour with 4.0–4.2 $\log_{10}$ ffu of the indicated virus. Serum was collected on day 7 and virus titre was determined on C6/36 cells.

b   Country and year virus strain was isolated.

c   'Endemic' denotes human or *Ae. aegypti* isolates associated with peridomestic transmission.

d   Serum collected from mice at peak viraemia on day 7 post infection. Virus titres in sera were determined by focus forming immunoassay (FFA) on C6/36 cells.

e   The limit of detection of the assay is 0.5 $\log_{10}$ ffu/ml.

Because of the great genetic diversity of the human population at risk for DENV infection and epidemiologic evidence of a possible role of host genetics in the development of DENV viraemia and progression to severe disease (Bravo *et al.*, 1987; Chaturvedi *et al.*, 2006; Halstead *et al.*, 2001; Loke *et al.*, 2002) (see Chapter 5), the *ex vivo* moDCs model also provided a unique opportunity to investigate whether host genetics (i.e. different racial and/or ethnic backgrounds) influences the outcome of DENV infection. The moDCs, which were obtained from a small number of human donors representing different ethnicities and racial backgrounds (Caucasian, West African, and South-East Asian), demonstrated interdonor variation in their infection rates, in agreement with previous reports (Cologna *et al.*, 2005; Sanchez *et al.*, 2006), but no differences in their ability to support DENV-2 replication (Table 11.4) (Vasilakis *et al.*, 2007b). However, the small sample size ($n = 3$) of donors from each ethnic group may have limited the power to detect meaningful inferences about the role of host genetics.

More recently, the replication profiles of a subset of endemic and sylvatic DENV-2 strains that included sylvatic strains isolated from humans in west Africa were evaluated in both mammalian (Huh-7 and Vero) and mosquito cells (C6/36) *in vitro* (Vasilakis *et al.*, 2008b). In this system, significant inter-genotypic variation in mean virus outputs was observed among strains of both endemic and sylvatic DENV-2 genotypes in all cell types (Table 11.5). Moreover, when the replication profiles of the two ecotypes (endemic and sylvatic) were compared, endemic DENV strains produced a significantly higher output of progeny in human liver (Huh-7) cells, but not in monkey kidney (Vero) or mosquito cells (Fig. 11.5). Interestingly, in Huh-7 cells, the endemic strains reached significantly higher maximum titres than the sylvatic strains and their replication dynamics were significantly different. This difference between endemic and sylvatic DENV-2 strains in both rapidity of replication and peak titre in Huh-7 cells *in vitro* (Fig. 11.5), differs from the previous findings described above with SCID mice engrafted with Huh-7 cells (SCID-Huh-7 mice) (Vasilakis *et al.*, 2007b). However, it may be inappropriate to compare the two studies directly, because different sets of DENV-2 isolates were used to represent the sylvatic and endemic ecotypes

DENV emergence and the role of adaptation to new hosts and vectors are important issues for arbovirology and have enormous public health

Table 11.4 Replication profile of endemic and sylvatic DENV-2 *ex vivo* in monocyte-derived dendritic cells (MoDCs)

Virus[a]	Location/year[b]	Epidemiological type[c]	Mean peak virus titre[d,e] ($\log_{10}$ ffu/ml $\pm$ SE)	GenBank accession no.
16681	Thailand/1964	Endemic	5.4 ± 0.1	U87411
1349	Burkina Faso/1982	Endemic	5.0 ± 0.3	EU056810
1328	Puerto Rico/1977	Endemic	3.1 ± 0.6	EU056812
IQT-1950	Peru/1995	Endemic	4.3 ± 0.2	EU056811
P8-1407	Malaysia/1970	Sylvatic	3.2 ± 0.0	EF105379
PM33974	Republic of Guinea/1981	Sylvatic	0.5 ± 0.0	EF105378
DakAr A510	Côte d'Ivoire/1980	Sylvatic	4.4 ± 0.2	EF105381
DakAr A1247	Côte d'Ivoire/1980	Sylvatic	0.5 ± 0.0	EF105383
DakAr 2022	Burkina Faso/1980	Sylvatic	2.6 ± 0.1	EF105386
DakAr 2039	Burkina Faso/1980	Sylvatic	4.6 ± 0.2	EF105382

a   moDCs (2.5×10^5 per sample) from two healthy human volunteers were infected with an MOI = 2 with the indicated virus. Supernatants were collected 48 p.i., and viral outputs were determined on C6/36 cells.

b   Country and year virus strain was isolated.

c   'Endemic' denotes human or *Ae. aegypti* isolates associated with peridomestic transmission.

d   Virus progeny in cell-free supernatants was collected at day 2 p.i. (peak viral output), and virus titres were determined by focus forming immunoassay (FFA) on C6/36 cells.

e   The limit of detection of the assay is 0.5 $\log_{10}$ ffu/ml.

implications (see Potential for sylvatic DENV re-emergence: epidemics and human contact section), especially considering the potential for eradicating the human transmission cycle with effective vaccines now under development. Overall, the experimental findings from entomological, human model and *in vitro* studies presented above fail to support the hypothesis that emergence of endemic DENV-2 was facilitated by adaptation of sylvatic DENV-2 to either peridomestic mosquito vectors or to humans as reservoir and amplification hosts. This implies that the potential for re-emergence of sylvatic DENV strains from Africa or Asia into the endemic cycle is quite high. Additional studies are required to extend these observations to the other DENV serotypes, although very few sylvatic DENV-4 strains are available and to date sylvatic DENV-3 and possibly -1 have not been isolated. In addition, characterization of human disease resulting from sylvatic DENV infections and, most importantly, human viraemia levels and duration, are needed

to more completely assess their public health threat. This critical information will require active surveillance in enzootic foci such as eastern Senegal and Malaysia.

Selection pressures

The introduction of South-East Asian DENV-2 genotypes to the Americas 30 years ago spawned a series of epidemics with increased pathogenicity relative to the existing American strains, as first documented with the appearance of DHF/DSS cases in Cuba (Kouri *et al.*, 1983) and subsequently throughout the Caribbean, Central and South America (Alvarez *et al.*, 2006; Cabrera-Batista *et al.*, 2005; Guzman *et al.*, 2002; Harris *et al.*, 2000; Kouri *et al.*, 1991; Kouri *et al.*, 1989; Rigau-Perez and Laufer, 2006; Uzcategui *et al.*, 2001) (Chapter 5). These epidemiological observations were corroborated by phylogenetic analyses (Rico-Hesse, 1990; Rico-Hesse *et al.*, 1997) indicating that the introduction of South-East Asian genotypes in the Americas coincided

Table 11.5 Replication profile of endemic and sylvatic DENV-2 in vitro

Virus[a]	Location/year[b]	Epidemiological type[c]	Mean peak virus titre[d,f] (Huh-7) (log$_{10}$ ffu/ml ± SE)	Mean peak virus titre[d,f] (Vero) (log$_{10}$ ffu/ml ± SE)	Mean peak virus titre[e,f] (C6/36) (log$_{10}$ ffu/ml ± SE)
16681	Thailand/1964	Endemic	7.2±0.1	6.6±0.0	6.8±0.0
1349	Burkina Faso/1982	Endemic	4.6±0.1	4.8±0.1	7.3±0.1
TVP-1915	Burkina Faso/1986	Endemic	6.6±0.1	5.8±0.0	6.2±0.1
P8-1407	Malaysia/1970	Sylvatic	4.4±0.2	3.8±0.0	6.8±0.0
DakAr A510	Côte d'Ivoire/1980	Sylvatic	4.7±0.1	4.5±0.0	6.9±0.0
DakAr D75505	Senegal/1991	Sylvatic	5.3±0.0	6.8±0.1	7.1±0.0
DakArHD10674	Senegal/1970	Sylvatic	4.5±0.1	5.3±0.1	7.5±0.1
IBH11234	Nigeria/1966	Sylvatic	3.7±0.1	3.1±0.1	5.3±0.1
IBH11208	Nigeria/1966	Sylvatic	4.8±0.0	4.3±0.1	6.1±0.0

a   Vero, Huh-7 and C6/36 cells, at 2.5×10^5, 2.5×10^5 and 5.0×10^5 cells/well respectively, were infected with a MOI=0.01 ffu/ml, in triplicate. Supernatants were collected 1, 2, 3, 4, 5 and 6h p.i., and viral output was determined on C6/36 cells.

b   Country and year virus strains were isolated.

c   'Endemic' denotes human or Ae. aegypti isolates associated with peridomestic transmission.

d   Virus titre represents cell-free supernatants that were collected at day 4 p.i. (peak viral output of Huh-7 or Vero cells), and virus titre was determined by focus forming immunoassay (FFA) on C6/36 cells.

e   Virus titre represents cell-free supernatants that were collected at day 5 p.i. (peak viral output of C6/36 cells), and virus titre was determined by focus forming immunoassay (FFA) on C6/36 cells.

f   The limit of detection of the assay is 0.9 log$_{10}$ ffu/ml.

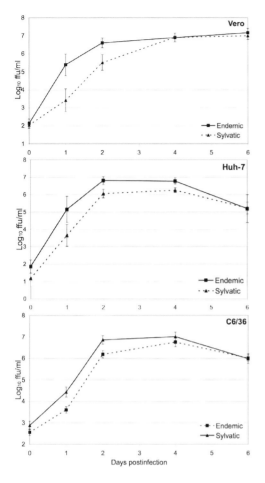

Figure 11.5 Comparative replication curves of sylvatic and endemic DENV-2 strains. Mean virus output of endemic and sylvatic DENV-2 genotypes on vertebrate cells lines Vero and Huh-7, and mosquito C6/36 cell line up to 6 days post infection. All infections were initiated at an MOI of 0.01 and titres are expressed as focus forming units.

with the origin and spread of DHF in the New World, whereas the American DENV genotypes are not associated with severe disease (Leitmeyer et al., 1999; Watts et al., 1999). In retrospect, these observations confirm the hypothesis that some DENV genotypes have a greater virulence and epidemic potential than others (Barnes and Rosen, 1974; Gubler et al., 1981; Rosen, 1977). However it is still not known whether the tendency of certain genotypes to cause severe disease results from greater intrinsic virulence of those genotypes, or a greater tendency to be enhanced in the presence of heterologous antibody, or some combination of the two traits.

To date, the selective forces that influence the occurrence of dengue outbreaks and disease severity remain poorly characterized. Detailed phylogenetic analyses based on diverse strains of endemic DENV have revealed evidence of limited, localized adaptive evolution, with the selective targets varying among serotypes, genotypes and viral proteins in the endemic ecotype. Utilizing molecular evolutionary analyses based on a ML method that measures rates of synonymous and non-synonymous substitution codon-by-codon (Yang et al., 2000) on the E gene of Vietnamese DENV isolates, as well as a collection of isolates representing all circulating DENV genotypes on a global scale, Twiddy et al. demonstrated that the E gene of DENV-2 is subject to strong purifying selection (expressed as the extremely low ratio of d_N/d_S) (Twiddy et al., 2002a). A limitation of studies that focus on codons affected convergently is that unique mutations, involving codons selected only once during the course of evolution, cannot be identified. Nonetheless, this study demonstrated evidence of weak positive selection in 1 and 17 amino acids within the Cosmopolitan and lineage 2 of the Asian genotype, respectively (Table 11.6). Furthermore, a higher level of positive selection was exhibited within the Cosmopolitan genotypes than the Asian genotypes (0.190 vs. 0.056). Interestingly, the positively selected E-390 site of the Cosmopolitan genotypes has been previously identified as a key virulence determinant (Sanchez and Ruiz, 1996), and maps within the distal face of domain III, a region associated with viral attachment to host cells.

In a parallel study, Twiddy et al. performed expanded analyses with larger datasets of E gene sequences (for DENV-2 only) in order to obtain a measure of the selection pressures acting upon all DENV serotypes (Twiddy et al., 2002b). The limited sample size of DENV-1 sequences did not allow for the detection of any positively selected sites, whereas weak positive selection was detected in both DENV-3 and DENV-4, and in the Cosmopolitan and lineage 2 of the Asian DENV-2 genotype. Most of the selected sites were located within or near potential B- (Aaskov et al., 1989; Innis et al., 1989; Wu et al., 2003) or T-cell epitopes (Kutubuddin et al., 1991; Leclerc et al., 1993; Megret et al., 1992; Roehrig et al.,

Table 11.6 Summary of select studies examining the selection pressures acting upon DENV

Serotype (gene) (epidemiological type)[a]	Dataset[b]	Selection	Selected sites[c]	Reference
DENV-2 (E) (endemic and sylvatic)	141 endemic, 4 sylvatic	Strong purifying selective constraints ($d_N/dS \ll 1$); weak positive selection in Cosmopolitan and Asian lineage 2 genotypes; no positive selection was detected to act upon sylvatic DENV-2	*Cosmopolitan:* 390; *Asian:* 52, 85, 90, 98, 100, 105, 112, 113, 122, 131, 144, 170, 330, 334, 342, 378, 392	Twiddy *et al.* (2002)
DENV-1,-2,-3,-4 (E) (endemic)	109 DENV-1–4, 171 DENV-2 only	Weak positive selection in: DENV-3 (two sites), DENV-4 (five sites) and DENV-2 Cosmopolitan lineage (two sites), Asian lineage 2 (17 sites)	DENV-2: *Cosmopolitan:* 52, 390; *Asian:* 52, 85, 90, 98, 100, 105, 112, 113, 122, 131, 144, 170, 330, 334, 342, 378, 392 DENV-3: 169, 380 DENV-4: 108, 131, 357, 429, 494	Twiddy *et al.* (2003)
DENV-2 (ORF)	36	No evidence of positive selection in the structural C and prM genes, as well as in non-structural NS1, NS2A and NS3 genes. Weak positive selection in NS2B (two sites), NS5 (two sites)	NS2B: 57, 63; NS5: 135, 637	Twiddy *et al.* (2003)
DENV-4 (C-prM-E, NS1–NS2A, NS4B 3'-UTR) (endemic)	82	Eight potentially positively selected sites in E, NS1, NS2A and NS4B genes. Only sites in NS2B (three sites) are suggestive of positive selection	E: 163, 351; NS2A: 14, 54, 101	Bennett *et al.* (2003)
DENV-1, -3 (ORF) (endemic)	10 DENV-1, 9 DENV-3	No evidence of positive selection acting upon any gene, site or lineage of DENV-1 and 3. Extremely weak positive selection in E, NS2A and NS4A (DENV-1) and NS3 (DENV-3)	None	Zhang *et al.* (2005)
DENV-1 (ORF) (endemic)	35	No evidence of strong positive selection contributing to the extinction of pre-1998 DENV-1. Weak positive selection in NS5 (three sites)	NS5: 127, 131, 669	Myat Thu *et al.* (2005)
DENV-2 (ORF)	14	Strong selective constraints. Weak positive selection unevenly distributed across the genome. Positively selected sites in E (two sites), NS2A (one site) and NS2B (13 sites)	E: 177, 329; NS2A: 195; NS4B: 12, 19, 21, 24, 26, 49, 96, 113, 116, 123, 197, 242, 246	Vasilakis *et al.* (2007a)

a   Endemic denotes human or *Ae. aegypti aegypti* isolates or strains that are associated with peridomestic transmission.

b   Dataset denotes the number of DENV sequences used for the analyses.

c   Selected sites listed by Bennett *et al.* (2003) represent only a subset of all selected sites that are relevant to the data presented.

1994; Simmons *et al.*, 2005), an association that suggests immune evasion as a selective factor (Table 11.6). Furthermore, some of the selected sites also mapped within functional domains involved in cellular tropism (Chen *et al.*, 1996; Sanchez and Ruiz, 1996) or fusion (Falconar, 2008; Megret *et al.*, 1992) (Table 11.6) (Twiddy *et al.*, 2002b). Within the same study, a dataset of 36 DENV-2 complete genomes was also used to examine the nature of the selection pressures acting across the DENV genome. There was no evidence of positive selection in the other structural proteins (prM or C). Interestingly, despite of extensive evidence for the existence of T- or B- cell epitopes in the non-structural NS1 (Falconar *et al.*, 1994; Garcia *et al.*, 1997; Henchal *et al.*, 1987; Huang *et al.*, 1999; Wu *et al.*, 2001), NS2A (Green *et al.*, 1997; Mathew *et al.*, 1998) and NS3 (Garcia *et al.*, 1997; Kurane *et al.*, 1991; Simmons *et al.*, 2005; Spaulding *et al.*, 1999; Wen *et al.*, 2008) proteins, no evidence of positive selection was observed within these genes. Localized positive selection was observed only in NS2B (2 sites) and NS5 (two sites) (Table 11.6). The selected NS2B sites are located within a 40 amino acid segment that has been shown to be essential for NS2B/NS3 protease activity (Falgout *et al.*, 1993), suggesting that these sites may play a role in the efficiency of the polyprotein processing. Of the NS5 selected sites, one is located within the conserved S-adenosylmethionine-utilizing methyltransferase (SAM) domain (135), suggesting am involvement in capping of virus genomes, whereas the other (637) is located near the GDD motif within the RNA-dependent RNA polymerase (RdRp) domain and may be involved in functional polymerase activity of NS5.

Bennett *et al.*, utilizing similar methods, attempted to determine whether positive selection plays a role in DENV-4 evolution and lineage turnover (Bennett *et al.*, 2003). In this study, 40% of the genome (C-prM-E, NS1–NS2A, NS4B and 3'-UTR) of 82 DENV-4 isolates collected in Puerto Rico over a 20 year period, was used to assess the role of viral molecular evolution in the changing patterns of DENV-4 incidence in Puerto Rico (CDC, 1998; De Jesus *et al.*, 1982; Dietz *et al.*, 1996). Eight positively selected sites were identified in the E, NS1, NS2A and NS4B genes, although only the 3 sites in the NS2A gene (Table

11.6) were consistently identified with all models used. Intriguingly, the NS2A positively selected sites were associated with isolates from the 1998 Puerto Rican epidemic (which evolved from lineages of the 1992 epidemic), as well as with subsequent epidemics throughout the Greater and Lesser Antilles (Foster *et al.*, 2003) (Chapter 9). This work suggests that selection pressures on non-structural genes may lead to lineage turnover, which could spawn epidemics as has been observed for other viruses (Knowles *et al.*, 2001).

Similarly, Zhang *et al.* attempted to determine whether lineage replacements in Thai DENV-1 and DENV-3 were the result of differences in viral fitness by examining the selection pressures acting upon complete viral genomes and lineages (Zhang *et al.*, 2005). Surprisingly, there was no evidence of strong positive selection accompanying lineage replacement events in either DENV-1 or DENV-3. Only weak evidence of positive selection (d_N/d_S ratio < 0.05 to < 0.200) was observed on E, NS2A and NS4A genes of DENV-1 and E, NS2A, NS3 and NS4A genes of DENV-3 (Table 11.6), suggesting that purifying selection was the dominant evolutionary process acting upon these lineages (Zhang *et al.*, 2005).

The role of selection or adaptive evolution acting upon genes other than E, which may be responsible for the lineage extinction of DENV-1 circulating in Myanmar, was examined by Myat Thu *et al.* (Myat Thu *et al.*, 2005). Almost all of the DENV-1 isolates sampled after 1998 were of genotype I, in contrast to genotype III isolates that were most commonly sampled earlier (Thu *et al.*, 2004). Analysis of 35 DENV-1 isolates, including 13 (nine human and four mosquito) collected around Yangon between 1971 and 2002, failed to demonstrate any evidence of strong positive selection that contributed to either the putative competitive displacement of genotype III by genotype I isolates. However, weak positive selection was observed in three sites within the NS5 gene (Table 11.6) (Myat Thu *et al.*, 2005). Interestingly, this study suggested that DENV evolution may be attributed to seasonal variations in the size of mosquito vector populations.

Collectively, the studies described above examined the selective forces and processes that shape the occurrence of DEN outbreaks, lineage extinctions and replacements, disease severity and evolution. Phylogenetic analyses suggest that endemic DENV strains have their ancestry in the sylvatic viruses (Wang *et al.*, 2000), but little is known about the evolutionary processes that characterize sylvatic DENV evolution. Twiddy *et al.* performed analyses based on a limited number of sylvatic DENV-2 E sequences (P8–1407, DakAr A578, PM339474 and DakArHD10674), but detected no evidence of positive selection (Table 11.6) (Twiddy *et al.*, 2002). A more recent and extensive study based on 14 complete sylvatic DENV-2 sequences demonstrated gene-specific d_N/d_S ratios characterized by strong purifying selection ($d_N/d_S << 1$) (Vasilakis *et al.*, 2007b). However, 16 amino acid sites unevenly distributed across the DENV genome showed evidence of weak positive selection. Of these, two were located in the E gene, one in NS2A, and 13 in NS4B. The functional importance of these changes is not known and none of these amino acid residues has previously been shown to mediate any DENV phenotype (Twiddy *et al.*, 2002). Intriguingly, the analysis indicated no evidence of positive selection acting upon NS4B in endemic DENV-2, and no positive selection was detected in the sylvatic lineage at amino acid E390, a key virulence determinant that has previously been reported to have responded to positive selection in endemic DENV-2 (Sanchez and Ruiz, 1996; Twiddy *et al.*, 2002; Vasilakis *et al.*, 2007a).

In conclusion, the studies described above imply that DENV-2 have not undergone extensive adaptive evolution during emergence of the endemic transmission cycles. This suggests that emergence into the endemic transmission cycle was not mediated by natural selection or host-specific (vertebrate and/or vector) adaptations, but rather by changing ecological conditions. This implies that sylvatic DENV currently circulating in the forests of West Africa and South-East Asia could readily reinitiate an endemic transmission cycle should the proper conditions arise.

Potential for sylvatic DENV re-emergence: epidemics and human contact

As is evident from the previous sections, our knowledge of vector – virus – host interactions in the sylvatic transmission DENV cycles range from dated to sporadic to speculative. The paucity of ecological and epidemiological data and the increasing deforestation occurring in South-East Asia for agricultural expansion of economic crop acreage (palm, coconut and rubber plantations) prohibit the formulation of any meaningful inferences about vector–host interactions of sylvatic DENV.

The last comprehensive ecologic and serologic studies of sylvatic DENV in Asia were those of Rudnick and colleagues that took place 40 years ago in Malaysia (Furumizo and Rudnick, 1979; Rudnick, 1965, 1984b, 1986; Rudnick and Lim, 1986; Rudnick *et al.*, 1967), which were a continuation of Gordon Smith's pioneering studies of the 1950s (Smith, 1956a–c, 1957, 1958). Smith's initial serologic surveys of inhabitants of diverse rural habitats (Table 11.1) had demonstrated complete DENV exposure (as evidenced by the presence of DENV neutralizing antibodies) by the age of 30 (Smith, 1956b). Although it is impossible even today to distinguish between sylvatic and endemic DENV exposure on the basis of clinical disease description or serological assays, Smith hypothesized that these sporadic and local outbreaks of 'rural' DEN may be due to forest spillover (Smith, 1956b).

In West Africa a more unambiguous understanding of sylvatic DENV circulation has emerged in the past 30 years. Ecologic and serologic research sponsored by the Institute Pasteur has documented multiple sylvatic amplification cycles occurring at roughly eight-year intervals since 1980 (Cornet, 1993; Diallo *et al.*, 2003; Saluzzo *et al.*, 1986a,b). Nonetheless, the degree and mechanism of human contact with sylvatic transmission cycles during these zoonotic outbreaks remains poorly understood. The field studies cited above suggest that the sylvatic DENV strains are confined to forest habitats or, if they indeed spill over into the adjacent human population, resulted in undetected subclinical and/or relatively mild illness (DF). However, the recent description of DHF due to sylvatic DENV infection suggests that either the enhanced pathogenic potential is an inherent trait of South-East Asian strains, irrespective of ecotype, or that this event represents an unusual manifestation due to yet undetermined aetiologies.

Overall the available data from Africa and South-East Asia suggest that sylvatic DENV come into regular contact with humans but undergo little or no secondary transmission. The apparent limitations on the ability of sylvatic DENV strains to generate significant outbreaks in humans may be attributable to different mechanisms in different sylvatic foci. In West Africa, the absence of efficient, peridomestic vectors, *Ae. aegypti* and *Ae. albopictus*, may prevent subsequent transmission from human infections. Recent studies from Senegal have shown that the gallery forest-dwelling mosquito, *Ae. furcifer*, is highly susceptible to sylvatic DENV infection (Diallo *et al.*, 2005), and disperses from the forest into villages (Diallo *et al.*, 2003). Thus, *Ae. furcifer*, like *Ae. albopictus* in Asia, may act as a bridge vector for exchange between forest and peridomestic habitats and may support limited spillover epidemics but may not be capable of generating large-scale outbreaks. In South-East Asia, in contrast, the high force of infection of endemic DENV, coupled with the ability of antibodies raised against endemic DENV to neutralize sylvatic strains of the same serotype, may restrict the ability of sylvatic strains to spread. The effect of human immunity to natural, endemic DENV infection or vaccination on infection by sylvatic DENV strains was recently evaluated by testing the neutralization capacity of sera from homotypic, DENV-2 vaccinees or naturally infected persons (Vasilakis *et al.*, 2008a). Comparisons between reactions with endemic and sylvatic DENV-2 strains indicated no consistent differences, indicating that immunity from endemic infection should protect against reinfection by sylvatic strains of the same serotype. Although additional testing with a focus on the other three serotypes is needed, these data suggest that tetravalent vaccination with adequate coverage could prevent DENV re-emergence from sylvatic cycles. However, unlike smallpox virus that no longer requires vaccination due to the lack of a zoonotic reservoir (Stewart and Devlin, 2006), the continued risk from the sylvatic cycles, which probably cannot be controlled in an environmentally acceptable manner, would necessitate sustained DEN vaccination even after eradication of the human cycle.

As mentioned earlier, the lack of any comprehensive ecologic and serologic Asian surveys in the past 40 years presents several obstacles for the establishment of an accurate assessment of the true extent of human exposure to sylvatic DENV. Even with functional reporting and surveillance systems, clinical diagnosis of dengue in both Africa and Asia is complicated by co-circulation of several other viruses that cause clinically indistinguishable febrile diseases (chikungunya, Japanese encephalitis, o'nyong-nyong and Zika viruses) (Moore *et al.*, 1975; Rudnick, 1986; Rudnick and Lim, 1986). Other obstacles to accurate assessment of the public health impact of sylvatic DENV in remote areas in Africa and Asia include (1) limited access of the population to health care facilities, (2) lack of access to viral and serological diagnostics, (3) popular beliefs that discourage standard medical treatment unless illness is severe, (4) a lack of veterinary (i.e. non-human primate) surveillance and (5) political and social instability in some areas of interest. These obstacles must be overcome to assess the roles of non-human primates and other vertebrate hosts in the maintenance of sylvatic DENV, the degree and routes of ecological contact between humans and sylvatic DENV, and the replication and immunological dynamics of sylvatic DENV in non-human primates.

Prevention of DENV re-emergence into the human cycle: implications for public health

The control of DENV, like that of most other arboviruses, currently relies on preventing exposure of people to mosquito vectors, reducing vector populations using insecticides, and reducing the presence of or access to vector larval habitats. All of these strategies are particularly essential for controlling DENV because of the ecology and behaviour of the principal vector, *Ae. aegypti*.

Preventing exposure to the infectious adult females of this species is complicated by their tendency to enter homes where there is ready access to human hosts. In many tropical regions of the world, homes do not include any screening on windows because air conditioning is not affordable to residents and screens limit airflow. *Ae. aegypti* as well as *Ae. albopictus* actively seek human hosts during the daytime when tropical temperatures demand maximum ventilation.

The endophilicity of adult females as well as the partitioning of larval populations in small containers near homes make both adulticide and larvicide control more difficult than for most mosquito species that develop in larger bodies of water more accessible for larvicide applications and whose adults remain outdoors where large-scale adulticide applications are more effective. The only widely effective control of *Ae. aegypti* occurred during the 1950s to 1970s in the Americas, where concerted efforts to control yellow fever resulted in the eradication of *Ae. aegypti* from many tropical and subtropical regions. However, these efforts relied heavily on the application of persistent insecticides such as DDT inside of homes, and this strategy was later widely found unacceptable due to negative environmental impacts.

Both *Ae. aegypti* and *Ae. albopictus* rely on artificial aquatic habitats for their larval habitats, and human habitations often provide ideal sites such as discarded containers that hold rainwater as well as large vessels used to store rainwater or water carried from a distance when municipal, piped water is not available. Unless properly protected by screening or larvicides, these containers can produce thousands of mosquitoes on a regular basis. For all of these reasons as well as the poor understanding in many locations of the transmission mechanisms of DENV and the risks for infection, control measures have had little or no effect in most locations on the incidence of disease.

Currently, control strategies for DEN are focusing heavily on the development of human vaccines (see Chapter 12). Because DENV produces only an acute infection, traditional approaches such as live-attenuated strains or inactivated preparations would be expected to succeed as they have for other arboviruses such as yellow fever or Japanese encephalitis. Newer approaches for virus attenuation such as chimerization, the development of replicon-based approaches that limit replication to initial target cells, and subunit vaccines that focus on immunogenic viral proteins also offer promise for improved balances between risk and efficacy. However, DEN vaccination is complicated by the increased risk of DHF/DSS following a secondary infection. This risk implies that vaccine-induced, protective immunity against only one or even only three of the four serotypes could increase the risk of life-threatening disease following natural infection. Therefore,

DEN vaccine development is focusing on tetravalent preparations that can protect against all four serotypes. This requirement presents a unique challenge because immunogenicity must be balanced for all vaccinees.

Despite these challenges, several DEN vaccine candidates show promise for prevention of both disease and transmission. Like smallpox (Stewart and Devlin, 2006) and polioviruses (Rey and Girard, 2008) that use humans as their only reservoir and amplification hosts, effective DEN vaccination could eventually eradicate human DENV strains. However, the risk for re-emergence posed by sylvatic DENV strains in Asia and Africa must be considered in long-term control strategies.

Acknowledgements

We dedicate this work to the memory of C.E. Gordon Smith, Albert Rudnick and all those whose trailblazing work contributed to the discovery and understanding of sylvatic DENV. The authors would like to thank Shannan Rossi for expert graphic design of the 'transmission cycles' figure. NV was supported by a grant from the Fine Foundation. SW's dengue research is supported by NIH grant AI069145. KH was supported by NIH-NM-INBRE (P20 RR016480-05) during preparation of this chapter.

References

CDC (1998). Dengue outbreak associated with multiple serotypes – Puerto Rico, 1998. MMWR. *47*, 952–956.

Aaskov, J.G., Geysen, H.M., and Mason, T.J. (1989). Serologically defined linear epitopes in the envelope protein of dengue 2 (Jamaica strain 1409). Arch. Virol. *105*, 209–221.

Alvarez, M., Rodriguez-Roche, R., Bernardo, L., Vazquez, S., Morier, L., Gonzalez, D., Castro, O., Kouri, G., Halstead, S.B., and Guzman, M.G. (2006). Dengue hemorrhagic Fever caused by sequential dengue 1–3 virus infections over a long time interval: Havana epidemic, 2001–2002. Am. J. Trop. Med. Hyg. *75*, 1113–1117.

Anonymous (1969). Group B arboviruses. In 1969 Annual Report (Ibadan, Nigeria, University of Ibadan Arbovirus Research Project), pp. 152–162.

Anonymous (1971–1972). Summary of Arbovirus isolates by type, 1971 and 1972. In Virus Research Laboratory (Ibadan, Nigeria, University of Ibadan), pp. 121.

Armstrong, P.M., and Rico-Hesse, R. (2001). Differential susceptibility of *Aedes aegypti* to infection by the American and South-East Asian genotypes of dengue type 2 virus. Vector Borne Zoonotic Dis. *1*, 159–168.

Armstrong, P.M., and Rico-Hesse, R. (2003). Efficiency of dengue serotype 2 virus strains to infect and disseminate in *Aedes aegypti*. Am. J. Trop. Med. Hyg. *68*, 539–544.

Balfour, A. (1914). The wild monkey as a reservoir for the virus of yellow fever. Lancet *1*, 1176–1178.

Barnes, W.J., and Rosen, L. (1974). Fatal hemorrhagic disease and shock associated with primary dengue infection on a Pacific island. Am. J. Trop. Med. Hyg. *23*, 495–506.

Barraud, P.J. (1928). The distribution of '*Stegomyia fasciata*' in India, with remarks on dengue and yellow fever. Indian J. Med. Res. *16*, 377–398.

Baudon, D., Robert, V., Roux, J., Lhuillier, M., Saluzzo, J.F., Sarthou, J.L., Cornet, M., Stanghellini, A., Gazin, P., Molez, J.F., Some, L., Darriet, F., Soudret, B.R., Guiguende, T.R., and Hennequin, M. (1986). [The 1983 yellow fever epidemic in Burkina Faso]. Bull. World Health Org. *64*, 873–882.

Bennett, S.N., Holmes, E.C., Chirivella, M., Rodriguez, D.M., Beltran, M., Vorndam, V., Gubler, D.J., and McMillan, W.O. (2003). Selection-driven evolution of emergent dengue virus. Mol. Biol. Evol. *20*, 1650–1658.

Bosio, C.F., Beaty, B.J., and Black, W.C.t. (1998). Quantitative genetics of vector competence for dengue-2 virus in *Aedes aegypti*. Am. J. Trop. Med. Hyg. *59*, 965–970.

Bravo, J.R., Guzman, M.G., and Kouri, G.P. (1987). Why dengue haemorrhagic fever in Cuba? 1. Individual risk factors for dengue haemorrhagic fever/dengue shock syndrome (DHF/DSS). Trans. R. Soc. Trop. Med. Hyg. *81*, 816–820.

Bryant, J., Wang, H., Cabezas, C., Ramirez, G., Watts, D., Russell, K., and Barrett, A. (2003). Enzootic transmission of yellow fever virus in Peru. Emerg. Infect. Dis. *9*, 926–933.

Bryant, J.E., Holmes, E.C., and Barrett, A.D. (2007). Out of Africa: a molecular perspective on the introduction of yellow fever virus into the Americas. PLoS Pathog *3*, e75.

Burke, D.S., Nisalak, A., Johnson, D.E., and Scott, R.M. (1988). A prospective study of dengue infections in Bangkok. Am. J. Trop. Med. Hyg. *38*, 172–180.

Burns, E.M., Ralph, P.L., Lerner, R.E., and Meacham, S. (1986). World Civilization, Vol. A. (New York, W. W. Norton & Co.).

Bylon, D. (1780). Korte Aantekening, wegens eene Algemeene Ziekte, Doorg aans Gen aamd de Knokkel-Koorts. In Verhandelungen van het batavi aasch Genootschop der Konsten in Wetenschappen (Batavia), pp. 17–30.

Cabrera-Batista, B., Skewes-Ramm, R., Fermin, C.D., and Garry, R.F. (2005). Dengue in the Dominican Republic: epidemiology for 2004. Microscopy Res Tech *68*, 250–254.

Calisher, C.H., Karabatsos, N., Dalrymple, J.M., Shope, R.E., Porterfield, J.S., Westaway, E.G., and Brandt, W.E. (1989). Antigenic relationships between flaviviruses as determined by cross-neutralization tests with polyclonal antisera. J. Gen. Virol. *70 (Pt 1)*, 37–43.

Cardosa, J., Ooi, M.H., Tio, P.H., Perera, D., Holmes, E.C., Bibi, K., and Abdul Manap, Z. (2009). Dengue virus serotype 2 from a sylvatic lineage isolated from a patient with dengue hemorrhagic fever. PLoS Negl. Trop. Dis. *3*, e423.

Carey, D.E., Causey, O.R., Reddy, S., and Cooke, A.R. (1971). Dengue viruses from febrile patients in Nigeria, 1964–68. Lancet *1*, 105–106.

Carey, D.E., Kemp, G.E., Troup, J.M., White, H.A., Smith, E.A., Addy, R.F., Fom, A.L., Pifer, J., Jones, E.M., Bres, P., and Shope, R.E.(1972). Epidemiological aspects of the 1969 yellow fever epidemic in Nigeria. Bull. World Health Org. *46*, 645–651.

Chaturvedi, U., Nagar, R., and Shrivastava, R. (2006). Dengue and dengue haemorrhagic fever: implications of host genetics. FEMS Immunol. Med. Microbiol *47*, 155–166.

Chen, Y., Maguire, T., and Marks, R.M. (1996). Demonstration of binding of dengue virus envelope protein to target cells. J. Virol. *70*, 8765–8772.

Chippaux, A., Deubel, V., Moreau, J.P., and Reynes, J.M. (1993). [Current situation of yellow fever in Latin America]. Bull. Soc. Path. Exot. *86*, 460–464.

Christie, J. (1881). On epidemics of dengue fever: their diffusion and etiology. Glasgow Med. J. *16*, 161–176.

Christophers, S.R. (1960). Systematic. In *Aedes aegypti*, The Yellow Fever Mosquito:Its life History, Bionomics and Structure (London, Cambridge University Press), pp. 21–53.

Clark, D.V., Mammen, M.P., Jr., Nisalak, A., Puthimethee, V., and Endy, T.P. (2005). Economic impact of dengue fever/dengue haemorrhagic fever in Thailand at the family and population levels. Am. J. Trop. Med. Hyg. *72*, 786–791.

Cologna, R., Armstrong, P.M., and Rico-Hesse, R. (2005). Selection for virulent dengue viruses occurs in humans and mosquitoes. J. Virol. *79*, 853–859.

Cologna, R., and Rico-Hesse, R. (2003). American genotype structures decrease dengue virus output from human monocytes and dendritic cells. J. Virol. *77*, 3929–3938.

Cordellier, R., Bouchite, B., Roche, J.C., Monteny, N., Diaco, B., and Akoliba, P. (1983). Circulation selvatique du virus Dengue 2, en 1980, dans les savanes sub-soudaniennes de Cote d'Ivoire. Cah ORSTOM ser Ent. Med. Parasitol. *21*, 165–179.

Cornet, M. (1993). Dengue virus in Africa. In Monograph on dengue/dengue haemorrhagic fever, P. Thongcharoen, ed. (New Delhi, India, WHO regional publication, South-East Asia. World Health Organization, Regional Office for South-East Asia), pp. 39–47.

Cornet, M., Robin, Y., Hannoun, C., Corniou, B., Bres, P., and Causse, G. (1968). An epidemic ofyellow fever in Senegal in 1965: Epidemiological studies. Bull. World Health Org. *39*, 845–858.

Cornet, M., Saluzzo, J.F., Hervy, J.P., Digoutte, J.P., Germain, M., Chauvancy, M.F., Eyraud, M., Ferrara, L., Heme, G., and Legros, F. (1984a). Dengue 2 au Senegal oriental: une poussee epizootique en milieu selvatique; isolements du virus a partir de moustiques et d'un singe et considerations epidemiologiques. Cah ORSTOM Ser. Ent. Med. Parasitol. *22*, 313–323.

Cornet, M., Saluzzo, J.F., Hervy, J.P., Digoutte, J.P., Germain, M., Chauvancy, M.F., Eyraud, M., Ferrara,

L., Heme, G., and Legros, F. (1984b). Dengue 2 au Senegal oriental: une poussee epizootique en milieu selvatique; isolements du virus a partir des moustiques et d'un singe et considerations epidemiologiques. Cah ORSTOM ser Ent. Med. Parasitol. *22*, 313–323.

Daggy, R.H. (1944). *Aedes scutellaris hebrideus* Edwards. A probable vector of dengue in the New Hebrides. War Med. *5*, 292.

De Jesus, M., Gubler, D.J., and Sather, G.E. (1982). A clinical survey of reported cases of dengue-like illness during the outbreak of dengue 1981 in Puerto Rico. Bol. Asoc. Med. P R *74*, 76–78.

de Thoisy, B.D., Dussart, P., and Kazanji, M. (2004). Wild terrestrial rainforest mammals as potential reservoirs for flaviviruses (yellow fever, dengue 2 and St Louis encephalitis viruses) in French Guiana. Trans. R. Soc. Trop. Med. Hyg. *98*, 409–412.

de Thoisy, B.D., Lacoste,V., Germain, A., Munoz-Jordan, J., Colon, C., Mauffrey, J.F., Delaval, M., Catzeflis, F., Kazanji, M., Matheus, S., Dussart, P., Morvan, J., Setién, A.A., Deparis, X., and Lavergne, A. (2008). Dengue infection in neotropical forest mammals. Vector Borne Zoonotic Dis [Epub ahead of print]

Diallo, M., Ba, Y., Faye, O., Soumare, M.L., Dia, I., and Sall, A.A. (2008). Vector competence of *Aedes aegypti* populations from Senegal for sylvatic and epidemic dengue 2 virus isolated in West Africa. Trans. R. Soc. Trop. Med. Hyg. *102*, 493–498.

Diallo, M., Ba, Y., Sall, A.A., Diop, O.M., Ndione, J.A., Mondo, M., Girault, L., and Mathiot, C. (2003). Amplification of the sylvatic cycle of dengue virus type 2, Senegal, 1999–2000: entomologic findings and epidemiologic considerations. Emerg. Infect. Dis. *9*, 362–367.

Diallo, M., Sall, A.A., Moncayo, A.C., Ba, Y., Fernandez, Z., Ortiz, D., Coffey, L.L., Mathiot, C., Tesh, R.B., and Weaver, S.C. (2005). Potential role of sylvatic and domestic african mosquito species in dengue emergence. Am. J. Trop. Med. Hyg. *73*, 445–449.

Dietz, V., Gubler, D.J., Ortiz, S., Kuno, G., Casta-Velez, A., Sather, G.E., Gomez, I., and Vergne, E. (1996). The 1986 dengue and dengue hemorrhagic fever epidemic in Puerto Rico: epidemiologic and clinical observations. PR Health Sci. J *15*, 201–210.

Edington, A.D. (1927). 'Dengue', as seen in the recent epidemic in Durban. J. Med. Assoc. South Africa *1*, 446–448.

Edwards, F.W. (1932). Diptera.Family Culicidae. In Genera Insectorum, P. Wystman, ed. (Brussels, Belgium, Desmet-Verteneuil).

Fagbami, A.H., Monath, T.P., and Fabiyi, A. (1977). Dengue virus infections in Nigeria: a survey for antibodies in monkeys and humans. Trans. R. Soc. Trop. Med. Hyg. *71*, 60–65.

Falconar, A.K. (2008). Use of synthetic peptides to represent surface-exposed epitopes defined by neutralizing dengue complex- and flavivirus group-reactive monoclonal antibodies on the native dengue type-2 virus envelope glycoprotein. J. Gen. Virol. *89*, 1616–1621.

Falconar, A.K., Young, P.R., and Miles, M.A. (1994). Precise location of sequential dengue virus subcomplex and complex B-cell epitopes on the nonstructural-1 glycoprotein. Arch. Virol. *137*, 315–326.

Ferguson, N., Anderson, R., and Gupta, S. (1999). The effect of antibody-dependent enhancement on the transmission dynamics and persistence of multiple-strain pathogens. Proc. Natl. Acad. Sci. U.S.A. *96*, 790–794.

Foster, J.E., Bennett, S.N., Vaughan, H., Vorndam, V., McMillan, W.O., and Carrington, C.V. (2003). Molecular evolution and phylogeny of dengue type 4 virus in the Caribbean. Virology *306*, 126–134.

Freier, J.E., and Grimstad, P.R. (1983). Transmission of dengue virus by orally infected *Aedes triseriatus*. Am. J. Trop. Med. Hyg. *32*, 1429–1434.

Furumizo, R.T., and Rudnick, A. (1979). Laboratory observations on the life history of two species of the *Aedes* (*Finlaya*) *niveus* subgroup (Diptera: Culicidae) in Malaysia. J. Med. Entomol. *15*, 573–575.

Garcia, G., Vaughn, D.W., and Del Angel, R.M. (1997). Recognition of synthetic oligopeptides from nonstructural proteins NS1 and NS3 of dengue-4 virus by sera from dengue virus-infected children. Am. J. Trop. Med. Hyg. *56*, 466–470.

Garg, P., Nagpal, J., Khairnar, P., and Seneviratne, S.L. (2008). Economic burden of dengue infections in India. Trans. R. Soc. Trop. Med. Hyg. *102*, 570–577.

Germain, A. (1984). Etiologie du virus de la fievre jeune en Afrique: l'apport de recherches recentes. Bull. Soc. Entomol. France *89*, 760–768.

Germain, M., Cornet, M., Mouchet, J., Herve, J.P., Robert, V., Camicas, J.L., Cordellier, R., Hervy, J.P., Digoutte, J.P., Monath, T.P., Salaun, J.J., Deubel, V., Robin, Y., Coz, J., Taufflieb, R., Saluzzo, J.F., and Gonzalez, J.P. (1981). La Fievre jeune selvatique en Afrique: donnees recentes et conceptions actuelles. Med. Trop. *41*, 31–43.

Germain, M., Francy, D.B., Monath, T.P., Ferrara, L., Bryan, J., Salaun, J.J., Heme, G., Renaudet, J., Adam, C., and Digoutte, J.P. (1980). Yellow fever in the Gambia, 1978 – 1979: entomological aspects and epidemiological correlations. Am. J. Trop. Med. Hyg. *29*, 929–940.

Germain, M., Sureau, P., Herve, J.P., Fabre, J., Mouchet, J., Robin, Y., and Geoffroy, B. (1976). Isolements du virus de la fievre jaune a partir d' Aedes du groupe A. *africanus* (Theobald) en Republique Centralafricaine. importance des savanes humides et semi-humides en tant que zone d'emergence du virus amaril. Cah ORSTOM ser Ent. Med. Parasitol *14*, 125–139.

Green, S., Kurane, I., Pincus, S., Paoletti, E., and Ennis, F.A. (1997). Recognition of dengue virus NS1–NS2a proteins by human CD4+ cytotoxic T lymphocyte clones. Virology *234*, 383–386.

Gubler, D.J. (1997). Dengue and dengue hemorrhagic fever: its history and resurgence as a global public health problem. In Dengue and Dengue Hemorrhagic Fever, D.J. Gubler, and G. Kuno, eds. (New York, CAB International), pp. 1–22.

Gubler, D.J., Novak, R.J., Vergne, E., Colon, N.A., Velez, M., and Fowler, J. (1985). *Aedes* (*Gymnometopa*) *mediovittatus* (Diptera: Culicidae), a potential maintenance vector of dengue viruses in Puerto Rico. J. Med. Entomol. *22*, 469–475.

Gubler, D.J., Suharyono, W., Lubis, I., Eram, S., and Gunarso, S. (1981). Epidemic dengue 3 in central Java, associated with low viremia in man. Am. J. Trop. Med. Hyg. *30*, 1094–1099.

Guilarde, A.O., Turchi, M.D., Siqueira, J.B., Jr., Feres, V.C., Rocha, B., Levi, J.E., Souza, V.A., Boas, L.S., Pannuti, C.S., and Martelli, C.M. (2008). Dengue and dengue hemorrhagic fever among adults: clinical outcomes related to viremia, serotypes, and antibody response. J. Infect. Dis. *197*, 817–824.

Guzman, M.G., Kouri, G., Valdes, L., Bravo, J., Vazquez, S., and Halstead, S.B. (2002). Enhanced severity of secondary dengue-2 infections: death rates in 1981 and 1997 Cuban outbreaks. Pan Am. J. Public Health *11*, 223–227.

Haddow, A.J. (1969). The natural history of yellow fever in Africa. Proc. R. Soc. Edinburgh B *70*, 191–227.

Halstead, S.B., Streit, T.G., Lafontant, J.G., Putvatana, R., Russell, K., Sun, W., Kanesa-Thasan, N., Hayes, C.G., and Watts, D.M. (2001). Haiti: absence of dengue hemorrhagic fever despite hyperendemic dengue virus transmission. Am. J. Trop. Med. Hyg. *65*, 180–183.

Hammon, W.M., Schrack, W.D., and Sather, G.E. (1958). Serological evidence for arthropod-borne viruses in the Philippines. Am. J. Trop. Med. Hyg. 7, 323–328.

Hanafusa, S., Chanyasanha, C., Sujirarat, D., Khuankhunsathid, I., Yaguchi, A., and Suzuki, T. (2008). Clinical features and differences between child and adult dengue infections in Rayong Province, southeast Thailand. SA J Trop. Med. Public Health *39*, 252–259.

Hanley, K.A., Nelson, J.T., Schirtzinger, E.E., Whitehead, S.S., and Hanson, C.T. (2008). Superior infectivity for mosquito vectors contributes to competitive displacement among strains of dengue virus. BMC Ecology *8*, 1.

Harrington, L.C., Edman, J.D., and Scott, T.W. (2001). Why do female *Aedes aegypti* (Diptera: Culicidae) feed preferentially and frequently on human blood? J. Med. Entomol. *38*, 411–422.

Harris, E., Videa, E., Perez, L., Sandoval, E., Tellez, Y., Perez, M.L., Cuadra, R., Rocha, J., Idiaquez, W., Alonso, R.E., Delgado, M.A., Campo, L.A., Acevedo, F., Gonzalez, A., Amador, J.J., and Balmaseda, A. (2000). Clinical, epidemiologic, and virologic features of dengue in the 1998 epidemic in Nicaragua. Am. J. Trop. Med. Hyg. *63*, 5–11.

Hayes, C.G., Phillips, I.A., Callahan, J.D., Griebenow, W.F., Hyams, K.C., Wu, S.J., and Watts, D.M. (1996). The epidemiology of dengue virus infection among urban, jungle, and rural populations in the Amazon region of Peru. Am. J. Trop. Med. Hyg. *55*, 459–463.

Henchal, E.A., Henchal, L.S., and Thaisomboonsuk, B.K. (1987). Topological mapping of unique epitopes on the dengue-2 virus NS1 protein using monoclonal antibodies. J. Gen. Virol. *68 (Pt 3)*, 845–851.

Hervy, J.P., Legros, F., Roche, J.C., Monteny, N., and Diaco, B. (1984). Circulation du Dengue 2 dans plusieurs milieux boises des savanes soudaniennes de la region de Bobo-Dioulasso (Burkina Faso). Considerations entomologiques et epidemiologiques. Cah ORSTOM ser Ent Med. Parasitol. *22*, 135–143.

Hirsch, A. (1883). Dengue, a comparatively new desease: its symptoms. In Handbook of Geographical and Historical Pathology (Syndenham Society), pp. 55–81.

Holmes, E.C. (2004). The phylogeography of human viruses. Mol Ecol *13*, 745–756.

Holmes, E.C., and Twiddy, S.S. (2003). The origin, emergence and evolutionary genetics of dengue virus. Infect. Genet. Evol. 3, 19–28.

Hota, S. (1952). Experimental studies on dengue. I. Isolation, identification and modification of the virus. J. Infect. Dis. *90*, 1–9.

Huang, J.H., Wey, J.J., Sun, Y.C., Chin, C., Chien, L.J., and Wu, Y.C. (1999). Antibody responses to an immunodominant nonstructural 1 synthetic peptide in patients with dengue fever and dengue hemorrhagic fever. J. Med. Virol *57*, 1–8.

Hugues, T.P. (1943). The reaction of the African grivet monkey (*Cercopithecus aethiops centralis*) to yellow fever virus. Trans. R. Soc. Trop. Med. Hyg. *36*, 339–346.

I.M.R. (1956). Annual Report 1955 (Kuala Lampur: Institute for Medical Research).

Innis, B.L., Thirawuth, V., and Hemachudha, C. (1989). Identification of continuous epitopes of the envelope glycoprotein of dengue type 2 virus. Am. J. Trop. Med. Hyg. *40*, 676–687.

Jumali, Sunarto, Gubler, D.J., Nalim, S., Eram, S., and Sulianti Saroso, J. (1979). Epidemic dengue hemorrhagic fever in rural Indonesia. III Entomological studies. Am. J. Trop. Med. Hyg. *28*, 717–724.

Kalayanarooj, S., and Nimmannitya, S. (2003). Clinical presentations of dengue hemorrhagic fever in infants compared to children. J Med Assoc Thai *86 Suppl 3*, S673–680.

Kimura, R., and Hotta, S. (1944). Research on dengue. Report 6: inoculation of mice with dengue virus Nippon Igaku *3379*, 629–633.

Knowles, N.J., Davies, P.R., Henry, T., O'Donnell, V., Pacheco, J.M., and Mason, P.W. (2001). Emergence in Asia of foot-and-mouth disease viruses with altered host range: characterization of alterations in the 3A protein. J. Virol. *75*, 1551–1556.

Koh, B.K., Ng, L.C., Kita, Y., Tang, C.S., Ang, L.W., Wong, K.Y., James, L., and Goh, K.T. (2008). The 2005 dengue epidemic in Singapore: epidemiology, prevention and control. Ann. Acad. Med. Singapore *37*, 538–545.

Kongsomboon, K., Singhasivanon, P., Kaewkungwal, J., Nimmannitya, S., Mammen, M.P., Jr., Nisalak, A., and Sawanpanyalert, P. (2004). Temporal trends of dengue fever/dengue hemorrhagic fever in Bangkok, Thailand from 1981 to 2000: an age-period-cohort analysis. SA J. Trop. Med. Public Health *35*, 913–917.

Kouri, G., Mas, P., Guzman, M.G., Soler, M., Goyenechea, A., and Morier, L. (1983). Dengue hemorrhagic fever in Cuba, 1981: rapid diagnosis of the etiologic agent. Bull Pan. Am. Health Org. *17*, 126–132.

Kouri, G., Valdez, M., Arguello, L., Guzman, M.G., Valdes, L., Soler, M., and Bravo, J. (1991). [Dengue epidemic in Nicaragua, 1985]. Rev. Inst. Med. Trop. Sao Paulo *33*, 365–371.

Kouri, G.P., Guzman, M.G., Bravo, J.R., and Triana, C. (1989). Dengue haemorrhagic fever/dengue shock syndrome: lessons from the Cuban epidemic, 1981. Bull. World Health Org. *67*, 375–380.

Kuno, G. (1997). Factors influencing the transmission of dengue viruses. In Dengue and Dengue Hemorrhagic Fever, D.J. Gubler, and G. Kuno, eds. (New York, CAB International), pp. 61–88.

Kurane, I., Brinton, M.A., Samson, A.L., and Ennis, F.A. (1991). Dengue virus-specific, human CD4+ CD8- cytotoxic T-cell clones: multiple patterns of virus cross-reactivity recognized by NS3-specific T-cell clones. J. Virol. 65, 1823–1828.

Kutubuddin, M., Kolaskar, A.S., Galande, S., Gore, M.M., Ghosh, S.N., and Banerjee, K. (1991). Recognition of helper T-cell epitopes in envelope (E) glycoprotein of Japanese encephalitis, west Nile and Dengue viruses. Mol. Immunol. 28, 149–154.

Leclerc, C., Deriaud, E., Megret, F., Briand, J.P., Van Regenmortel, M.H., and Deubel, V. (1993). Identification of helper T-cell epitopes of dengue virus E-protein. Mol. Immunol. 30, 613–625.

Lee, I.K., Liu, J.W., and Yang, K.D. (2008). Clinical and laboratory characteristics and risk factors for fatality in elderly patients with dengue hemorrhagic fever. Am. J. Trop. Med. Hyg. 79, 149–153.

Lee, V.H. (1969). *Aedes aegypti* surveillance in Ibadan. Bull. Entomol. Soc. Nigeria 2, 87–89.

Lee, V.H., Monath, T.P., Tomori, O., Fagbami, A., and Wilson, D.C. (1974). Arbovirus studies in Nupeko forest, a possible natural focus of yellow fever virus in Nigeria. I.I. Entomological investigations and viruses isolated. Trans. R. Soc. Trop. Med. Hyg. 68, 39–43.

Leichtenstern, O. (1896). Influenza and Dengue. In Specielle Pathologie und Therapie, H. Nothnagel, ed. (Wien, Alfred Holder), pp. 133–226.

Leitmeyer, K.C., Vaughn, D.W., Watts, D.M., Salas, R., Villalobos, I., de, C., Ramos, C., and Rico-Hesse, R. (1999). Dengue virus structural differences that correlate with pathogenesis. J. Virol. 73, 4738–4747.

Loke, H., Bethell, D., Phuong, C.X., Day, N., White, N., Farrar, J., and Hill, A. (2002). Susceptibility to dengue hemorrhagic fever in vietnam: evidence of an association with variation in the vitamin d receptor and Fc gamma receptor IIa genes. Am. J. Trop. Med. Hyg. 67, 102–106.

Mackerras, I.M. (1946). Transmission of dengue fever by *Aedes* (*Stegomyia*) *scutellaris* Walk. in New Guinea. Trans. R. Soc. Trop. Med. Hyg. 40, 295–312.

Mathew, A., Kurane, I., Green, S., Stephens, H.A., Vaughn, D.W., Kalayanarooj, S., Suntayakorn, S., Chandanayingyong, D., Ennis, F.A., and Rothman, A.L. (1998). Predominance of HLA-restricted cytotoxic T-lymphocyte responses to serotype-cross-reactive epitopes on nonstructural proteins following natural secondary dengue virus infection. J. Virol. 72, 3999–4004.

Megret, F., Hugnot, J.P., Falconar, A., Gentry, M.K., Morens, D.M., Murray, J.M., Schlesinger, J.J., Wright, P.J., Young, P., Van Regenmortel, M.H., and Deubel, V. (1992). Use of recombinant fusion proteins and monoclonal antibodies to define linear and discontinuous antigenic sites on the dengue virus envelope glycoprotein. Virology 187, 480–491.

Messer, W.B., Gubler, D.J., Harris, E., Sivananthan, K., and de Silva, A.M. (2003). Emergence and global spread of a dengue serotype 3, subtype III virus. Emerg. Infect. Dis. 9, 800–809.

Meunier, D.M., Aron, N., and Mazzariol, M.J. (1988). The 1987 yellow fever epidemic in Mali: viral and immunological diagnosis. Trans. R. Soc. Trop. Med. Hyg. 82, 767.

Monath, T.P. (1994). Yellow fever and dengue – the interactions of virus, vector and host in the re-emergence of epidemic disease. Sem Virol 5, 133–145.

Monath, T.P. (2001). Yellow Fever. In The Encyclopedia of Arthropod-transmitted Infections, M.W. Service, ed. (Wallingford, U.K., CAB International), pp. 571–577.

Moncayo, A.C., Fernandez, Z., Ortiz, D., Diallo, M., Sall, A., Hartman, S., Davis, C.T., Coffey, L., Mathiot, C.C., Tesh, R.B., and Weaver, S.C. (2004). Dengue emergence and adaptation to peridomestic mosquitoes. Emerg. Infect. Dis. 10, 1790–1796.

Monlun, E., Zeller, H., Traore-Lamizana, M., Hervy, J.P., Adam, F., Mondo, M., and Digoutte, J.P. (1992). Caracteres cliniques et epidemiologiques de la dengue 2 au Senegal. Med. Mal. Infect. 22, 718–721.

Moore, D.L., Causey, O.R., Carey, D.E., Reddy, S., Cooke, A.R., Akinkugbe, F.M., David-West, T.S., and Kemp, G.E. (1975). Arthropod-borne viral infections of man in Nigeria, 1964–1970. Ann Trop Med Parasitol 69, 49–64.

More, F.W. (1904). Observations on dengue fever in Singapore. J Malaya Branch Br Med Assoc 1, 24–29.

Myat Thu, H., Lowry, K., Jiang, L., Hlaing, T., Holmes, E.C., and Aaskov, J. (2005). Lineage extinction and replacement in dengue type 1 virus populations are due to stochastic events rather than to natural selection. Virology 336, 163–172.

Olaleye, O.D., Omilabu, S.A., Faseru, O., and Fagbami, A.H. (1988). 1987 yellow fever epidemics in Oyo State, Nigeria: a survey for yellow fever virus haemagglutination inhibiting antibody in residents of two communities before and after the epidemics. Virologie 39, 261–266.

Pepper, P. (1941). A note on David Bylon and dengue. Ann. Med. Hist. 3, 363–368.

Powell, J.R., Tabachnick, W.J., and Arnold, J. (1980). Genetics and the origin of a vector population: *Aedes aegypti*, a case study. Science (New York, NY 208, 1385–1387.

Rey, M., and Girard, M.P. (2008). The global eradication of poliomyelitis: progress and problems. Comparative immunology, microbiology and infectious diseases 31, 317–325.

Rico-Hesse, R. (1990). Molecular evolution and distribution of dengue viruses type 1 and 2 in nature. Virology 174, 479–493.

Rico-Hesse, R., Harrison, L.M., Salas, R.A., Tovar, D., Nisalak, A., Ramos, C., Boshell, J., de Mesa, M.T., Nogueira, R.M., and da Rosa, A.T. (1997). Origins of dengue type 2 viruses associated with increased pathogenicity in the Americas. Virology 230, 244–251.

Rigau-Perez, J.G. (1998). The early use of break-bone fever (Quebranta huesos, 1771) and dengue (1801) in Spanish. Am. J. Trop. Med. Hyg. 59, 272–274.

Rigau-Perez, J.G., and Laufer, M.K. (2006). Dengue-related deaths in Puerto Rico, 1992–1996: diagnosis and clinical alarm signals. Clin. Infect. Dis. *42*, 1241–1246.

Roberts, D.R., Peyton, E.L., Pinheiro, F.P., Balderrama, F., and Vargas, R. (1984). Associations of arbovirus vectors with gallery forests and domestic environments in southeastern Bolivia. Bull. Pan Am. Health Org. *18*, 337–350.

Robin, Y., Cornet, M., Heme, G., and Le Gonidec, G. (1980a). Isolement du virus de la dengue au Senegal. Ann. Virol. (Institut Pasteur) *131*, 149–154.

Robin, Y., Cornet, M., Heme, G., and Le Gonidec, G. (1980b). Isolement du virus dela dengue au Senegal. Ann. Virol. (Institut Pasteur) *131*, 149–154.

Roche, J.C., Cordellier, R., Hervy, J.P., Digoutte, J.P., and Monteny, N. (1983). Isolement de 96 souches de virus Dengue 2 a partir de moustiques captures en Cote d'Ivoire et en Haute Volta. Ann Virol (Institut Pasteur) *134E.*, 233–244.

Rodhain, F. (1991). The role of monkeys in the biology of dengue and yellow fever. Comp. Immunol. Microbiol. Infect. Dis. *14*, 9–19.

Roehrig, J.T., Risi, P.A., Brubaker, J.R., Hunt, A.R., Beaty, B.J., Trent, D.W., and Mathews, J.H. (1994). T-helper cell epitopes on the E-glycoprotein of dengue 2 Jamaica virus. Virology *198*, 31–38.

Rosen, L. (1958). Experimental infection of New World monkeys with dengue and yellow fever viruses. Am. J. Trop. Med. Hyg. *7*, 406–410.

Rosen, L. (1977). The Emperor's new clothes revisited, or reflections on the pathogenesis of dengue hemorrhagic fever. Am. J. Trop. Med. Hyg. *26*, 337–343.

Rosen, L., Roseboom, L.E., Gubler, D.J., Lien, J.C., and Chaniotis, B.N. (1985). Comparative susceptibility of mosquito species and strains to oral and parenteral infection with dengue and Japanese encephalitis viruses. Am. J. Trop. Med. Hyg. *34*, 603–615.

Rosen, L., Rozeboom, L.E., Sweet, B.H., and Sabin, A.B. (1954). The transmission of dengue by *Aedes polynesiensis* Marks. Am. J. Trop. Med. Hyg. *3*, 878–882.

Ross, R.W., and Gillett, J.D. (1950). The cyclical transmission of yellow fever virus through the grivet monkey, *Cercopithecus aethiops centralis* Neumann and the mosquito *Aedes africanus* Theobald. Ann. Trop. Med. Parasitol. *44*, 351–356.

Rothman, A.L. (2003). Immunology and immunopathogenesis of dengue disease. Adv. Virus Res. *60*, 397–419.

Rudnick, A. (1965). Studies of the ecology of dengue in Malaysia: a preliminary report. J. Med. Entomol. *2*, 203–208.

Rudnick, A. (1978). Ecology of dengue virus. Asian J. Infect. Dis. *2*, 156–160.

Rudnick, A. (1984a). The ecology of the dengue virus complex in Peninsular Malaysia. In Proceedings of the International Conference on Dengue/D.H.F., T. Pang, and R. Pathmanathan, eds. (Kuala Lumpur, University of Malaysia Press), p. 7.

Rudnick, A. (1984b). The ecology of the dengue virus complex in peninsular Malaysia. Paper presented at: Proceedings of the International Conference on dengue/DHF (Kuala Lampur, Malaysia, University of Malaysia Press).

Rudnick, A. (1986). Dengue virus ecology in Malaysia. Inst. Med. Res. Malays. Bull. *23*, 51–152.

Rudnick, A., and Lim, T.W. (1986). Dengue fever studies in Malaysia. Inst. Med. Res. Malays. Bull. *23*, 51–152.

Rudnick, A., Marchette, N.J., and Garcia, R. (1967). Possible jungle dengue – recent studies and hypotheses. Jpn J Med Sci Biol *20*, 69–74.

Rush, A.B. (1789). An account of the bilious remitting fever, as it appeared in philadelphia in the summer and autumn of the year 1980. In Medical Inquiries and Observations (Philadelphia, Prichard and Hall), pp. 89–100.

Saluzzo, J.F., Cornet, M., Adam, C., Eyraud, M., and Digoutte, J.P. (1986a). [Dengue 2 in eastern Senegal: serologic survey in simian and human populations. 1974–85]. Bull Soc Pathol Exot Filiales *79*, 313–322.

Saluzzo, J.F., Cornet, M., Castagnet, P., Rey, C., and Digoutte, J.P. (1986b). Isolation of dengue 2 and dengue 4 viruses from patients in Senegal. Trans. R. Soc. Trop. Med. Hyg. *80*, 5.

Sanchez, I.J., and Ruiz, B.H. (1996). A single nucleotide change in the E protein gene of dengue virus 2 Mexican strain affects neurovirulence in mice. J. Gen. Virol. *77 (Pt 10)*, 2541–2545.

Sanchez, V., Hessler, C., DeMonfort, A., Lang, J., and Guy, B. (2006). Comparison by flow cytometry of immune changes induced in human monocyte-derived dendritic cells upon infection with dengue 2 live-attenuated vaccine or 16681 parental strain. FEMS Immunol Med Microbiol *46*, 113–123.

Serie, C., Lindrec, A., Poirier, A., Andral, L., and Neri, P. (1968). Studies on yellow fever in Ethiopia. I. Introduction: clinical symptoms of yellow fever. Bull World Health Organ *38*, 835–841.

Siler, J.F., Hall, M.W., and Hitchens, A.P. (1926). Dengue: Its history, epidemiology, mechanism of transmission, etiology, clinical manifestations, immunity, and prevention. Philippine J Sci *29*, 1–252.

Simmons, C.P., Dong, T., Chau, N.V., Dung, N.T., Chau, T.N., Thao le, T.T., Dung, N.T., Hien, T.T., Rowland-Jones, S., and Farrar, J. (2005). Early T-cell responses to dengue virus epitopes in Vietnamese adults with secondary dengue virus infections. J. Virol. *79*, 5665–5675.

Simmons, J.S., St John, J.H., and Reynolds, F.H.K. (1931). Experimental studies of dengue. Philippine J Sci *44*, 1–252.

Siqueira, J.B., Jr., Martelli, C.M., Coelho, G.E., Simplicio, A.C., and Hatch, D.L. (2005). Dengue and dengue hemorrhagic fever, Brazil, 1981–2002. Emerg. Infect. Dis. *11*, 48–53.

Skae, F.M.T. (1902). Dengue fever in Penang. Brit Med J *2*, 1581–1582.

Smith, C.E. (1956a). Isolation of three strains of type 1 dengue virus from a local outbreak of the disease in Malaya. J. Hyg. (Lond) *54*, 569–580.

Smith, C.E. (1956b). The history of dengue in tropical Asia and its probable relationship to the mosquito *Aedes aegypti*. J. Trop. Med. Hyg. *59*, 243–251.

Smith, C.E. (1957). A localized outbreak of dengue fever in Kuala Lumpur: serological aspects. J. Hyg. (Lond.) *55*, 207–223.

Smith, C.E. (1958). The distribution of antibodies to Japanese Encephalitis, dengue, and yellow fever viruses in five rural communities in Malaya. Trans. R. Soc. Trop. Med. Hyg. *52*, 237–252.

Smith, C.E.G. (1956c). A localized outbreak of dengue fever in Kuala Lumpur: epidemiological and clinical aspects. Med. J. Malaya *10*, 289–303.

Smithburn, K.C., and Haddow, A.J. (1949). The susceptibility of African wild animals to yellow fever. 1. Monkeys. Am. J. Trop. Med. Hyg. *29*, 414–423.

Soper, F.L., Penna, H., Cardoso, E., Serafim, J.J., Frobisher, M., and Pinheiro, J. (1933). Yellow Fever without *Aedes aegypti*. Study of a rural epidemic in the Valley of Chan aan, Espiritu Santo, Brasil. Am. J. Hyg. *18*, 555–587.

Spaulding, A.C., Kurane, I., Ennis, F.A., and Rothman, A.L. (1999). Analysis of murine CD8(+) T-cell clones specific for the Dengue virus NS3 protein: flavivirus cross-reactivity and influence of infecting serotype. J. Virol. *73*, 398–403.

Stanton, A.T. (1919). The mosquitoes of Far Eastern ports with special reference to the prevalence of *Stegomyia fasciata*. Bull. Ent. Res. *10*, 333–344.

Steadman, G.W. (1828). Some account of an anomalous disease which raged in the islands of St. Thomas and Santa Cruz, in the West Indies, during the months of September, October, November, December, and January 1827–8. Edinb. Med. Surg. J. *30*, 227–248.

Stewart, A.J., and Devlin, P.M. (2006). The history of the smallpox vaccine. J. Infect. *52*, 329–334.

Tabachnick, W.J., and Powell, J.R. (1979). A world-wide survey of genetic variation in the yellow fever mosquito, *Aedes aegypti*. Genet. Re.s *34*, 215–229.

Theobald, F.V. (1901). A monograph of the Culicidae of the World Vol. 1 (London, British Museum of Natural History).

Thu, H.M., Lowry, K., Myint, T.T., Shwe, T.N., Han, A.M., Khin, K.K., Thant, K.Z., Thein, S., and Aaskov, J. (2004). Myanmar dengue outbreak associated with displacement of serotypes 2, 3, and 4 by dengue 1. Emerg. Infect. Dis. *10*, 593–597.

Traore-Lamizana, M., Zeller, H., Monlun, E., Mondo, M., Hervy, J.P., Adam, F., and Digoutte, J.P. (1994). Dengue 2 outbreak in southeastern Senegal during 1990: virus isolations from mosquitoes (Diptera: Culicidae). J. Med. Entomol. *31*, 623–627.

Twiddy, S.S., Farrar, J.J., Vinh Chau, N., Wills, B., Gould, E.A., Gritsun, T., Lloyd, G., and Holmes, E.C. (2002a). Phylogenetic relationships and differential selection pressures among genotypes of dengue-2 virus. Virology *298*, 63–72.

Twiddy, S.S., Holmes, E.C., and Rambaut, A. (2003). Inferring the rate and time-scale of dengue virus evolution. Mol. Biol. Evol. *20*, 122–129.

Twiddy, S.S., Woelk, C.H., and Holmes, E.C. (2002b). Phylogenetic evidence for adaptive evolution of dengue viruses in nature. J. Gen. Virol. *83*, 1679–1689.

Uzcategui, N.Y., Camacho, D., Comach, G., Cuello de Uzcategui, R., Holmes, E.C., and Gould, E.A. (2001). Molecular epidemiology of dengue type 2 virus in Venezuela: evidence for in situ virus evolution and recombination. J. Gen. Virol. *82*, 2945–2953.

Vasconcelos, P.F., Costa, Z.G., Travassos Da Rosa, E.S., Luna, E., Rodrigues, S.G., Barros, V.L., Dias, J.P., Monteiro, H.A., Oliva, O.F., Vasconcelos, H.B., Oliveira, R.C., Sousa, M.R., Barbosa Da Silva, J., Cruz, A.C., Martins, E.C., and Travasos Da Rosa, J.F. (2001). Epidemic of jungle yellow fever in Brazil, 2000: implications of climatic alterations in disease spread. J. Med. Virol. *65*, 598–604.

Vasconcelos, P.F., Rodrigues, S.G., Degallier, N., Moraes, M.A., da Rosa, J.F., da Rosa, E.S., Mondet, B., Barros, V.L., and da Rosa, A.P. (1997). An epidemic of sylvatic yellow fever in the southeast region of Maranhao State, Brazil, 1993–1994: epidemiologic and entomologic findings. Am. J. Trop. Med. Hyg. *57*, 132–137.

Vasilakis, N., Durbin, A., Travassos da Rosa, A.P.A., Munoz-Jordan, J.L., Tesh, R.B., and Weaver, S.C. (2008a). Antigenic relationships between sylvatic and endemic dengue viruses. Am. J. Trop. Med. Hyg. *79*, 128–132.

Vasilakis, N., Fokam, E.B., Hanson, C.T., Weinberg, E., Sall, A.A., Whitehead, S.S., Hanley, K.A., and Weaver, S.C. (2008b). Genetic and phenotypic characterization of sylvatic dengue virus type 2 strains. Virology *377*, 296–307.

Vasilakis, N., Holmes, E.C., Fokam, E.B., Faye, O., Diallo, M., Sall, A.A., and Weaver, S.C. (2007a). Evolutionary processes among sylvatic dengue type 2 viruses. J. Virol. *81*, 9591–9595.

Vasilakis, N., Shell, E.J., Fokam, E.B., Mason, P.W., Hanley, K.A., Estes, D.M., and Weaver, S.C. (2007b). Potential of ancestral sylvatic dengue-2 viruses to re-emerge. Virology *358*, 402–412.

Vasilakis, N., Tesh, R.B., and Weaver, S.C. (2008c). Sylvatic dengue virus type 2 activity in humans, Nigeria, 1966. Emerg. Infect. Dis. *14*, 502–504.

Vasilakis, N., and Weaver, S.C. (2008). The History and Evolution of Human Dengue Emergence. Adv. Vir. Res. *72*, 1–76.

Vaughn, D.W., Green, S., Kalayanarooj, S., Innis, B.L., Nimmannitya, S., Suntayakorn, S., Endy, T.P., Raengsakulrach, B., Rothman, A.L., Ennis, F.A., and Nisalak, A. (2000). Dengue viremia titer, antibody response pattern, and virus serotype correlate with disease severity. J. Infect. Dis. *181*, 2–9.

Vazeille-Falcoz, M., Failloux, A.B., Mousson, L., Elissa, N., and Rodhain, F. (1999). [Oral receptivity of *Aedes aegypti formosus* from Franceville (Gabon, central Africa) for type 2 dengue virus]. Bull. Soc. Pathol. Exot. (1990) *92*, 341–342.

Wang, E., Ni, H., Xu, R., Barrett, A.D., Watowich, S.J., Gubler, D.J., and Weaver, S.C. (2000). Evolutionary relationships of endemic/epidemic and sylvatic dengue viruses. J. Virol. *74*, 3227–3234.

Watts, D.M., Porter, K.R., Putvatana, P., Vasquez, B., Calampa, C., Hayes, C.G., and Halstead, S.B. (1999). Failure of secondary infection with American genotype dengue 2 to cause dengue haemorrhagic fever. Lancet *354*, 1431–1434.

Wen, J.S., Jiang, L.F., Zhou, J.M., Yan, H.J., and Fang, D.Y. (2008). Computational prediction and identification of dengue virus-specific CD4(+) T-cell epitopes. Virus Res. *132*, 42–48.

WHO (1996a). Yellow fever. Wkly Epidemiol. Rec. *71*, 232.

WHO (1996b). Yellow fever. Wkly Epidemiol. Rec. *71*, 342–343.

WHO (1997). Dengue haemorrhagic fever: diagnosis, treatment prevention and control (W.H.O., Geneva).

Wichmann, O., Hongsiriwon, S., Bowonwatanuwong, C., Chotivanich, K., Sukthana, Y., and Pukrittayakamee, S. (2004). Risk factors and clinical features associated with severe dengue infection in adults and children during the 2001 epidemic in Chonburi, Thailand. Trop. Med. Int. Health 9, 1022–1029.

Witayathawornwong, P. (2005). DHF in infants, late infants and older children: a comparative study. SA J. Trop. Med. Public Health *36*, 896–900.

Wolfe, N.D., Kilbourn, A.M., Karesh, W.B., Rahman, H.A., Bosi, E.J., Cropp, B.C., Andau, M., Spielman, A., and Gubler, D.J. (2001). Sylvatic transmission of arboviruses among Bornean orangutans. Am. J. Trop. Med. Hyg. *64*, 310–316.

Wu, H.C., Huang, Y.L., Chao, T.T., Jan, J.T., Huang, J.L., Chiang, H.Y., King, C.C., and Shaio, M.F. (2001). Identification of B-cell epitope of dengue virus type 1 and its application in diagnosis of patients. J. Clin. Microbiol *39*, 977–982.

Wu, H.C., Jung, M.Y., Chiu, C.Y., Chao, T.T., Lai, S.C., Jan, J.T., and Shaio, M.F. (2003). Identification of a dengue virus type 2 (DEN-2) serotype-specific B-cell epitope and detection of DEN-2-immunized animal serum samples using an epitope-based peptide antigen. J. Gen. Virol. *84*, 2771–2779.

Zhang, C., Mammen, M.P., Jr., Chinnawirotpisan, P., Klungthong, C., Rodpradit, P., Monkongdee, P., Nimmannitya, S., Kalayanarooj, S., and Holmes, E.C. (2005). Clade replacements in dengue virus serotypes 1 and 3 are associated with changing serotype prevalence. J. Virol. *79*, 15123–15130.

Part V

Strategies for Disease Control

Prospects and Challenges for Dengue Virus Vaccine Development

12

Stephen S. Whitehead and Anna P. Durbin

Abstract

A safe and effective vaccine for the control of dengue virus disease is urgently needed and long overdue. Because each of the four dengue virus serotypes can cause the full spectrum of dengue disease, vaccination must protect against each serotype. An unprecedented number of vaccine candidates are in development and under clinical evaluation, with live attenuated vaccines being the most advanced. Considerable effort is also being made in the development of inactivated, subunit protein, virus vectored, and DNA vaccine candidates. The need to elicit protective immunity without predisposing for antibody-mediated enhanced disease, the need for rapid and tetravalent protection, and the need for an economical vaccine have presented challenges in the development pathway. Nevertheless, innovative research and development continues to provide solutions to these obstacles.

Introduction

Dengue virus (DENV) infection continues to cause significant morbidity and mortality throughout the tropical and subtropical regions of the world, with children bearing the greatest burden of disease (Halstead *et al.*, 2007). Unlike other flaviviruses such as yellow fever virus, Japanese encephalitis virus and tick-borne encephalitis virus, no licensed vaccine exists for dengue. This is despite nearly eighty years of vaccine-related research and development, which have intensified in the last 30 years. Interest in and support for dengue vaccine development

has been created by the generation of numerous commissions, initiatives, organizations, and partnerships, all emphasizing the continued urgency for a vaccine and fostering a community that is capable of overcoming the many obstacles in the path of development. Notwithstanding the many challenges facing the development of a successful DENV vaccine, the most recent decade has produced two live attenuated tetravalent vaccine candidates currently in phase II clinical evaluation, several live attenuated vaccine candidates in phase I clinical evaluation, and many subunit, DNA, and vectored vaccines in pre-clinical stages of development (Whitehead *et al.*, 2007). The candidate vaccines that have been most extensively evaluated and are furthest along in development are live attenuated DENV vaccines, which will be discussed in the greatest detail below.

The ideal vaccine

Dengue vaccine development proceeds without the benefit of a full understanding of the pathogenesis of dengue disease or a suitable animal disease model, and unlike other polyvalent vaccines where components can be introduced for use as they are developed, four dengue vaccines must be developed simultaneously to avoid the theoretical risk of antibody-mediated immune enhancement of disease severity of subsequent heteroserotypic infections (see Chapters 5 and 14). This point illustrates the first characteristic of an ideal vaccine: safety. To be safe, a dengue vaccine must not only be weakly or non-reactogenic, sterile, and pure, but it must also be functionally tetravalent, eliciting simultaneous protection against all four dengue virus serotypes. Second, vaccine efficacy must be achieved at an affordable cost, which is dependent not only on the cost of research and development and the cost of production, but also on the dosing strategy (number of doses required) and method of vaccine administration. This is particularly challenging since most countries where dengue disease burden is greatest are also countries with developing economies. Although dengue vaccination remains a cost-effective activity (Shepard et al., 2004), development of a dengue vaccine is influenced by the lack of disease incidence in developed countries, where pricing strategies could help to offset the cost of vaccination in less affluent areas. In addition to efforts to decrease the overall cost of vaccine manufacture, attention must be paid to strategies to maximize vaccine potency and stability, while decreasing the need for repeated doses or expensive delivery methods. Currently, live attenuated vaccine candidates promise to be the most economical vaccine option.

Live attenuated dengue vaccines

Live attenuated virus vaccines can induce durable humoral and cellular immune responses, mimicking natural infection. However, the replication level of a live DENV vaccine needs to be sufficiently low to preclude the development of significant illness. High levels of viraemia following natural infection, $>10^5$ infectious units/ml, can be associated with dengue fever; levels 10- to 100-fold greater are detected in patients with severe dengue disease (Vaughn et al., 2000). A reasonable range of viraemia for a live attenuated DENV vaccine would be approximately 10^1–10^2 infectious units/ml, based on the level observed following administration of the live attenuated yellow fever vaccine (Guirakhoo et al., 2006). Although symptoms should be minimal following DENV vaccination, some mild reactogenicity in an overall safe vaccine is to be expected as part of the normal response to a replicating DENV and may be a reasonable trade-off in an endemic region where DENV causes severe disease. For a live attenuated vaccine, this is the nature of the delicate balance between vaccine reactogenicity and its subsequent immunogenicity.

Early mouse-brain derived dengue viruses

The earliest reported efforts to create a live attenuated dengue vaccine were published in 1929 by Blanc and Caminopetros (1929), who attempted to attenuate dengue virus by treating it with ox bile. During the 1940s, both American and Japanese investigators initiated vaccine development programmes (Hotta, 1952; Sabin, 1952; Sabin and Schlesinger, 1945; Schlesinger et al., 1956). These investigators serially passaged DENV-1 or DENV-2 by intracerebral inoculation of mice and noted that as the virus became adapted to mice, it became less pathogenic for humans. After seven mouse brain passages, Sabin (1952) showed in human volunteers that the virus was attenuated and resulted in only mild symptoms of dengue, usually accompanied by a macular rash. Subsequently, volunteers were protected from challenge with wild-type dengue virus (Hotta, 1952; Sabin, 1952; Sabin and Schlesinger, 1945; Schlesinger et al., 1956), and this protection was generally attributed to neutralizing antibody (Wisseman et al., 1963). Interestingly, Sabin and others noted decreased protection and/or lower neutralizing antibody titres induced by the DENV-1 vaccine when it was mixed with the live attenuated yellow fever 17D vaccine (Fujita et al., 1969; Sabin, 1952). Nevertheless, these early vaccines were not pursued further due to concerns regarding impurities in mouse-brain derived preparations.

Tissue culture-derived live attenuated dengue viruses

In response to the Virus Commission of the United States Armed Forces Epidemiology Board, researchers at Walter Reed Army Institute of Research (WRAIR) in the United States and the Center for Vaccine Development at Mahidol University in Thailand independently began programmes to develop a live attenuated dengue vaccine in the 1970s. Candidates from both groups were developed from viruses isolated from dengue patients and then attenuated by sequential passage in primary dog kidney (PDK) cells or primary green monkey kidney (PGMK) cells.

The Mahidol University live attenuated DENV vaccine candidates identified for inclusion in a tetravalent formulation were DENV-1 (16007, PDK 13); DENV-2 (16681, PDK-53), DENV-3

(16562, PGMK-30, FRhL-3); and DENV-4 (1036, PDK-48). These vaccine candidates were tested as monovalent, bivalent, and trivalent formulations in adult volunteers (Bhamarapravati and Yoksan, 1989, 1997; Bhamarapravati et al., 1987; Vaughn et al., 1996) and reported to be well tolerated with fever, rash, and mild liver enzyme elevations being the most commonly reported side effects. Seroconversion rates to these candidates were as high as 90–100%. (Bhamarapravati and Yoksan, 1989, 2000). Importantly, volunteers with pre-existing dengue antibody who were vaccinated with a monovalent vaccine did not show enhanced vaccine reactions, and serious dengue illness was not reported during follow-up. The safety profile of the vaccine was further strengthened by the observation that monotypic antibody remained detectable for up to three years. After three years heterotypic antibody was detected, suggesting that these vaccinees may have experienced a subsequent asymptomatic natural dengue infection (Bhamarapravati and Yoksan, 1997). The tetravalent vaccine was eventually licensed to Aventis Pasteur for further development. It was soon noted that reactogenicity appeared to be greater for the tetravalent formulation than previously observed for the monovalent vaccines (Kanesa-thasan et al., 2001). This was attributed to the increased replication of the DENV-3 component, which also induced the highest neutralizing antibody titre. To diminish the apparent interference of the DENV-3 component and improve seroconversion rates to the other serotypes, different dose formulations of the vaccine were studied in adult volunteers, and a two-dose vaccination schedule was evaluated for the first time (Sabcharoen et al., 2002). Varying the concentration of the DENV-3 component in the tetravalent formulation coupled with a second vaccination six months after the first improved the immunogenicity of the tetravalent vaccine and overcame the immunodominance of the DENV-3 component. The vaccine was then studied in 82 Thai children aged 5–12 years (Sabcharoen et al., 2004). The vaccine was administered at three time points: day 0, 3–5 months later, and again at 12 months after initial vaccination. The first dose of vaccine did not appear to enhance the severity of reactions with the second and third doses, despite low seroconversion rates to all four serotypes after

the first vaccination. The most frequently reported adverse events were fever, rash, and headache. Mild increases in liver enzymes and neutropenia were also reported. The frequency and severity of systemic reactions were reduced with subsequent doses of vaccine. Following re-derivation of the DENV-3 component, further trials of the Aventis Pasteur vaccine candidate were halted due to unacceptably levels of reactogenicity and formulation issues associated with the DENV-3 component (Kitchener et al., 2006; Sanchez et al., 2006). The spectrum of clinical symptoms, clinical chemistries, and haematological abnormalities seen in the vaccinees were similar to those seen during natural infection, albeit milder and less frequent indicating that the vaccine candidate was incompletely attenuated.

WRAIR has generated numerous live attenuated vaccine candidates derived from DENV isolates sequentially passaged in PDK cells, followed by production in FRhL (fetal rhesus lung) cells. Several of these early candidates, although attenuated in preclinical studies, were subsequently found to be unacceptably reactogenic in human trials (Bancroft et al., 1984; Bancroft et al., 1981; Eckels et al., 1984; Innis et al., 1988; McKee et al., 1987; Scott et al., 1983). The vaccine candidates selected for initial inclusion in the WRAIR tetravalent formulation were DENV-1 45AZ5 PDK-20, DENV-2 S16803 PDK-50, DENV-3 CH53489 PDK-20, and DENV-4 341750 PDK-20. These candidate viruses were tested first as monovalent formulations and then evaluated in several different tetravalent formulations to determine the ideal passage number and dose of each candidate (Edelman et al., 1994; Edelman et al., 2003; Kanesa-Thasan et al., 2003; Sun et al., 2003). The most common side effects noted in recipients of the tetravalent formulations were low-grade fever, rash, and transient neutropenia (Edelman et al., 2003; Sun et al., 2003). In an effort to reduce reactogenicity, the DENV-1 component was replaced with a higher passage DENV-1 virus (PDK-27 rather than PDK-20) and to improve immunogenicity, the DENV-4 component was replaced with a lower passage virus (PDK-6 rather than PDK-20). During the course of its development, the WRAIR dengue vaccine was licensed to GlaxoSmithKline. A phase I trial evaluating this tetravalent vaccine in a small number of flavivirus-naïve children (6–7 years of age) was completed in Thailand (Simasathien et al., 2008). The tetravalent formulation was composed of DENV-1 PDK-27, DENV-2 PDK-50, DENV-3 PDK-20, and DENV-4 PDK-6 at doses ranging from 5.1 to 6.3 $\log_{10}$ (plaque-forming units, pfu). Two doses of the vaccine were administered six months apart. Although the vaccine was poorly immunogenic after dose 1, with the lowest antibody responses to DENV-1, a tetravalent neutralizing antibody response was induced in almost all volunteers following the second dose of vaccine (Simasathien et al., 2008). The results of a phase II trial in flavivirus-naïve adults in the US was recently published (Sun et al., 2008). The tetravalent vaccine (formulation 17) consisted of the same viruses tested previously in children, but at doses ranging from 4.4 to 5.6 $\log_{10}$ pfu, and was administered on day 0 and 180 of the study. Side effects after the primary dose were mild and consisted of low-grade headache, fever, and rash. Side effects were less frequent after the second dose and consisted primarily of headache. Viraemia following primary vaccination was detectable in 75% of the volunteers, yet undetectable after the second dose. Although trivalent and tetravalent neutralizing antibody responses were elicited in only 26% and 4% of volunteers, respectively, following the primary vaccination, immunogenicity following the second dose improved with trivalent and tetravalent neutralizing antibody responses in 25% and 63% of volunteers, respectively. This formulation was selected for further clinical evaluation in the USA, Thailand, and Puerto Rico. Recently, each of the vaccine components has been re-derived in Vero cells and a phase IIb trial in Thailand is planned.

Genetically engineered live attenuated dengue vaccines

During the last fifteen years, the development of molecular genetic techniques for the manipulation of the virus genome, and the ability to recover viruses via transfection of tissue culture cells has lead to the generation of many genetically engineered vaccine candidates. These vaccine viruses have been created by numerous methods, including: chimerization in which the structural protein coding region of a flavivirus is replaced by that from a specific DENV serotype;

introduction of attenuating point mutations into the viral genome; deletion of nucleotides from the untranslated regions (UTRs) of the genome; and exchange of the UTRs between different dengue virus serotypes.

In a vaccine strategy developed at the Laboratory of Infectious Diseases (NIAID, NIH, Bethesda, Maryland) (Blaney et al., 2006), a cDNA clone of DENV-4 Dominica/81 was used to engineer deletion mutations into the 3′-UTR of DENV-4. This conferred varying levels of attenuation in rhesus monkeys compared to the wild-type parent virus (Men et al., 1996). Of particular interest was the 3′ 172–143 deletion mutation, later referred to as Δ30, which achieved a desirable balance between level of attenuation and immunogenicity in monkeys. The DENV-4 virus containing the Δ30 mutation (DEN4Δ30) has been evaluated in flavivirus-naïve adult volunteers and was shown to be safe, asymptomatic, and immunogenic at all doses administered ($10^1 – 10^5$ pfu) (Durbin et al., 2001; Durbin et al., 2005). Clinical signs following vaccination consisted primarily of an asymptomatic rash; low-grade neutropenia and liver enzyme elevations were also observed. Viraemia, less than 10 pfu/ml serum, was present in about half of the volunteers. Vaccine virus was not transmitted from infected volunteers to mosquitoes. This lack of transmission is probably attributable to the low level of viraemia in vaccines. However the Δ30 mutation confers an additional safeguard against transmission due to its negative impact on infection and dissemination of DENV-4 in mosquitoes (Troyer et al., 2001). Seroconversion rates were consistently 95–100%. A derivative of DEN4Δ30 bearing additional mutations in the NS5 gene was shown to be further attenuated and to manifest decreased liver reactogenicity relative to its parent and evaluation in humans has shown it to be a satisfactory vaccine candidate, although slightly less immunogenic than the DEN4Δ30 parental virus (McArthur et al., 2008). The success of the DEN4Δ30 vaccine in humans launched a unique strategy to create vaccine candidates for the other three DENV serotypes: nucleotides analogous to the Δ30 mutation of DEN4Δ30 were deleted from the UTR of each of the remaining DENV serotypes. Introduction of the Δ30 mutation into DENV-1 resulted in a vaccine candidate attenuated to a level similar to that observed in monkeys for DEN4Δ30 (Whitehead *et al.*, 2003) and well tolerated and immunogenic in humans (Durbin *et al.*, 2006a), with a clinical profile very similar to that observed with DEN4Δ30. However, introduction of the Δ30 mutation into DENV-2 conferred only a modest level of attenuation in monkeys (Blaney *et al.*, 2004b), and introduction into DENV-3 failed to attenuate the resulting virus (Blaney *et al.*, 2004a). To create attenuated vaccine candidates for DENV-2 and DENV-3, a separate, chimeric strategy was developed in which the prM and E genes of DEN4Δ30 were replaced with those derived from DENV-2 or DENV-3. The Δ30 mutation has been shown to contribute to the attenuation observed for chimeric viruses DEN2/4Δ30 and DENV-3/4Δ30 (Blaney *et al.*, 2004a; Whitehead *et al.*, 2003), and it is clear that the level of attenuation associated with the background virus, in this case DEN4Δ30, is maintained in the chimeric virus without the need for additional attenuating mutations in the structural genes. Tetravalent formulations containing these chimeric viruses along with DEN1Δ30 and DEN4Δ30 have been tested in monkeys and shown to be attenuated and to elicit balanced antibody responses (Blaney *et al.*, 2005). Vaccine candidate DEN2/4Δ30 has been shown to be safe and immunogenic in adult volunteers at a dose of 10^3 pfu (Durbin *et al.*, 2006b). Results of the phase I evaluation of DEN3/4Δ30 have shown the vaccine candidate to be safe, but to have an unacceptably low level of infectivity at doses of 10^3 or 10^5 pfu (A.P. Durbin, J.H. McArthur, J.A. Marron,, K.A. Wanionek, B. Thumar, D.J. Pierro, A.C. Schmidt, B.R. Murphy, S.S. Whitehead, manuscript in preparation). Alternate DENV-3 vaccine candidates have been generated using full-length DENV-3 in which the 3′-UTR contains two deletion mutations (Δ30/31), or in which the entire 3′-UTR has been exchanged with that derived from DEN4Δ30 (Blaney *et al.*, 2008). Phase I evaluation of these new vaccine candidates has been initiated. Once a suitable DENV-3 component is identified, a tetravalent formulation consisting of that component along with DEN1Δ30, DEN2/4Δ30, and DEN4Δ30 will be evaluated in human volunteers. The NIAID vaccine candidates, which have been licensed to the Butantan Foundation (Brazil),

Biological E. Ltd. (India), Panacea Biotec Ltd. (India), and VaBiotech (Vietnam), should be economical to produce since the DENV-1, -2, and -4 components can be given at 10^3 pfu.

Similar to the strategy described above, a live attenuated chimeric DENV vaccine has been developed utilizing the yellow fever 17D vaccine virus as the genetic background. The ChimeriVax platform, developed by Acambis, Inc. has been used to create chimeric vaccine candidates in which the prM and E proteins of the yellow fever 17D virus are replaced with those derived from each DENV serotype (Guirakhoo *et al.*, 2001; Guirakhoo *et al.*, 2000). The prM and E genes used in these chimeric viruses were from virus strains isolated from dengue cases. Although the ChimeriVax-DEN2 vaccine virus component appeared to be immunodominant in early tetravalent formulations, modification of the doses of the individual chimeric viruses in the tetravalent formulation were able to overcome this problem in monkeys (Guirakhoo *et al.*, 2004; Guirakhoo *et al.*, 2000). All 24 monkeys inoculated with one of four different tetravalent ChimeriVax formulations became viraemic, with mean peak titres ranging from 2.1 $\log_{10}$ to 2.6 $\log_{10}$. These titres were similar to those induced by the yellow fever 17D vaccine virus and lower than those induced by wild-type DENV. Neutralizing antibody titres induced by the tetravalent formulation in monkeys were similar to those induced by the monovalent ChimeriVax-DEN viruses, and the vaccine was highly protective against challenge (Guirakhoo *et al.*, 2004). In the only published phase I trial in humans, the ChimeriVax-DEN2 vaccine was found to be safe and highly immunogenic at both low dose (10^3 pfu) and high dose (10^5 pfu) in both yellow fever immune and non-immune individuals (Guirakhoo *et al.*, 2006). The most frequent side effects associated with the vaccine were headache, myalgia, and fatigue. Although a higher percentage of ChimeriVax-DEN2 recipients were viraemic than those who received yellow fever vaccine, the peak level of viraemia induced by the vaccine was lower than that induced by the yellow fever 17D vaccine. ChimeriVax-DEN2 was highly immunogenic in all vaccinees, all of whom seroconverted to DENV-2 and remained seropositive at 6 and 12 months post vaccination. Cross-reactivity to the other DENV serotypes was minimal in YF-naïve individuals but did increase in those with pre-existing antibody to YF. This technology has been licensed by Sanofi Pasteur and the ChimeriVax-DEN tetravalent formulation at a dose of 5 $\log_{10}$pfu of each component is currently being evaluated as a three-dose vaccine (0, 3–4, and 12 months) in phase II studies in the USA, Mexico, and the Philippines.

A third chimeric strategy has been developed by the US Centers for Disease Control and Prevention (CDC) and is based on the Mahidol University DENV-2 PDK-53 vaccine candidate (Huang et al., 2000; Huang et al., 2003). The major genetic determinants of attenuation of the DENV-2 PDK-53 virus are located in the non-structural and untranslated regions of the genome (Butrapet et al., 2000) and the E protein is wild-type with the exception of a single amino acid substitution. Chimeric viruses were created in which the prM and E genes of DENV-2 PDK-53 were replaced with those derived from wild-type DENV-1, -3, or -4. These viruses were shown to be immunogenic and protective in the AG129 mouse model (Huang et al., 2003). In 2006, these vaccine candidates were licensed to InViragen Inc., and further evaluation of the tetravalent formulation in non-human primates is ongoing and a phase I clinical trial is planned.

Inactivated dengue vaccines

Inactivated or killed virus vaccines have been used successfully to control disease caused by flaviviruses such as Japanese encephalitis virus and tick-borne encephalitis virus. It is likely that a similar approach would work for dengue virus. Inactivated dengue vaccines would have three theoretical advantages over live attenuated dengue vaccines: they would not revert to more virulent viruses because they do not replicate, the immune response should be balanced in a tetravalent formulation, and they can be given to persons who may be immunocompromised. Although this vaccine approach may be possible, it is probably not tenable since the cost of an effective vaccine may be prohibitive. Inactivated vaccines are generally more expensive than live vaccines to produce and require multiple doses. Unfortunately, the cost issue is compounded for dengue virus due to the lack of robust virus yields

compared to other flaviviruses, the need for four vaccine components, and the limited financial resources available for vaccination in dengue endemic regions.

The earliest attempts to develop inactivated DENV vaccines were initiated over 80 years ago (Blanc and Caminopetros, 1929; Simmons et al., 1931). Subcutaneous vaccination of volunteers with saline suspensions of either filtrates of infected mosquitoes or dried blood which had been previously collected from dengue-infected patients failed to protect volunteers from subsequent DENV challenge (Simmons et al., 1931). Blanc and Caminopetros reported that DENV in blood was killed when mixed with one-fifteenth volume of ox bile (Blanc and Caminopetros, 1929). However, volunteers inoculated with this mixture were not protected against DENV challenge, probably due the relatively low concentration of antigen and the fact the only a single dose was given.

A purified, formalin-inactivated DENV-2 vaccine (PIV) developed by WRAIR was immunogenic in animal models (Putnak et al., 1996a; Putnak et al., 1996b; Putnak et al., 2005). This vaccine was prepared from the DENV-2 strain S16803 and propagated in Vero cells. The virus was concentrated by ultracentrifugation and purified over a sucrose gradient resulting in an approximately 10 to 20-fold virus concentration. The virus was then inactivated with formalin and mixed with aluminium hydroxide or other adjuvants. The PIV induced neutralizing antibody responses in all macaques after two doses of vaccine and was highly protective against challenge, most notably in those macaques that received higher doses of the PIV or PIV given with one of several adjuvants (Putnak et al., 1996a; Putnak et al., 2005).

Virus-vectored DENV vaccines

Recombinant poxviruses and adenoviruses expressing foreign proteins have been demonstrated to induce strong humoral and cellular responses in humans against various pathogens (Catanzaro et al., 2006; Liniger et al., 2007; Moss, 1996). These viruses infect cells, express their proteins de novo within the cell, and then the processed, glycosylated antigens can associate with the cell membranes or be secreted into the medium. In addition, because of intracellular translation and processing of the gene products, MHC class I dependent immune responses can be induced. Despite these properties seen with other vectored antigens, early studies of recombinant vaccinia viruses (VV) expressing the structural proteins of DENV-2 or DENV-4 were disappointing (Bray *et al.*, 1989; Deubel *et al.*, 1988). These constructs expressed prM and a full-length E protein but failed to induce neutralizing antibody and failed to protect monkeys from wild-type challenge. Since recombinant VV vectors expressing full-length E protein have been shown to express only intracellular E protein, VV expressing C-terminal truncated E proteins lacking the hydrophobic membrane anchor region were constructed. The truncated E protein expressed by VV was secreted from the cell and was more immunogenic and efficacious in mice and in rhesus macaques than VV expressing full length E (Men *et al.*, 1991; Men *et al.*, 2000). In an updated approach, the replication-deficient modified vaccinia Ankara (MVA) virus was used as a vector to express the C-terminal truncated E protein of DENV-2 (Men *et al.*, 2000). Two doses of this vector conferred relatively low levels of neutralizing antibody in monkeys and only partial protection upon virus challenge, while three doses resulted in higher levels of neutralizing antibody and complete protection following challenge. Construction of MVA vectors expressing E protein of the other three dengue serotypes has not been reported.

Because of their safety and low pathogenicity in humans, replication-defective adenoviruses vectors have been used to express the prM and E protein of DENV-2 (Jaiswal *et al.*, 2003). Mice were immunized at 0, 1, and 2 months with 10^7 pfu of the recombinant adenovirus (rAd). After the third injection, neutralizing antibody titres induced by the rAd against DENV-2 were comparable to those induced by a purified inactivated DENV-2 vaccine. Because of the ability of the adenovirus genome to accommodate lengthy insertions, two bivalent rAd vectors have been developed that expresses the prM and E proteins of DENV-1 and DENV-2 [cAdVaxD(1–2)] or the prM and E of DENV-3 and DENV-4 [cAdVaxD(3–4)] (Holman *et al.*, 2007; Raja *et al.*, 2007). In mice, each of the bivalent vectors induced neutralizing antibody to the respective

DENV serotypes and a cellular immune response was detected in vaccinated groups 4 – 10 weeks following primary vaccination (Holman *et al.*, 2007). Subsequently, the bivalent vectors were co-administered to rhesus monkeys as a two dose tetravalent vaccine and induced significant protection against challenge with wild-type DENV (Raviprakash *et al.*, 2008). However, the usefulness of such vectors in humans immune to adenovirus remains to be demonstrated.

Venezuelan equine encephalitis virus (VEE) has been utilized as a vector for a DENV-1 vaccine (Chen *et al.*, 2007) and DENV-2 vaccine (White *et al.*, 2007). VEE expressing non-propagating virus replicon particles (VRP) from DENV-2 prM and E genes was shown to be immunogenic and protective in weanling mice and was able to induce neutralizing antibody even in the presence of maternally derived DENV-2 antibody. VEE expressing replicon particles from DENV-1 prM and E were evaluated in cynomolgus macaques. The DENV-1 VRP was given at 0, 1, and 4 months and was compared with a DENV-1 DNA vaccine. The DENV-1 VRP induced comparable neutralizing antibody titres to three doses of the naked DNA vaccine in immunized macaques and both vaccines were partially protective against wild-type DENV-1 challenge (Chen *et al.*, 2007). In a prime-boost regimen, the DENV-1 DNA vaccine was given at times 0 and 1 month, followed by the DENV-1 VRP at 4 months. This regimen induced complete protection against DENV-1 challenge in all vaccinated macaques more than four months after the last vaccination (Chen *et al.*, 2007).

An innovative strategy has been used to produce subviral particles in cells infected with virus containing C-deleted genomes produced by a trans-encapsidating cell system (Widman *et al.*, 2008). Because the infecting virus particle lacks a C gene, it cannot produce infectious progeny, but still encodes immunogenic subviral particles. The system has been used successfully to produce a West Nile virus vaccine candidate, RepliVax WN, capable of protecting mice and hamsters. Replacement of the prM and E genes of RepliVax WN with those derived from DENV-2 and encapsidation in WNV C-expressing cells was only partially successful and required the accumulation of growth adaptation mutations (Suzuki *et al.*, 2008). Although immunized mice were only partially protected after a single inoculation, the study served as a proof of concept for future work.

Recombinant subunit protein vaccines

Recombinant subunit protein vaccines are being evaluated as alternative vaccine strategies to avoid some of these difficulties associated with live attenuated and inactivated vaccines. Recombinant DENV proteins can be expressed in baculovirus, yeast, *Escherichia coli*, vaccinia virus, insect cells, or mammalian cells, and then purified for use as non-replicating subunit vaccines. Research has shown that several important factors are essential for the successful expression of useful antigens. The E protein contains the major antigenic epitopes of DENV and requires a properly folded conformation to elicit neutralizing antibodies. This generally requires the co-expression of prM (Allison *et al.*, 1995; Guirakhoo *et al.*, 1992), since full-length E protein, without the co-expression of prM, is not secreted, thereby failing to induce neutralizing antibody (Feighny *et al.*, 1994; Fonseca *et al.*, 1994). For this reason, E protein co-expressed with prM has been most extensively studied in subunit protein vaccines, and the baculovirus expression system has been most widely utilized for the expression of these proteins (Bielefeldt-Ohmann *et al.*, 1997; Delenda *et al.*, 1994a; Delenda *et al.*, 1994b; Deubel *et al.*, 1991; Eckels *et al.*, 1994; Feighny *et al.*, 1994; Kelly *et al.*, 2000; Putnak *et al.*, 1991; Staropoli *et al.*, 1997). Co-expression of prM and E induces the formation of viral-like particles (VLPs) that are secreted into the medium, and the VLPs expressed from baculovirus, yeast, or mammalian cells are quite immunogenic, inducing both neutralizing antibody and partial or full protection from wild-type DENV challenge in mice (Kelly *et al.*, 2000; Konishi and Fujii, 2002; Sugrue *et al.*, 1997). An alternative strategy to improve secretion and solubility of the E protein, and consequently its immunogenicity, is the removal of the carboxy-terminal end of the E protein containing the membrane anchor region. (Delenda *et al.*, 1994a; Delenda *et al.*, 1994b; Deubel *et al.*, 1991; Men *et al.*, 1991; Staropoli *et al.*, 1996). Replacement of the 100 carboxy-terminal amino acids of the E protein with six histidine residues also provided a convenient method of purification via

metal affinity chromatography (Staropoli *et al.*, 1997). This vaccine preparation, when given with aluminium hydroxide as an adjuvant, induced neutralizing antibody and was highly protective in mice.

A recombinant subunit DENV-2 E protein developed by Hawaii Biotech, Inc., comprising the N-terminal 80% of E and prM (r80E) expressed in *Drosophila* cells, was evaluated in rhesus macaques (Putnak *et al.*, 2005). The r80E was administered with one of five different adjuvants on days 0 and 30. Control groups included monkeys immunized with saline, purified inactivated DENV-2 given with adjuvants, or a single dose of a live attenuated DENV-2 vaccine candidate. Although the live attenuated vaccine candidate provided the best overall protective efficacy against wild-type challenge, the subunit protein vaccines given with adjuvant induced robust neutralizing antibody titres and provided a high degree of protection against wild-type challenge. It should also be noted that vaccination with r80E also primed for a much more robust immune response following challenge with wild-type DENV-2 compared to monkeys receiving live attenuated vaccine or saline.

In another study using truncated DENV-4 E protein expressed in *Pichia pastoris*, cynomolgus macaques immunized with four doses of antigen mixed with aluminium hydroxide (Guzman *et al.*, 2003) developed barely detectable ELISA and haemagglutination inhibition antibody; at the time of virus challenge neutralizing antibody titres were only 1:30. Although partial protection against wild-type DENV-2 challenge viraemia was demonstrated, the researchers considered the results insufficient to recommend the subunit preparation as a vaccine candidate. Moreover they suggested that conformational differences between recombinant and native E protein, antigen presentation, and immunization schedule could have influenced the lack of full protection.

Numerous other strategies for the expression of subunit antigens have been used. Recombinant fusion proteins comprised of the DENV-2 E protein fused to either the maltose binding protein (MBP) of *E. coli*, the Staphylococcal A protein, or the meningococcal P64K protein were expressed in *E. coli* (Hermida *et al.*, 2004; Simmons *et al.*, 1998; Srivastava *et al.*, 1995). These fusion proteins induced high levels of neutralizing antibody and protection against DENV challenge in mice. When the MBP was cleaved from the E protein, immunogenicity of the protein was greatly decreased. A hybrid protein consisting of the carboxy-truncated DENV-2 E fused in frame with the hepatitis B surface antigen was expressed in *Pichia pastoris*. Although the construct formed stable VLPs and induced antibody, the resulting antibodies were not capable of neutralizing DENV-2 (Bisht *et al.*, 2002). A recombinant hybrid DENV-2/DENV-3 protein comprised of the N-terminal two-thirds of DENV-2 and truncated carboxy-terminal one-third of DENV-3 was expressed in the baculovirus system (Bielefeldt-Ohmann *et al.*, 1997). Although the antigen induced a strong cross-reactive T-cell response and virus-specific antibodies to both DENV-2 and DENV-3, neutralizing antibody was specific only for DENV-2.

Many studies in mice have demonstrated the ability of E proteins to induce neutralizing antibody and to fully or partially protect against challenge with wild-type DENV (Delenda *et al.*, 1994a; Men *et al.*, 1991; Putnak *et al.*, 1991; Simmons *et al.*, 1998; Srivastava *et al.*, 1995). Despite generally promising results in mice, the subunit protein vaccines described above were inferior to live DENV vaccines when tested in non-human primates, providing only partial protection against challenge with wild-type DENV (Eckels *et al.*, 1994; Guzman *et al.*, 2003; Velzing *et al.*, 1999). However, in all of the reported macaque studies, there was no evidence of enhanced challenge virus replication, even in the presence of minimally protective antibody levels, confirming the safety advantage of this vaccination strategy. To date, a DENV subunit vaccine has not been tested in humans, although Hawaii Biotech has begun a phase I evaluation of an E protein subunit vaccine for West Nile virus.

DNA vaccines

A significant amount of effort has been spent on the generation of DNA vaccines for flaviviruses (Putnak *et al.*, 2003), with some progress in the last 5 years towards the development of a DENV DNA vaccine. Overall, DNA vaccines are thought to have several advantages over more traditional methods (Whalen, 1996). DNA is stable for long periods of time and is resistant to extremes of temperature, overcoming cold-chain restrictions. Because the proteins produced by DNA vaccines are translated and processed within the host cell, they are able to induce class I MHC-dependent immune responses. In addition, DNA vaccines may cause less reactogenicity than live vaccines (Rhodes *et al.*, 1994) and may be useful to vaccinate against multiple pathogens in a single vaccination (Gurunathan *et al.*, 2000). Nevertheless, the DNA vaccine approach in general carries its own theoretical risks (Klinman *et al.*, 1997), which include the possibility of nucleic acid integration into the recipient's genome with the potential to activate oncogenes or inactivate tumour suppressor genes and the possibility of inducing anti-DNA antibodies in the host leading to autoimmune disease. Neither concern has been validated in an experimental model (Martin *et al.*, 1999; Parker *et al.*, 1999). In addition, DNA vaccines may also induce immune responses in tissues not normally infected by the pathogen.

Preclinical evaluation of the first DENV DNA vaccine candidate expressing the prM and E genes was published in 1997 using the plasmid expression vector VR1 012 containing the prM gene and 92% of the E gene of DENV-2 (Kochel *et al.*, 1997). Neutralizing antibody was induced in all vaccinated mice and a higher survival rate was observed in vaccinated mice following challenge with wild-type DENV-2 virus (Kochel *et al.*, 1997; Porter *et al.*, 1998). Neutralizing antibody titres in mice were significantly increased when the DNA construct was engineered such that the expressed DENV-2 E protein was fused to lysosome-associated membrane protein (LAMP) to enhance trafficking of the protein to the lysosome for improved MHC class II expression (Raviprakash *et al.*, 2001). Four different DENV-1 DNA constructs were evaluated in mice, and it was determined that the plasmid expressing the prM and full-length DENV-1 E protein (DENV-1ME) elicited the highest antibody response (Raviprakash *et al.*, 2000a). DENV-1ME was further studied in rhesus macaques and *Aotus* monkeys to evaluate its protective efficacy against challenge (Kochel *et al.*, 2000; Raviprakash *et al.*, 2000b). In these studies, the DENV-1ME vaccine only partially protected monkeys against challenge with wild-type DENV-1. Studies were then performed in *Aotus* monkeys to evaluate whether the co-administration of DENV-1ME with plasmids expressing multiple copies of human immunostimulatory sequences (ISS) or *Aotus* granulocyte macrophage colony-stimulating factor (GM-CSF) could improve the immunogenicity and protective efficacy of the vaccine (Raviprakash *et al.*, 2003). Vaccination with DENV-1ME along with expression of ISS and GM-CSF led to complete protection against DENV-1 challenge in over 85% of monkeys. A DENV-3 DNA vaccine was recently tested in *Aotus*, but only some of the monkeys were partially protected against DENV-3 challenge (Blair *et al.*, 2006). Prime-boost strategies using combinations of DENV DNA vaccines, subunit vaccines, and inactivated vaccines have been evaluated in rhesus monkeys and have proven to be only minimally protective (Simmons *et al.*, 2006). The DENV-1ME DNA vaccine is currently being evaluated in healthy human adult volunteers in the US.

Konishi and colleagues successfully immunized mice with a tetravalent dengue DNA vaccine consisting of four plasmids expressing the prM and E of each DENV serotype (Konishi *et al.*, 2006). The tetravalent DNA vaccine induced comparable antibody titres to those induced by the individual plasmids when given as monovalent vaccines (Konishi *et al.*, 2006). Colleagues at Maxygen Inc., and the NRMC have utilized DNA 'shuffling', where DNA sequence for E protein epitopes from all four DENV serotypes are combined randomly to create a panel of novel chimeric antigens which are screened for their ability to induce multivalent T-cell and neutralizing antibody responses to all four DENV serotypes. (Apt *et al.*, 2006; Raviprakash *et al.*, 2006). Although the constructs evaluated were capable of inducing neutralizing antibody against all four serotypes, vaccinated macaques were not protected against DENV-2 challenge (Raviprakash *et al.*, 2006). The authors surmise that the DENV-2 neutralizing antibody titres induced by the DNA shuffle vaccines were not sufficient to protect against DENV-2 infection and that strategies to enhance this immune response should be evaluated.

The protective efficacy of a DNA vaccine expressing the non-structural protein NS1 has been evaluated in mouse models (Costa *et al.*,

2007; Wu *et al.*, 2003). Because NS1 antibody is non-neutralizing and does not bind to the virus, it is proposed that an NS1 vaccine would eliminate the risk of antibody dependent enhancement in vaccine recipients, even as antibody titres waned over time. Mice immunized with the NS1 vaccine were partially protected from wild-type DENV challenge (Costa *et al.*, 2007; Wu *et al.*, 2003), and protection was enhanced by administration of a plasmid expressing IL-12 (Wu *et al.*, 2003).

Obstacles and challenges to vaccine development

By the end of 2009, over 25 unique DENV vaccine candidates will have been tested in clinical trials during the previous decade, yet a licensed vaccine is still not available. Although progress is continual, and many vaccine candidates have progressed to phase II studies, obstacles remain and progress is slower than hoped. Almost all vaccine candidates tested to date have been live attenuated virus. This is due in part to the success of the live attenuated yellow fever vaccine and the initial expectation that such an approach would require a single dose, as in the case of yellow fever vaccine, and would elicit the same type of immune response achieved following natural infection. However, while natural immunity to dengue virus is eventually very effective, it is acquired by the sequential infection of dengue virus serotypes in a manner that transiently increases the risk for subsequent severe dengue disease. Following infection by two or more DENV serotypes, an individual is generally protected against severe dengue disease (Gibbons *et al.*, 2007). Because vaccination cannot proceed safely in an analogous sequential manner, immunity to all four dengue virus serotypes must be elicited simultaneously by a tetravalent vaccine, and herein lies the greatest obstacle. Currently studied dengue vaccines are actually tetravalent, not unlike the multi-component live vaccines for poliovirus or rotavirus. Unfortunately, the tetravalent requirement of a DENV vaccine increases the development burden more than just four-fold when one considers that combination of vaccine components can have unexpected and undesirable results. Thus, the multi-component nature of DENV live vaccines demands that special attention be given to developing each of the four components to achieve a balanced level of infectivity and replication in the human host, thereby inducing a balanced and protective immune response. Unfortunately, infectivity and immunogenicity in non-human primates models has not always clearly predicted the outcome of human trials. In the absence of other animal models for dengue virus, phase I and II clinical trials become pivotal in assessing the suitability of vaccine components and formulations.

At this time, an immune marker that correlates with protection has not been firmly established, although it is clear from passively acquired immunity in infants that humoral immunity, presumably neutralizing antibody, is sufficient for protection (Kliks et al., 1988; Pengsaa et al., 2006), and this has guided most pre-clinical and clinical studies to date. It is hoped that as vaccines move into phase III evaluation, data will be available to determine the quantitative relationship between pre-infection neutralizing antibody level and response to natural challenge with wild type dengue virus. Based on the experience of the ongoing clinical trials using tetravalent formulations of live vaccine candidates, and the assumption that neutralizing antibody levels are a surrogate marker of efficacy, it is likely that a single vaccine dose will be insufficient to elicit protective immunity to all four DENV serotypes. Thus booster immunizations may be required. The challenge then becomes: when and how to boost? Protection against all four DENV is sought in the shortest time interval. However, data in monkeys (Blaney et al., 2005) and in humans have indicated that it is difficult to boost the DENV immune response by a second dose of live vaccine in an interval shorter than 4–6 months, and that vaccinees may not be fully immune until after a third dose given at 12 months. It may be possible to devise prime-boost strategies using other types of vaccines to shorten this interval, but the number of required vaccine components quickly becomes unwieldy from a development perspective, and vaccination compliance may be difficult to maintain. Nevertheless, in special circumstances, such as vaccination of travellers or military personnel, a rapid prime-boost strategy may be warranted. Once again, carefully designed phase I and II studies become essential to establish immune correlates of protection and to optimize immunization schedules.

Live attenuated DENV vaccines have been most extensively evaluated in clinical trials and are furthest along in the development pathway. The insufficient potency of dengue DNA and subunit protein vaccines in non-human primate models is problematic, as is the concern that pre-existing and acquired immunity to poxvirus and adeno-virus vectors may diminish their effectiveness as vaccines. Novel adjuvants and prime-boost strategies have had some success in improving the performance of these candidates and should be further evaluated (Chen et al., 2007; Imoto and Konishi, 2007; Putnak et al., 2005; Wu et al., 2003; Yang et al., 2003). It is anticipated that clinical trials evaluating novel recombinant subunit protein, DNA, and vectored vaccines will be initiated in the coming years, either alone or as part of a prime-boost strategy. In addition, clinical trials of several new live attenuated DENV vaccine candidates are planned.

The control of dengue disease by vaccination may necessitate the use of different types of vaccines depending on the vaccination goal and the disease setting. In endemic regions, routine immunization against dengue is targeted to infants and young children 1 – 3 years of age and should be coordinated with current childhood immunization schedules. As discussed above, this type of vaccine will need to inexpensive. Alternatively, there is a need to protect travellers, seasonal labourers, and military personnel who visit or work in dengue endemic regions. Vaccination in this case will need to be rapid, with closely spaced booster doses if needed, and may be more tolerant to increased cost.

As in all vaccine development programmes, ongoing support for manufacture of clinical trial materials and clinical evaluation are critical. Since its formal organization in 2001 (Almond et al., 2002), the Pediatric Dengue Vaccine Initiative (www.pdvi.org) has been instrumental in moving dengue vaccine development forward through its establishment of partnerships, funding of research and development, and establishment of several field trial sites in endemic regions. In addition, the World Health Organization has recently published Guidelines for the Clinical Evaluation of Dengue Vaccines in Endemic Areas (Edelman and Hombach, 2008) and sponsors numerous workshops and consultations designed to address many issues of DENV vaccine development. Activities by these organizations have created a positive momentum during the last decade.

Despite the numerous obstacles in its path, the development of safe and effective vaccines to protect against dengue is pressing forward at a remarkable rate and represents contributions from the pharmaceutical industry, private sector, non-profit organizations, academia, and government. Innovative and novel approaches to overcoming the challenges of vaccine development and implementation are forthcoming, and there is a general optimism that this long-overdue and urgently needed vaccine will be become a reality.

Acknowledgements

Preparation of this manuscript was supported in part by the Intramural Research Program of the NIH, National Institute of Allergy and Infectious Diseases. The authors wish to thank Dr Brian Murphy for critical review of this manuscript.

References

Allison, S.L., Stadler, K., Mandl, C.W., Kunz, C., and Heinz, F.X. (1995). Synthesis and secretion of recombinant tick-borne encephalitis virus protein E in soluble and particulate form. J. Virol. 69, 5816–5820.

Almond, J., Clemens, J., Engers, H., Halstead, S., Khiem, H., Pablos-Mendez, A., Pervikov, Y., and Tram, T. (2002). Accelerating the development and introduction of a dengue vaccine for poor children, 5–8 December 2001, Ho Chi Minh City, VietNam. Vaccine 20, 3043.

Apt, D., Raviprakash, K., Brinkman, A., Semyonov, A., Yang, S., Skinner, C., Diehl, L., Lyons, R., Porter, K., and Punnonen, J. (2006). Tetravalent neutralizing antibody response against four dengue serotypes by a single chimeric dengue envelope antigen. Vaccine 24, 335–344.

Bancroft, W.H., Scott, R.M., Eckels, K.H., Hoke, C.H., Jr., Simms, T.E., Jesrani, K.D., Summers, P.L., Dubois, D.R., Tsoulos, D., and Russell, P.K. (1984). Dengue virus type 2 vaccine: reactogenicity and immunogenicity in soldiers. J. Infect. Dis. 149, 1005–1010.

Bancroft, W.H., Top, F.H., Jr., Eckels, K.H., Anderson, J.H., Jr., McCown, J.M., and Russell, P.K. (1981). Dengue-2 vaccine: virological, immunological, and clinical responses of six yellow fever-immune recipients. Infect. Immun. 31, 698–703.

Bhamarapravati, N., and Yoksan, S. (1989). Study of bivalent dengue vaccine in volunteers. Lancet 1, 1077.

Bhamarapravati, N., and Yoksan, S. (1997). Live attenuated tetravalent dengue vaccine. In Dengue and dengue hemorrhagic fever, D.J. Gubler, and G. Kuno, eds. (New York, CAB International), pp. 367–377.

Bhamarapravati, N., and Yoksan, S. (2000). Live attenuated tetravalent dengue vaccine. Vaccine 18 Suppl. 2, 44–47.

Bhamarapravati, N., Yoksan, S., Chayaniyayothin, T., Angsubphakorn, S., and Bunyaratvej, A. (1987). Immunization with a live attenuated dengue-2-virus candidate vaccine (16681-PDK 53): clinical, immunological and biological responses in adult volunteers. Bull World Health Organ 65, 189–195.

Bielefeldt-Ohmann, H., Beasley, D.W., Fitzpatrick, D.R., and Aaskov, J.G. (1997). Analysis of a recombinant dengue-2 virus-dengue-3 virus hybrid envelope protein expressed in a secretory baculovirus system. J. Gen. Virol. 78, 2723–2733.

Bisht, H., Chugh, D.A., Raje, M., Swaminathan, S.S., and Khanna, N. (2002). Recombinant dengue virus type 2 envelope/hepatitis B surface antigen hybrid protein expressed in – can function as a bivalent immunogen. J. Biotechnol. 99, 97–110.

Blair, P.J., Kochel, T.J., Raviprakash, K., Guevara, C., Salazar, M., Wu, S.J., Olson, J.G., and Porter, K.R. (2006). Evaluation of immunity and protective efficacy of a dengue-3 premembrane and envelope DNA vaccine in Aotus nancymae monkeys. Vaccine 24, 1427–1432.

Blanc, G., and Caminopetros, J. (1929). Bull. Acad. Med. 102, 47–40.

Blaney, J.E., Jr., Durbin, A.P., Murphy, B.R., and Whitehead, S.S. (2006). Development of a live attenuated dengue virus vaccine using reverse genetics. Viral Immunol. 19, 10–32.

Blaney, J.E., Jr., Hanson, C.T., Firestone, C.Y., Hanley, K.A., Murphy, B.R., and Whitehead, S.S. (2004a). Genetically modified, live attenuated dengue virus type 3 vaccine candidates. Am. J. Trop. Med. Hyg. 71, 811–821.

Blaney, J.E., Jr., Hanson, C.T., Hanley, K.A., Murphy, B.R., and Whitehead, S.S. (2004b). Vaccine candidates derived from a novel infectious cDNA clone of an American genotype dengue virus type 2. BMC Infect. Dis. 4, 39.

Blaney, J.E., Jr., Matro, J.M., Murphy, B.R., and Whitehead, S.S. (2005). Recombinant, live-attenuated tetravalent dengue virus vaccine formulations induce a balanced, broad, and protective neutralizing antibody response against each of the four serotypes in rhesus monkeys. J. Virol. 79, 5516–5528.

Blaney, J.E., Jr., Sathe, N.S., Goddard, L., Hanson, C.T., Romero, T.A., Hanley, K.A., Murphy, B.R., and Whitehead, S.S. (2008). Dengue virus type 3 vaccine candidates generated by introduction of deletions in the 3' untranslated region (3'-UTR) or by exchange of the DENV-3 3'-UTR with that of DENV-4. Vaccine 26, 817–828.

Bray, M., Zhao, B.T., Markoff, L., Eckels, K.H., Chanock, R.M., and Lai, C.J. (1989). Mice immunized with recombinant vaccinia virus expressing dengue 4 virus structural proteins with or without nonstructural protein NS1 are protected against fatal dengue virus encephalitis. J. Virol. 63, 2853–2856.

Butrapet, S., Huang, C.Y., Pierro, D.J., Bhamarapravati, N., Gubler, D.J., and Kinney, R.M. (2000). Attenuation markers of a candidate dengue type 2 vaccine virus, strain 16681 (PDK-53), are defined by mutations in the 5′ noncoding region and nonstructural proteins 1 and 3. J. Virol. *74*, 3011–3019.

Catanzaro, A.T., Koup, R.A., Roederer, M., Bailer, R.T., Enama, M.E., Moodie, Z., Gu, L., Martin, J.E., Novik, L., Chakrabarti, B.K., Butman, B.T., Gall, J.G., King, C.R., Andrews, C.A., Sheets, R., Gomez, P.L., Mascola, J.R., Nabel, G.J., and Graham, B.S. (2006). Phase 1 safety and immunogenicity evaluation of a multiclade HIV-1 candidate vaccine delivered by a replication-defective recombinant adenovirus vector. J. Infect. Dis. *194*, 1638–1649.

Chen, L., Ewing, D., Subramanian, H., Block, K., Rayner, J., Alterson, K.D., Sedegah, M., Hayes, C., Porter, K., and Raviprakash, K. (2007). A heterologous DNA prime-Venezuelan equine encephalitis virus replicon particle boost dengue vaccine regimen affords complete protection from virus challenge in cynomolgus macaques. J. Virol. *81*, 11634–11639.

Costa, S.M., Azevedo, A.S., Paes, M.V., Sarges, F.S., Freire, M.S., and Alves, A.M. (2007). DNA vaccines against dengue virus based on the ns1 gene: the influence of different signal sequences on the protein expression and its correlation to the immune response elicited in mice. Virology *358*, 413–423.

Delenda, C., Frenkiel, M.P., and Deubel, V. (1994a). Protective efficacy in mice of a secreted form of recombinant dengue-2 virus envelope protein produced in baculovirus infected insect cells. Arch Virol *139*, 197–207.

Delenda, C., Staropoli, I., Frenkiel, M.P., Cabanie, L., and Deubel, V. (1994b). Analysis of C-terminally truncated dengue 2 and dengue 3 virus envelope glycoproteins: Processing in insect cells and immunogenic properties in mice. J. Gen. Virol. *75*, 1569–1578.

Deubel, V., Bordier, M., Megret, F., Gentry, M.K., Schlesinger, J.J., and Girard, M. (1991). Processing, secretion, and immunoreactivity of carboxy terminally truncated dengue-2 virus envelope proteins expressed in insect cells by recombinant baculoviruses. Virology *180*, 442–447.

Deubel, V., Kinney, R.M., Esposito, J.J., Cropp, C.B., Vorndam, A.V., Monath, T.P., and Trent, D.W. (1988). Dengue 2 virus envelope protein expressed by a recombinant vaccinia virus fails to protect monkeys against dengue. J. Gen. Virol. *69*, 1921–1929.

Durbin, A.P., Karron, R.A., Sun, W., Vaughn, D.W., Reynolds, M.J., Perreault, J.R., Thumar, B., Men, R., Lai, C.J., Elkins, W.R., Chanock, R.M., Murphy, B.R., and Whitehead, S.S. (2001). Attenuation and immunogenicity in humans of a live dengue virus type-4 vaccine candidate with a 30 nucleotide deletion in its 3′-untranslated region. Am. J. Trop. Med. Hyg. *65*, 405–413.

Durbin, A.P., McArthur, J., Marron, J.A., Blaney, J.E., Jr., Thumar, B., Wanionek, K., Murphy, B.R., and Whitehead, S.S. (2006a). The live attenuated dengue serotype 1 vaccine rDEN1Delta30 is safe and highly immunogenic in healthy adult volunteers. Hum. Vaccin. *2*, 167–173.

Durbin, A.P., McArthur, J.H., Marron, J.A., Blaney, J.E., Thumar, B., Wanionek, K., Murphy, B.R., and Whitehead, S.S. (2006b). rDEN2/4Delta30(ME), a

live attenuated chimeric dengue serotype 2 vaccine is safe and highly immunogenic in healthy dengue-naive adults. Hum. Vaccin. *2*, 255–260.

Durbin, A.P., Whitehead, S.S., McArthur, J., Perreault, J.R., Blaney, J.E., Jr., Thumar, B., Murphy, B.R., and Karron, R.A. (2005). rDEN4 Delta 30, a Live attenuated dengue virus type 4 vaccine candidate, is safe, immunogenic, and highly infectious in healthy adult volunteers. J. Infect. Dis. *191*, 710–718.

Eckels, K.H., Dubois, D.R., Summers, P.L., Schlesinger, J.J., Shelly, M., Cohen, S., Zhang, Y.M., Lai, C.J., Kurane, I., Rothman, A., Hasty, S., and Howard, B. (1994). Immunization of monkeys with baculovirus-dengue type-4 recombinants containing envelope and nonstructural proteins: Evidence of priming and partial protection. Am. J. Trop. Med. Hyg. *50*, 472–478.

Eckels, K.H., Scott, R.M., Bancroft, W.H., Brown, J., Dubois, D.R., Summers, P.L., Russell, P.K., and Halstead, S.B. (1984). Selection of attenuated dengue 4 viruses by serial passage in primary kidney cells. V. Human response to immunization with a candidate vaccine prepared in fetal rhesus lung cells. Am. J. Trop. Med. Hyg. *33*, 684–689.

Edelman, R., and Hombach, J. (2008). 'Guidelines for the clinical evaluation of dengue vaccines in endemic areas': summary of a World Health Organization Technical Consultation. Vaccine *26*, 4113–4119.

Edelman, R., Tacket, C.O., Wasserman, S.S., Vaughn, D.W., Eckels, K.H., Dubois, D.R., Summers, P.L., and Hoke, C.H., Jr. (1994). A live attenuated dengue-1 vaccine candidate (45AZ5) passaged in primary dog kidney cell culture is attenuated and immunogenic for humans. J. Infect. Dis. *170*, 1448–1455.

Edelman, R., Wasserman, S.S., Bodison, S.A., Putnak, R.J., Eckels, K.H., Tang, D., Kanesa-Thasan, N., Vaughn, D.W., Innis, B.L., and Sun, W. (2003). Phase I trial of 16 formulations of a tetravalent live-attenuated dengue vaccine. Am. J. Trop. Med. Hyg. *69*, 48–60.

Feighny, R., Burrous, J., and Putnak, R. (1994). Dengue type-2 virus envelope protein made using recombinant baculovirus protects mice against virus challenge. Am. J. Trop. Med. Hyg. *50*, 322–328.

Fonseca, B.A.L., Pincus, S., Shope, R.E., Paoletti, E., and Mason, P.W. (1994). Recombinant vaccinia viruses co-expressing dengue-1 glycoproteins prM and E induce neutralizing antibodies in mice. Vaccine *12*, 279–285.

Fujita, N., Oda, K., Yasui, Y., and Hotta, S. (1969). Research on dengue in tissue culture: I.V. Serologic responses of human beings to combined inoculations of attenuated, tissue-cultured type 1 dengue virus and yellow fever vaccine. Kobe J. Med. Sci. *15*, 163–180.

Gibbons, R.V., Kalanarooj, S., Jarman, R.G., Nisalak, A., Vaughn, D.W., Endy, T.P., Mammen, M.P., Jr., and Srikiatkhachorn, A. (2007). Analysis of repeat hospital admissions for dengue to estimate the frequency of third or fourth dengue infections resulting in admissions and dengue hemorrhagic fever, and serotype sequences. Am. J. Trop. Med. Hyg. *77*, 910–913.

Guirakhoo, F., Arroyo, J., Pugachev, K.V., Miller, C., Zhang, Z.X., Weltzin, R., Georgakopoulos, K., Catalan, J., Ocran, S., Soike, K., Ratterree, M., and Monath, T.P. (2001). Construction, safety, and immunogenicity in nonhuman primates of a chimeric yellow fever-dengue virus tetravalent vaccine. J. Virol. 75, 7290–7304.

Guirakhoo, F., Bolin, R.A., and Roehrig, J.T. (1992). The Murray Valley encephalitis virus prM protein confers acid resistance to virus particles and alters the expression of epitopes within the R2 domain of E glycoprotein. Virology 191, 921–931.

Guirakhoo, F., Kitchener, S., Morrison, D., Forrat, R., McCarthy, K., Nichols, R., Yoksan, S., Duan, X., Ermak, T.H., Kanesa-Thasan, N., Bedford, P., Lang, J., Quentin-Millet, M.J., and Monath, T.P. (2006). Live attenuated chimeric yellow fever dengue type 2 (ChimeriVax-DEN2) vaccine: Phase I clinical trial for safety and immunogenicity: effect of yellow fever preimmunity in induction of cross neutralizing antibody responses to all 4 dengue serotypes. Hum. Vaccin. 2, 60–67.

Guirakhoo, F., Pugachev, K., Zhang, Z., Myers, G., Levenbook, I., Draper, K., Lang, J., Ocran, S., Mitchell, F., Parsons, M., Brown, N., Brandler, S., Fournier, C., Barrere, B., Rizvi, F., Travassos, A., Nichols, R., Trent, D., and Monath, T. (2004). Safety and efficacy of chimeric yellow fever-dengue virus tetravalent vaccine formulations in nonhuman primates. J. Virol. 78, 4761–4775.

Guirakhoo, F., Weltzin, R., Chambers, T.J., Zhang, Z., Soike, K., Ratterree, M., Arroyo, J., Georgakopoulos, K., Catalan, J., and Monath, T.P. (2000). Recombinant chimeric yellow fever-dengue type 2 virus is immunogenic and protective in nonhuman primates. J. Virol. 74, 5477–5485.

Gurunathan, S., Wu, C.Y., Freidag, B.L., and Seder, R.A. (2000). DNA vaccines: a key for inducing long-term cellular immunity. Curr. Opin. Immunol. 12, 442–447.

Guzman, M.G., Rodriguez, R., Hermida, L., Alvarez, M., Lazo, L., Mune, M., Rosario, D., Valdes, K., Vazquez, S., Martinez, R., Serrano, T., Paez, J., Espinosa, R., Pumariega, T., and Guillen, G. (2003). Induction of neutralizing antibodies and partial protection from viral challenge in Macaca fascicularis immunized with recombinant dengue 4 virus envelope glycoprotein expressed in Pichia pastoris. Am. J. Trop. Med. Hyg. 69, 129–134.

Halstead, S.B., Suaya, J.A., and Shepard, D.S. (2007). The burden of dengue infection. Lancet 369, 1410–1411.

Hermida, L., Rodri, guez, R., Lazo, L., Silva, R., Zulueta, A., Chinea, G., Lopez, C., Guzman, M.G., and Guillen, G. (2004). A dengue-2 Envelope fragment inserted within the structure of the P64k meningococcal protein carrier enables a functional immune response against the virus in mice. J. Virol. Methods 115, 41–49.

Holman, D.H., Wang, D., Raviprakash, K., Raja, N.U., Luo, M., Zhang, J., Porter, K.R., and Dong, J.Y. (2007). Two complex, adenovirus-based vaccines that together induce immune responses to all four dengue virus serotypes. Clin. Vaccin. Immunol. 14, 182–189.

Hotta, S. (1952). Experimental studies on dengue. I. Isolation, identification and modification of the virus. J. Infect. Dis. 90, 1–9.

Huang, C.Y., Butrapet, S., Pierro, D.J., Chang, G.J., Hunt, A.R., Bhamarapravati, N., Gubler, D.J., and Kinney, R.M. (2000). Chimeric dengue type 2 (vaccine strain PDK-53)/dengue type 1 virus as a potential candidate dengue type 1 virus vaccine. J. Virol. 74, 3020–3028.

Huang, C.Y., Butrapet, S., Tsuchiya, K.R., Bhamarapravati, N., Gubler, D.J., and Kinney, R.M. (2003). Dengue 2 PDK-53 virus as a chimeric carrier for tetravalent dengue vaccine development. J. Virol. 77, 11436–11447.

Imoto, J., and Konishi, E. (2007). Dengue tetravalent DNA vaccine increases its immunogenicity in mice when mixed with a dengue type 2 subunit vaccine or an inactivated Japanese encephalitis vaccine. Vaccine 25, 1076–1084.

Innis, B.L., Eckels, K.H., Kraiselburd, E., Dubois, D.R., Meadors, G.F., Gubler, D.J., Burke, D.S., and Bancroft, W.H. (1988). Virulence of a live dengue virus vaccine candidate: A possible new marker of dengue virus attenuation. J. Infect. Dis. 158, 876–880.

Jaiswal, S., Khanna, N., and Swaminathan, S. (2003). Replication-defective adenoviral vaccine vector for the induction of immune responses to dengue virus type 2. J. Virol. 77, 12907–12913.

Kanesa-Thasan, N., Edelman, R., Tacket, C.O., Wasserman, S.S., Vaughn, D.W., Coster, T.S., Kim-Ahn, G.J., Dubois, D.R., Putnak, J.R., King, A., Summers, P.L., Innis, B.L., Eckels, K.H., and Hoke, C.H. Jr.(2003). Phase 1 studies of Walter Reed Army Institute of Research candidate attenuated dengue vaccines: selection of safe and immunogenic monovalent vaccines. Am. J. Trop. Med. Hyg. 69, 17–23.

Kanesa-thasan, N., Sun, W., Kim-Ahn, G., Van Albert, S., Putnak, J.R., King, A., Raengsakulrach, B., Christ-Schmidt, H., Gilson, K., Zahradnik, J.M., Vaughn, D.W., Innis, B.L., Saluzzo, J.F., and Hoke, C.H. Jr (2001). Safety and immunogenicity of attenuated dengue virus vaccines (Aventis Pasteur) in human volunteers. Vaccine 19, 3179–3188.

Kelly, E.P., Greene, J.J., King, A.D., and Innis, B.L. (2000). Purified dengue 2 virus envelope glycoprotein aggregates produced by baculovirus are immunogenic in mice. Vaccine 18, 2549–2559.

Kitchener, S., Nissen, M., Nasveld, P., Forrat, R., Yoksan, S., Lang, J., and Saluzzo, J.F. (2006). Immunogenicity and safety of two live-attenuated tetravalent dengue vaccine formulations in healthy Australian adults. Vaccine 24, 1238–1241.

Kliks, S.C., Nimmannitya, S., Nisalak, A., and Burke, D.S. (1988). Evidence that maternal dengue antibodies are important in the development of dengue hemorrhagic fever in infants. Am. J. Trop. Med. Hyg. 38, 411–419.

Klinman, D.M., Takeno, M., Ichino, M., Gu, M., Yamshchikov, G., Mor, G., and Conover, J. (1997). DNA vaccines: safety and efficacy issues. Springer Semin. Immunopathol. 19, 245–256.

Kochel, T., Wu, S.J., Raviprakash, K., Hobart, P., Hoffman, S., Porter, K., and Hayes, C. (1997). Inoculation of plasmids expressing the dengue-2 envelope gene elicit neutralizing antibodies in mice. Vaccine 15, 547–552.

Kochel, T.J., Raviprakash, K., Hayes, C.G., Watts, D.M., Russell, K.L., Gozalo, A.S., Phillips, I.A., Ewing, D.F., Murphy, G.S., and Porter, K.R. (2000). A dengue virus serotype-1 DNA vaccine induces virus neutralizing antibodies and provides protection from viral challenge in Aotus monkeys. Vaccine 18, 3166–3173.

Konishi, E., and Fujii, A. (2002). Dengue type 2 virus subviral extracellular particles produced by a stably transfected mammalian cell line and their evaluation for a subunit vaccine. Vaccine 20, 1058–1067.

Konishi, E., Kosugi, S., and Imoto, J. (2006). Dengue tetravalent DNA vaccine inducing neutralizing antibody and anamnestic responses to four serotypes in mice. Vaccine 24, 2200–2207.

Liniger, M., Zuniga, A., and Naim, H.Y. (2007). Use of viral vectors for the development of vaccines. Expert review of vaccines 6, 255–266.

Martin, T., Parker, S.E., Hedstrom, R., Le, T., Hoffman, S.L., Norman, J., Hobart, P., and Lew, D. (1999). Plasmid DNA malaria vaccine: the potential for genomic integration after intramuscular injection. Hum Gene Ther 10, 759–768.

McArthur, J.H., Durbin, A.P., Marron, J.A., Wanionek, K.A., Thumar, B., Pierro, D.J., Schmidt, A.C., Blaney, J.E., Jr., Murphy, B.R., and Whitehead, S.S. (2008). Phase I clinical evaluation of rDEN4Delta30–200,201: a live attenuated dengue 4 vaccine candidate designed for decreased hepatotoxicity. Am. J. Trop. Med. Hyg. 79, 678–684.

McKee, K.T., Jr., Bancroft, W.H., Eckels, K.H., Redfield, R.R., Summers, P.L., and Russell, P.K. (1987). Lack of attenuation of a candidate dengue 1 vaccine (45AZ5) in human volunteers. Am. J. Trop. Med. Hyg. 36, 435–442.

Men, R., Bray, M., Clark, D., Chanock, R.M., and Lai, C.J. (1996). Dengue type 4 virus mutants containing deletions in the 3′ noncoding region of the RNA genome: analysis of growth restriction in cell culture and altered viremia pattern and immunogenicity in rhesus monkeys. J. Virol. 70, 3930–3937.

Men, R., Bray, M., and Lai, C.J. (1991). Carboxy-terminally truncated dengue virus envelope glycoproteins expresssed on the cell surface and secreted extracellularly exhibit increased immunogenicity in mice. J. Virol. 65, 1400–1407.

Men, R., Wyatt, L., Tokimatsu, I., Arakaki, S., Shameem, G., Elkins, R., Chanock, R., Moss, B., and Lai, C.J. (2000). Immunization of rhesus monkeys with a recombinant of modified vaccinia virus Ankara expressing a truncated envelope glycoprotein of dengue type 2 virus induced resistance to dengue type 2 virus challenge. Vaccine 18, 3113–3122.

Moss, B. (1996). Genetically engineered poxviruses for recombinant gene expression, vaccination, and safety. Proc. Natl. Acad. Sci. U.S.A. 93, 11341–11348.

Parker, S.E., Borellini, F., Wenk, M.L., Hobart, P., Hoffman, S.L., Hedstrom, R., Le, T., and Norman, J.A. (1999). Plasmid DNA malaria vaccine: tissue distribution and safety studies in mice and rabbits. Hum. Gene Ther. 10, 741–758.

Pengsaa, K., Luxemburger, C., Sabcharoen, A., Limkittikul, K., Yoksan, S., Chambonneau, L., Chaovarind, U., Sirivichayakul, C., Lapphra, K., Chanthavanich, P., and Lang, J. (2006). Dengue virus infections in the first 2 years of life and the kinetics of transplacentally transferred dengue neutralizing antibodies in thai children. J. Infect. Dis. 194, 1570–1576.

Porter, K.R., Kochel, T.J., Wu, S.J., Raviprakash, K., Phillips, I., and Hayes, C.G. (1998). Protective efficacy of a dengue 2 DNA vaccine in mice and the effect of CpG immuno-stimulatory motifs on antibody responses. Arch. Virol. 143, 997–1003.

Putnak, R., Barvir, D.A., Burrous, J.M., Dubois, D.R., D'Andrea, V.M., Hoke, C.H., Jr., Sadoff, J.C., and Eckels, K.H. (1996a). Development of a purified, inactivated, dengue-2 virus vaccine prototype in Vero cells: immunogenicity and protection in mice and rhesus monkeys. J. Infect. Dis. 174, 1176–1184.

Putnak, R., Cassidy, K., Conforti, N., Lee, R., Sollazzo, D., Truong, T., Ing, E., Dubois, D., Sparkuhl, J., Gastle, W., and Hoke, C.H. Jr (1996b). Immunogenic and protective response in mice immunized with a purified, inactivated, Dengue-2 virus vaccine prototype made in fetal rhesus lung cells. Am. J. Trop. Med. Hyg. 55, 504–510.

Putnak, R., Coller, B.A., Voss, G., Vaughn, D.W., Clements, D., Peters, I., Bignami, G., Houng, H.S., Chen, R.C., Barvir, D.A., Seriwatana, J., Cayphas, S., Garcon, N., Gheysen, D., Kanesa-Thasan, N., McDonell, M., Humphreys, T., Eckels, K.H., Prieels, J.P., and Innis, B.L. (2005). An evaluation of dengue type-2 inactivated, recombinant subunit, and live-attenuated vaccine candidates in the rhesus macaque model. Vaccine 23, 4442–4452.

Putnak, R., Feighny, R., Burrous, J., Cochran, M., Hackett, C., Smith, G., and Hoke, C.H., Jr. (1991). Dengue-1 virus envelope glycoprotein gene expressed in recombinant baculovirus elicits virus-neutralizing antibody in mice and protects them from virus challenge. Am. J. Trop. Med. Hyg. 45, 159–167.

Putnak, R., Porter, K., and Schmaljohn, C. (2003). DNA vaccines for flaviviruses. Adv. Virus Res. 61, 445–468.

Raja, N.U., Holman, D.H., Wang, D., Raviprakash, K., Juompan, L.Y., Deitz, S.B., Luo, M., Zhang, J., Porter, K.R., and Dong, J.Y. (2007). Induction of bivalent immune responses by expression of dengue virus type 1 and type 2 antigens from a single complex adenoviral vector. Am. J. Trop. Med. Hyg. 76, 743–751.

Raviprakash, K., Apt, D., Brinkman, A., Skinner, C., Yang, S., Dawes, G., Ewing, D., Wu, S.J., Bass, S., Punnonen, J., and Porter, K. (2006). A chimeric tetravalent dengue DNA vaccine elicits neutralizing antibody to all four virus serotypes in rhesus macaques. Virology 353, 166–173.

Raviprakash, K., Ewing, D., Simmons, M., Porter, K.R., Jones, T.R., Hayes, C.G., Stout, R., and Murphy, G.S. (2003). Needle-free Biojector injection of a dengue virus type 1 DNA vaccine with human immunostimulatory sequences and the GM-CSF gene increases immunogenicity and protection from virus challenge in Aotus monkeys. Virology 315, 345–352.

Raviprakash, K., Kochel, T.J., Ewing, D., Simmons, M., Phillips, I., Hayes, C.G., and Porter, K.R. (2000a). Immunogenicity of dengue virus type 1 DNA vaccines expressing truncated and full length envelope protein. Vaccine 18, 2426–2434.

Raviprakash, K., Marques, E., Ewing, D., Lu, Y., Phillips, I., Porter, K.R., Kochel, T.J., August, T.J., Hayes, C.G., and Murphy, G.S. (2001). Synergistic neutralizing antibody response to a dengue virus type 2 DNA vaccine by incorporation of lysosome-associated membrane protein sequences and use of plasmid expressing GM-C.S.F. Virology 290, 74–82.

Raviprakash, K., Porter, K.R., Kochel, T.J., Ewing, D., Simmons, M., Phillips, I., Murphy, G.S., Weiss, W.R., and Hayes, C.G. (2000b). Dengue virus type 1 DNA vaccine induces protective immune responses in rhesus macaques. J. Gen. Virol. 81 Pt 7, 1659–1667.

Raviprakash, K., Wang, D., Ewing, D., Holman, D.H., Block, K., Woraratanadharm, J., Chen, L., Hayes, C., Dong, J.Y., and Porter, K. (2008). A tetravalent dengue vaccine based on a complex adenovirus vector provides significant protection in rhesus monkeys against all four serotypes of dengue virus. J. Virol. 82, 6927–6934.

Rhodes, G.H., Abai, A.M., Margalith, M., Kuwahara-Rundell, A., Morrow, J., Parker, S.E., and Dwarki, V.J. (1994). Characterization of humoral immunity after DNA injection. Dev. Biol. Standard. 82, 229–236.

Sabchareon, A., Lang, J., Chanthavanich, P., Yoksan, S., Forrat, R., Attanath, P., Sirivichayakul, C., Pengs aa, K., Pojjaroen-Anant, C., Chambonneau, L., Saluzzo, J.F., and Bhamarapravati, N. (2004). Safety and immunogenicity of a three dose regimen of two tetravalent live-attenuated dengue vaccines in five- to twelve-year-old Thai children. Pediatr. Infect. Dis. J. 23, 99–109.

Sabchareon, A., Lang, J., Chanthavanich, P., Yoksan, S., Forrat, R., Attanath, P., Sirivichayakul, C., Pengs aa, K., Pojjaroen-Anant, C., Chokejindachai, W., Jagsudee, A., Saluzzo, J.F., and Bhamarapravati, N. (2002). Safety and immunogenicity of tetravalent live-attenuated dengue vaccines in Thai adult volunteers: role of serotype concentration, ratio, and multiple doses. Am. J. Trop. Med. Hyg. 66, 264–272.

Sabin, A.B. (1952). Research on dengue during World War I.I. Am. J. Trop. Med. Hyg. 1, 30–50.

Sabin, A.B., and Schlesinger, R.W. (1945). Production of immunity to dengue with virus modified by propagation in mice. Science 101, 640–642.

Sanchez, V., Gimenez, S., Tomlinson, B., Chan, P.K., Thomas, G.N., Forrat, R., Chambonneau, L., Deauvieau, F., Lang, J., and Guy, B. (2006). Innate and adaptive cellular immunity in flavivirus-naive human recipients of a live-attenuated dengue serotype 3 vaccine produced in Vero cells (VDV3). Vaccine 24, 4914–4926.

Schlesinger, R.W., Gordon, I., Frankel, J.W., Winter, J.W., Patterson, P.R., and Dorrance, W.R. (1956). Clinical and serologic response of man to immunization with attnuated dengue and yellow fever viruses. J. Immunol. 77, 352–364.

Scott, R.M., Eckels, K.H., Bancroft, W.H., Summers, P.L., McCown, J.M., Anderson, J.H., and Russell, P.K. (1983). Dengue 2 vaccine: dose response in volunteers in relation to yellow fever immune status. J. Infect. Dis. 148, 1055–1060.

Shepard, D.S., Suaya, J.A., Halstead, S.B., Nathan, M.B., Gubler, D.J., Mahoney, R.T., Wang, D.N., and Meltzer, M.I. (2004). Cost-effectiveness of a pediatric dengue vaccine. Vaccine 22, 1275–1280.

Simasathien, S., Thomas, S.J., Watanaveeradej, V., Nisalak, A., Barberousse, C., Innis, B.L., Sun, W., Putnak, J.R., Eckels, K.H., Hutagalung, Y., Gibbons, R.V., Zhang,

C., De La Barrera, R., Jarman, R.G., Chawachalasai, W., and Mammen, M.P. Jr (2008). Safety and Immunogenicity of a Tetravalent Live-attenuated Dengue Vaccine in Flavivirus Naive Children. Am. J. Trop. Med. Hyg. 78, 426–433.

Simmons, J.S., St John, J.H., and Reynolds, F.H.K. (1931). Experimental studies of dengue. Philippine J. Sci. 44, 1–252.

Simmons, M., Nelson, W.M., Wu, S.J., and Hayes, C.G. (1998). Evaluation of the protective efficacy of a recombinant dengue envelope B domain fusion protein against dengue 2 virus infection in mice. Am. J. Trop. Med. Hyg. 58, 655–662.

Simmons, M., Porter, K.R., Hayes, C.G., Vaughn, D.W., and Putnak, R. (2006). Characterization of antibody responses to combinations of a dengue virus type 2 DNA vaccine and two dengue virus type 2 protein vaccines in rhesus macaques. J. Virol. 80, 9577–9585.

Srivastava, A.K., Putnak, J.R., Warren, R.L., and Hoke, C.H., Jr. (1995). Mice immunized with a dengue type 2 virus E and NS1 fusion protein made in Escherichia coli are protected against lethal dengue virus infection. Vaccine 13, 1251–1258.

Staropoli, I., Clement, J.M., Frenkiel, M.P., Hofnung, M., and Deubel, V. (1996). Dengue virus envelope glycoprotein can be secreted from insect cells as a fusion with the maltose-binding protein. J. Virol. Methods 56, 179–189.

Staropoli, I., Frenkiel, M.P., Megret, F., and Deubel, V. (1997). Affinity-purified dengue-2 virus envelope glycoprotein induces neutralizing antibodies and protective immunity in mice. Vaccine 15, 1946–1954.

Sugrue, R.J., Fu, J., Howe, J., and Chan, Y.C. (1997). Expression of the dengue virus structural proteins in Pichia pastoris leads to the generation of virus-like particles. J. Gen. Virol. 78, 1861–1866.

Sun, W., Cunningham, D., Wasserman, S.S., Perry, J., Putnak, J.R., Eckels, K.H., Vaughn, D.W., Thomas, S.J., Kanesa-Thasan, N., Innis, B.L., and Edelman, R. (2008). Phase 2 clinical trial of three formulations of tetravalent live-attenuated dengue vaccine in flavivirus-naive adults. Hum. Vaccin. 5.

Sun, W., Edelman, R., Kanesa-Thasan, N., Eckels, K.H., Putnak, J.R., King, A.D., Houng, H.S., Tang, D., Scherer, J.M., Hoke, C.H. Jr., and Innis, B.L. (2003). Vaccination of human volunteers with monovalent and tetravalent live-attenuated dengue vaccine candidates. Am. J. Trop. Med. Hyg. 69, 24–31.

Suzuki, R., Winkelmann, E.R., and Mason, P.W. (2009). Construction and characterization of a single-cycle chimeric flavivirus vaccine candidate that protects mice against lethal challenge with dengue virus type 2. J. Virol. 83, 1870–1880.

Troyer, J.M., Hanley, K.A., Whitehead, S.S., Strickman, D., Karron, R.A., Durbin, A.P., and Murphy, B.R. (2001). A live attenuated recombinant dengue-4 virus vaccine candidate with restricted capacity for dissemination in mosquitoes and lack of transmission from vaccinees to mosquitoes. Am. J. Trop. Med. Hyg. 65, 414–419.

Vaughn, D.W., Green, S., Kalayanarooj, S., Innis, B.L., Nimmannitya, S., Suntayakorn, S., Endy, T.P., Raengsakulrach, B., Rothman, A.L., Ennis, F.A., and

Nisalak, A. (2000). Dengue viremia titer, antibody response pattern, and virus serotype correlate with disease severity. J. Infect. Dis. *181*, 2–9.

Vaughn, D.W., Hoke, C.H., Jr., Yoksan, S., LaChance, R., Innis, B.L., Rice, R.M., and Bhamarapravati, N. (1996). Testing of a dengue 2 live-attenuated vaccine (strain 16681 PDK 53) in ten American volunteers. Vaccine *14*, 329–336.

Velzing, J., Groen, J., Drouet, M.T., van Amerongen, G., Copra, C., Osterhaus, A.D., and Deubel, V. (1999). Induction of protective immunity against Dengue virus type 2: comparison of candidate live attenuated and recombinant vaccines. Vaccine *17*, 1312–1320.

Whalen, R.G. (1996). DNA vaccines for emerging infectious diseases: what if? Emerg. Infect. Dis. *2*, 168–175.

White, L.J., Parsons, M.M., Whitmore, A.C., Williams, B.M., de Silva, A., and Johnston, R.E. (2007). An immunogenic and protective alphavirus replicon particle-based dengue vaccine overcomes maternal antibody interference in weanling mice. J. Virol. *81*, 10329–10339.

Whitehead, S.S., Blaney, J.E., Durbin, A.P., and Murphy, B.R. (2007). Prospects for a dengue virus vaccine. Nat. Rev. *5*, 518–528.

Whitehead, S.S., Falgout, B., Hanley, K.A., Blaney Jr, J.E., Jr., Markoff, L., and Murphy, B.R. (2003). A live, attenuated dengue virus type 1 vaccine candidate with a 30-nucleotide deletion in the 3′ untranslated region is highly attenuated and immunogenic in monkeys. J. Virol. *77*, 1653–1657.

Widman, D.G., Ishikawa, T., Fayzulin, R., Bourne, N., and Mason, P.W. (2008). Construction and characterization of a second-generation pseudoinfectious West Nile virus vaccine propagated using a new cultivation system. Vaccine *26*, 2762–2771.

Wisseman, C.L., Jr., Sweet, B.H., Rosenzweig, E.C., and Eylar, O.R. (1963). Attenuated living type 1 dengue vaccines. Am. J. Trop. Med. Hyg. *12*, 620–623.

Wu, S.F., Liao, C.L., Lin, Y.L., Yeh, C.T., Chen, L.K., Huang, Y.F., Chou, H.Y., Huang, J.L., Shaio, M.F., and Sytwu, H.K. (2003). Evaluation of protective efficacy and immune mechanisms of using a non-structural protein NS1 in DNA vaccine against dengue 2 virus in mice. Vaccine *21*, 3919–3929.

Yang, Z.Y., Wyatt, L.S., Kong, W.P., Moodie, Z., Moss, B., and Nabel, G.J. (2003). Overcoming immunity to a viral vaccine by DNA priming before vector boosting. J. Virol. *77*, 799–803.

Novel Therapeutic Approaches for Dengue Disease

13

Mayuri, Elisa La Bauve and Richard J. Kuhn

Abstract

Dengue has emerged as the most common mosquito-borne viral disease of humans in the past three decades. There are no available vaccines or antivirals against DENV. Currently, vector control is the only method for prevention of the disease. Development of a successful vaccine would require for it to be effective against all four DENV serotypes, economical, and provide long-term protection. Antivirals directed against one or more stages of the virus life cycle are likely to form an important part of dengue disease therapeutics. The strategies that have been used in the past towards development of an effective antiviral against dengue, as well as those being employed currently are discussed in light of information from structural biology, computational biology and molecular virology, highlighting the potential opportunities and obstacles to their use.

Introduction

The virus and the disease

In recent years dengue has become a major international public health concern, and the World Health Organization (WHO) has declared dengue the most common mosquito-borne viral disease of humans. About 2.5 billion people (including 1 billion children) reside in the tropical and subtropical areas of the world where dengue virus (DENV) can be transmitted and are therefore exposed to infection (Gubler, 2002a). More than 30 years ago the disease was limited to mainly five South-East Asian countries, but the virus is now endemic in over 100 countries including South and Central America (Whitehead et al., 2007; WHO, 1997). Transmitted primarily by the *Aedes aegypti* mosquito, there are four distinct, but closely related viruses (DENV 1–4) that cause dengue. Unplanned urbanization, lack of mosquito control, population growth in urban centres of tropical countries, and increased movement of viruses through infected humans due to modern transportation have all contributed to the marked increase in epidemic occurrence of the disease (Gubler, 2006). Due to lack of available therapy for the disease, control of the disease is limited to efforts at reducing and eradicating mosquito vectors, which has proved to be expensive and ineffective (Gubler, 1998; Halstead, 2002).

Each of the four DENV serotypes is capable of causing all of the disease signs and symptoms ranging from the mild dengue fever (DF) to dengue haemorrhagic fever (DHF) or dengue shock syndrome (DSS). Dengue fever is a usually mild but debilitating viral fever that seldom causes death. The clinical features of dengue fever vary according to the age of the patient. Infants and young children may have a fever with rash. Older children and adults may have either a mild fever or the incapacitating disease with abrupt onset and high fever, severe headache, arthralgia and rash. Dengue haemorrhagic fever (DHF) and dengue shock syndrome (DSS) are the more severe complications of the disease that can be potentially fatal and are characterized by high fever, often with enlargement of the liver, and in severe cases circulatory failure and haemorrhagic manifestations (Gubler, 1998; WHO, 1997). DHF has been classified into four grades according to the severity of shock and bleeding (WHO, 1993). There are approximately 50–100 million cases of dengue fever, 500,000 cases of DHF/DSS and more than 20,000 deaths each year as a result of DENV infection (Gubler, 1998, 2002b; Halstead, 1999; Kurane and Takasaki, 2001; WHO, 1997).

It has been suggested that DHF/DSS arise in a small percentage of the patients suffering from DF as a result of antibody-dependent enhancement or ADE. The presence of neutralizing antibodies from a primary DENV infection, during subsequent infection with DENV of a different serotype has been suggested to contribute to ADE (Halstead, 1988, 1990; Halstead and O'Rourke, 1977; Kliks et al., 1989; Morens and Halstead, 1990). This is because, while the neutralizing antibodies are effective in long-term protection against the first serotype, they are incapable of providing life-long protection against the other DENV serotypes and enhance the disease severity (Guzman et al., 2007; Vaughn et al., 2000). It has been suggested that binding of the antibodies produced in response to the first serotype with the virus of a different serotype allows the Fc portion of the antibodies to recognize and bind to the Fc-γ receptors on macrophages and monocytes. This would thereby increase uptake of the virus by these cells, leading to increased viral replication, virus release, viraemia, and increased disease intensity (Brown et al., 2006; Green and Rothman, 2006; Gubler, 1998; Mady et al., 1993). Another hypothetical scenario is that cells with bound virus-antibody complexes release cytokines that suppress immune response and inflammation and lead to down-regulation of the host antiviral response. This would allow increased virus replication and virus production eventually leading to the severe forms of DHF/DSS (Suhrbier and La Linn, 2003). While secondary infection with a distinct DENV serotype from that causing the primary infection has been suggested to increase the risk for DHF, it has also been found to occur in patients that have not been previously infected, suggesting that ADE might not be solely responsible for the more severe forms of the disease. Factors such as age, sex, race, nutritional status and genetic predisposition are believed to play a significant role in disease outcome (Bravo et al., 1987; Halstead and Simasthien, 1970; Kabra et al., 1999). In addition, viral factors like genotypic differences between strains have also been shown to affect the severity of the disease and these differences were found to occur in both the structural and non-structural coding regions of the viral genome as well as the 5' and the 3' non-translated regions (NTRs) (Leitmeyer et al., 1999; Mangada and Igarashi, 1998; Pandey and Igarashi, 2000; Watts et al., 1999). Therefore, the diverse manifestations of infection with DENV could be arising from an even more diverse and complex interplay of factors. All these factors that contribute to the disease need to be considered for the development of an effective, safe, economical therapeutic approach that also provides life-long immunity against dengue disease.

Treatment

There is no specific treatment for dengue fever. A dengue vaccine has not yet been licensed, but research on several dengue vaccines for human use is under way. Currently vector control is the available method for dengue and DHF prevention and control. There are several factors that make vaccine development against dengue disease (both mild and severe forms) a difficult undertaking, especially in the case of live-attenuated vaccines. Firstly, any of the four DENV serotypes can elicit all the known symptoms of the clinical disease (Whitehead et al., 2007) and there is no cross-protection between the four serologically distinct DENV serotypes. Therefore, there is the possibility of ADE by monotypic antibody leading to DHF/DSS with subsequent natural infections (Halstead, 2003). Secondly, the possibility of recombination of attenuated strains with wild-type virus and reversion to the virulent strain requires that vaccine strategies be carefully evaluated to address these risks (Rawlinson et al., 2006; Twiddy et al., 2002). The control of dengue will be possible only after an efficient tetravalent vaccine has been developed that is safe for use in children 9–12 months of age, especially in dengue endemic countries, is economical, and provides long-term immunity (Gubler, 2006). Progress in vaccine development has also been slow because these viruses grow poorly in cell culture and there is no reliable animal model. The development and clinical evaluation of dengue vaccine candidates is additionally challenging because the progression of the disease and the pathology are not well understood (WHO, 1997). It is therefore desirable to develop alternative therapeutic approaches. Insights into the DENV life cycle and structural information about several viral proteins obtained in recent years should facilitate drug design and the development of specific antivirals to combat the disease.

Antiviral therapy against flaviviruses and DENV in particular, suffered initially due to dearth of information about the virus and its life cycle. The development of efficient cell culture systems has allowed more insight into each stage of the virus life cycle – entry, replication, assembly, and maturation. Rapid progress in structural studies on both the virus as well as viral proteins has linked efforts and information from the fields of structural biology [X-ray crystallography, nuclear magnetic resonance (NMR), cryo-electron Microscopy (cryo-EM)], computational biology (virtual drug screening), medicinal chemistry (structure-based drug design and synthesis) and molecular virology (drug efficacy). The recent developments in the field of dengue antiviral therapy are discussed in detail, with an emphasis on the properties of the viral structural proteins that make them attractive targets for therapeutic approaches as well as other antiviral strategies that target different aspects of the virus life cycle.

Virus life cycle

Flaviviruses have a ~10.7 kb plus-sense genome that codes for three structural and seven non-structural proteins (Lindenbach and Rice, 2001). Upon release into cellular cytoplasm, the RNA is translated as a single polyprotein. The N-terminal one-third of this polyprotein consists of the structural proteins – capsid (C), membrane (M, which is encoded as the precursor, prM) and envelope (E). The capsid protein is involved in the packaging of the viral RNA. The RNA and C protein are surrounded by a host-derived lipid bilayer into which the E and M proteins are embedded by means of the two trans-membrane helices present in each protein. Flaviviruses enter host cells by receptor-mediated endocytosis (Lindenbach and Rice, 2001). The E protein contains the binding site for the receptor (Crill and Roehrig, 2001; Hung *et al.*, 1999). The low pH environment within the endosome triggers conformational changes in E that bring about fusion of the viral and endosomal membranes (Allison *et al.*, 1995b; Heinz and Allison, 2000; Stiasny *et al.*, 1996). This conformation of E protein (referred to as fusogenic or fusion-activated state) drives membrane fusion and results in the release of the viral genomic RNA into the cytoplasm and its subsequent translation into a single polyprotein. The signal sequences on the polyprotein allow translocation to the endoplasmic reticulum (ER) membrane (Lindenbach and Rice, 2001). The polyprotein is proteolytically cleaved by viral (NS2B-3) and cellular proteases into the individual proteins on the ER membrane, and replication occurs in vesicle packets containing non-structural proteins and cellular proteins (Fig. 13.1) (Mackenzie and Westaway, 2001).

The C protein packages the viral genome in the membrane bound vesicles and virus assembly takes place on the ER membranes driven by interactions between C, RNA, and the membrane embedded homodimer complexes of prM and E proteins (Lindenbach and Rice, 2001; Lorenz *et al.*, 2002; Mackenzie and Westaway, 2001; Zhang *et al.*, 2003b; Zhang *et al.*, 2004). This results in formation of non-infectious immature virus particles that are transported through the cellular exocytotic pathway (Fig. 13.1). During this transport, the prM–E protein complex undergoes a conformational transition, triggered by the low pH in the trans-Golgi network (TGN), that allows prM to be cleaved by furin, for the formation of the mature virion with surface exposed homodimers of E and M (Elshuber *et al.*, 2003; Kuhn *et al.*, 2002; Mukhopadhyay *et al.*, 2005; Stadler *et al.*, 1997). The role of prM protein in immature virions is to prevent premature fusion of the virion in the TGN due to induction of low pH-induced conformational changes in E (Heinz *et al.*, 1994; Stadler *et al.*, 1997; Yu *et al.*, 2008). The prM cleavage is critical for formation and release of mature virus particles; when this cleavage is prevented, immature virus particles are released from the infected cell (Elshuber *et al.*,

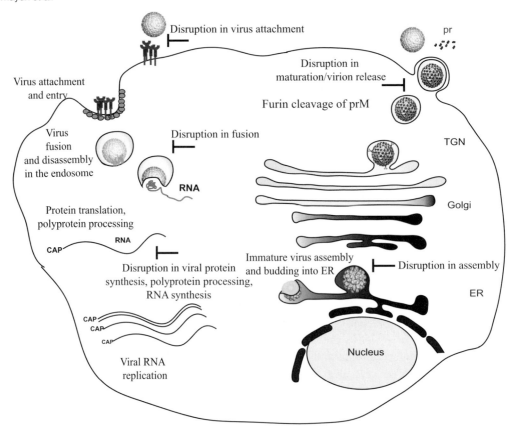

Figure 13.1 Overview of the flavivirus life cycle. Multiple structural changes occur during the life cycle, all of which have been targets for antiviral drug design. Viral entry into host cells occurs by virus attachment and subsequent entry via receptor-mediated endocytosis. Once inside the endosome, an acidic drop causes fusion of endosomal and viral membranes, resulting in release of the viral genome and subsequent virus replication and assembly. Following virus assembly in the ER, viral maturation occurs via the secretory pathway. The low pH environment of the trans-Golgi network (TGN) brings about a conformational change in the E proteins, causing them to rearrange. In the late Golgi, furin cleaves pr from M, which subsequently remains associated with the immature particle until it is released from the cell. Once released from the cell the pr dissociates from E and the particle becomes infectious. Potential points in the life cycle that can be targeted for disruption by antiviral therapy are highlighted.

2003). In the mature virus particles released from the cell, the E protein is conformationally primed to mediate low pH-triggered membrane fusion in the endosome after virus uptake.

This sequence of events highlights that structural changes in E drive the virus infection through three main conformational states: mature virus (entry), fusogenic, and immature virus (Fig. 13.2). The immature virus undergoes further transitions to produce the infectious virus. The extensive structural studies on immature and mature DENV by cryo-EM imaging and reconstruction, and X-ray crystallographic studies on E and prM proteins have allowed generation of 'pseudo-

atomic' structures of the virus (Fig. 13.3) (Kuhn *et al.*, 2002; Li *et al.*, 2008a; Modis *et al.*, 2003, 2004; Yu *et al.*, 2008; Zhang *et al.*, 2003b; Zhang *et al.*, 2007; Zhang *et al.*, 2004). Information from these studies has permitted careful delineation of the structural rearrangements that these proteins undergo during the virus life cycle. Each of these structural rearrangements or conformational changes requires significant flexibility of the E protein and its interaction with prM or M. These conformational transitions therefore present opportunities for antiviral targeting of the entry, assembly or maturation steps of the virus life cycle.

A   E Protein Ectodomain Monomer

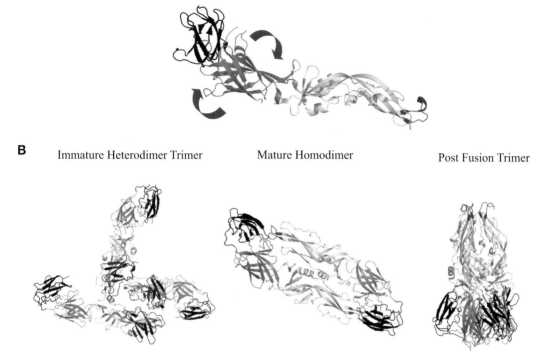

B   Immature Heterodimer Trimer   Mature Homodimer   Post Fusion Trimer

Figure 13.2 Flavivirus E protein and its oligomeric states. **A** Ribbon diagram of the E protein ectodomain. Domains I, II and III are shown in dark grey, light grey, and black, respectively. The fusion loop is the darker region at the distal tip of domain II. Hinge rotations between DII and DI and DI and DII are illustrated with block arrows. **B** Ribbon diagrams of the prM-E immature heterodimer trimer, mature homodimer and post-fusion E homotrimer. The pr peptide is highlighted in dark grey at the distal tip of domain II of the trimer and in the immature particle, protecting the fusion peptide from undergoing premature fusion during maturation.

Virus structural proteins as targets for antiviral therapy

Flavivirus E protein and early steps in virus replication and entry

The flavivirus E protein, 495 aa in length in DENV, is a class II fusion protein (Heinz and Allison, 2001; Schibli and Weissenhorn, 2004) and is structurally different from the class I viral fusion proteins for which influenza virus haemagglutinin is the prototype. There are obvious structural differences between class I and class II viral fusion proteins, but they share some common principles that are important for the fusion process. These include the synthesis of a fusion-incompetent precursor, the generation of a metastable and fusion-competent state by a proteolytic cleavage event, and an irreversible conversion to a lower-energy state during fusion (Schibli and Weissenhorn, 2004). The atomic structures of E proteins from the ectodomain (residues 1–395) of tick-borne encephalitis virus (TBEV), West Nile virus (WNV) DENV-2 and DENV-3 have been solved using X-ray crystallography (Fig. 13.2A) (Kanai et al., 2006; Modis et al., 2003, 2005; Rey et al., 1995). The ectodomain is the truncated E protein that lacks the double trans-membrane anchor region (made up of two antiparallel coiled coils) and the 'stem' region which is 50 aa in length and is arranged as two helices. The stem-anchor region is located between the lipid bilayer and the exposed ectodomain. This ectodomain consists of three domains – domain I (DI), the central eight-stranded β-barrel domain containing the N-terminus; domain II (DII), the elongated, finger-like 'dimerization' domain that is respon-

Noninfectious immature particle   Infectious mature particle

A   **B**   **C**

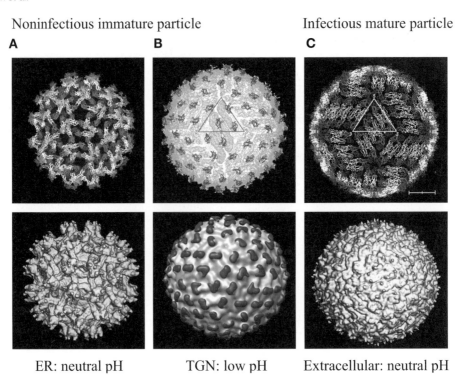

ER: neutral pH   TGN: low pH   Extracellular: neutral pH

Figure 13.3 Structures of immature and mature DENV. **A** Pseudo-atomic structure of the immature virus particle, at neutral pH (ER) conformation, with prM highlighted in light blue (upper panel) (Zhang *et al.* 2004). The three domains of E, I, II, and III are highlighted in red, yellow and blue respectively. Surface-shaded representation of immature particles at neutral pH (lower panel). **B** Pseudo-atomic structure of the immature virus particle in the low pH conformation. One icosahedral asymmetric unit and a representative E protein raft is highlighted (upper panel). Surface-shaded representation of immature particles at the trans-Golgi network – TGN low pH conformation (lower panel). prM is shaded in blue (Yu *et al.* 2008). **C** Pseudo-atomic structure of mature DENV particle at extracellular neutral pH, with each monomer shown in red, yellow and blue for domains I, II and III, respectively. One icosahedral asymmetric unit is outlined (upper panel) (Kuhn *et al.* 2002). Surface shaded electron micrograph of the mature infectious virus particle following disassociation of pr from the particle (lower panel). A colour version of this figure is located in the plate section at the back of the book.

sible for the dimeric contacts and is flanked by DI on one side; and domain III (DIII), the immunoglobulin-like domain that is suggested to be involved in receptor binding (Bhardwaj *et al.*, 2001; Rey *et al.*, 1995) and antibody neutralization (Beasley and Barrett, 2002; Crill and Chang, 2004; Crill and Roehrig, 2001; Halstead *et al.*, 2005; Kaufmann *et al.*, 2006). At the distal tip of DII is a small hydrophobic sequence, called the fusion peptide (Allison *et al.*, 2001). Domains I and II are connected by four polypeptide chains, whereas domains I and III are connected by a single polypeptide chain. The relative orientations of each of these domains during entry, fusion and maturation are characteristic for the corresponding E protein structural states (Fig. 13.2B). Also,

the motion at the hinge regions between these domains (DI-DII and DI-DIII) has been suggested to drive the local conformational changes in E that is translated into structural transformation throughout the viral surface during virus entry, fusion, and maturation.

The mature flavivirus particle has a spikeless, smooth surface and a diameter of ~500 Å, where the E and M proteins form a glycoprotein shell, almost completely covering the lipid bilayer and making it inaccessible from the exterior. The pseudo-atomic structures of the DENV, obtained by fitting the X-ray structure of E into the cryo-EM density of the mature virus (Fig. 13.3C), have shown that the E protein homodimers lie flat on the viral surface and lack the conventional $T = 3$

icosahedral symmetry (Kuhn *et al.*, 2002). These E protein homodimers consist of antiparallel arrangement of two E monomers where DII contributes to most of the contacts for dimerization and these contacts are primarily hydrophobic (Modis *et al.*, 2003, 2005; Zhang *et al.*, 2004). Cryo-EM reconstructions of mature DENV have revealed that within the glycoprotein shell, there are 180 copies of the E protein arranged as 90 homodimers to make 30 'rafts', each composed of three E homodimers oriented in parallel to give rise to a herringbone pattern (Kuhn *et al.*, 2002; Mukhopadhyay *et al.*, 2003). The E protein DIII is the only domain that protrudes from the relatively smooth surface, which possibly contributes to its role in receptor binding. The fusion loop at the distal end of DII is protected in a cavity between DI and DIII in the mature virus (Fig. 13.2B). In mature and immature virus structures, the E proteins differ by an angle of 27° between DI and DII, indicating that the flexibility in this hinge region may contribute significantly to the different conformational states of the virus. Compared to the relative positions in the cryo-EM density of the virus particle, the hinge angle between E protein domains I and II differs by 21° in the X-ray structure. This angular difference has been suggested to allow the E protein to structurally adapt to the viral surface, and position the protein in a metastable state for transition into the fusogenic trimer form during virus entry (Stiasny *et al.*, 2001).

Virus entry into a host cell requires attachment of the virus to a cellular receptor. Several different cell surface proteins have been proposed to function as the cellular receptors for DENV. These include the glycosaminoglycan heparan sulphate, which is expressed on many cell types (Chen *et al.*, 1997; Germi *et al.*, 2002; Hilgard and Stockert, 2000; Lee *et al.*, 2006); the dendritic cell-specific intracellular adhesion molecule 3-grabbing non-integrin (DC-SIGN) which is expressed on immature dendritic cells (Navarro-Sanchez *et al.*, 2003; Tassaneetrithep *et al.*, 2003); and C-type lectin, the mannose receptor (MR), that has recently been shown to bind DENV, Japanese encephalitis virus (JEV) and TBEV (Miller *et al.*, 2008). DC-SIGN is also a mannose-specific lectin receptor, thought to interact with carbohydrate moieties on the E protein (Navarro-

Sanchez *et al.*, 2003). The glycosylated residues spatially map near DIII on the E homodimer structure (Rey *et al.*, 1995; Zhang *et al.*, 2003b; Zhang *et al.*, 2004). It has recently been suggested that MR, DC-SIGN and heparan sulphate may function as the initial attachment factors for DENV and entry may require interaction with as yet unidentified receptors (Lozach *et al.*, 2005; Miller *et al.*, 2008; Pokidysheva *et al.*, 2006). In addition, the carbohydrate moieties on the virus surface have been speculated to modulate receptor-specificity. This is because the number and position of glycosylated residues are not conserved even among different strains of the same virus (Navarro-Sanchez *et al.*, 2003). Despite disparity in receptor specificity, inhibition of virus entry is has been one of the areas where research is being focused on for development of antivirals against dengue.

Inhibition of virus entry
Antiviral peptides
For more than two decades, peptides directed against virus surface proteins and domains critical for virus entry have been researched for viruses displaying class I fusion mechanism. The best example comes from work on human immunodeficiency virus-1 (HIV-1). Peptides derived from one of the predicted helical domains of the HIV transmembrane protein gp41 was found to block virus entry and was later approved by FDA as an anti-HIV drug (Qureshi *et al.*, 1990; Wild *et al.*, 1994). These results encouraged similar strategies directed against not just HIV but also several other viruses. Antiviral peptides have been designed and tested for blocking of both DENV and WNV entry. In one such study, physiochemical algorithms were used in combination with structural information on E protein to identify potential peptides that could interfere with protein–protein interactions during virus entry (Hrobowski *et al.*, 2005). Of all the designed peptides, two peptides from WNV E DII region and one from DENV representing a pre-anchor stem region displayed effective inhibition of virus in plaque reduction assays at micromolar concentrations (29.0 µM for the DENV peptide). These peptides were speculated to inhibit a step following the virus–receptor interaction. The DENV peptide showed antiviral activity against WNV

as well, suggesting that it could be developed and refined as a broad range antiviral. In another study WNV E protein was used to screen a murine brain cDNA phage display library to identify peptides that could inhibit WNV and DENV infection in cell culture and mice. One of the peptides showed inhibitory activity towards both WNV and DENV-2. Another peptide, when tested in mice infected with WNV, reduced viraemia and fatality and was detected in the brain, suggesting that it had the ability to cross the blood–brain barrier (Bai *et al.*, 2007). These studies demonstrate the use of structural information of E protein in targeting viral entry into cells, and suggest that antiviral peptides could be a promising form of DENV therapy.

Sulphated polysaccharides

Sulphated polysaccharides have been demonstrated to possess in vitro antiviral activity against several RNA viruses (Baba *et al.*, 1988). The sulphated glycosaminoglycans, heparin and polyanion- suramin, have been found to inhibit DENV infection in Vero cells (Chen *et al.*, 1997). It was suggested that the inhibitory activity of several polyanionic compounds varied with the molecular size and level of sulphation (Chen *et al.*, 1997; Marks *et al.*, 2001). In another study, sulphated polysaccharides extracted from seaweeds were found to possess antiviral activity against DENV-2 but were not very effective against the other serotypes (Talarico *et al.*, 2005). Also, the inhibitory activity of these compounds differed based on the cell type used, such that they were ineffective in mosquito cells. Lee *et al.* tested the anti-viral activity of three heparan sulphate analogues or mimetics that differed in size, structure and sulphation content, in vitro and in vivo (Lee *et al.*, 2006). These polysaccharides, heparin, suramin, and pentosan polysulphate (PPS) were tested for inhibition of virus infectivity in BHK cells and therapeutic activity in mouse models. While in vitro, heparin showed the best inhibitory activity, this in vitro anti-flavivirus activity did not correlate with the in vivo therapeutic value. Only one compound, PI-88, led to increased survival in mice infected with DEN and other flaviviruses. This inconsistency in the reported antiviral activity of sulphated polysaccharides indicates

that they need to be extensively tested to obtain reproducible results not only in vitro but also in animal models. Another class of compounds, the polyoxotungstates, that have been less extensively studied, were shown to possess broad-spectrum antiviral activities against several RNA viruses including DENV-2 and were non-cytotoxic to Vero cells (Shigeta *et al.*, 2003). However these compounds have not been characterized in detail for effects on DENV and need to be investigated further.

Structure based targeting of virus entry

The E protein rafts consisting of E homodimers provide another basis for drug design targeting protein–protein interactions within the E dimer interface (Kuhn *et al.*, 2002; Mukhopadhyay *et al.*, 2003) (Figs. 13.2B and 13.3C). Computational methods have been employed to perform an *in silico* screen of a small molecule NIH compound library based on the available structures of DENV E protein (Zhou *et al.*, 2008). These compounds were selected based on their ability to fit into two proposed binding pockets, thought to be involved in interactions between the E monomers within the dimer. These compounds could therefore potentially inhibit viral maturation or host cell entry by blocking or destabilizing protein–protein interactions between individual E proteins in the mature DENV. Compounds targeting these binding pockets could potentially disrupt interactions between E proteins, or alternatively stabilize interactions in one conformation, preventing further downstream rearrangements. Disruptions of these types would inhibit virus entry or assembly. Therefore antivirals that target the mature virus during entry into host cells and thereby block the virus at the onset are extremely promising candidates because delivery of target compounds into the host cell during stages of fusion and maturation is significantly more challenging.

Low pH-mediated fusion

Upon receptor-mediated uptake of the mature virus, the low pH environment of the late endosome triggers the fusion cascade. During fusion, E undergoes tremendous conformational changes where its homodimers dissociate into monomers and then re-associate to form fusion-competent (fusogenic) homotrimers (Fig. 13.2B). The fusion peptides presumably penetrate the membrane and initiate the fusion process. The rearrangement of the E homodimers into the homotrimers results in the exposure of the fusion loop at the distal tip of DII, allowing its interaction with the host cell membrane (Allison *et al.*, 1995a; Modis *et al.*, 2004; Stiasny *et al.*, 2002). This trimeric state of the E protein is called the post-fusion state and the structure was determined by X-ray crystallography for DENV-2 as well as TBEV (Bressanelli *et al.*, 2004; Modis *et al.*, 2004). The transition from E homodimers to homotrimers does not affect the structural integrity of the E protein domains, but requires extensive reorientation of the domains relative to each other at the DI-DII and DI-DIII hinges. Domain II rotates by an angle of 30° relative to DI. This rotation occurs about the DI-DII hinge, which is near the kl hairpin (residues 270–279), and this interface is critical for the pH-dependent fusion mechanism. Mutation of residues in this region has been shown to change the pH threshold of fusion (Modis *et al.*, 2003). In contrast, DIII is folded back by about 36 Å towards DII and rotates by ~70°. This folding over of DIII brings its C-terminus closer to the fusion loop and the homotrimeric structure acquires a hairpin like orientation, where helix 1 of the stem region is directed towards DII of the same E monomer (Bressanelli *et al.*, 2004). The conserved hydrophobic residues in the fusion loop (Trp101, Leu107 and Phe108) are fully exposed on the trimer surface and form a concave region. The fusion loop penetrates the lipid bilayer membrane up to the outer leaflet and is held there by aromatic residues Trp101 and Phe108, while the hydrophobic side chains of Leu107 and Phe108 can accommodate the lipid head groups (Bressanelli *et al.*, 2004; Modis *et al.*, 2004; Stiasny *et al.*, 2004). The hemifusion stage is an intermediate state where only the outer leaflets of the two membranes have mixed, and this eventually leads to formation of the final post-fusion trimer and opening of the fusion pore (Modis *et al.*, 2004; Stiasny and Heinz, 2006). Therefore each step during the fusion event (the large rotation and folding over of DIII, the exposure and outward projection of DII, and the rotation of DI and DII around the hairpin) involves structural

rearrangements that provide opportunities for drug-design directed towards inhibiting fusion.

Inhibition of fusion

The structural rearrangements that take place during low-pH-triggered fusion in a DENV-infected cell have been the basis for antiviral design. Among the three domains of E protein, DIII forms a continuous polypeptide segment that can fold independently (Bhardwaj *et al.*, 2001). A recent study reported the expression of recombinant DIII protein from DENV-2 which was found to be capable of inhibiting low-pH induced virus fusion at the cell surface of BHK cells as well as within the endosomal pathway (Liao and Kielian, 2005). Bacterially expressed DIII or DIII + helixI (DIIIH1) was tested for inhibitory effects on DENV-1 and DENV-2 fusion in BHK cells. The DIIIH1 peptide strongly inhibited both DENV-1 and DENV-2 fusion (~70% inhibition of DENV-2 at a concentration of 50 μM) but DIII alone did not inhibit DENV fusion. This indicated that helix 1, which is an NH_2-terminal region of the stem and has been shown to be involved in E protein trimerization (Allison *et al.*, 1999), could have a role in the fusion mechanism. It was proposed that this inhibition was due to DIII binding to the trimeric conformation that prevented hairpin formation and therefore fusion. Similarly in another report it was shown that recombinant WNV DIII could inhibit DENV-2 entry into mosquito cells but not into Vero cells as determined by plaque neutralization assays. On the other hand, DENV-1 and DENV-2 DIII proteins inhibited the respective viruses into both HepG2 and mosquito cells (Chu *et al.*, 2005). This suggested that despite different mosquito vector ranges for WNV and DENV-2, the DIII of both these viruses utilized the same receptors for entry into mosquito cells, but this is not the case for mammalian cells. These examples underscore the scope of development of an antiviral strategy that targets the E structural intermediates during fusion.

Immature virus

Cryo-EM reconstruction of the immature DENV-2 and YFV has allowed comparison and contrast with the mature virus (Zhang *et al.*, 2003a). The immature particles are different from the mature native virus in size, appearance and the arrangement of the prM and E proteins (Fig. 13.3). The immature virus is bigger (~600 Å) and is spiky in appearance. The surface projections are due to the 60 trimers of prM-E heterodimers. Similar to the mature virion, these immature particles lack classical *T*=3 symmetry. The dengue virus prM glycoprotein consists of 166 amino acids. Furin cleavage releases the N-terminal 91 residues constituting 'pr' during maturation, leaving the ectodomain (residues 92–130) and C-terminal transmembrane region (residues 131–166) of 'M' in the virion. The pr peptide protects immature virions against premature fusion with the host membrane (Li *et al.*, 2008a; Zhang *et al.*, 2003b). The recent X-ray structure of the DENV-2 prM-E heterodimer (Fig. 13.2B and 3B) provides valuable insight on prM and its location relative to E at neutral and low pH (Li *et al.*, 2008a). The pr peptide consists of seven antiparallel β strands, three disulphide bonds and one glycosylated Asn residue (N69). The conformational state of E was similar to that found in the immature virus, since the angle between DI and DII differed only by 5° from that seen in the immature virus, but by 23° in comparison to the mature virus. A pseudo-atomic structure of the immature DENV generated by fitting the prM-E crystal structure into the cryo-EM density of the immature virus (Zhang *et al.*, 2004) revealed that the loops and carbohydrate side chains associated with residue Asn69 on the pr peptide completely covered the hydrophobic fusion loop in DII of the E protein at both neutral and low pH, where it prevents premature fusion of the immature virus. Also, the relative orientation and conformation of pr and traces of M density in the immature virus at neutral pH rendered the furin cleavage site almost inaccessible. This suggested that immature virus at low pH must have a different conformation and relative orientation of prM, to allow furin cleavage to occur and structural transitions in the prM-E heterodimer to form the E homodimers during maturation.

In an accompanying report, the low-pH-dependent furin cleavage of pr from the prM-E heterodimer in the immature virus was carefully analysed in combination with structural studies on the immature virus (Yu *et al.*, 2008). The over-all structure of the low pH immature particle was more similar to the cleaved mature virion than to the immature neutral-pH virion (Fig. 13.3A and C). Cryo-EM reconstruction of immature virus under different pH conditions showed that the immature virus undergoes a reversible conformational change at low pH (6.0) where the particle appears relatively smooth, whereas exposure to higher pH (8.0) causes the particles to become spiky. These particles have a diameter of ~530 Å, which is smaller than the immature virus particle at neutral pH (600 Å). Furin cleavage occurs only when immature particles are exposed to pH in the range of 5.0–6.0. The pseudo-atomic structures of the immature virus obtained by fitting the crystal structure of the prM-E heterodimer into the cryo-EM structure of the immature virus at low pH revealed that, as reported earlier, the E protein is arranged as three parallel E homodimers, and that two pr peptides are associated with each dimer. The fusion loop is buried in the pr and E interface and the pr peptide interacts with DII from one E monomer and DI of the other monomer. These extensive interactions presumably contribute to the stability of the immature viruses at low pH. These observations indicate that the immature virus particle with trimeric prM-E spikes undergoes a conformational change due to acidification in the trans-Golgi network (TGN), which exposes the hidden furin cleavage site. This study also revealed that the pr peptide remains associated with the virion after furin cleavage in the acidic environment of the TGN to prevent membrane fusion in the TGN, and is released only when the virus encounters neutral pH (Figs. 13.1 and 13.3).

The structure of pr and its interaction with E at different stages of the virus life cycle has provided more information about virus maturation and more targets for drug design. Compounds or peptides that mimic the pr peptide could be designed to prevent membrane fusion by retaining the E protein in the metastable, homodimeric state and therefore inhibiting virus growth and infectivity. Although this has yet to be achieved, the structural details about the E ectodomain dimer conformation have been exploited by several groups to identify antiviral compounds targeting the mature virus. These efforts and the preliminary results are discussed below.

Ligand binding pocket in DENV-2 E protein and its implications for antiviral drug design

The DENV-2 E protein structure with bound detergent *n*-octyl-β-d-glucoside (β-OG) or in the unbound form (Modis *et al.*, 2003), provided tremendous information for design of viral inhibitors (Fig. 13.4). The main difference between the two structures was the local rearrangement at the '*kl*' β-hairpin region (residues 268–280) in DII of the homodimers and the opening of a hydrophobic pocket between domains I and II, which was occupied by a single β-OG molecule in selected molecules. In the presence of β-OG, the kl loop was observed to be shifted towards the dimer con-

tact region, forming a salt bridge and a hydrogen bond with the i and j loops of the other monomer in the dimer. This shift led to closing of the 'cavities' along the dimer contact and opening of a hydrophobic channel at the DI–DII interface that accommodated β-OG. This β-OG occupied state was referred to as the open state, while the structure in the absence of the β-OG molecule was the closed state where the hydrophobic residues on the kl loop were buried. Mutations of residues that participate in this DI–DII interface alter the required pH threshold and affect virulence (Rey *et al.*, 1995). Since the virus fusion event involves dissociation of E homodimers and rearrangement into homotrimers, it was suggested that the movement at DI–DII interface might contribute

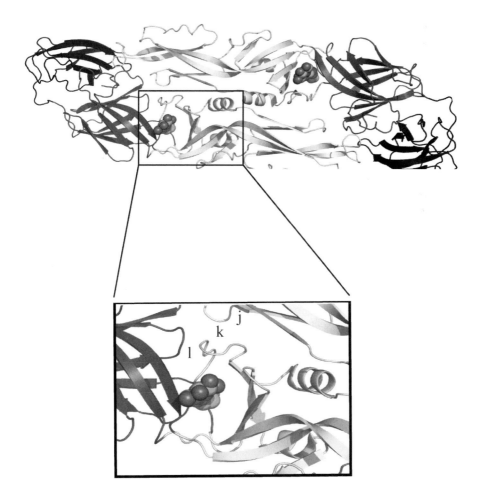

Figure 13.4 Ligand binding pocket in E protein. Structure of E protein dimer showing β-OG bound within the pocket and an enlarged view of bound β-OG. The binding of β-OG confers a new conformation within the k1 loop in an open position. This is an ideal target for small molecules, as blocking the flexing of this hinge region could inhibit the assembly, maturation and entry of virus particles.

to these conformational changes during fusion. It was proposed that the shift in the kl loop opens the hydrophobic pocket and allows DII to hinge away from its dimer partner and display the fusion peptide at its distal tip towards the target cell membrane.

The DI–DII interface, and specifically the kl loop, was identified as a critical determinant for initiation of the low-pH-induced conformational changes during fusion that result in formation of the fusogenic E trimers. The presence of a hydrophobic pocket occupied by β-OG in this pre-fusion structural conformation of E and the role of this interface in structural transitions during fusion and maturation of the virus make it a potential target for antivirals. The compounds designed to bind in this ligand-binding pocket could act in two ways: (1) they might inhibit the structural changes that drive the fusion event or (2) the hydrophobic pocket could also be exploited by designing smaller molecules that would 'open' the closed state kl hairpin and inhibit infection by premature triggering of a conformational change that mirrors the low pH-induced structural rearrangements. The ligand-binding pocket in E has therefore been an appealing target for design of antiviral compounds against DENV and other pathogenic flaviviruses.

Antiviral therapy targeting the ligand binding pocket in DENV E protein

Structure based design of antivirals targeting structural proteins in antivirus drug discovery is not a novel concept and has been used previously in the case of picornaviruses [enteroviruses (Padalko *et al.*, 2004; Rossmann *et al.*, 2000) and rhinoviruses (Badger *et al.*, 1988; Hadfield *et al.*, 1999; Heinz *et al.*, 1989; Rossmann and McKinlay, 1992)], as well as influenza virus (Hsieh and Hsu, 2007) and HIV (Copeland, 2006; Veiga *et al.*, 2006). The structure of the first human rhinovirus, solved in 1985 by Rossmann and coworkers, revealed the presence of a 25-Å-deep depression or canyon at the junction of viral structural proteins VP1 and VP2/3, and was found to encircle each fivefold axis of symmetry on the surface of the virion (Rossmann *et al.*, 1985). A hydrophobic pocket beneath the canyon on the floor of the outer coat protein VP1was the site of binding by several antiviral 'WIN' compounds that raised the

floor of the canyon as a consequence of binding. These compounds were found to inhibit virus attachment to host cells as well as uncoating (Badger *et al.*, 1988; Kim *et al.*, 1993; Reisdorph *et al.*, 2003; Smith *et al.*, 1986) and were developed by structure-based compound selection and refinement. Similarly, structural information about the DENV E protein and the flexibility at the DI–DII interface has prompted several groups to initiate drug design based on the hydrophobic pocket and subsequent screening to test the efficiency of virus inhibition. The basic layout of this drug discovery effort involves (1) *in silico* screening of compound databases or libraries for promising candidates by molecular docking into the β-OG pocket based on structural similarities to this detergent molecule; (2) subjecting the selected compounds to a biological validation assay to assess the inhibition of DENV growth in mammalian cell culture systems; and (3) testing compounds that show promise in cell culture assays for efficacy in animal models before they can be considered for clinical trials in humans.

We have initiated a drug discovery programme where compounds designed on the basis of the DENV E protein structure were tested for efficacy in inhibition against flaviviruses. An initial *in silico* screening of three compound libraries (143,000 compounds) from the National Cancer Institute was performed against the β-OG binding site in the hydrophobic pocket of DENV E (Zhou *et al.*, 2008). A hierarchical strategy was used to select 23 compounds after eliminating the high-energy conformers and visual inspection of the remaining compounds for structural similarities to β-OG for binding to the β-OG pocket and drug-like characteristics. These compounds were assayed for cytotoxicity in mammalian cells and inhibitory effect on flavivirus infectivity, specifically virus entry, spread, and viral replication (Li *et al.*, 2008b; Zhou *et al.*, 2008). Nine of 23 compounds tested showed inhibitory effect on virus growth with inhibitory concentrations (IC50) in the range of 20–500 μM. As a more direct approach to evaluate the inhibitory effect of the compounds on virus assembly and release, a replicon system was used to generate pseudo-infectious particles, or PIPS, in the presence or absence of compounds (Jones *et al.*, 2003). Of the nine compounds identified above, five reduced replication of PIPS, indicating disruption at the level of entry/assembly/release. These initial results with the selected compounds led to structural studies of protein-compound complexes. One of the compounds was found to display direct interaction with the hydrophobic pocket of DENV E protein in NMR studies (Zhou *et al.*, 2008). The initial lead compounds described above were used as templates in the design of second generation compounds that were tested for effect on virus growth. The docking models of these leads were examined to obtain more information about the binding of these compounds in the hydrophobic pocket of E. This computational analysis revealed that the ligand-binding pocket is most hydrophobic in the middle, while the end region is hydrophilic and can interact with ligands through stronger and electrostatically favourable interactions. Therefore the compounds were designed such that they could account for the limited steric space in the middle of the pocket and more accommodating ends. Of 36 compounds tested, two were found to be extremely effective at inhibiting virus growth at significantly low concentrations (1–5 μM) (Li *et al.*, 2008b). Future work with these compounds will include further modification of the two promising targets to generate third generation compounds and investigation into the stage of virus infection being inhibited.

A similar strategy was used by Yang *et al.* to screen for tetracycline derivatives for anti-DENV-2 effects by binding in the hydrophobic pocket of E (Yang *et al.*, 2007). To avoid the issue of lead compound toxicity, the screening was based on molecular docking using structural databases of medical compounds (CMC database) with molecular weights in the range of 200–800 Daltons and not consisting of multiple components. Out of 5331 compounds, 10 were finally selected for validation by their ability to inhibit DENV plaque formation assay on BHK cells. Of these 10 compounds, two tetracycline derivatives, rolitetracycline and doxycycline, significantly inhibited plaque formation with IC_{50} of ~ 67.1 μM and 55.6 μM, respectively. These compounds were docked into the ligand-binding pocket of E and displayed common hydrophobic interactions with critical residues that have been shown to affect membrane fusion during viral entry.

These reports demonstrate that structural information has greatly improved and facilitated antiviral drug design. Development of high-throughput systems for rapid screening in biological assays and suitable animal models to study the effect of these drugs during DENV infection or disease should be the focus of future endeavours (Holbrook and Gowen, 2008). While viral structural proteins are promising candidates for antiviral drug design, therapeutic strategies involving diverse viral targets and novel approaches are being pursued and have shown encouraging preliminary results. Some of these schemes are discussed further below.

Inhibitors of virus assembly

α-Glucosidase inhibitors

The first steps of DENV assembly take place in association with the membranes of the endoplasmic reticulum (ER) (Chambers *et al.*, 1990; Lindenbach and Rice, 2001). In flavivirus-infected mammalian cells, a 14-residue oligosaccharide, (Glc)3(Man)9(GlcNAc)2, is added to specific asparagine residues on the prM and E proteins in the ER lumen. This oligosaccharide precursor is sequentially trimmed at the glucose residues in the ER by resident α-glucosidases I and II (Lobigs, 1993). α-Glucosidase I removes the terminal α(1,2)-linked glucose from Glc3Man9GlcNAc2, and α-glucosidase II removes the second and the third α(1,3)-linked glucose residues (Hebert *et al.*, 1995). The processing of the N-linked oligosaccharide in the ER has been found to be important for secretion of several enveloped viruses, and inhibition of the glucosidase mediated trimming affects the life cycle of several viruses including HIV and HBV (Mehta *et al.*, 1998). Similarly, two ER α-glucosidase inhibitors, Castanospermine (CST) and deoxynojirimycin (DNJ), both of which inhibit the early stages of glycoprotein processing, were found to affect virus production in mouse neuronal cells infected with DENV-1 (Courageot *et al.*, 2000). A related ER α-glucosidase inhibitor, *N*-nonyl-deoxynojirimycin (NN-DNJ), was found to suppress production of infectious DENV-2 from BHK cells (Wu *et al.*, 2002). In fact, CST was later found to be effective against all four DENV serotypes in human hepatoma cells and prevented mortality in mice infected with DENV-2. This effect was specific for DENV, as opposed to other flaviviruses like West Nile virus and Yellow fever virus (Whitby *et al.*, 2005). These inhibitors would indirectly affect the N-linked oligosaccharide cleavage of the structural proteins prM and E, as well as the non-structural protein NS1, therefore affecting post-translational modification of these proteins. This could result in misfolded proteins and a reduction in secretion of glycoproteins and intracellular NS1 levels (Ray and Shi, 2006). Virus assembly, as well as replication, is therefore affected. These results are encouraging and provide candidate antivirals that can be tested further for efficacy in animal models for potential use in humans.

Nucleic acid-based therapy

RNA interference or silencing

RNA interference (RNAi) is a highly conserved gene silencing mechanism that plays an important role in post-transcriptional regulation of gene expression in a wide variety of species including plants, fungi, flies and even mammals. RNAi is thought to contribute to preservation of genomic integrity in these organisms and has also been suggested to protect the host from viral infections by degrading the extraneous genetic material like viral RNA (Ma *et al.*, 2007). In recent years, RNAi has become a powerful tool in functional analysis of genes and in rational drug design. It has been used in therapeutic approaches for several infectious diseases, tumours and metabolic disorders. Several studies have shown that RNAi technology has several advantages over traditional anti-viral drugs and vaccines, because of its ease of use, rapid action, and high level of efficiency when applied to the different stages of virus–host interactions (Tan and Yin, 2004). Strategies for delivery of RNAi reagents into mammalian cells can be divided into two types, the transient RNAi (for acute virus infections) and the stable/inducible RNAi (for chronic virus infections). The expression of endogenous host genes that are critical for viral replication can be targeted using siRNAs (Hannon and Rossi, 2004). Alternatively, the virus can be directly targeted by pre-treating cells with virus-specific siRNAs. This allows cleavage and removal of viral RNA after being targeted to the RNA induced silencing complex, RISC (Tan and Yin, 2004).

Use of RNAi against flaviviruses

Initial reports with flaviviruses tested RNAi-based strategies to target WNV (McCown *et al.*, 2003) and DENV-2 (Adelman *et al.*, 2002). In case of DENV-2, the study was carried out in mosquito C6/36 cells transformed with a plasmid that could transcribe inverted-repeat RNA (irRNA) derived from the DENV-2 genome and form dsRNA. It was demonstrated that the mosquito cell line expressing this plasmid was rendered resistant to DENV-2 infection and immune to viral RNA replication. However this work was not followed up by studies in mammalian cells and animal models. In the WNV study, siRNAs against the capsid and NS5 genes were expressed under the control of an RNA polymerase I promoter in mammalian cells, and led to reduction in viral RNA, protein and virus production. In another study, siRNAs directed against WNV E protein were administered to mice pre-infected with the virus. The siRNA pre-treatment led to reduction in viral load in the mice and protected the animals from lethal infection (Bai *et al.*, 2005).

Problems associated with RNAi

There are several challenges facing the widespread use of RNAi in antiviral therapy. The most pertinent is the viral escape strategies. One of the main issues with flaviviruses is that viral RNA replication occurs in reorganized ER membrane packets, which makes the viral RNA resistant to RNAi (Geiss *et al.*, 2005). Therefore, efficient delivery of the siRNA to the target subcellular location can be challenging. Secondly, the long-term RNAi treatment for chronic viral infections leads to escape viral mutants (induced viral escape route) that tend to be resistant to siRNA, where a single nucleotide change in the target sequence can lead to resistance to the antiviral activity of the siRNA (Haasnoot *et al.*, 2007). Also, while the host tries to mount an antiviral siRNA response, there are increasing reports of viruses being able to regulate or exploit the RNAi machinery during infection. For example, the hepatitis C virus (HCV) replication was found to be stimulated by the cellular miRNA, miR-122 by unknown mechanism (Jopling *et al.*, 2005). In addition, various herpes viruses encode miRNAs that are thought to target specific cellular genes (Cui *et al.*, 2006; Pfeffer *et al.*, 2005). Simian virus 40 (SV40) encodes miRNAs that regulate viral gene expression, thus reducing recognition of infected cells by the immune system (Sullivan *et al.*, 2005). The HIV tat protein has been shown to inhibit Dicer activity *in vitro* (Bennasser *et al.*, 2005). Therefore this interplay or race between the host and viruses for the RNAi machinery is a critical point of interference for antiviral therapy.

Possible solutions

The common theme behind most of these obstacles is single-shot siRNA antiviral therapy. There is increasing concern that monotypic siRNA therapies might not be sufficient to address the issues of virus escape mutants and evasion mechanisms. Therefore, strategies aimed at overcoming viral resistance of RNAi by combining RNAi effectors with each other, or with gene expression inhibitors, are being developed (reviewed in (Grimm and Kay, 2007). This multi-prong RNAi strategy has been referred to as 'combinatorial RNAi' (coRNAi). This approach has been tested in HIV by using multiple short-hairpin (sh)RNA doses (ter Brake *et al.*, 2006). Further, targeting cellular proteins through RNAi, or designing RNAi effectors that can 'invade' stable RNA structures that protect viral RNA, and even targeting viral mRNA that code for viral proteins which act as RNAi antagonists, are some of the measures that can be taken to improve the efficacy of RNAi as an antiviral therapy. RNAi has been used successfully in significant inhibition or elimination of several viruses, including HIV, HCV, HBV, coronavirus, influenza A virus (IAV), and human papillomavirus (HPV) (Ma *et al.*, 2007). These results are an encouraging demonstration that RNAi may prove to be effective in treatment and reduction of the dengue disease.

Antisense therapy: morpholino oligomers

Nucleic acid-based antiviral candidates have long included antisense DNA or RNA decoys. However, the development of these antisense effectors was severely hampered by issues such as toxicity, instability in serum, ineffective delivery to the target subcellular compartment and insufficient efficacy (Summerton and Weller, 1997). Development of chemically modified nucleic acids has addressed some of these issues (Kurreck, 2003).

One such modification is morpholino antisense oligomers that have been used in antiviral therapy against several human viruses and are being further evaluated in clinical trials (Ma *et al.*, 2000).

Phosphorodiamidate morpholino oligomers (PMOs) are one such class of morpholino antisense compounds that contain purine or pyrimidine bases attached to a backbone composed of a six-member morpholine ring joined by phosphorodiamidate intersubunit linkages and are usually about 20 subunits in length (Summerton and Weller, 1997). PMOs are water soluble, nuclease resistant (Hudziak *et al.*, 1996), and bind to RNA by Watson–Crick base pairing (Summerton and Weller, 1997). These compounds act by forming a stable, sequence-specific duplex with RNA, blocking the availability of the target RNA to biomolecules for any biosynthetic process requiring RNA–RNA or protein–RNA interaction. They most significantly affect translation or interfere with pre-mRNA splicing. PMOs have been found to be effective against viral RNA replication of several viruses (Neuman *et al.*, 2004; Stein *et al.*, 2001). Peptide-conjugated PMOs (P-PMOs) are used preferentially over non-conjugated PMOs. The most effective P-PMO in terms of cellular uptake and antiviral activity was the conjugation of arginine-rich peptides to PMOs (Moulton *et al.*, 2004).

In an initial study on use of PMOs against DENV-2, phophorothioate oligonucleotides containing C-5 propyne-substituted pyrimidines were found to be most effective (in contrast with unmodified phosphodiester oligonucleotides as well as the corresponding phosphorothioate oligonucleotides) against the virus when monkey kidney cells (LLC-MK2) microinjected with this PMO were infected with DENV-2 (Raviprakash *et al.*, 1995). There was a 50–75% reduction in viral titre with compounds directed against the translation initiation site of DENV RNA as well as the 3'-untranslated (UTR). Another study once again with monkey kidney (Vero) cells demonstrated that, out of five different DENV-specific Arg P-PMOs, those targeting the 5' stem–loop (5' SL) region and the 3' cyclization sequence (3' CS) in the DENV-2 genome were most effective in reducing the viral load (5 log reduction). The PMO targeting the 3' CS was effective against all of four DENV serotypes, reducing the viral titre by up to 4 logs during the course of the infection (Kinney *et al.*, 2005). However, in both these studies, the mechanism of actual viral inhibition was not determined or explained. A more recent report studied the role of the 3'-UTR more extensively, using arginine-rich P-PMOs targeting several conserved elements (pseudo-knots, CS and the 3' SL) (Holden *et al.*, 2006). The study was carried out in BHK cells infected with DENV-2 and identified an additional region in the 3'-UTR, the 3' SLT region (including the pentanucleotide loop), that could be targeted for reduction in viral replication and viral titre. A more detailed investigation of the stage of inhibition by these antisense oligos revealed that the 3' SLT P-PMO led to inhibition of both translation of viral RNA as well as synthesis of viral RNA, suggesting that this region (in addition to 5' CS and 3' CS) was important both for the viral RNA translation early during infection as well as stabilizing RNA–RNA or RNA–protein interactions during viral RNA replication.

These DENV UTR targets and the corresponding P-PMOs have not yet been tested in animal models, but provide a promising start for more focused efforts to evaluate them for antiviral therapy. The approval of the antisense phosphorothioate oligomer that blocks the translation of CMV immediate early 2 (IE2) mRNA by the FDA as an antiviral for the treatment of CMV retinitis (Marwick, 1998) is an example of a PMO being successfully used in therapeutics. PMOs and peptide-PMOs meet most of the requirements for an anti-DENV therapeutic: non-toxic, inexpensive, easy to administer, stable for months at variable temperatures (Iversen *et al.*, 2003; Kipshidze *et al.*, 2001), especially with more strategies being developed to improve delivery into cells (Dias and Stein, 2002). However, all the promising candidates need to be put through rigorous tests in animal models to confirm their suitability for use as a dengue antiviral.

Nucleoside analogues in antiviral therapy

Nucleoside analogues have been an important class of antiviral compounds in the treatment of several viral diseases. These compounds are predominantly prodrugs that need to be converted to their antiviral nucleotide metabolite forms

(Parker, 2005). Viruses, for the most part, do not express the enzymes that are necessary for activation of nucleoside analogues and therefore rely on the host purine or pyrimidine metabolic enzymes for the activation of nucleotides that can inhibit viral replication. Ribavirin (1-β-d-ribofuranosyl-1*H*-1,2,4-triazole-3-carboxamide) is one such commonly used nucleoside analogue. First synthesized in 1972, ribavirin was then demonstrated to possess broad spectrum antiviral activity (Sidwell *et al.*, 1972; Streeter *et al.*, 1973). Ribavirin has been shown to be an effective drug with *in vivo* activity against several RNA viruses including influenza (Gilbert *et al.*, 1985), respiratory syncytial (RSV) (Taber *et al.*, 1983), Lassa fever (McCormick *et al.*, 1986), and Hanta virus infections (Huggins, 1989). It is an approved inhaled drug for treatment of respiratory syncytial virus and together with interferon alpha in the treatment of HCV infections (Davis *et al.*, 1990; Taber *et al.*, 1983).

Mode of action

Upon entry into the cell, ribavirin is phosphorylated by cellular adenosine kinase at its 5' position to ribavirin monophosphate (RMP) (Balzarini *et al.*, 1993; Willis *et al.*, 1978), which is then converted to ribavirin triphosphate (RTP) by the activity of nucleoside mono- and diphosphate kinases, respectively (Gallois-Montbrun *et al.*, 2003). Ribavirin monophosphate (RMP) acts as a competitive inhibitor of inosine monophosphate dehydrogenase (IMPDH), one of the key enzymes involved in the biosynthesis of guanine nucleotides (Parker, 2005). This inhibition results in depletion of the cellular guanine nucleotide pool, which indirectly affects capping and polymerase activities of both cellular and viral proteins. The antiviral effect of ribavirin is therefore exerted due to the relative inability of the viral enzymes to compete with cellular enzymes for the depleted nucleotide pool (Benarroch *et al.*, 2004). Besides reducing the intracellular pool of nucleotides available for viral polymerase and methyltranferases, it has also been shown that ribavirin causes a more error-prone replication of several viral genomes (Contreras *et al.*, 2002; Crotty *et al.*, 2000; Eigen, 2002; Lanford *et al.*, 2001; Zhou *et al.*, 2003). Inhibition of IMP dehydrogenase leads to imbalances in the intracellular nucleotide pool which could cause the viral polymerases to replace GTP with alternative nucleotides (ATP, UTP, or CTP). This phenomenon may lead to introduction of lethal mutations in viral genomes and could account for the antiviral activity of ribavirin.

There is increasing evidence that there is another mechanism of the antiviral activity of ribavirin: interaction of RTP with viral RNA polymerases. In mammalian cells, RTP accumulates to very high intracellular concentrations (greater than $100 \mu M$), similar to the normal intracellular concentrations of ATP and GTP (Smee et al., 2001) and therefore can effectively compete with the natural nucleotides. Recent studies have been done to evaluate the interaction of RTP with the RNA polymerases of poliovirus (Crotty et al., 2000) and HCV (Bougie and Bisaillon, 2003; Vo et al., 2003). These investigators concluded that both the inhibition of RNA synthesis and the induction of mutations could contribute to the activity of ribavirin. More specifically, it has been demonstrated that RTP inhibits the 2'-O-methyltranferase (MTase) activity of the DENV NS5 MTase domain in vitro (Benarroch et al., 2004). Structural studies further revealed that RTP occupied the GTP/RNA cap binding region on the NS5 MTase domain, leading to suggestions that ribavirin might target the viral methyltransferases to contribute to the overall antiviral activity of the compound (Benarroch et al., 2004). In addition, ribavirin has also been shown to cause reduction of the intracellular viral replicase activity as assayed in vitro using viral replicase complexes obtained from infected cells treated with the drug (Takhampunya et al., 2006b). Since both capping and RNA-dependent RNA polymerase activities for flaviviruses are carried out by the same non-structural protein, NS5, these results strongly hint at a more direct role for ribavirin against DENV.

Problems

Although ribavirin has been shown to possess significant in vivo activity against several RNA viruses, in DENV infected cells, it has a cytostatic effect and has not been effective in animal models (Crance et al., 2003). For instance, in one of the animal studies, ribavirin treatment of rhesus monkeys infected with DENV had little or no ef-fect on viraemia (Malinoski et al., 1990). In mice, intraperitoneal administration of ribavirin had no effect on survival following intracerebral inoculation with DENV (Koff et al., 1983). Therefore, it has found limited use in dengue disease therapy and needs to be further evaluated for efficacy in non-human primates. In addition, analogues of ribavirin that are more effective antivirals against DENV may be pursued and developed further.

Mycophenolic acid (MPA) is a non-nucleoside analogue used as an immunosuppressant drug. It is also an IMPDH inhibitor like ribavirin, and has been shown to reduce DENV replication in infected human hepatoma cells (Diamond et al., 2002) and in rhesus monkey kidney cells (Takhampunya et al., 2006b). But once again, this drug needs to be tested for efficacy in animal models of dengue.

Nitric oxide

Nitric oxide (NO) is a versatile molecule with varied biological functions. It acts as a biological messenger in a variety of biological processes. In vertebrate systems, NO is biosynthesized endogenously from arginine and oxygen by various nitric oxide synthase (NOS) enzymes and by reduction of inorganic nitrate. A biologically important reaction of nitric oxide is S-nitrosylation, the conversion of thiol groups, including cysteine residues in proteins, to form S-nitrosothiols (RSNOs). Nitric oxide plays a role in smooth muscle relaxation and therefore vasodilation, which is important in neurotransmission and inhibition of platelet activation, among other pathways. Nitric oxide is also generated by macrophages, monocytes, dendritic cells and neutrophils as part of the human immune response. These cells are the main sites of replication for several RNA viruses, including DENV (Lin et al., 2002; Wu et al., 2000), and S-Nitrosylation is a critical modification of several viral proteins. For example, nitration of the VP1 capsid protein or S-nitrosylation of the cysteine protease 3C by NO has been shown to contribute to the inhibition of coxsackievirus infection (Padalko et al., 2004; Saura et al., 1999). Similarly, S-nitrosylation of the viral protease by NO leads to suppression of HIV infection (Persichini et al., 1998).

Studies on the effect of endogenously synthesized NO in macrophages or monocytes on

DENV did not yield consistent results. One study reported that, in macrophages and monocytes, it led to decreased viral replication for DENV (Neves-Souza *et al.*, 2005) while other studies showed that there was no significant difference between the NO levels in uninfected versus DENV-infected monocytes and macrophages (Espina *et al.*, 2003). However, in several studies with cultured cells other than macrophages or monocytes, and an external source of NO, DENV replication and production was inhibited. Studies on C6/36 mosquito cells (Neves-Souza *et al.*, 2005) and N18 neuroblastoma cells (Charnsilpa *et al.*, 2005) demonstrated that NO donors lead to a reduction in viral RNA replication and in viral protein synthesis. Another study reported that NO (provided by donor *S*-nitroso-*N*-acetylpenicillamine, SNAP) in LLC-MK2 cells infected with DENV-2 resulted in reduced virus RNA replication. *In vitro* assays revealed that NO specifically affects the *de novo* activity of the RNA-dependent RNA polymerase, NS5, while the helicase activity of NS3 remained unaffected even at high concentrations of the NO donor (Takhampunya *et al.*, 2006a). These results shed some light on the possible viral targets of NO in a DENV infection.

There are caveats associated with the use of NO in anti-DENV therapy. Most of the above mentioned experiments required the NO donor to be added to the cells almost immediately after infection. Experiments examining the temporal relationship between NO treatment and DENV infection need to be performed to explore the effectiveness of NO as an antiviral during later stages of infection. Also, the possibility that NO may have other viral targets besides NS5 has not yet been completely addressed. Once again, the actual effect of NO as an antiviral during infection of a non-human primate will help cement the possibility of its use in antiviral therapy.

Inhibition of host gene functions

Viruses depend on host cellular machinery at almost every stage of the virus infection: entry, translation of viral genome, replication, polyprotein processing, assembly and maturation. Therefore, one approach to antiviral therapy is the inhibition of host cell proteins or enzymes

critical for the virus life cycle. The ideal scenario for this approach would be to target a host protein that is critical for the virus but is redundant for the host (Monath, 2008). Kinases are enzymes that phosphorylate proteins and are critical for the transmission of biochemical signals in many signal transduction pathways. Phosphorylation-dependent signal transduction pathways are involved in DENV infection where they contribute to cell survival (Chang *et al.*, 2006; Lee *et al.*, 2005) and evasion of the immune response (Ho *et al.*, 2005; Munoz-Jordan *et al.*, 2003). Also, phosphorylation of viral proteins such as DENV NS5 (Kapoor *et al.*, 1995) by cellular kinases is known to regulate their subcellular localization, and presumably their functions. In one such effort, a c-Src kinase inhibitor (dasatinib) was shown to inhibit the growth and spread of DENV-1–4 in Vero, Huh-7, and C6/36 cells (Chu and Yang, 2007). This inhibitor was chosen from a library of kinase inhibitors for its ability to inhibit DENV infection with both pre- and post-infection exposure. In addition, siRNA-based silencing of cSrc kinase lead to reduction in DENV replication. Immunofluorescence and transmission EM showed that this inhibition of c-Src kinase leads to inhibition of DENV assembly in the ER. These results indicate that targeting host cell functions required for virus survival might be a promising approach to antiviral therapy.

Future perspectives

The main objectives driving dengue antiviral research today are that the drug should be safe, economical, and effective against all four DENV serotypes, and should be able to provide long-term protection. All of the strategies explained and discussed above seem promising and have shown partial success against DENV infection in mammalian cell culture systems and sometimes even in mouse models. However the main concerns with most of these therapeutic approaches are that they have not been examined for inhibitory effects on all four DENV serotypes. Initial reports for some of the antiviral approaches have not been followed up, raising questions about the reproducibility of the antiviral phenomena. Several reported antivirals have been tested at only one time point, pre- or post- infection in tissue culture systems, and therefore need to be subjected to more rigorous and diverse regimes, different cell types (mosquito, mammalian) and against all four DENV serotypes and perhaps even against other pathogenic flaviviruses to monitor broad-spectrum antiviral activity.

The development of effective antiviral approaches is complicated further because of the issue of their suitability and safety for human use. Another challenge is to ensure that the therapeutic measures tested experimentally in mammalian cells will be as effective in humans. Suitable animal models can help address this question as they can recapitulate the human disease. Animal models permit studies of infectivity, pathogenesis, humoral and cellular immunity, cytokine responses, and antiviral and vaccine efficacy. They also allow detailed analysis throughout the course of the viral infection. The availability of samples from the initial stages of infection provides a better assessment of the time course of the infection and insight into the early events of subclinical infection (Holbrook and Gowen, 2008). Non-human primate animal models serve as the most relevant animal systems, as they mimic the human host responses well. However they are very expensive experimental systems and DENV generally does not reproduce human-like disease in non-human primates. Therefore, smaller animal models like mice, hamsters and guinea pigs may serve as good hosts to study the efficacy of an antiviral before moving on to a larger host, assuming that the dengue disease model can be established in them.

Advances towards the development of therapeutic approaches against dengue have been possible because of greater insights into the virus life cycle, availability of structural details about viral proteins, and the ability to utilize structural information for effective antiviral design. The strides made in the last 30 years towards understanding the virus and the disease, as well as progress in the field of vaccine design and immunomodulatory molecules, indicates that development of an effective therapy against dengue, although challenging, is not improbable.

References

Adelman, Z.N., Sanchez-Vargas, I., Travanty, E.A., Carlson, J.O., Beaty, B.J., Blair, C.D., and Olson, K.E. (2002). RNA silencing of dengue virus type 2 replication in transformed C6/36 mosquito cells transcribing an inverted-repeat RNA derived from the virus genome. J. Virol. 76, 12925–12933.

Allison, S.L., Schalich, J., Stiasny, K., Mandl, C.W., and Heinz, F.X. (2001). Mutational evidence for an internal fusion peptide in flavivirus envelope protein E. J. Virol. 75, 4268–4275.

Allison, S.L., Schalich, J., Stiasny, K., Mandl, C.W., Kunz, C., and Heinz, F.X. (1995a). Oligomeric rearrangement of tick-borne encephalitis virus envelope proteins induced by an acidic pH. J. Virol. 69, 695–700.

Allison, S.L., Stadler, K., Mandl, C.W., Kunz, C., and Heinz, F.X. (1995b). Synthesis and secretion of recombinant tick-borne encephalitis virus protein E in soluble and particulate form. J. Virol. 69, 5816–5820.

Allison, S.L., Stiasny, K., Stadler, K., Mandl, C.W., and Heinz, F.X. (1999). Mapping of functional elements in the stem-anchor region of tick-borne encephalitis virus envelope protein E. J. Virol. 73, 5605–5612.

Baba, M., Snoeck, R., Pauwels, R., and de Clercq, E. (1988). Sulfated polysaccharides are potent and selective inhibitors of various enveloped viruses, including herpes simplex virus, cytomegalovirus, vesicular stomatitis virus, and human immunodeficiency virus. Antimicrob. Agents Chemother. 32, 1742–1745.

Badger, J., Minor, I., Kremer, M.J., Oliveira, M.A., Smith, T.J., Griffith, J.P., Guerin, D.M., Krishnaswamy, S., Luo, M., Rossmann, M.G., McKinlay, M.A., Diana, G.D., Dutko, F.J., Fancher, M., Rueckert, R.R., and Heinz, B.A. (1988). Structural analysis of a series of antiviral agents complexed with human rhinovirus 14. Proc. Natl. Acad. Sci. U.S.A. 85, 3304–3308.

Bai, F., Town, T., Pradhan, D., Cox, J., Ashish, Ledizet, M., Anderson, J.F., Flavell, R.A., Krueger, J.K., Koski, R.A., and Fikrig, E. (2007). Antiviral peptides targeting the west nile virus envelope protein. J. Virol. 81, 2047–2055.

Bai, F., Wang, T., Pal, U., Bao, F., Gould, L.H., and Fikrig, E. (2005). Use of RNA interference to prevent lethal murine west nile virus infection. J. Infect. Dis. 191, 1148–1154.

Balzarini, J., Karlsson, A., Wang, L., Bohman, C., Horska, K., Votruba, I., Fridland, A., Van Aerschot, A., Herdewijn, P., and De Clercq, E. (1993). Eicar (5-ethynyl-1-beta-D-ribofuranosylimidazole-4-carboxamide). A novel potent inhibitor of inosinate dehydrogenase activity and guanylate biosynthesis. J. Biol. Chem. 268, 24591–24598.

Beasley, D.W., and Barrett, A.D. (2002). Identification of neutralizing epitopes within structural domain III of the West Nile virus envelope protein. J. Virol. 76, 13097–13100.

Benarroch, D., Egloff, M.P., Mulard, L., Guerreiro, C., Romette, J.L., and Canard, B. (2004). A structural basis for the inhibition of the NS5 dengue virus mRNA 2'-O-methyltransferase domain by ribavirin 5'-triphosphate. J. Biol. Chem. 279, 35638–35643.

Bennasser, Y., Le, S.Y., Benkirane, M., and Jeang, K.T. (2005). Evidence that HIV-1 encodes an siRNA and a suppressor of RNA silencing. Immunity 22, 607–619.

Bhardwaj, S., Holbrook, M., Shope, R.E., Barrett, A.D., and Watowich, S.J. (2001). Biophysical characterization and vector-specific antagonist activity of domain

III of the tick-borne flavivirus envelope protein. J. Virol. *75*, 4002–4007.

Bougie, I., and Bisaillon, M. (2003). Initial binding of the broad spectrum antiviral nucleoside ribavirin to the hepatitis C virus RNA polymerase. J. Biol. Chem. *278*, 52471–52478.

Bravo, J.R., Guzman, M.G., and Kouri, G.P. (1987). Why dengue haemorrhagic fever in Cuba? 1. Individual risk factors for dengue haemorrhagic fever/dengue shock syndrome (DHF/DSS). Trans. R. Soc. Trop. Med. Hyg. *81*, 816–820.

Bressanelli, S., Stiasny, K., Allison, S.L., Stura, E.A., Duquerroy, S., Lescar, J., Heinz, F.X., and Rey, F.A. (2004). Structure of a flavivirus envelope glycoprotein in its low-pH-induced membrane fusion conformation. EMBO J. *23*, 728–738.

Brown, M.G., King, C.A., Sherren, C., Marshall, J.S., and Anderson, R. (2006). A dominant role for FcgammaRII in antibody-enhanced dengue virus infection of human mast cells and associated CCL5 release. J. Leukoc. Biol. *80*, 1242–1250.

Chambers, T.J., Hahn, C.S., Galler, R., and Rice, C.M. (1990). Flavivirus genome organization, expression, and replication. Annu. Rev. Microbiol. *44*, 649–688.

Chang, T.H., Liao, C.L., and Lin, Y.L. (2006). Flavivirus induces interferon-beta gene expression through a pathway involving RIG-I-dependent IRF-3 and PI3K-dependent NF-kappaB activation. Microbes Infect. *8*, 157–171.

Charnsilpa, W., Takhampunya, R., Endy, T.P., Mammen, M.P., Jr., Libraty, D.H., and Ubol, S. (2005). Nitric oxide radical suppresses replication of wild-type dengue 2 viruses *in vitro*. J. Med. Virol. *77*, 89–95.

Chen, Y., Maguire, T., Hileman, R.E., Fromm, J.R., Esko, J.D., Linhardt, R.J., and Marks, R.M. (1997). Dengue virus infectivity depends on envelope protein binding to target cell heparan sulfate. Nat. Med. *3*, 866–871.

Chu, J.J., Rajamanonmani, R., Li, J., Bhuvanakantham, R., Lescar, J., and Ng, M.L. (2005). Inhibition of West Nile virus entry by using a recombinant domain III from the envelope glycoprotein. J. Gen. Virol. *86*, 405–412.

Chu, J.J., and Yang, P.L. (2007). c-Src protein kinase inhibitors block assembly and maturation of dengue virus. Proc. Natl. Acad. Sci. U.S.A. *104*, 3520–3525.

Contreras, A.M., Hiasa, Y., He, W., Terella, A., Schmidt, E.V., and Chung, R.T. (2002). Viral RNA mutations are region specific and increased by ribavirin in a full-length hepatitis C virus replication system. J. Virol. *76*, 8505–8517.

Copeland, K.F. (2006). Inhibition of HIV-1 entry into cells. Recent Patents Anti-Infect. Drug Disc. *1*, 107–112.

Courageot, M.P., Frenkiel, M.P., Dos Santos, C.D., Deubel, V., and Despres, P. (2000). Alpha-glucosidase inhibitors reduce dengue virus production by affecting the initial steps of virion morphogenesis in the endoplasmic reticulum. J. Virol. *74*, 564–572.

Crance, J.M., Scaramozzino, N., Jouan, A., and Garin, D. (2003). Interferon, ribavirin, 6-azauridine and glycyrrhizin: antiviral compounds active against pathogenic flaviviruses. Antiviral Res. *58*, 73–79.

Crill, W.D., and Chang, G.J. (2004). Localization and characterization of flavivirus envelope glycoprotein cross-reactive epitopes. J. Virol. *78*, 13975–13986.

Crill, W.D., and Roehrig, J.T. (2001). Monoclonal antibodies that bind to domain III of dengue virus E glycoprotein are the most efficient blockers of virus adsorption to Vero cells. J. Virol. *75*, 7769–7773.

Crotty, S., M aag, D., Arnold, J.J., Zhong, W., Lau, J.Y., Hong, Z., Andino, R., and Cameron, C.E. (2000). The broad-spectrum antiviral ribonucleoside ribavirin is an RNA virus mutagen. Nat. Med. *6*, 1375–1379.

Cui, C., Griffiths, A., Li, G., Silva, L.M., Kramer, M.F., G aasterland, T., Wang, X.J., and Coen, D.M. (2006). Prediction and identification of herpes simplex virus 1-encoded microRNAs. J. Virol. *80*, 5499–5508.

Davis, G.L., Balart, L.A., Schiff, E.R., Lindsay, K., Bodenheimer, H.C., Jr., Perrillo, R.P., Carey, W., Jacobson, I.M., Payne, J., Dienstag, J.L., *et al.* (1990). Treatment of chronic hepatitis C with recombinant alpha-interferon. A multicentre randomized, controlled trial. The Hepatitis Interventional Therapy Group. J. Hepatol. 11 Suppl 1, S31–35.

Diamond, M.S., Zachariah, M., and Harris, E. (2002). Mycophenolic acid inhibits dengue virus infection by preventing replication of viral RNA Virology *304*, 211–221.

Dias, N., and Stein, C.A. (2002). Antisense oligonucleotides: basic concepts and mechanisms. Mol. Cancer Ther. *1*, 347–355.

Eigen, M. (2002). Error catastrophe and antiviral strategy. Proc. Natl. Acad. Sci. U.S.A. *99*, 13374–13376.

Elshuber, S., Allison, S.L., Heinz, F.X., and Mandl, C.W. (2003). Cleavage of protein prM is necessary for infection of BHK-21 cells by tick-borne encephalitis virus. J. Gen. Virol. *84*, 183–191.

Espina, L.M., Valero, N.J., Hernandez, J.M., and Mosquera, J.A. (2003). Increased apoptosis and expression of tumor necrosis factor-alpha caused by infection of cultured human monocytes with dengue virus. Am. J. Trop. Med. Hyg. *68*, 48–53.

Gallois-Montbrun, S., Chen, Y., Dutartre, H., Sophys, M., Morera, S., Guerreiro, C., Schneider, B., Mulard, L., Janin, J., Veron, M., Deville-Bonne, D., and Canard, B. (2003). Structural analysis of the activation of ribavirin analogs by NDP kinase: comparison with other ribavirin targets. Mol. Pharmacol. *63*, 538–546.

Geiss, B.J., Pierson, T.C., and Diamond, M.S. (2005). Actively replicating West Nile virus is resistant to cytoplasmic delivery of siR.N.A. Virol. J. *2*, 53.

Germi, R., Crance, J.M., Garin, D., Guimet, J., Lortat-Jacob, H., Ruigrok, R.W., Zarski, J.P., and Drouet, E. (2002). Heparan sulfate-mediated binding of infectious dengue virus type 2 and yellow fever virus. Virology *292*, 162–168.

Gilbert, B.E., Wilson, S.Z., Knight, V., Couch, R.B., Quarles, J.M., Dure, L., Hayes, N., and Willis, G. (1985). Ribavirin small-particle aerosol treatment of infections caused by influenza virus strains A/Victoria/7/83 (H1N1) and B/Texas/1/84. Antimicrob. Agents Chemother. *27*, 309–313.

Green, S., and Rothman, A. (2006). Immunopathological mechanisms in dengue and dengue hemorrhagic fever. Current opinion in infectious diseases *19*, 429–436.

Grimm, D., and Kay, M.A. (2007). RNAi and Gene Therapy: A Mutual Attraction. Hematology Am Soc Hematol Educ Program *2007*, 473–481.

Gubler, D.J. (1998). Dengue and dengue hemorrhagic fever. Clin. Microbiol. Rev. *11*, 480–496.

Gubler, D.J. (2002a). Epidemic dengue/dengue hemorrhagic fever as a public health, social and economic problem in the 21st century. Trends Microbiol. *10*, 100–103.

Gubler, D.J. (2002b). The global emergence/resurgence of arboviral diseases as public health problems. Arch. Med. Res. *33*, 330–342.

Gubler, D.J. (2006). Dengue/dengue haemorrhagic fever: history and current status. Novartis Found. Symp. *277*, 3–16; discussion 16–22, 71–13, 251–253.

Guzman, M.G., Alvarez, M., Rodriguez-Roche, R., Bernardo, L., Montes, T., Vazquez, S., Morier, L., Alvarez, A., Gould, E.A., Kouri, G., and Halstead, S.B. (2007). Neutralizing antibodies after infection with dengue 1 virus. Emerg. Infect. Dis. *13*, 282–286.

Haasnoot, J., Westerhout, E.M., and Berkhout, B. (2007). RNA interference against viruses: strike and counter-strike. Nat. Biotechnol. *25*, 1435–1443.

Hadfield, A.T., Diana, G.D., and Rossmann, M.G. (1999). Analysis of three structurally related antiviral compounds in complex with human rhinovirus 16. Proc. Natl. Acad. Sci. U. S. A. *96*, 14730–14735.

Halstead, S.B. (1988). Pathogenesis of dengue: challenges to molecular biology. Science *239*, 476–481.

Halstead, S.B. (1990). Global epidemiology of dengue hemorrhagic fever. South-East Asian J. Trop. Med. Public Health *21*, 636–641.

Halstead, S.B. (1999). Is there an inapparent dengue explosion? Lancet *353*, 1100–1101.

Halstead, S.B. (2002). Dengue. Curr. Opin. Infect. Dis. *15*, 471–476.

Halstead, S.B. (2003). Neutralization and antibody-dependent enhancement of dengue viruses. Adv. Virus Res. *60*, 421–467.

Halstead, S.B., Heinz, F.X., Barrett, A.D., and Roehrig, J.T. (2005). Dengue virus: molecular basis of cell entry and pathogenesis, 25–27 June 2003, Vienna, Austria. Vaccine *23*, 849–856.

Halstead, S.B., and O'Rourke, E.J. (1977). Dengue viruses and mononuclear phagocytes. I. Infection enhancement by non-neutralizing antibody. J. Exp. Med. *146*, 201–217.

Halstead, S.B., and Simasthien, P. (1970). Observations related to the pathogenesis of dengue hemorrhagic fever. I.I. Antigenic and biologic properties of dengue viruses and their association with disease response in the host. Yale J. Biol. Med. *42*, 276–292.

Hannon, G.J., and Rossi, J.J. (2004). Unlocking the potential of the human genome with RNA interference. Nature *431*, 371–378.

Hebert, D.N., Foellmer, B., and Helenius, A. (1995). Glucose trimming and reglucosylation determine glycoprotein association with calnexin in the endoplasmic reticulum. Cell *81*, 425–433.

Heinz, B.A., Rueckert, R.R., Shepard, D.A., Dutko, F.J., McKinlay, M.A., Fancher, M., Rossmann, M.G., Badger, J., and Smith, T.J. (1989). Genetic and molecular analyses of spontaneous mutants of human rhi-

novirus 14 that are resistant to an antiviral compound. J. Virol. *63*, 2476–2485.

Heinz, F.X., and Allison, S.L. (2000). Structures and mechanisms in flavivirus fusion. Adv. Virus Res. *55*, 231–269.

Heinz, F.X., and Allison, S.L. (2001). The machinery for flavivirus fusion with host cell membranes. Curr. Opin. Microbiol. *4*, 450–455.

Heinz, F.X., Auer, G., Stiasny, K., Holzmann, H., Mandl, C., Guirakhoo, F., and Kunz, C. (1994). The interactions of the flavivirus envelope proteins: implications for virus entry and release. Arch. Virol. Suppl. *9*, 339–348.

Hilgard, P., and Stockert, R. (2000). Heparan sulfate proteoglycans initiate dengue virus infection of hepatocytes. Hepatology *32*, 1069–1077.

Ho, L.J., Hung, L.F., Weng, C.Y., Wu, W.L., Chou, P., Lin, Y.L., Chang, D.M., Tai, T.Y., and Lai, J.H. (2005). Dengue virus type 2 antagonizes IFN-alpha but not IFN-gamma antiviral effect via down-regulating Tyk2-STAT signaling in the human dendritic cell. J. Immunol. *174*, 8163–8172.

Holbrook, M.R., and Gowen, B.B. (2008). Animal models of highly pathogenic RNA viral infections: encephalitis viruses. Antiviral Res. *78*, 69–78.

Holden, K.L., Stein, D.A., Pierson, T.C., Ahmed, A.A., Clyde, K., Iversen, P.L., and Harris, E. (2006). Inhibition of dengue virus translation and RNA synthesis by a morpholino oligomer targeted to the top of the terminal 3' stem–loop structure. Virology *344*, 439–452.

Hrobowski, Y.M., Garry, R.F., and Michael, S.F. (2005). Peptide inhibitors of dengue virus and West Nile virus infectivity. Virol. J. *2*, 49.

Hsieh, H.P., and Hsu, J.T. (2007). Strategies of development of antiviral agents directed against influenza virus replication. Curr. Pharm. Des. *13*, 3531–3542.

Hudziak, R.M., Barofsky, E., Barofsky, D.F., Weller, D.L., Huang, S.B., and Weller, D.D. (1996). Resistance of morpholino phosphorodiamidate oligomers to enzymatic degradation. Antisense Nucleic Acid Drug Dev. *6*, 267–272.

Huggins, J.W. (1989). Prospects for treatment of viral hemorrhagic fevers with ribavirin, a broad-spectrum antiviral drug. Rev. Infect. Dis. 11 Suppl 4, S750–761.

Hung, S.L., Lee, P.L., Chen, H.W., Chen, L.K., Kao, C.L., and King, C.C. (1999). Analysis of the steps involved in Dengue virus entry into host cells. Virology *257*, 156–167.

Iversen, P.L., Arora, V., Acker, A.J., Mason, D.H., and Devi, G.R. (2003). Efficacy of antisense morpholino oligomer targeted to c-myc in prostate cancer xenograft murine model and a Phase I safety study in humans. Clin. Cancer Res. 9, 2510–2519.

Jones, C.T., Ma, L., Burgner, J.W., Groesch, T.D., Post, C.B., and Kuhn, R.J. (2003). Flavivirus capsid is a dimeric alpha-helical protein. Journal of virology 77, 7143–7149.

Jopling, C.L., Yi, M., Lancaster, A.M., Lemon, S.M., and Sarnow, P. (2005). Modulation of hepatitis C virus RNA abundance by a liver-specific MicroR.N.A. Science *309*, 1577–1581.

Kabra, S.K., Jain, Y., Singhal, T., and Ratageri, V.H. (1999). Dengue hemorrhagic fever: clinical manifestations and management. Indian J. Pediatr. *66*, 93–101.

Kanai, R., Kar, K., Anthony, K., Gould, L.H., Ledizet, M., Fikrig, E., Marasco, W.A., Koski, R.A., and Modis, Y. (2006). Crystal structure of west nile virus envelope glycoprotein reveals viral surface epitopes. J. Virol. *80*, 11000–11008.

Kapoor, M., Zhang, L., Ramachandra, M., Kusukawa, J., Ebner, K.E., and Padmanabhan, R. (1995). Association between NS3 and NS5 proteins of dengue virus type 2 in the putative RNA replicase is linked to differential phosphorylation of NS5. J. Biol. Chem. *270*, 19100–19106.

Kaufmann, B., Nybakken, G.E., Chipman, P.R., Zhang, W., Diamond, M.S., Fremont, D.H., Kuhn, R.J., and Rossmann, M.G. (2006). West Nile virus in complex with the Fab fragment of a neutralizing monoclonal antibody. Proc. Natl. Acad. Sci. U.S.A. *103*, 12400–12404.

Kim, K.H., Willingmann, P., Gong, Z.X., Kremer, M.J., Chapman, M.S., Minor, I., Oliveira, M.A., Rossmann, M.G., Andries, K., Diana, G.D., Dutko, F.J., McKinlay, M.A., and Pevear, D.C. (1993). A comparison of the anti-rhinoviral drug binding pocket in HRV14 and HRV1A. J. Mol. Biol. *230*, 206–227.

Kinney, R.M., Huang, C.Y., Rose, B.C., Kroeker, A.D., Dreher, T.W., Iversen, P.L., and Stein, D.A. (2005). Inhibition of dengue virus serotypes 1 to 4 in vero cell cultures with morpholino oligomers. J. Virol. *79*, 5116–5128.

Kipshidze, N., Moses, J., Shankar, L.R., and Leon, M. (2001). Perspectives on antisense therapy for the prevention of restenosis. Curr. Opin. Mol. Ther. *3*, 265–277.

Kliks, S.C., Nisalak, A., Brandt, W.E., Wahl, L., and Burke, D.S. (1989). Antibody-dependent enhancement of dengue virus growth in human monocytes as a risk factor for dengue hemorrhagic fever. Am. J. Trop. Med. Hyg. *40*, 444–451.

Koff, W.C., Pratt, R.D., Elm, J.L., Jr., Venkateshan, C.N., and Halstead, S.B. (1983). Treatment of intracranial dengue virus infections in mice with a lipophilic derivative of ribavirin. Antimicrob. Agents Chemother. *24*, 134–136.

Kuhn, R.J., Zhang, W., Rossmann, M.G., Pletnev, S.V., Corver, J., Lenches, E., Jones, C.T., Mukhopadhyay, S., Chipman, P.R., Strauss, E.G., Baker, T.S., and Strauss, J.H. (2002). Structure of dengue virus: implications for flavivirus organization, maturation, and fusion. Cell 108,Kurane, I., and Takasaki, T. (2001). Dengue fever and dengue haemorrhagic fever: challenges of controlling an enemy still at large. Rev. Med. Virol. *11*, 301–311.

Kurreck, J. (2003). Antisense technologies. Improvement through novel chemical modifications. Eur. J. Biochem. *270*, 1628–1644.

Lanford, R.E., Chavez, D., Guerra, B., Lau, J.Y., Hong, Z., Brasky, K.M., and Beames, B. (2001). Ribavirin induces error-prone replication of GB virus B in primary tamarin hepatocytes. J. Virol. *75*, 8074–8081.

Lee, C.J., Liao, C.L., and Lin, Y.L. (2005). Flavivirus activates phosphatidylinositol 3-kinase signaling to block caspase-dependent apoptotic cell death at the early stage of virus infection. J. Virol. *79*, 8388–8399.

Lee, E., Pavy, M., Young, N., Freeman, C., and Lobigs, M. (2006). Antiviral effect of the heparan sulfate mimetic, PI-88, against dengue and encephalitic flaviviruses. Antiviral Res. *69*, 31–38.

Leitmeyer, K.C., Vaughn, D.W., Watts, D.M., Salas, R., Villalobos, I., de, C., Ramos, C., and Rico-Hesse, R. (1999). Dengue virus structural differences that correlate with pathogenesis. J. Virol. *73*, 4738–4747.

Li, L., Lok, S.M., Yu, I.M., Zhang, Y., Kuhn, R.J., Chen, J., and Rossmann, M.G. (2008a). The flavivirus precursor membrane-envelope protein complex: structure and maturation. Science *319*, 1830–1834.

Li, Z., Khaliq, M., Zhou, Z., Post, C.B., Kuhn, R.J., and Cushman, M. (2008b). Design, synthesis, and biological evaluation of antiviral agents targeting flavivirus envelope proteins. J. Med. Chem. *51*, 4660–4671.

Liao, M., and Kielian, M. (2005). The conserved glycine residues in the transmembrane domain of the Semliki Forest virus fusion protein are not required for assembly and fusion. Virology *332*, 430–437.

Lin, Y.W., Wang, K.J., Lei, H.Y., Lin, Y.S., Yeh, T.M., Liu, H.S., Liu, C.C., and Chen, S.H. (2002). Virus replication and cytokine production in dengue virus-infected human B lymphocytes. J. Virol. *76*, 12242–12249.

Lindenbach, B.D., and Rice, C.M. (2001). Flaviviridae: The Viruses and Their Replication. In Fields Virology, D.M. Knipe, and P.M. Howley, eds. (Lippincott Williams & Wilkins, Philadelphia), pp. 991–1041.

Lobigs, M. (1993). Flavivirus premembrane protein cleavage and spike heterodimer secretion require the function of the viral proteinase NS3. Proc. Natl. Acad. Sci. U. S. A. *90*, 6218–6222.

Lorenz, I.C., Allison, S.L., Heinz, F.X., and Helenius, A. (2002). Folding and dimerization of tick-borne encephalitis virus envelope proteins prM and E in the endoplasmic reticulum. J. Virol. *76*, 5480–5491.

Lozach, P.Y., Burleigh, L., Staropoli, I., Navarro-Sanchez, E., Harriague, J., Virelizier, J.L., Rey, F.A., Despres, P., Arenzana-Seisdedos, F., and Amara, A. (2005). Dendritic cell-specific intercellular adhesion molecule 3-grabbing non-integrin (DC-SIGN)-mediated enhancement of dengue virus infection is independent of DC-SIGN internalization signals. J. Biol. Chem. *280*, 23698–23708.

Ma, D.D., Rede, T., Naqvi, N.A., and Cook, P.D. (2000). Synthetic oligonucleotides as therapeutics: the coming of age. Biotechnol. Annu. Rev. *5*, 155–196.

Ma, Y., Chan, C.Y., and He, M.L. (2007). RNA interference and antiviral therapy. World J. Gastroenterol. *13*, 5169–5179.

Mackenzie, J.M., and Westaway, E.G. (2001). Assembly and maturation of the flavivirus Kunjin virus appear to occur in the rough endoplasmic reticulum and along the secretory pathway, respectively. J. Virol. *75*, 10787–10799.

Mady, B.J., Kurane, I., Erbe, D.V., Fanger, M.W., and Ennis, F.A. (1993). Neuraminidase augments Fc gamma receptor II-mediated antibody-dependent enhancement of dengue virus infection. The Journal of general virology 74 (Pt 5), 839–844.

Malinoski, F.J., Hasty, S.E., Ussery, M.A., and Dalrymple, J.M. (1990). Prophylactic ribavirin treatment of dengue type 1 infection in rhesus monkeys. Antiviral Res. *13*, 139–149.

Mangada, M.N., and Igarashi, A. (1998). Molecular and *in vitro* analysis of eight dengue type 2 viruses isolated from patients exhibiting different disease severities. Virology *244*, 458–466.

Marks, R.M., Lu, H., Sundaresan, R., Toida, T., Suzuki, A., Imanari, T., Hernaiz, M.J., and Linhardt, R.J. (2001). Probing the interaction of dengue virus envelope protein with heparin: assessment of glycosaminoglycan-derived inhibitors. J. Med. Chem. *44*, 2178–2187.

Marwick, C. (1998). First 'antisense' drug will treat CMV retinitis. JAMA 280, 871.

McCormick, J.B., King, I.J., Webb, P.A., Scribner, C.L., Craven, R.B., Johnson, K.M., Elliott, L.H., and Belmont-Williams, R. (1986). Lassa fever. Effective therapy with ribavirin. N. Engl. J. Med. *314*, 20–26.

McCown, M., Diamond, M.S., and Pekosz, A. (2003). The utility of siRNA transcripts produced by RNA polymerase i in down regulating viral gene expression and replication of negative- and positive-strand RNA viruses. Virology *313*, 514–524.

Mehta, A., Zitzmann, N., Rudd, P.M., Block, T.M., and Dwek, R.A. (1998). Alpha-glucosidase inhibitors as potential broad based anti-viral agents. FEBS Lett. *430*, 17–22.

Miller, J.L., deWet, B.J., Martinez-Pomares, L., Radcliffe, C.M., Dwek, R.A., Rudd, P.M., and Gordon, S. (2008). The mannose receptor mediates dengue virus infection of macrophages. PLoS Pathog. *4*, e17.

Modis, Y., Ogata, S., Clements, D., and Harrison, S.C. (2003). A ligand-binding pocket in the dengue virus envelope glycoprotein. Proc. Natl. Acad. Sci. U. S. A. *100*, 6986–6991.

Modis, Y., Ogata, S., Clements, D., and Harrison, S.C. (2004). Structure of the dengue virus envelope protein after membrane fusion. Nature *427*, 313–319.

Modis, Y., Ogata, S., Clements, D., and Harrison, S.C. (2005). Variable surface epitopes in the crystal structure of dengue virus type 3 envelope glycoprotein. J. Virol. *79*, 1223–1231.

Monath, T.P. (2008). Treatment of yellow fever. Antiviral Res. *78*, 116–124.

Morens, D.M., and Halstead, S.B. (1990). Measurement of antibody-dependent infection enhancement of four dengue virus serotypes by monoclonal and polyclonal antibodies. J. Gen. Virol. 71 (Pt 12), 2909–2914.

Moulton, H.M., Nelson, M.H., Hatlevig, S.A., Reddy, M.T., and Iversen, P.L. (2004). Cellular uptake of antisense morpholino oligomers conjugated to arginine-rich peptides. Bioconjug. Chem. *15*, 290–299.

Mukhopadhyay, S., Kim, B.S., Chipman, P.R., Rossmann, M.G., and Kuhn, R.J. (2003). Structure of West Nile virus. Science *302*, 248.

Mukhopadhyay, S., Kuhn, R.J., and Rossmann, M.G. (2005). A structural perspective of the flavivirus life cycle. Nat. Rev. Microbiol. 3, 13–22.

Munoz-Jordan, J.L., Sanchez-Burgos, G.G., Laurent-Rolle, M., and Garcia-Sastre, A. (2003). Inhibition of interferon signaling by dengue virus. Proc. Natl. Acad. Sci. U. S. A. *100*, 14333–14338.

Navarro-Sanchez, E., Altmeyer, R., Amara, A., Schwartz, O., Fieschi, F., Virelizier, J.L., Arenzana-Seisdedos, F., and Despres, P. (2003). Dendritic-cell-specific ICAM3-grabbing non-integrin is essential for the productive infection of human dendritic cells by mosquito-cell-derived dengue viruses. EMBO Rep. *4*, 723–728.

Neuman, B.W., Stein, D.A., Kroeker, A.D., Paulino, A.D., Moulton, H.M., Iversen, P.L., and Buchmeier, M.J. (2004). Antisense morpholino-oligomers directed against the 5′ end of the genome inhibit coronavirus proliferation and growth. J. Virol. *78*, 5891–5899.

Neves-Souza, P.C., Azeredo, E.L., Zagne, S.M., Valls-de-Souza, R., Reis, S.R., Cerqueira, D.I., Nogueira, R.M., and Kubelka, C.F. (2005). Inducible nitric oxide synthase (iNOS) expression in monocytes during acute Dengue Fever in patients and during *in vitro* infection. BMC Infect. Dis. *5*, 64.

Padalko, E., Verbeken, E., De Clercq, E., and Neyts, J. (2004). Inhibition of coxsackie B3 virus induced myocarditis in mice by 2-(3,4-dichlorophenoxy)-5-nitrobenzonitrile. J. Med. Virol. 72, 263–267.

Pandey, B.D., and Igarashi, A. (2000). Severity-related molecular differences among nineteen strains of dengue type 2 viruses. Microbiol. Immunol. *44*, 179–188.

Parker, W.B. (2005). Metabolism and antiviral activity of ribavirin. Virus Res. *107*, 165–171.

Persichini, T., Colasanti, M., Lauro, G.M., and Ascenzi, P. (1998). Cysteine nitrosylation inactivates the HIV-1 protease. Biochem. Biophys. Res. Commun. *250*, 575–576.

Pfeffer, S., Sewer, A., Lagos-Quintana, M., Sheridan, R., Sander, C., Grässer, F.A., van Dyk, L.F., Ho, C.K., Shuman, S., Chien, M., Russo, J.J., Ju, J., Randall, G., Lindenbach, B.D., Rice, C.M., Simon, V., Ho, D.D., Zavolan, M., and Tuschl, T. (2005). Identification of microRNAs of the herpesvirus family. Nat. Methods 2, 269–276.

Pokidysheva, E., Zhang, Y., Battisti, A.J., Bator-Kelly, C.M., Chipman, P.R., Xiao, C., Gregorio, G.G., Hendrickson, W.A., Kuhn, R.J., and Rossmann, M.G. (2006). Cryo-EM reconstruction of dengue virus in complex with the carbohydrate recognition domain of DC-SIGN. Cell *124*, 485–493.

Qureshi, N.M., Coy, D.H., Garry, R.F., and Henderson, L.A. (1990). Characterization of a putative cellular receptor for HIV-1 transmembrane glycoprotein using synthetic peptides. AIDS 4, 553–558.

Raviprakash, K., Liu, K., Matteucci, M., Wagner, R., Riffenburgh, R., and Carl, M. (1995). Inhibition of dengue virus by novel, modified antisense oligonucleotides. J. Virol. *69*, 69–74.

Rawlinson, S.M., Pryor, M.J., Wright, P.J., and Jans, D.A. (2006). Dengue virus RNA polymerase NS5: a potential therapeutic target? Curr. Drug Targets 7, 1623–1638.

Ray, D., and Shi, P.Y. (2006). Recent advances in flavivirus antiviral drug discovery and vaccine development. Recent Patents Anti-Infect. Drug Disc. *1*, 45–55.

Reisdorph, N., Thomas, J.J., Katpally, U., Chase, E., Harris, K., Siuzdak, G., and Smith, T.J. (2003). Human rhinovirus capsid dynamics is controlled by canyon flexibility. Virology *314*, 34–44.

Rey, F.A., Heinz, F.X., Mandl, C., Kunz, C., and Harrison, S.C. (1995). The envelope glycoprotein from tick-borne encephalitis virus at 2 A resolution. Nature 375, 291–298.

Rossmann, M.G., Arnold, E., Erickson, J.W., Frankenberger, E.A., Griffith, J.P., Hecht, H.J., Johnson, J.E., Kamer, G., Luo, M., Mosser, A.G., Rueckert, R.R., Sherry, S., and Vriend, G. (1985). Structure of a human common cold virus and functional relationship to other picornaviruses. Nature 317, 145–153.

Rossmann, M.G., Bella, J., Kolatkar, P.R., He, Y., Wimmer, E., Kuhn, R.J., and Baker, T.S. (2000). Cell recognition and entry by rhino- and enteroviruses. Virology 269, 239–247.

Rossmann, M.G., and McKinlay, M.A. (1992). Application of crystallography to the design of antiviral agents. Infect. Agents Dis. 1, 3–10.

Saura, M., Zaragoza, C., McMillan, A., Quick, R.A., Hohenadl, C., Lowenstein, J.M., and Lowenstein, C.J. (1999). An antiviral mechanism of nitric oxide: inhibition of a viral protease. Immunity 10, 21–28.

Schibli, D.J., and Weissenhorn, W. (2004). Class I and class II viral fusion protein structures reveal similar principles in membrane fusion. Mol. Membr. Biol. 21, 361–371.

Shigeta, S., Mori, S., Kodama, E., Kodama, J., Takahashi, K., and Yamase, T. (2003). Broad spectrum anti-RNA virus activities of titanium and vanadium substituted polyoxotungstates. Antiviral Res. 58, 265–271.

Sidwell, R.W., Huffman, J.H., Khare, G.P., Allen, L.B., Witkowski, J.T., and Robins, R.K. (1972). Broad-spectrum antiviral activity of Virazole: 1-beta-D-ribofuranosyl-1,2,4-triazole-3-carboxamide. Science 177, 705–706.

Smee, D.F., Bray, M., and Huggins, J.W. (2001). Antiviral activity and mode of action studies of ribavirin and mycophenolic acid against orthopoxviruses in vitro. Antivir. Chem. Chemother. 12, 327–335.

Smith, T.J., Kremer, M.J., Luo, M., Vriend, G., Arnold, E., Kamer, G., Rossmann, M.G., McKinlay, M.A., Diana, G.D., and Otto, M.J. (1986). The site of attachment in human rhinovirus 14 for antiviral agents that inhibit uncoating. Science 233, 1286–1293.

Stadler, K., Allison, S.L., Schalich, J., and Heinz, F.X. (1997). Proteolytic activation of tick-borne encephalitis virus by furin. J. Virol. 71, 8475–8481.

Stein, D.A., Skilling, D.E., Iversen, P.L., and Smith, A.W. (2001). Inhibition of Vesivirus infections in mammalian tissue culture with antisense morpholino oligomers. Antisense Nucleic Acid Drug Dev. 11, 317–325.

Stiasny, K., Allison, S.L., Mandl, C.W., and Heinz, F.X. (2001). Role of metastability and acidic pH in membrane fusion by tick-borne encephalitis virus. J. Virol. 75, 7392–7398.

Stiasny, K., Allison, S.L., Marchler-Bauer, A., Kunz, C., and Heinz, F.X. (1996). Structural requirements for low-pH-induced rearrangements in the envelope glycoprotein of tick-borne encephalitis virus. J. Virol. 70, 8142–8147.

Stiasny, K., Allison, S.L., Schalich, J., and Heinz, F.X. (2002). Membrane interactions of the tick-borne encephalitis virus fusion protein E at low pH. J. Virol. 76, 3784–3790.

Stiasny, K., Bressanelli, S., Lepault, J., Rey, F.A., and Heinz, F.X. (2004). Characterization of a membrane-associated trimeric low-pH-induced Form of the class II viral fusion protein E from tick-borne encephalitis virus and its crystallization. J. Virol. 78, 3178–3183.

Stiasny, K., and Heinz, F.X. (2006). Flavivirus membrane fusion. J. Gen. Virol. 87, 2755–2766.

Streeter, D.G., Witkowski, J.T., Khare, G.P., Sidwell, R.W., Bauer, R.J., Robins, R.K., and Simon, L.N. (1973). Mechanism of action of 1- -D-ribofuranosyl-1,2,4-triazole-3-carboxamide (Virazole), a new broad-spectrum antiviral agent. Proc. Natl. Acad. Sci. U. S. A. 70, 1174–1178.

Suhrbier, A., and La Linn, M. (2003). Suppression of antiviral responses by antibody-dependent enhancement of macrophage infection. Trends Immunol. 24, 165–168.

Sullivan, C.S., Grundhoff, A.T., Tevethia, S., Pipas, J.M., and Ganem, D. (2005). SV40-encoded microRNAs regulate viral gene expression and reduce susceptibility to cytotoxic T-cells. Nature 435, 682–686.

Summerton, J., and Weller, D. (1997). Morpholino antisense oligomers: design, preparation, and properties. Antisense Nucleic Acid Drug Dev. 7, 187–195.

Taber, L.H., Knight, V., Gilbert, B.E., McClung, H.W., Wilson, S.Z., Norton, H.J., Thurson, J.M., Gordon, W.H., Atmar, R.L., and Schlaudt, W.R. (1983). Ribavirin aerosol treatment of bronchiolitis associated with respiratory syncytial virus infection in infants. Pediatrics 72, 613–618.

Takhampunya, R., Padmanabhan, R., and Ubol, S. (2006a). Antiviral action of nitric oxide on dengue virus type 2 replication. J. Gen. Virol. 87, 3003–3011.

Takhampunya, R., Ubol, S., Houng, H.S., Cameron, C.E., and Padmanabhan, R. (2006b). Inhibition of dengue virus replication by mycophenolic acid and ribavirin. J. Gen. Virol. 87, 1947–1952.

Talarico, L.B., Pujol, C.A., Zibetti, R.G., Faria, P.C., Noseda, M.D., Duarte, M.E., and Damonte, E.B. (2005). The antiviral activity of sulfated polysaccharides against dengue virus is dependent on virus serotype and host cell. Antiviral Res. 66, 103–110.

Tan, F.L., and Yin, J.Q. (2004). RNAi, a new therapeutic strategy against viral infection. Cell Res. 14, 460–466.

Tassaneetrithep, B., Burgess, T.H., Granelli-Piperno, A., Trumpfheller, C., Finke, J., Sun, W., Eller, M.A., Pattanapanyasat, K., Sarasombath, S., Birx, D.L., Steinman, R.M., Schlesinger, S., and Marovich, M.A. (2003). DC-SIGN (CD209) mediates dengue virus infection of human dendritic cells. J. Exp. Med. 197, 823–829.

ter Brake, O., Konstantinova, P., Ceylan, M., and Berkhout, B. (2006). Silencing of HIV-1 with RNA interference: a multiple shRNA approach. Mol. Ther. 14, 883–892.

Twiddy, S.S., Farrar, J.J., Vinh Chau, N., Wills, B., Gould, E.A., Gritsun, T., Lloyd, G., and Holmes, E.C. (2002). Phylogenetic relationships and differential selection pressures among genotypes of dengue-2 virus. Virology 298, 63–72.

Vaughn, D.W., Green, S., Kalayanarooj, S., Innis, B.L., Nimmannitya, S., Suntayakorn, S., Endy, T.P., Raengsakulrach, B., Rothman, A.L., Ennis, F.A., and Nisalak, A. (2000). Dengue viremia titer, antibody response pattern, and virus serotype correlate with disease severity. J. Infect. Dis. *181*, 2–9.

Veiga, A.S., Santos, N.C., and Castanho, M.A. (2006). An insight on the leading HIV entry inhibitors. Recent Patents Anti-Infect. Drug Disc. *1*, 67–73.

Vo, N.V., Young, K.C., and Lai, M.M. (2003). Mutagenic and inhibitory effects of ribavirin on hepatitis C virus RNA polymerase. Biochemistry (Mosc.) *42*, 10462–10471.

Watts, D.M., Porter, K.R., Putvatana, P., Vasquez, B., Calampa, C., Hayes, C.G., and Halstead, S.B. (1999). Failure of secondary infection with American genotype dengue 2 to cause dengue haemorrhagic fever. Lancet *354*, 1431–1434.

Whitby, K., Pierson, T.C., Geiss, B., Lane, K., Engle, M., Zhou, Y., Doms, R.W., and Diamond, M.S. (2005). Castanospermine, a potent inhibitor of dengue virus infection *in vitro* and *in vivo*. J. Virol. *79*, 8698–8706.

Whitehead, S.S., Blaney, J.E., Durbin, A.P., and Murphy, B.R. (2007). Prospects for a dengue virus vaccine. Nat. Rev. Microbiol. *5*, 518–528.

WHO (1993). Vaccine prospects for dengue haemorrhagic fever. In World Health Forum, p. 203.

WHO (1997). Dengue haemorrhagic fever: diagnosis, treatment, prevention and control. 1997. In Geneva: World Health Organization (WHO).

Wild, C.T., Shugars, D.C., Greenwell, T.K., McDanal, C.B., and Matthews, T.J. (1994). Peptides corresponding to a predictive alpha-helical domain of human immunodeficiency virus type 1 gp41 are potent inhibitors of virus infection. Proc. Natl. Acad. Sci. U. S. A. *91*, 9770–9774.

Willis, R.C., Carson, D.A., and Seegmiller, J.E. (1978). Adenosine kinase initiates the major route of ribavirin activation in a cultured human cell line. Proc. Natl. Acad. Sci. U. S. A. *75*, 3042–3044.

Wu, S.F., Lee, C.J., Liao, C.L., Dwek, R.A., Zitzmann, N., and Lin, Y.L. (2002). Antiviral effects of an iminosugar derivative on flavivirus infections. J. Virol. *76*, 3596–3604.

Wu, S.J., Grouard-Vogel, G., Sun, W., Mascola, J.R., Brachtel, E., Putvatana, R., Louder, M.K., Filgueira, L., Marovich, M.A., Wong, H.K., Blauvelt, A., Murphy, G.S., Robb, M.L., Innes, B.L., Birx, D.L., Hayes, C.G., and Frankel, S.S. (2000). Human skin Langerhans cells are targets of dengue virus infection. Nat. Med. *6*, 816–820.

Yang, J.M., Chen, Y.F., Tu, Y.Y., Yen, K.R., and Yang, Y.L. (2007). Combinatorial computational approaches to identify tetracycline derivatives as flavivirus inhibitors. PLoS ONE 2, e428.

Yu, I.M., Zhang, W., Holdaway, H.A., Li, L., Kostyuchenko, V.A., Chipman, P.R., Kuhn, R.J., Rossmann, M.G., and Chen, J. (2008). Structure of the immature dengue virus at low pH primes proteolytic maturation. Science *319*, 1834–1837.

Zhang, W., Chipman, P.R., Corver, J., Johnson, P.R., Zhang, Y., Mukhopadhyay, S., Baker, T.S., Strauss, J.H., Rossmann, M.G., and Kuhn, R.J. (2003a). Visualization of membrane protein domains by cryo-electron microscopy of dengue virus. Nat. Struct. Biol. *10*, 907–912.

Zhang, Y., Corver, J., Chipman, P.R., Zhang, W., Pletnev, S.V., Sedlak, D., Baker, T.S., Strauss, J.H., Kuhn, R.J., and Rossmann, M.G. (2003b). Structures of immature flavivirus particles. EMBO J. *22*, 2604–2613.

Zhang, Y., Kaufmann, B., Chipman, P.R., Kuhn, R.J., and Rossmann, M.G. (2007). Structure of immature West Nile virus. J. Virol. *81*, 6141–6145.

Zhang, Y., Zhang, W., Ogata, S., Clements, D., Strauss, J.H., Baker, T.S., Kuhn, R.J., and Rossmann, M.G. (2004). Conformational changes of the flavivirus E glycoprotein. Structure *12*, 1607–1618.

Zhou, S., Liu, R., Baroudy, B.M., Malcolm, B.A., and Reyes, G.R. (2003). The effect of ribavirin and IMPDH inhibitors on hepatitis C virus subgenomic replicon RNA Virology *310*, 333–342.

Zhou, Z., Khaliq, M., Suk, J.E., Patkar, C., Li, L., Kuhn, R.J., and Post, C.B. (2008). Antiviral compounds discovered by virtual screening of small-molecule libraries against dengue virus E protein. ACS Chem. Biol. *3*, 765–775.

Progress in Passive Immunotherapy

Ana P. Goncalvez, Robert H. Purcell and Ching-Juh Lai

Abstract

Dengue is currently endemic in more than one hundred countries around the world. It causes approximately 50–100 million infections annually, including 250,000–500,000 cases of dengue haemorrhagic fever/dengue shock syndrome (DHF/DSS). According to the World Health Organization (WHO), two-fifths of the world population is at risk of dengue virus (DENV) infection. It has been suggested that globalization and climate change have had a significant impact on the emergence of DENV in new areas. No vaccine or therapy against DENV is currently approved for use in humans, and alternative strategies to control DENV infection are urgently needed, particularly because the design of such strategies may also inform efforts in vaccine design. This chapter outlines the prophylaxis/therapeutic potential of monoclonal antibodies (MAbs) against DENV and highlights the challenges to implementation of this strategy, including antibody-dependent enhancement (ADE), genetic variability of DENV strains, potential for selection of MAb escape variants, and financial cost. Moreover, we describe recent immunologic and structural studies that have provided a new understanding of antibody-mediated neutralization mechanisms and protection against DENV and other flavivirus infections. These insights are having an important impact on the development of vaccines and antibody-based therapies.

Overview

Dengue is one of the most important resurgent tropical diseases of the past 30 years, with an expanding geographic distribution of both the viruses and its mosquito vectors, increased frequency of epidemics, co-circulation of multiple virus serotypes and emergence of dengue haemorrhagic fever/dengue shock syndrome (DHF/DSS) in new areas (Kyle and Harris, 2008). Four serotypes of dengue virus (DENV) are transmitted among humans by *Aedes aegypti* and *Aedes albopictus* mosquitoes. These account for approximately 100 million infections leading to dengue fever (DF) and up to 500,000 cases of DHF/DSS and 25,000 deaths annually (Gubler, 1998). Development of a safe and effective vaccine against dengue has proven challenging (see Chapter 12) and alternative strategies for prevention of DF/DHF are needed.

Passive immunization with immune sera from humans or animals has been widely used for prophylaxis and therapy for viral and bacterial diseases since the 1890s (Casadevall *et al.*, 2004), with clinical success in many cases. However, disadvantages, which include the risk of allergic reactions and non-standard manufacturing, have limited their use. Although immunoglobulins are generally safe (Pennington, 1990), side-effects have been associated with high doses, including renal failure (Pasatiempo *et al.*, 1994) and aseptic meningitis (Sekul *et al.*, 1994). Monoclonal antibodies (MAbs) represent an attractive alternative to animal and human derived sera for several reasons. MAbs are homogeneous immunoglobulins that recognize one epitope and have higher specific activity than polyclonal sera (Lang *et al.*, 1993). Additionally, MAbs are not derived from human blood, and the risk for contamination with blood-borne viruses is markedly reduced. In contrast to polyclonal sera, MAbs are reproducible reagents that can be produced in large quantities, and titres can be adjusted to generate standardized products (Farid, 2006).

In spite of their advantages, few MAbs for infectious agents, including rabies, WNV, CMV and EBV, are in clinical trials [Reviewed in (Marasco and Sui, 2007; Xiao and Dimitrov, 2007)], and only one humanized MAb, against respiratory syncytial virus (RSV), has been marketed (http://www.fda.gov/cder/biologics/products/palimed102302.htm). Nevertheless, human or humanized monoclonal antibodies might find relevant applications, especially against infectious agents for which vaccines and traditional therapies do not exist, such as DENV.

Several examples document the effectiveness of serotherapy against flaviviruses in humans. During the West Nile virus (WNV) outbreak in Israel in 2000, a woman with chronic lymphocytic leukemia became comatose after contracting WNV encephalitis. She recovered after treatment with intravenous immunoglobulin (IVIG) (Shimoni *et al.*, 2001). A second patient, a lung transplant recipient, also recovered from WNV encephalitis after treatment with Israeli IVIG (Hamdan *et al.*, 2002). Six other patients, who were subsequently treated, had variable outcomes: two improved, two had no improvement, and two died (Agrawal and Petersen, 2003; Haley *et al.*, 2003). Although specific anti-tick-borne encephalitis virus (TBEV) immune globulin preparations (Encegam®, FSME-Bulin®) for post-exposure treatment have been available since the 1970s, their effectiveness for disease prevention has never been demonstrated (Roehrig *et al.*, 2001). Currently, Latvia is the only European country where TBE-specific immunoglobulin preparations, imported from Russia, are still in use (Broker and Kollaritsch, 2008). Although inconclusive, these reports have stimulated interest in the use of passive immunization for treating disease caused by flaviviruses.

Studies of MAbs in mice, non-human primates and other species have paved the way for development of human therapies. The prophylactic and therapeutic potential of MAbs against DENV and other flaviviruses has now been demonstrated (Henchal *et al.*, 1988; Kaufman *et al.*, 1989; Kaufman *et al.*, 1987; Kimura-Kuroda and Yasui, 1988; Lai *et al.*, 2007; Mathews and Roehrig, 1984; Oliphant *et al.*, 2005; Throsby *et al.*, 2006; Zhang *et al.*, 1989a). However substantial challenges to the deployment of MAb therapy for flavivirus disease remain. Most prominently, DHF/DSS is thought to result from antibody-dependent enhancement (ADE), increased uptake of virus when it is complexed to pre-existing subneutralizing or non-neutralizing antibodies resulting from a prior infection with a heterologous DENV serotype. In theory, ADE of DENV infection could complicate the prophylactic/therapeutic administration of antibodies to humans. Improvement in antiviral activity without the risk of ADE could potentially be achieved by modification of Fc effector functions.

Besides ADE, additional challenges remain for MAb therapy against DENV. The high costs of production, storage, and administration of antibodies are disadvantages of antibody-based therapies. For example, the cost of Synagis® for the prevention of RSV disease is estimated at approximately $4000.00 per child per season, which affects the cost–benefit ratio (Reeve *et al.*, 2006). The efficacy of MAb-based therapies may be diminished by the genetic variability of DENV and the emergence of MAb escape mutants. DENV has a significant genetic diversity, demonstrable in its four distinct serotypes (see Chapters 9 and 11). Moreover, antibody-resistant mutants can be generated *in vitro*, and it is likely that the same can occur in patients. High diversity of MAbs may overcome these problems.

In the remainder of this chapter, we describe the development of MAbs against flaviviruses, and the recent advances in our understanding of neutralizing and non-neutralizing antibody responses against DENV and other flaviviruses, including the examination of key neutralization epitopes on flavivirus envelope (E) protein and MAb-associated antiviral mechanisms of action. We also discuss in this section the role of Fc effector functions in protection against dengue, the implications of prophylactic use of MAbs with a modified Fc region, and the overall potential of passive prophylaxis/therapy with anti-DENV MAbs.

Development of MAbs against flaviviruses

Flavivirus neutralizing MAbs have been isolated in a variety of species using a broad range of approaches.

Mouse hybridomas

This was the first technology developed for the isolation of neutralizing MAbs. Mouse hybridomas are generated from the stable fusion of immortalized myeloma cells with B-cells from immunized mice (Kohler and Milstein, 1975). This technology was widely applied to flaviviruses with the purpose of studying their antigenic structures and replication cycles [Reviewed in (Roehrig, 2003)]. Studies with mouse MAbs have revealed three antigenic domains of flavivirus envelope (E) glycoprotein, with the identification of flavivirus group-, complex-, and type-specific epitopes. It has been shown with these studies that the E-glycoprotein is responsible for two critical functions: virus attachment to target cells and fusion of virus and cell membranes. In addition, mouse MAbs have been extremely important for demonstration of antibody-mediated passive protection against flavivirus diseases (Chiba et al., 1999; Gupta et al., 2003; Henchal et al., 1988; Iacono-Connors et al., 1996; Kaufman et al., 1989; Kaufman et al., 1987; Kimura-Kuroda and Yasui, 1988; Mathews and Roehrig, 1984; Niedrig et al., 1994; Phillpotts et al., 1987; Schlesinger et al., 1985, 1987; Schlesinger and Chapman, 1995; Tan et al., 1990; Yamauchi, 1989; Zhang et al., 1989a).

In spite of the great utility of mouse MAbs as tools for basic research, they have certain properties that have limited their clinical use. First, mouse MAbs can induce an immunogenic response in humans, so-called human anti-mouse antibodies (HAMA) or human anti-globulin antibodies (HAGA). The HAMA response can compromise the murine MAb efficacy because of immunologically mediated rapid clearance as well as the potential for allergic reactions (Mirick et al., 2004). Furthermore, mouse antibodies do not bind the human neonatal Fc receptor (hFcRn) (Ober et al., 2001), also resulting in a relatively faster clearance in humans (Frodin et al., 1990). In addition, mouse MAbs exhibit inefficient antibody effector functions [e.g. antibody-dependent cellular cytotoxicity (ADCC), complement-dependent cytotoxicity (CDC) and phagocytosis] (Carter, 2006; Nimmerjahn and Ravetch, 2008). Nevertheless, the limitations of mouse antibodies can be solved, in part, by chimerization or humanization (Gonzales et al., 2005).

Chimerization or humanization of murine MAbs

This is the oldest methodology for generating therapeutic humanized MAbs, and it is particularly useful for extensively characterized murine MAbs with antiviral properties. Initially, the development of chimeric murine MAbs involved the replacement of the mouse constant region with the human constant region of the antibody (Morrison et al., 1984). Subsequently, this approach was extended to the construction of humanized antibodies by joining the variable (V; antigen binding) domains of a mouse MAb to the constant domains of a human antibody (Riechmann et al., 1988). Nowadays, all of the therapeutic antibodies on the market are human/mouse chimeric or humanized antibodies. The generation of an antibody against WNV with potential use as a therapeutic is the only example of the utility of this technology in the flavivirus field (Oliphant et al., 2005). This mouse MAb, E16, is specific for domain III (DIII) of the WNV Envelope (E) glycoprotein and has a potent neutralizing activity in vitro and in vivo. The humanized version of E16 successfully retained its antigen specificity, avidity, and neutralizing activity in vitro (Oliphant et al., 2005). Moreover, a single dose of humanized E16 protected mice against lethal WNV infection when administered after WNV challenge. Therefore, this MAb is a potential option for treatment against WNV infection in humans.

Human antibodies from transgenic mice

The generation of human antibodies by immunization of mice that are transgenic for human immunoglobulin (Ig) genes was first described in 1994 (Green et al., 1994; Lonberg, 2008; Lonberg et al., 1994). Briefly, mice with knocked out endogenous Ig genes are bred with mice transgenic for the human Ig loci. The resulting mice, unable to generate murine antibodies, rearrange the transgenic human Ig loci for the generation of human antibody repertoires. The binding affinity of these antibodies is often high, reflecting *in vivo* affinity maturation. However, the transgenic mouse system is limited to human pathogens that are able to infect mice. In addition, more immunizations or extensive antibody screening are needed with these mice, possibly due to differences in binding of human Igs to mouse Fc receptors, compared with their mouse counterparts (Nimmerjahn and Ravetch, 2008).

The transchromo (TC) mouse technology, which allows the introduction of DNA megabase-sized DNA segments into cells, represents the next step in the evolution of procedures involving transgenic mice. TC technology has been applied to introduce the entire human Ig loci in mice, which then express a fully diverse repertoire of human Igs. However, the instability of the Igκ locus resulted in inefficient hybridoma production. This was solved by cross-breeding the Kirin TC Mouse bearing the human Ig heavy chain (HCF14) with the Medarex YAC-transgenic mouse bearing approximately 50% of the human IgVκ gene segments, stably integrated into the mouse genome (Ishida *et al.*, 2002). The resulting mouse, the KM Mouse, performed similarly to the normal mouse in terms of efficiency of hybridoma production, antibody affinity constants, and antigen-specific titres of IgG. There has been little experience with antibody generation in TC mice infected with flaviviruses. A single report described antisera and purified γ-globulin from TC mice immunized with Japanese encephalitis virus (JEV) that had an *in vitro* neutralization titre 33-fold higher than immune human serum (Ishida *et al.*, 2002). Hence, more studies are needed to evaluate fully the utility of TC mouse technology.

Anti-flavivirus antibodies from phage display libraries

Display technologies utilize microorganisms, including phage, virus, bacteria, and yeast, as well as non-infectious entities (mammalian cells and ribosomes), to present repertoires of single-chain variable antibody fragments (scFvs), antigen-binding fragments (Fabs) or domain antibodies (Dabs) expressed on their surfaces. These can then be enriched and isolated through several cycles of selection against the desired target (Chanock *et al.*, 1993; Ho *et al.*, 2006; Hoogenboom, 2005; Huse *et al.*, 1989). Display technologies have become standard methods for isolating human MAbs since they can be applied to any antigen and can result in antibodies of high affinity through repetitive rounds of selection (Burton and Barbas, 1994; Hoogenboom, 2005).

Construction of an antibody library involves amplification of diverse immunoglobulin variable-region gene segments (as scFv or Fab fragments) from natural sources, such as from animal or human B-cells. The library is then cloned for expression or display on the surface of carrier particles that also encode the expressed gene. The antibody repertoire displayed on these particles is subjected to selection of clones binding to a desirable target; antibodies that do not bind are removed while those that do are amplified. After several rounds of selection, antibodies binding to antigen are screened, for example, using enzyme-linked immunosorbent assay (ELISA) or fluorescent-activated cell sorting (FACS). The genes encoding antibody variable regions with a particular specificity can then be cloned into whole human IgG expression vectors and transfected into cell lines to produce human MAbs (Sanna *et al.*, 1999; Sarantopoulos *et al.*, 1994).

Phage display technology has been used for selection of human MAbs against DENV and WNV from non-immunized (so-called naïve) individuals (Cabezas *et al.*, 2008; Gould *et al.*, 2005). Human scFv fragments that bind the envelope protein of WNV were selected from two large phage display libraries (Gould *et al.*, 2005; Sui *et al.*, 2004). These fragments were able to protect mice from death when given prior to infection with virus. Moreover, some of these antibody fragments had therapeutic effect when given at days 1 and 4 after infection. A second study reported the selection of scFv fragments from a human naïve library that were predominantly specific to DENV serotypes 1, 3 or 4. Two of those fragments were cross-reactive with at least two DENV serotypes (Cabezas *et al.*, 2008). The neutralizing activity, *in vitro* and *in vivo*, of these scFv fragments was not investigated. Because these libraries were created from naïve humans, the phage-displayed scFv fragments do not necessarily represent the same antibodies that would be found in an immune individual. For instance, MAbs selected from immune chimpanzees infected with DENV were predominantly cross-reactive to all four serotypes of DENV and other flaviviruses (Goncalvez *et al.*, 2004a; Goncalvez *et al.*, 2004b; Men *et al.*, 2004). In contrast, the scFv fragments against DENV selected from the naïve human library were predominantly serotype-specific (Cabezas *et al.*, 2008).

The isolation of antibodies from immune humans offers the advantage of fully exploiting the strength of the human antibody response to a human pathogen. The human immune system is exposed to all of the relevant antigens displayed by an infecting agent and can consequently have a broad response to most of these epitopes. Antibody phage libraries from humans who were previously infected with yellow fever virus (YFV), WNV or JEV have been generated in recent years (Arakawa *et al.*, 2007; Daffis *et al.*, 2005; Throsby *et al.*, 2006). Only MAbs recovered from convalescent WNV-infected humans were tested for their protective ability *in vivo* (Throsby *et al.*, 2006). One hundred twenty-one MAbs specifically bound to the WNV E protein and four MAbs to the premembrane (prM) protein. Neutralizing activity was demonstrated *in vitro* for 12 MAbs, and three of these antibodies protected mice from lethal WNV challenge.

Due to their close genetic relationship with humans, chimpanzees have also shown utility for the selection of novel and potentially therapeutic MAbs. Chimpanzees can be infected and/or immunized to obtain optimal neutralizing antibody titres and other immune responses. Therefore, they represent a valuable source of MAbs that can identify epitopes likely to be functionally important in humans. Furthermore, chimpanzee immunoglobulins are ~98% homologous to those of humans, so that MAbs from these animals are potentially useful for passive immunotherapy (Ehrlich *et al.*, 1990; Ueda *et al.*, 1988). Phage display-based selection has identified chimpanzee MAbs against a variety of infectious agents, including HIV (Vijh-Warrier *et al.*, 1995; Warrier *et al.*, 1994), RSV (Crowe *et al.*, 1994), hepatitis viruses (Schofield *et al.*, 2000; Schofield *et al.*, 2002), anthrax (Chen *et al.*, 2006b) and vaccinia virus (Chen *et al.*, 2007b; Chen *et al.*, 2006a). Notably, chimpanzee MAbs have also been generated against flaviviruses, including DENV (Goncalvez *et al.*, 2004a; Men *et al.*, 2004) and JEV (Goncalvez *et al.*, 2008).

Generally speaking, phage display-based selection is a more widely used method than yeast display-based selection for development of human MAbs. However, these are complementary methods, since selection biases may occur during panning. For instance, a recent study has shown that antibodies selected from an HIV-1 immune scFv cDNA library displayed on yeast were not only similar to many antibodies isolated from the same library displayed on phage, but yeast yielded twice as many novel antibodies (Bowley *et al.*, 2007). In addition, the method of presentation of antigen for panning may affect the availability of neutralizing epitopes. Other limitations in the use of phage or yeast display technology are that target antigens must be accessed *a priori*, functional assays (i.e. viral neutralization) cannot be used in the initial step of selection and antibodies isolated in *Escherichia coli* or yeast may be expressed suboptimally in mammalian cells.

Memory B-cell immortalization

Characteristically, memory B-cells persist throughout a human lifetime and therefore represent an excellent resource of specific antibodies generated after infection or vaccination. This source is available even several years after antigenic stimulation by taking a small sample of peripheral blood (Bernasconi *et al.*, 2002). In addition, memory B-cells conserve the potential of considerable growth and can be immortalized by Epstein Barr virus (EBV). The EBV-transformed B-cells secrete antibodies in large quantities, thereby facilitating screening of specific antibodies in culture supernatants (Bernasconi *et al.*, 2002; Simmons *et al.*, 2007; Traggiai *et al.*, 2004). Another advantage of EBV-immortalized B-cells is the expression of surface Ig, which allows the antigen-based selection of these cells (Leyendeckers *et al.*, 1999).

In this approach, memory B-cells are isolated from peripheral blood mononuclear cells (PBMCs) of an individual recovering from a viral infection. They are transformed by EBV in the presence of polyclonal stimuli, such as CpG oligodeoxynucleotides and irradiated allogeneic PBMCs. CpG oligonucleotides stimulate B-cells via Toll-like receptor 9 (TLR9) ligands that can increase more than 100-fold the efficiency of transformation (Krieg, 2002; Traggiai *et al.*,

2004). Positive cultures, identified by binding and neutralization assays, are cloned by limiting dilution and human MAbs are produced from the cloned memory B-cells (Bernasconi *et al.*, 2002). Memory B-cells isolated from the blood of patients who recovered from SARS-CoV and H5N1 avian influenza infections have been immortalized to isolate MAbs specific for different viral proteins (Simmons *et al.*, 2007; Traggiai *et al.*, 2004). At present, this approach is being used to develop human MAbs against other infectious agents, including HCV, HIV-1, CMV and DENV (Lanzavecchia *et al.*, 2007).

Identification of flavivirus neutralizing epitopes with mouse, chimpanzee and human MAbs

Specific antibodies are essential components of the adaptive immune response to flaviviruses and contribute to virus clearance and maintenance of long-lasting immunity. Although antibodies are believed to protect through Fc-γ-receptor-dependent viral clearance and complement-mediated lysis of virus or infected cells, the most critical function of antibodies seems to be interfering with virus entry into cells through neutralization of receptor binding and inhibition of fusion. MAbs generated through the approaches outlined above are valuable tools to analyse the antigenic structure of the flavivirus proteins at the molecular level. A variety of *in vitro* methodologies have been used for epitope mapping of flavivirus MAbs (Table 14.1), and have provided much information about mechanisms of neutralization.

The positive-stranded RNA genome of flaviviruses encodes a single polyprotein that is processed to produce three structural proteins, i.e. capsid (C), precursor membrane (prM) and envelope (E) proteins, and seven non-structural (NS) proteins (Lindenbach and Rice, 2003) (see Chapter 2). The prM protein is further cleaved by the cellular enzyme furin before release of the mature virion. Two transmembrane proteins are exposed on the surface of the resulting mature and infectious virion, the antiparallel homodimer E-protein and the M-protein (Kuhn *et al.*, 2002; Zhang *et al.*, 2003a). Antibody responses have been described against both surface proteins, E

and M. However, the primary target for neutralizing antibodies is the E protein. Another target of protective antibodies is the NS1 protein (Chung *et al.*, 2006; Henchal *et al.*, 1988), which is secreted from infected cells (Winkler *et al.*, 1989). Antibodies against NS3 (Tan *et al.*, 1990) and NS5 (Wong *et al.*, 2003) are also produced during infection but their role is unknown.

The envelope glycoprotein (E), the major target of neutralizing and protective antibodies

In the replication cycle of flaviviruses, the E protein mediates two important functions: virus attachment to a putative cell receptor(s) and viral fusion with the cellular membranes (Mukhopadhyay *et al.*, 2005). The molecular structure of the E protein is reviewed in Chapter 13. However, because the E protein is the major target of flavivirus neutralizing antibodies, a brief comment on its structure is presented in this chapter. The structures of the E protein ectodomains from DENV-2 and DENV-3 have been determined by X-ray crystallography (Modis *et al.*, 2003, 2005). The proteins resemble their homologs from WNV and tick-borne encephalitis (TBE) viruses in their flat, elongated, dimeric architecture and in their folding into three structural domains (I, II, and III; Fig. 14.1) (Kanai *et al.*, 2006; Nybakken *et al.*, 2006; Rey *et al.*, 1995). The central domain, E protein Domain I (E DI), is a nine-stranded β-barrel that forms a flexible link between the other two E protein domains. In DENV, an N-linked glycosylation site is present at amino acid position 154–156 of E DI. E protein Domain II (E DII) is composed of two extended loops that project from DI and contains a highly conserved fusion loop at its tip. In contrast with E DI and DII, E protein domain III (E DIII) is formed by a continuous sequence that folds as an immunoglobulin-like motif (Kanai *et al.*, 2006; Modis *et al.*, 2003, 2005; Nybakken *et al.*, 2006; Nybakken *et al.*, 2005; Rey *et al.*, 1995; Volk *et al.*, 2004; Wu *et al.*, 2003b). Four peptide loops are present on the lateral ridge of E DIII. This region is predicted to contain the cellular receptor-binding site (Bhardwaj *et al.*, 2001; Chu *et al.*, 2005; Hung *et al.*, 2004).

Cryo-electron microscopy and pseudo-atomic modelling have defined the structure of

Table 14.1 Methods for mapping epitopes of flavivirus-neutralizing antibodies

Methods	References
Isolation of neutralization escape mutants	Beasley and Aaskov, 2001; Beasley and Barrett, 2002; Cecilia and Gould, 1991; Choi *et al.*, 2007; Daffis *et al.*, 2005; Gao *et al.*, 1994; Goncalvez *et al.*, 2008; Goncalvez *et al.*, 2004b; Hasegawa *et al.*, 1992; Holzmann *et al.*, 1989; Holzmann *et al.*, 1997; Jiang *et al.*, 1993; Lai *et al.*, 2007; Lin *et al.*, 1994; Lobigs *et al.*, 1987; Lok *et al.*, 2001; Mandl *et al.*, 1989; McMinn *et al.*, 1995; Ryman *et al.*, 1997a; Serafin and Aaskov, 2001
Binding of antibodies to linear peptides or fragments from E protein	Falconar, 1999, 2008; Mandl *et al.*, 1989; Razumov *et al.*, 2005; Roehrig *et al.*, 1998; Roehrig *et al.*, 1989; Roehrig *et al.*, 1990; Sanchez *et al.*, 2005; Thullier *et al.*, 2001
Binding of antibodies to linear peptides displayed on phage	Chen *et al.*, 2007a; Lin and Wu, 2004; Thullier *et al.*, 2001; Wu *et al.*, 2003a; Yao *et al.*, 1995
Site-directed or random mutagenesis of E protein and forward genetic screens with	
(a) Direct coating of E proteins or domains in ELISA plates	Crill and Chang, 2004; Gromowski and Barrett, 2007; Gromowski *et al.*, 2008; Hiramatsu *et al.*, 1996; Lin and Wu, 2003; Lisova *et al.*, 2007; Morita *et al.*, 2001; Sanchez *et al.*, 2005
(b) Display of E proteins or domains on yeast	Oliphant *et al.*, 2005; Oliphant *et al.*, 2006; Sukupolvi-Petty *et al.*, 2007
Nuclear magnetic resonance	Volk *et al.*, 2004; Wu *et al.*, 2003b
X-ray crystallography	Lok *et al.*, 2008; Nybakken *et al.*, 2005

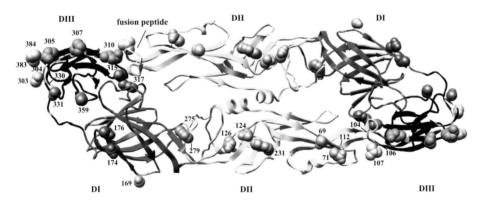

Figure 14.1 Mapping of representative MAb epitopes onto the DENV-2 E protein dimer. The three domains of DENV-2 E glycoprotein are shown: Domain I is red, domain II is yellow, and domain III is blue. The amino acid residues that significantly affect binding of MAbs, and/or were responsible for escape mutants, are in space-filling representation. Domain III: type-specific MAb 1F1 (residues 303, 304, 330, 383 and 384), subcomplex-specific MAb 1A1D-2 (residues 304, 305, 307 and 310) and cross-reactive MAb 13D4-1 (residues 315, 317, 331 and 359) (Sukupolvi-Petty *et al.*, 2007). Domain II: flavivirus cross-reactive MAb 1A5 (residues 106 and 107) (Goncalvez *et al.*, 2004b); flavivirus cross-reactive MAb 6B6C-1 (residues 104, 106, 126 and 231) and flavivirus group-reactive MAb 1B7-5 (residues 104, 126 and 231) (Crill and Chang, 2004); and MAb 10F2 (residues 69, 71, 112 and 124) (Lok *et al.*, 2001). Domain I: DENV-4-specific MAb 5H2 (residues 174 and 176) (Lai *et al.*, 2007); MAb 4G2 (residues 169 and 275) (Serafin and Aaskov, 2001); and MAb M10 (residue 279) (Beasley and Aaskov, 2001). A colour version of this figure is located in the plate section at the back of the book.

the mature DENV and WNV virions (Kuhn *et al.*, 2002; Mukhopadhyay *et al.*, 2003; Zhang *et al.*, 2003a). These viruses are relatively smooth particles composed of 90 E protein dimers and 180 M protein monomers. The mature flavivirus virion has a quasi-icosahedral symmetry, resulting in three distinct chemical environments in which the E protein exists. When the E protein is in its homodimeric form, the fusion loop is buried in a pocket at the DI–DIII interface of the adjacent E protein.

In immature virus particles, the E protein assumes a distinct conformation. There are 60 trimeric spikes, each consisting of three prM/E heterodimers (Long, 2008; Zhang *et al.*, 2003b; Zhang *et al.*, 2007). The prM protein covers the fusion peptide of each E protein. During the secretory process and through the acidic compartment of the trans-Golgi network, prM undergoes a furin-mediated cleavage event that promotes viral maturation (Stadler *et al.*, 1997). After attachment and endocytosis by a target cell, a reduction of pH in the early endosome prompts another structural transition of E protein on mature virions: the dimers dissociate and the exposed fusion loops insert into the target membrane, reclustering the E monomers into trimeric intermediates. DIII of the E protein then rotates towards the DII fusion loop, promoting the formation of a hairpin-like structure that results in the fusion of the viral membrane with the membrane of the target cell (Bressanelli *et al.*, 2004; Harrison, 2008; Modis *et al.*, 2004).

The specific binding epitopes of neutralizing antibodies from mice, chimpanzees and humans have been examined for several members of the *Flavivirus* genus. Different methods have served that purpose (Table 14.1). In general, these studies have identified flavivirus neutralizing antibodies that mapped to sites of the three domains of the E protein. These neutralizing antibodies can be either virus-specific or virus cross-reactive, and the virus-specific antibodies are the strongest neutralizers of flavivirus infection.

Epitopes on Domain III of the E protein

A significant level of information on the binding sites of neutralizing antibodies, mainly by the identification of mutations on E protein, and more recently by the co-crystallization of E protein DIII-Fab complexes, have suggested that the primary neutralizing epitopes localize on E DIII. In particular, one epitope on the lateral ridge of E DIII is recognized by many of the strongest neutralizing type-specific antibodies reported for flaviviruses (Beasley and Aaskov, 2001; Beasley and Barrett, 2002; Cecilia and Gould, 1991; Gromowski and Barrett, 2007; Gromowski *et al.*, 2008; Hiramatsu *et al.*, 1996; Holzmann *et al.*, 1997; Lin and Wu, 2003; Mandl *et al.*, 1989; Oliphant *et al.*, 2005; Roehrig *et al.*, 1998; Ryman *et al.*, 1998; Sanchez *et al.*, 2005; Seif *et al.*, 1995; Sukupolvi-Petty *et al.*, 2007; Throsby *et al.*, 2006; Volk *et al.*, 2004; Wu *et al.*, 2003b). Interestingly, this epitope is structurally conserved but highly variable at the sequence level among flavivirus species, which suggests a significant role of this dominant epitope in neutralization. An analogous WNV neutralizing type-specific epitope has been extensively characterized by error-prone PCR mutagenesis of WNV E DIII and expression on yeast cells (Oliphant *et al.*, 2005). These experiments, together with X-ray crystallographic studies of the structural interface between DIII and Fab E16 (Nybakken *et al.*, 2005), have demonstrated that the antibody binds to four discontinuous segments of E DIII, including the N-terminal linker region (residues 302 to 309), and the strand-connecting loops BC (residues 330 to 333), DE (residues 365 to 368) and FG (residues 389 to 391).

In recent studies, the contact residues of a panel of anti-DENV-2 E DIII-specific MAbs were mapped (Gromowski and Barrett, 2007; Gromowski *et al.*, 2008; Sukupolvi-Petty *et al.*, 2007). In general, these MAbs can be classified as follows: (a) serotype-specific MAbs with the strongest neutralizing activities localize to the epitopes on the lateral ridge of DIII and, as with the anti-WNV MAb E16, are centred on the FG loop (residues 383 and 384); (b) DENV subcomplex-specific neutralizing MAbs, which recognize several but not all of the four DENV-serotypes, bind an adjacent epitope centred on the A strand of DIII (residues 305, 307 and 310); (c) a third group of MAbs comprises poorly neutralizing MAbs against all DENV serotypes and other flaviviruses, which recognize an epitope with residues within the AB loop of DIII (residues 315 and 317) (Fig. 14.1). The epitopes recognized by the latter group of antibodies seem to have limited accessibility to the surface of the mature virion (Sukupolvi-Petty *et al.*, 2007). As seen for the DIII-specific poorly neutralizing anti-WNV MAbs (Oliphant *et al.*, 2005), a relative lack of accessibility might influence the stoichiometry of binding in such a way that antibody neutralization is poorly attained (Pierson *et al.*, 2007).

Epitopes on Domain II-fusion loop of the E protein

An important group of epitopes recognized by neutralizing antibodies has been mapped to the fusion loop (FL) at the tip of DII (Fig. 14.1). These epitopes are conserved among flaviviruses and elicit cross-reactive antibodies (Crill and Chang, 2004; Goncalvez *et al.*, 2004b; Oliphant *et al.*, 2006; Roehrig *et al.*, 1998; Stiasny *et al.*, 2006). Panels of mouse flavivirus group-reactive anti-E MAbs have identified three distinct overlapping epitopes on E glycoprotein containing conserved FL residues (Crill and Chang, 2004; Oliphant *et al.*, 2006; Stiasny *et al.*, 2006). Similarly, FL residues comprise a subset of the epitopes of broadly cross-reactive MAbs isolated from chimpanzees and humans (Goncalvez *et al.*, 2004b; Throsby *et al.*, 2006). The level of neutralization observed with this class of antibodies *in vitro* and *in vivo* is significantly lower, when compared with DIII MAbs (Goncalvez *et al.*, 2004a; Goncalvez *et al.*, 2004b; Oliphant *et al.*, 2006; Roehrig *et al.*, 1998; Stiasny *et al.*, 2006; Throsby *et al.*, 2006). A study of TBEV has shown that cross-reactive E DII-FL antibodies bind only with low avidity to native virions, but with high avidity to disintegrated virions, suggesting partial occlusion of these epitopes in infectious virions (Stiasny *et al.*, 2006). The cryptic character of these sites in native virions might explain the inefficient neutralization of broadly cross-reactive antibodies. However, several of the cross-reactive MAbs do bind and neutralize virus, suggesting that viral particles may be in motion, such that some MAbs may bind partially buried amino acid residues of the FL.

Epitopes on the E protein outside the DIII and DII-fusion loop

Flavivirus-neutralizing antibodies also mapped to other regions on the E protein, besides DIII and DII-FL epitopes. Studies with WNV and TBEV have shown that neutralization epitopes can be located in the dimer interface, central interface, and lateral ridge of DII, as well as in the hinge region between DI and DII and on the lateral ridge of DI (Guirakhoo *et al.*, 1989; Holzmann *et al.*, 1997; Oliphant *et al.*, 2006; Stiasny *et al.*, 2007a). Similarly, neutralization epitopes in DI and DII have also been reported for YFV (Daffis *et al.*, 2005; Lobigs *et al.*, 1987; Ryman *et al.*, 1997a), JEV (Goncalvez *et al.*, 2008; Morita *et al.*, 2001) and Murray Valley encephalitis virus (MVEV) (McMinn *et al.*, 1995).

For DENV, epitope mapping using fragments or peptides of the E glycoprotein have identified three antigenic sites in DI and DII [Reviewed by (Roehrig, 2003)]. DENV epitope mapping, by isolation of neutralization escape variants, has distinguished few DI and DII-nonFL neutralization epitopes (Fig. 14.1). A Phe-279 mutation at the DI–DII interface of DENV-1 E causes escape from neutralization by an IgM MAb (Beasley and Aaskov, 2001). Phe-279, conserved in all DENV serotypes, is located in the kl-hairpin region of DII, which has a role in the conformational rearrangement that drives membrane fusion (Modis *et al.*, 2003, 2004, 2005). A second IgM epitope on DENV-2 DII was characterized by the selection of a DENV-2 variant containing four amino acid changes at positions Thr-69, Glu-71, Glu-112 and Glu-124. This mutant lost its ability to cause fusion in mosquito cells (Lok *et al.*, 2001).

Only a few mouse MAbs that neutralize flaviviruses have been shown to react with epitope determinants on DI of the E glycoprotein (Holzmann *et al.*, 1997; Ryman *et al.*, 1997a; Ryman *et al.*, 1997b). Although mouse MAbs that map to DENV E DI are generally non-neutralizing (Roehrig *et al.*, 1998), a highly neutralizing DENV-4-specific MAb (5H2) recovered from a chimpanzee binds to an epitope within this domain (Lai *et al.*, 2007; Men *et al.*, 2004). Important sites for binding of MAb 5H2 localize to amino acids 174 and 176 in E, near or within the three-amino-acid loop between the H_0 and G_0 β-strands (Fig. 14.1). The sequence variability among DENVs and other flaviviruses in the region that includes the three-amino-acid loop 174 to 176 accounts for the exquisite specificity of 5H2. Another study has identified this position as an important epitope determinant of the JEV highly neutralizing MAb A3 (Goncalvez *et al.*, 2008). Humanized MAbs 5H2 and A3 proved to be protective against DENV-4 and JEV infections in monkeys and mice, respectively. These results support the assumption that a cluster of epitopes involving antigenic determinants of the H_0–G_0 loop may play an important role in inducing flavivirus type-specific antibodies in rodents and nonhuman primates. It remains to be determined the importance of antibody responses to these epitopes in humans.

An update on the antigenic model of DENV E glycoprotein derived from studies with chimpanzee and human MAbs

Although studies of the naturally and vaccine-induced antibody response in humans and non-human primates have shown that DENV-specific neutralizing antibodies control viraemia, the epitope specificities of the generated antibodies remain unclear. Recently, epitope analysis of antibodies recovered from DENV-infected chimpanzees have shown that DENV infections induce broadly reactive, but weakly to non-neutralizing antibodies that are reactive with major epitopes involving the FL (Goncalvez *et al.*, 2004a; Goncalvez *et al.*, 2004b; Men *et al.*, 2004). Unexpectedly, neutralizing antibodies that bind to DENV DIII-specific epitopes have not been recovered from chimpanzees, in spite of intense effort. Instead, potent DENV-4-neutralizing MAbs, which bind to sites in DI, were selected from the same libraries (Lai *et al.*, 2007; Men *et al.*, 2004). Most interestingly, MAbs isolated from chimpanzees and mice infected with other flaviviruses bind to analogous epitopes on E DI (Goncalvez *et al.*, 2008; Holzmann *et al.*, 1997; Ryman *et al.*, 1997a; Ryman *et al.*, 1997b). The significance of this cluster of E DI epitopes that induce flavivirus type-specific antibodies remains to be determined.

A recent analysis of the antibody response in humans with primary DENV-2 infection confirms the observations in chimpanzees (Lai *et al.*, 2008). In this report, more than 90% of anti-E antibodies were cross-reactive and non-neutralizing against heterologous viruses, and only a minor proportion was type specific. Moreover, the binding of polyclonal anti-E antibodies from these patients was almost completely abolished by single point mutations in the FL of DENV-2, but not by those outside the DII-FL. Based upon these observations, it is speculated that the cross-reactive DII-FL epitopes are immunodominant in humans. Type-specific highly neutralizing antibodies could be present as minor subsets.

Two studies analysing the human B-cell repertoire before and after WNV infection also showed a predominant number of antibodies reactive to DII, whereas DIII-specific antibodies were relatively rare or absent (Gould *et al.*, 2005; Throsby *et al.*, 2006). Kinetic studies of the antibody response in mice, horses and humans infected with WNV also support the notion that the immune response is reoriented to E DII of WNV and neutralizing E DIII antibodies are infrequent (Oliphant *et al.*, 2007; Sanchez *et al.*, 2007).

Neutralizing mechanisms of flavivirus antibodies

The primary function of an antibody is to bind the antigen, which in some cases can have a direct effect, for instance by preventing virus attachment and entry into target cells. This effect can be tested *in vitro* and, in general, correlates with animal protection *in vivo*. However, MAbs binding to virions or virus-infected cells recruit secondary effector functions involving Fc receptors and components of the complement cascade *in vivo*.

For flaviviruses, antibodies can inhibit viral infection directly by blocking cellular attachment, membrane fusion, and E protein-mediated internalization (Crill and Roehrig, 2001; Gollins and Porterfield, 1986; Nybakken *et al.*, 2005; Pierson *et al.*, 2008). Moreover, MAbs can prevent structural transitions of mature virions into fusion forms of the virus (Lok *et al.*, 2008).

The ability of a set of anti-DENV-2 E-specific MAbs to block virus adsorption to Vero cells was analysed by Crill and Roehrig in 2001. Similar experiments performed by us with mouse and chimpanzee MAbs, are shown in Fig. 14.2. In the experiments, most MAbs neutralized virus infectivity at least partially by blocking virus adsorption. As previously shown (Crill and Roehrig, 2001), MAbs specific for DIII were the strongest blockers of virus adsorption (Fig. 14.2A), consistent with the notion that DIII is responsible for

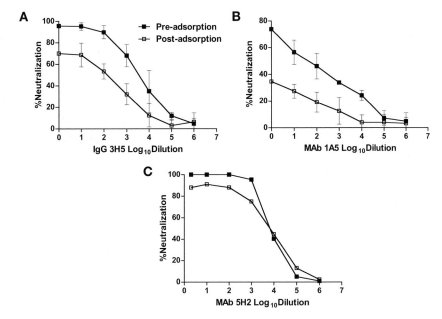

Figure 14.2 Comparison of neutralization of DENV before and after adsorption to Vero cells by representative MAbs for the three distinct E-glycoprotein domains. **A** Domain III, mouse MAb 3H5; **B** Domain II, chimpanzee MAb 1A5; **C** Domain I, chimpanzee MAb 5H2. In pre- and post-adsorption antibody neutralization assays, a constant amount of DENV-2 (A, B) or DENV-4 (C) (70 FFU) was tested against various dilutions of MAb.

viral attachment (Hung *et al.*, 2004; Modis *et al.*, 2003; Rey *et al.*, 1995). The DII-specific neutralizing MAbs blocked virus adsorption but to a lesser extent and at relatively high concentrations, when compared with those specific for E DIII, consistent with the results of Crill and Roehrig (Crill and Roehrig, 2001) (Fig. 14.2B). Additional evidence for DENV neutralization by blocking of virus binding to receptors has been reported for the anti-DENV E DIII MAb 4E11 (Thullier *et al.*, 2001). This antibody blocked the binding of a synthetic peptide DENV-1 $_{296-400}$ to heparin, a highly sulphated heparan sulphate (HSHS) molecule. The initial attachment of DENV and other flaviviruses to target cells *in vitro* has been shown to depend on cell surface glycosaminoglycans (Chen *et al.*, 1997; Hung *et al.*, 2004). Since the binding of 4E11 to a peptide consisting of the DIII 296 to 400 amino acid residues suppressed the interaction with heparin, it is postulated that the 4E11 epitope on DENV contains amino acids that are critical for HSHS binding.

It has been suggested recently that DENV-specific E DIII MAbs might mediate inhibition of viral attachment to host cells by changing the arrangement of the E glycoproteins on the viral surface (Lok *et al.*, 2008). The DENV complex-specific MAb 1A1D-2 strongly neutralizes DENV serotypes 1, 2 and 3, but does not bind to serotype 4 (Roehrig *et al.*, 1998). The co-crystallization of a complex of the 1A1D-2 Fab fragment and DIII of the E glycoprotein, shows that this antibody bound to an epitope partially occluded in the structure of the mature DENV virion (Lok *et al.*, 2008). However, MAb 1A1D-2 is able to bind DENV-2 at 37°C, suggesting that higher temperature promotes increased mobility of the E proteins on the surface of the virus, thereby making occluded epitopes available. Based upon these data, the authors proposed that structural transitions of the viral proteins exposing occluded epitopes could be trapped through antibody binding.

In addition to blocking attachment, flavivirus MAbs against E DIII also seem to mediate fusion inhibition (Nybakken *et al.*, 2005; Stiasny *et al.*, 2007a). According to docking studies of the anti-WNV Fab E16-DIII complex onto the structures of the pre-fusion DENV E glycoprotein dimmer and post-fusion trimer (Modis *et al.*, 2003, 2004; Nybakken *et al.*, 2005), it appears that E16 binding prevents the ~70° rotation of DIII towards

DII in the dimer to trimer transition (Bressanelli *et al.*, 2004; Modis *et al.*, 2004). Supporting the post-fusion inhibition mechanism of MAb E16 is the fact that this antibody appears to block low pH-dependent plasma membrane fusion. Moreover, pre-binding or post-binding of WNV with E16 significantly protects against infection (Nybakken *et al.*, 2005).

Significantly, neutralizing MAbs against DII-FL were able to block adsorption of DENV and WNV, although at relatively high concentration (Crill and Chang, 2004; Crill and Roehrig, 2001; Nybakken *et al.*, 2005; Oliphant *et al.*, 2006). An example of this type of experiment is shown in Fig. 14.2B using the humanized-chimpanzee MAb 1A5 (Goncalvez *et al.*, 2004b). Pre-incubation of DENV with MAb 1A5 protected Vero cells against infection. In contrast, the ability of MAb 1A5 to inhibit infection was significantly reduced when added after virus binding. Notably, MAbs specific for DII-FL also were shown to block virus-mediated cell membrane fusion (Allison *et al.*, 2001; Crill and Chang, 2004; Roehrig *et al.*, 1998), indicating that single flavivirus MAbs may have multiple mechanisms of virus inhibition.

New insights into the mechanism of antibody-mediated flavivirus fusion inhibition were recently reported using TBEV and MAbs that bind to epitopes distributed along the three domains of the E glycoprotein (Stiasny *et al.*, 2007a). Depending on the location of their binding sites, the MAbs impaired early or late stages of the fusion process, either by blocking the initial interaction with the target membrane or by interfering with the proper formation of the post-fusion structure of E, respectively. The effect of MAbs on the overall fusion process was investigated in a fusion assay with liposomes and fluorescence-labelled purified virions (Corver *et al.*, 2000). Most of the MAbs directed against the three E domains had a fusion inhibitory effect, with the exception of two of the four MAbs against DIII tested, B2 and B3, which did not have a significant effect on the fusion reaction. Thus, certain MAbs that react with E DIII demonstrate that in some instances, fusion may not be affected by the presence of MAbs bound to the E protein. In this study, the authors also investigated whether binding of the MAbs inhibited interaction of the FL with target membranes. Most of the MAbs had no effect or only minor effects on inhibition of FL binding to liposomes, suggesting that their binding sites in E are far from the FL. Surprisingly, MAbs that react directly with the FL exhibited an intermediate level of binding inhibition (Allison *et al.*, 2001; Stiasny *et al.*, 2006). The authors propose that the partial inhibition of fusion activity by these MAbs might be explained by the short exposure time of FL during the E dimer dissociation at low pH, just prior to becoming buried after insertion into the target membrane (Stiasny *et al.*, 2007b).

MAbs that bind to epitopes on DI and DII-non FL were able to strongly inhibit TBEV fusion to liposomes (Stiasny *et al.*, 2007a). Additionally, the humanized MAb 5H2, which binds an analogous DI epitope on DENV-4 E protein, neutralized equally efficiently before and after binding of the virus to Vero cells (Fig. 14.2C and Lai *et al.*, 2007). The locations of the MAbs' antigenic determinants in the pre- and post- fusion conformations of E suggest that these MAbs may prevent the conformational rearrangement of the E protein essential for the fusion process.

JEV has also provided a model for investigation of the mechanism of antibody-mediated inhibition of flavivirus infection. MAb 503 is a highly neutralizing and specific antibody that has been shown to have protective activity against JEV infection in mice (Butrapet *et al.*, 1998; Kimura-Kuroda and Yasui, 1988). The epitope reactive to MAb 503 appears to map to the junction of domains I and II (Morita *et al.*, 2001). Analysis of radiolabelled JEV and observations using confocal laser microscopy and electron microscopy suggest that the neutralization and protective activities of MAb 503 do not have an effect on the binding of virus to the cell surface. Treatment with MAb 503 strongly inhibits JEV-induced cell fusion and internalization of JEV into the cells (Butrapet *et al.*, 1998). Therefore, the neutralization activity of this antibody seems to be involved in the post-attachment steps of JEV infection.

Besides epitope specificity, an additional determinant that modulates viral inhibition is epitope accessibility. According to the multiple-hit theory and stoichiometric analysis of epitope occupancy for neutralization (Klasse and Sattentau, 2002), the most potent antibodies neutralize the virus at concentrations with low occupancy of the epitopes available for binding on the virion. The structure of the flavivirus mature virion is a pseudo-icosahedral particle with three E monomers per asymmetric unit. Consequently, the E protein is present in three different environments accessible for binding to antibodies or receptors (Kuhn *et al.*, 2002; Mukhopadhyay *et al.*, 2003). Recent studies have assessed the proportion of antibody to a given flavivirus E DIII epitope that must be bound to a virion for neutralization (Gromowski and Barrett, 2007; Gromowski *et al.*, 2008; Pierson *et al.*, 2007). MAbs that bind the lateral ridge of E DIII inhibit infection under conditions of low occupancy of the available epitopes on the virion. In contrast, weakly neutralizing MAbs that bind other epitopes on DIII require a saturation of occupancy of available epitopes to block infection. A similar approach to estimate the occupancy of MAbs that bind to DENV epitopes outside the E DIII is shown in Fig. 14.3. To reach 50% of virus neutralization, the occupancy for the DENV-4 highly neutralizing MAb 5H2 was approximately 30% of available sites on E DI (Fig. 14.3A), whereas a 90% occupancy of DII-FL epitopes was required for the poorly neutralizing MAb 1A5 (Fig. 14.3B).

It is generally accepted that besides antibody concentration, the strength of binding or affinity also modulates the relative occupancy of epitopes available on the virion for a specific antibody (Klasse and Sattentau, 2002). Thus, affinity does seem to correlate with neutralization potency and protection when MAbs that bind to a specific structural epitope are evaluated (Gromowski and Barrett, 2007; Gromowski *et al.*, 2008; Oliphant *et al.*, 2005; Pierson *et al.*, 2007).

In addition to occupancy and binding affinity, virion maturity seems to have an impact on the sensitivity of virions to antibody-mediated neutralization (Nelson *et al.*, 2008). The E proteins of immature virions are arranged into 60 trimeric spikes consisting of prM and E proteins (Zhang *et al.*, 2003b; Zhang *et al.*, 2007). Cleavage of prM results in a transition of the heterodimers into the dimeric E arrangement that distinguishes mature virions. Interestingly, preparations of flavivirus virions from mammalian cells were found to be a mixture of mature and immature particles, suggesting that viral maturation may not be an efficient process (Zhang *et al.*, 2007). In

2008, Nelson *et al.* showed that complete virion maturation results in a significant reduction in sensitivity to neutralization by antibodies specific for the poorly accessible DII-FL. However, the same antibodies can neutralize more efficiently virions that retain partially immature character. This could explain, in part, how antibodies that bind to poorly accessible epitopes on E protein are still able to neutralize virions (Goncalvez *et al.*, 2004b; Oliphant *et al.*, 2006; Stiasny *et al.*, 2006).

In vivo protection mediated by DENV MAbs

Anti-E protein MAbs

A pertinent evaluation of neutralization potency of antibodies to specific epitopes is their protective activity *in vivo* in a passive transfer model. There are a vast number of reports of neutralizing MAbs against flaviviruses being effective in immunotherapy and/or immunoprophylaxis. Most reports are associated with anti-E protein MAbs, since E is the major surface glycoprotein and the principal antigen that elicits protective neutralizing antibodies. Most research groups have performed prophylaxis and post-viral challenge MAb protection studies in rodents (Chiba *et al.*, 1999; Goncalvez *et al.*, 2008; Gupta *et al.*, 2003; Henchal *et al.*, 1988; Iacono-Connors *et al.*, 1996; Kaufman *et al.*, 1989; Kaufman *et al.*, 1987; Kimura-Kuroda and Yasui, 1988; Lai *et al.*, 2007; Mathews and Roehrig, 1984; Niedrig *et al.*, 1994; Oliphant *et al.*, 2005; Phillpotts *et al.*, 1987; Schlesinger *et al.*, 1985, 1987; Schlesinger and Chapman, 1995; Tan *et al.*, 1990; Throsby *et al.*, 2006; Yamauchi, 1989; Zhang *et al.*, 1989a), although a few studies have used non-human primates (Goncalvez *et al.*, 2007; Lai *et al.*, 2007; Zhang *et al.*, 1989a). Both protection and virus infection with improved survival have been seen in animals after virus challenge and administration of antibodies.

A significant difficulty in the assessment of the protective capacities of MAbs against DENV infections and correlates of protection *in vivo* is the lack of suitable animal models that mimic human dengue disease (see Chapter 6). Humans, non-human primates and mosquitoes are known to be natural hosts for DENV infections (Monath, 1994). Non-human primates, including chimpanzees, gibbons, and macaques,

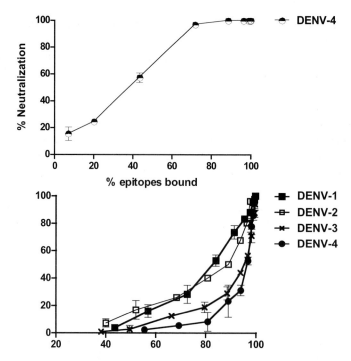

Figure 14.3 Occupancy for antibody-mediated neutralization of DENV. Percentage of epitopes bound for neutralization mediated by **A** MAb 5H2 and **B** MAb 1A5 were estimated from MAb dose-response curves against the percentage of epitopes bound by MAb at different concentrations. Percentage of epitopes bound was calculated using the binding avidity value obtained by ELISA and applying the following equation: percent of epitopes bound = [MAb]/([MAb] + KD) (Pierson *et al.*, 2007). [MAb]: MAb concentration (nM); KD: binding avidity.

have been experimentally infected, and, in some cases, have developed viraemia sufficient to infect mosquitoes (Schlesinger, 1977). However, DENV does not cause clinical illness and disease in non-human primates.

Mice often develop encephalitis after intracerebral challenge with highly adapted DENV strains (Boonpucknavig *et al.*, 1981). Although the disease presentation in mice differs from that in humans following DENV infections, this model is considered suitable for studying the protective capacity of MAbs that are passively transferred. Nevertheless, murine models are still limited when testing MAbs of human origin, since the human Fc region has a lower binding affinity for the human Fcγ receptors compared with their murine counterparts. In addition, there are structural differences in the Fcγ receptor-intracellular domains and the cellular expression pattern of these receptors (Nimmerjahn and Ravetch, 2008).

In mouse models, both polyclonal serum and monoclonal antibodies against E, prM, NS1 and NS3 proteins have provided *in vivo* protection against DENV infection (Henchal *et al.*, 1988; Johnson and Roehrig, 1999; Kaufman *et al.*, 1989; Kaufman *et al.*, 1987; Lai *et al.*, 2007; Lin *et al.*, 1998b; Schlesinger *et al.*, 1987; Tan *et al.*, 1990; Wu *et al.*, 2003c). In an early study, a panel of ͼ murine neutralizing MAbs against DENV E were shown to protect BALB/cJ mice against lethal intracranial (i.c.) DENV-2 challenge, administered 1 day after intraperitoneal (i.p.) inoculation of ascitic fluid containing the MAbs (Kaufman *et al.*, 1987). Contrarily, passive immunization with some of the MAbs used by Kaufman *et al.* (1987) did not protect mice lacking interferon (IFN)α/β and IFNγ receptors (AG129), which were challenged peripherally with a mouse-adapted DENV-2 strain (Johnson and Roehrig, 1999). The failure of MAbs to protect mice against DENV infection might be associated with differences between the two studies, including antibody doses, routes of virus infection and differences inherent to the mouse models.

SCID mice engrafted with human K562 cells have also been used to determine whether a neutralizing antibody against DENV *in vitro* was also protective *in vivo* (Lin *et al.*, 1998b). These mice, peripherally infected with DENV-2, developed paralysis and died. Pretreatment of 10^7 PFU of DENV-2 strain PL046 with MAb 56–3.1 at 37°C prolonged the survival of infected mice. Although not as effective as the pretreatment, a significant increase in the average survival time was still observed when the MAb was administered to the mice 1 day before DENV challenge. Characterization of this MAb in terms of effective dose and epitope mapping awaits further analysis.

Until recently, there has been a lack of human-derived antibodies for characterization of DENV protective epitopes discovered with mouse antibodies. However, DENV type-specific and cross-reactive antibodies have been recovered from infected chimpanzees by repertoire cloning (Goncalvez *et al.*, 2004a; Men *et al.*, 2004). The flavivirus-cross-reactive MAb 1A5, which has a low neutralization potency *in vitro* (Goncalvez *et al.*, 2004b), was tested *in vivo* with suckling and adult mice (Fig. 14.4A and B). Groups of 4-day-old Swiss mice were inoculated with IgG 1A5 by i.p. injection at a dose range of 40 to 1000 µg per mouse. One day later, mice were challenged with 25 50% lethal doses (LD_{50}) (74 FFU) of neurovirulent DENV-2, strain NGC, by i.c. inoculation. Nearly all mice in the control group and the group that received a low dose of antibody succumbed to infection, except the group that received 1000 µg of the antibody, which had a survival rate of 36% (Fig. 14.4A). The average survival time (AST) of this group was significantly different when compared with the PBS group (8.4 ± 0.2 and 6.9 ± 0.4 days, respectively; $P = 0.03$, t-test). However, an improvement in the survival rate was observed in adult mice infused with higher doses of IgG 1A5 (50% protective dose of ~200 µg per mouse; Fig. 14.4B).

The protective potency of MAb 1A5 was also tested in rhesus monkeys (Goncalvez *et al.*, 2007). Monkeys that received a high dose of IgG 1A5

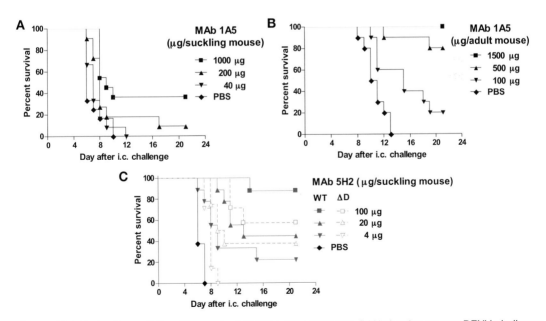

Figure 14.4 Protective activity of humanized DENV IgG1 antibodies (MAbs) using mouse DENV challenge models. Inbred BALBc mice ($n = 10$) were injected i.p. with MAb 1A5 (**A** and **B**) or MAb 5H2 (**C**) at various doses indicated. Unprotected control mice were administered PBS diluent. Twenty-four hours later mice were infected i.c. with DENV-2 strain NGB (**A** and **B**), or DENV-4 strain H241 (**C**). The animals were monitored daily and euthanized when clinical signs of infection appeared. Kaplan–Meier survival curves are shown. A colour version of this figure is located in the plate section at the back of the book.

(18 mg/Kg) were not protected from DENV-4 infection, despite having a mean serum IgG 1A5 concentration of 166 µg/ml on the day of virus challenge. However, a 2–3 day delay in peak viraemia was observed in this group, when compared to the peak viraemia detected in monkeys given lower antibody doses and the control group. One may speculate that the high antibody concentration resulted in an initial inhibition of DENV-4 replication in these monkeys. Nevertheless, the magnitude of the viral burden despite the presence of neutralizing antibodies indicated that MAb 1A5 was ineffective *in vivo*.

In a similar manner, the chimpanzee-derived highly neutralizing MAb 5H2 was evaluated for its protective ability (Lai *et al.*, 2007). Passive transfer of IgG 5H2 in 3- to 4-day-old BALB/c mice, at a dose of approximately 20 µg/mouse, afforded 50% protection against challenge with 25 LD_{50} of the mouse-neurovirulent DENV-4 strain H241. Rhesus monkeys that received 2 mg/kg of IgG 5H2 one day prior to virus challenge were completely protected, as indicated by the absence of viraemia and lack of seroconversion. Taken together, *in vitro* neutralizing activity of human–chimpanzee MAbs correlated with protective activity *in vivo*. The highly neutralizing MAb 5H2 that binds to an E DI epitope has a protective effect in mice and monkeys (Lai *et al.*, 2007). In contrast, MAb 1A5, which has a lower neutralizing activity and binds to an epitope that includes determinants of E DII-FL, protected adult mice at a high MAb dose, but did not protect suckling mice or monkeys (Goncalvez *et al.*, 2007). These studies are the first to evaluate protection of non-human primates against DENV infection by passive transfer of antibodies derived from an immunocompetent primate.

Demonstration of passive protection with MAbs against prM, NS1 and NS3 proteins

Besides the E glycoprotein, the prM structural protein also appears to contain protective epitopes. The neutralizing ability of anti-prM antibodies has been analysed for several flaviviruses (Falconar, 1999; Holbrook *et al.*, 2001; Iacono-Connors *et al.*, 1996; Kaufman *et al.*, 1989; Men *et al.*, 2004; Takegami *et al.*, 1982; Throsby *et al.*, 2006). Only three of these studies were performed with DENV. A panel of prM specific MAbs for DENV-3 and DENV-4 has been used to protect BALB/cJ mice against lethal challenge with homologous or heterologous DENV serotypes. Four of five MAbs protected mice against heterologous challenge (Kaufman *et al.*, 1989). Interestingly, some of the anti-prM MAbs, which lack neutralizing ability *in vitro*, have been shown to mediate passive protection in mice (Kaufman *et al.*, 1989; Young, 1990). Little is known regarding the mechanism of antibody-mediated protection by anti-prM MAbs. Nevertheless, since some of these antibodies are able to fix complement *in vitro*, a possible mechanism of protection may involve the activation of complement by the binding of MAbs to prM protein on the infected cells.

The non-structural (NS) proteins, particularly NS1, have also been shown to be immunogenic and to elicit high antibody titres. NS1 exhibits a high conservation of sequence among flaviviruses (Blitvich *et al.*, 2001; Wallis *et al.*, 2004). Interestingly, high levels of NS1 are detected in the sera of patients experiencing secondary DENV infections (Alcon *et al.*, 2002; Libraty *et al.*, 2002; Young *et al.*, 2000). It is believed that NS1 contributes significantly to the formation of immune complexes that may play a significant role in the pathogenesis of severe dengue disease.

Different approaches have been used to map antibody epitopes on NS1 of flaviviruses. These include competitive binding assays, NS1 protein-derived fragments or NS1 synthetic peptides, yeast surface display and bacterial expression of truncated NS1 fragments (Chung *et al.*, 2006; Falconar and Young, 1991; Falconar *et al.*, 1994; Garcia *et al.*, 1997; Hall *et al.*, 1990; Henchal *et al.*, 1987; Mason *et al.*, 1990; Putnak *et al.*, 1988; Roehrig, 2003). In general, three regions of the NS1 protein are targets for anti-NS1 antibodies (Chung *et al.*, 2006).

Immunization studies with NS1 and passive administration of anti-NS1 MAbs in mice have suggested that this protein may enhance protection against and/or clearance of flaviviruses (Chung *et al.*, 2006; Despres *et al.*, 1991; Falconar, 1989; Falgout *et al.*, 1990; Gould *et al.*, 1986; Hall *et al.*, 1996; Henchal *et al.*, 1988; Jacobs *et al.*, 1992, 1994; Lin *et al.*, 1998a; Qu *et al.*, 1993; Schlesinger *et al.*, 1986; Schlesinger *et al.*, 1985, 1987; Schlesinger *et al.*, 1993; Timofeev *et al.*, 2004). However, protection conferred by immunization with DENV NS1 protein could not be demonstrated in monkeys (Zhang *et al.*, 1989b). In another study, AG129 interferon receptor-deficient mice were immunized with chimeric (DENV-2/WNV) viruses containing prM and E proteins of WNV and C protein and non-structural proteins of DENV-2 (Calvert *et al.*, 2006). These viruses elicited high levels of anti-DENV NS1 antibodies, but no or very low levels of DENV neutralizing antibodies. Although the mean survival time increased significantly over non-immunized controls, only 15% of immunized mice survived after DENV-2 challenge. Thus, protection mediated by NS1 is not as effective as protection mediated by the E protein. The mechanisms by which NS1 elicits protection are not fully understood. An early study with MAbs against YFV NS1 suggested that protection *in vivo* was dependent on the Fc region of the antibody (Schlesinger *et al.*, 1993). In a more recent study, anti-NS1 MAbs against WNV were tested in Fcγ RI- and RIII-deficient mice, which are severely impaired in antibody-dependent phagocytic and cellular cytotoxic responses (Chung *et al.*, 2006). The protective effect of some MAbs was lost in these models, suggesting that the binding of Fc regions to Fcγ RI or RIII determined the protective effect of these antibodies. Moreover, only MAbs that recognize cell surface-associated NS1 triggered Fcγ R-mediated phagocytosis and clearance of WNV-infected cells, and this Fcγ receptor-dependent mechanism of NS1 MAb-mediated protection seemed to depend on the IgG subclass (Chung *et al.*, 2007).

The protective role of MAbs against other flavivirus non-structural proteins is less clear. NS3 is antigenic in humans, as antibodies to NS3 have been detected in sera of DENV-infected patients. The role of DENV NS3 protein in protection was examined in passive protection studies with anti-DENV-1 NS3 MAbs. These MAbs were able to increase the survival time of mice challenged with a lethal dose of DENV-1, although the survival rate was low (17–25%) (Tan *et al.*, 1990). The combination of protective MAbs against different proteins of DENV may be useful for a therapeutic strategy that prevents the emergence of resistant viral variants.

Challenges to passive immunotherapy and/or immunoprophylaxis of DENV infections

In spite of the fast commercial growth of therapeutic antibodies (Casadevall *et al.*, 2004), their application in prevention and/or treatment of infectious diseases has been significantly slower when compared with other types of diseases (Reichert, 2007; Xiao and Dimitrov, 2007). Scientific challenges specific to the development of MAbs as anti-infectious agents, and clinical and commercial challenges have had an impact on the interest in this area. Some of the potential obstacles include ADE, antibody escape variants and viral variability. Another challenge to the development of anti-viral MAbs is competition from a growing interest in the development of small-molecule inhibitors and antiviral vaccines.

Antibody-dependent enhancement (ADE)

Essential to any discussion of DENV immunotherapy and/or immunoprophylaxis is the effect of ADE on viral replication. ADE has been proposed as the underlying mechanism for DHF/DSS (Halstead, 1970), two clinical conditions frequently seen in patients infected with a second heterotypic DENV infection or in infants with maternally transferred anti-DENV antibodies (Kliks *et al.*, 1988). ADE is thought to occur when preexisting subneutralizing antibodies and the infecting DENV form complexes that bind to Fc receptor-bearing cells, leading to increased virus uptake and replication (Kliks *et al.*, 1988). ADE has been demonstrated *in vitro* using dengue immune sera or MAbs and cells of monocytic and B lymphocytic lineages bearing Fc receptors (Lin *et al.*, 2002; Littaua *et al.*, 1990; Morens *et al.*, 1987).

ADE has also been associated with the increased virulence, manifested as early death, in mice passively administered certain monoclonal antibodies followed by infection with JEV (Kimura-Kuroda and Yasui, 1988) or YFV (Gould and Buckley, 1989). ADE of DENV-2 replication *in vivo* was also studied by passive antibody transfer, in which monkeys infused with human dengue antiserum developed viraemia titres up to 50-fold higher than control monkeys (Halstead,

1979). More recently, the humanized flavivirus cross-reactive MAb 1A5 was shown to up-regulate DENV-4 replication by an ADE mechanism in juvenile monkeys (Goncalvez *et al.*, 2007). Monkeys in this study received a range of passively transferred subneutralizing concentrations of IgG 1A5 and were then infected with DENV-4. Infection enhancement of up to 100-fold in the viraemia titre was detected. Furthermore, peak viraemia titres were detected on days 5–6 in both studies, around the time when a subject's illness would be expected to progress to DHF. Despite having experienced ADE of DENV replication, monkeys did not generally become ill in either of the studies (Halstead *et al.*, 1973).

Prospective studies in humans have suggested a correlation between higher viraemia titres and increased risk of severe dengue in second, heterotypic DENV infections (Endy *et al.*, 2004; Vaughn *et al.*, 2000). This hypothesis, however, has been contradicted by recent work showing that the infection-enhancing activity detected with pre-illness sera *in vitro* does not correlate with increased viraemia titres and disease severity in subsequent infection with DENV-2 or DENV-3 (Endy *et al.*, 2004; Laoprasopwattana *et al.*, 2005). As discussed in Chapter 5, differences in DENV serotypes or even strains may influence the likelihood of ADE and resulting severe dengue disease in humans (Endy *et al.*, 2004; Vaughn *et al.*, 2000).

ADE is a possible adverse outcome of antibody-based therapies against DENV if the antibody titres generated decline below their neutralization threshold. Strategies to reduce ADE by mutational analysis of the key structures in the Fc of anti-DENV MAbs have been recently explored (Goncalvez *et al.*, 2007). A nine amino acid deletion at the N-terminus of Fc, designated ΔD mutation, in MAb 1A5 resulted in complete abrogation of DENV ADE *in vitro* (Fig. 14.5A). The authors have performed experiments to address whether the ΔD mutation in the antibody Fc region effects the functional activity of the chimpanzee-derived MAb 5H2 (Lai *et al.*, 2007; Men *et al.*, 2004). IgG elimination is likely dominated by affinity of the Fc region for the neonatal Fc receptor (FcRn), and the nature of affinity for the specific target of the antibody (Ghetie and Ward, 2002; Lobo *et al.*, 2004). The FcRn-Fc co-crystal structure reveals that FcRn binds to the C_H2–C_H3 hinge region of IgG antibodies (Martin *et al.*, 2001b), at a site that is distinct from the binding sites of the classical Fcγ Rs or the C1q component of complement. ΔD mutation is not close to the FcRn binding site responsible for the maintenance of IgG half-life; however, other factors may contribute to the rate of antibody elimination. The effect of ΔD on the pharmacokinetics of IgG 5H2 was recently investigated. Two chimpanzees were infused intravenously with full-length IgG 5H2 WT or IgG 5H2 ΔD, and serum samples were collected at various time points. Chimpanzees that received either IgG 5H2 WT or IgG 5H2 ΔD demonstrated prolonged serum antibody clearance. The elimination half-life for both antibodies was ~16 days (unpublished data), which corresponds to the half-life for the clearance of human IgG1 in humans.

Antibodies are natural components of the immune system and interact with immune cells and proteins to achieve antiviral effects (Nimmerjahn and Ravetch, 2008). For instance, the Fc region can mediate complement-dependent cytotoxicity (CDC), which can cause a direct virotoxic effect or mediate destruction of target cells. In addition, antibody-dependent cell-mediated cytotoxicity (ADCC) and phagocytosis are mediated through interaction of virus-bound MAbs with Fcγ receptors. Alterations of the sequence in the Fc region that affect the Fcγ receptor and complement binding would be expected to modulate effector cell functions. Experiments were performed to explore the ability of the variant ΔD to interact with human effector-triggering molecules *in vitro*. Human Fcγ RI, Fcγ RIIA and C1q were tested for binding to full-length IgGs 1A5 and its ΔD variant in an ELISA assay. The binding of the ΔD variant to these effector molecules was significantly reduced (Fig. 14.5B–D). Thus, abrogation of ADE of DENV replication *in vitro* is associated with the reduction of ΔD variant binding to the monocyte and macrophage Fcγ Rs.

Further experiments were performed to determine whether the reduced binding of the ΔD variants of IgG to effector molecules would diminish the protective ability of these anti-DENV antibodies *in vivo*. Suckling BALB/c mice were administered IgG 5H2 WT or ΔD intraperitoneally, 1 day prior to i.c. challenge with 40 LD_{50}

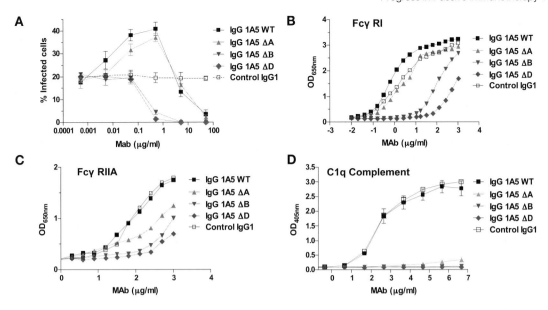

Figure 14.5 ADE of DENV-4 and binding to human Fcγ receptors and C1q by IgG 1A5 WT and its variants. **A** ADE of DENV-4 infection at 1 MOI in U937 cells mediated by IgG 1A5 Fc variants. IgG 1A5 variants ΔA and ΔB, which contain sequence substitutions of subclass IgG2 or IgG4 in the Fc, were described previously (Goncalvez *et al.*, 2007). These IgG subclasses have reduced Fcγ receptor-binding affinity (Armour *et al.*, 1999). IgG 1A5 ΔD contains a 9-aa deletion in the N-terminal Fc region. Dengue-negative human serum was used as control. **B**, **C** and **D** Binding of IgG 1A5 and variants to human Fcγ RI, Fcγ RIIA and C1q proteins. IgG 1A5 WT and the control IgG1 bind Fcγ RI, Fcγ RIIA, and C1q. ΔA ΔB and ΔD variants show no binding to C1q protein. The ΔA variant shows similar Fcγ RI binding to the IgG 1A5 WT, whereas the ΔB and ΔD variants show weak or no binding to human Fcγ Rs. A colour version of this figure is located in the plate section at the back of the book.

(220 FFU) of DENV-4. Although IgG 5H2 ΔD afforded protection, it was less effective than protection mediated by IgG 5H2 WT (Fig. 14.4C). Therefore, the Fc region seems to enhance the protective capacity of IgG 5H2 by virtue its ability to bind Fcγ receptors and/or C1q molecules. Nevertheless, the evaluation of anti-DENV IgG MAbs of human origin in a mouse model should be analysed cautiously for several reasons. First, the disease presentation in mice differs from that in humans following DENV infections, as well the site of viral inoculation. Second, differences in the intracellular domains and the cellular expression pattern between human and mouse Fcγ receptors have been described (Nimmerjahn and Ravetch, 2008). In support of these differences, a recent study showed that exchanging the human Fcγ1 of an anti-CD326 antibody with the murine Fcγ2a significantly improved ADCC in mouse effector cells and increased anti-tumour activity in a mouse model (Lutterbuese *et al.*, 2007).

The contributions of different antibody-mediated clearance mechanisms have been studied in various animal models. For example, a MAb against WNV had a similar protective activity in C1qa or C4-deficient mice, which cannot activate complement by the antibody-dependent classical pathway, when compared to wild-type mice in passive transfer experiments (Mehlhop and Diamond, 2006; Oliphant *et al.*, 2005). On the other hand, in Fcγ RI- and Fcγ RIII-deficient mice, lower antibody doses resulted in higher mortality rates, while high doses provided complete protection (Oliphant *et al.*, 2005). In an HIV-primate model, a dramatic decrease in the ability of a broadly neutralizing antibody to protect macaques against SHIV challenge was reported when Fc receptor activity, but not complement-binding activity was abrogated. Accordingly, the Fcγ receptor, but not complement binding, is important in antibody-mediated protection against HIV-1 *in vivo* (Hessell *et al.*,

2007). One interpretation of these results could be that after virus challenge, the antibody with full effector functions may neutralize infectious virions, thereby preventing infection. On the other hand, the lack of Fcγ-receptor-binding of the IgG variants may adversely affect the suppression of viral release from infected cells. Contributions of anti-DENV E specific antibodies to host defence through Fc receptors and complement pathways are less clear. Further analyses in a more suitable animal model, such as monkeys, are required to address the feasibility of engineering the Fc portion of IgG to abrogate ADE.

DENV variability and antibody escape variants

Many studies of DENV evolution have shown that the virus exists as four distinct serotypes.; see also Chapters 9 and 11), which contain well-defined phylogenetic clusters (i.e. subtypes or genotypes) causing human diseases (Rico-Hesse, 2003). Multiple factors are responsible for this diversity, for instance, high mutation rates in RNA replication, geographical migration by hosts and vectors, and the increasing size and density of the human population that might facilitate viral transmission (Holmes and Burch, 2000; Monath, 1994

The evolution of virulence has been a controversial aspect of extensive analysis in DENV biology. Phylogenetic studies of many DENV strains have proposed an association between specific genotypes and disease severity (Messer *et al.*, 2003; Rico-Hesse, 2003). It has been reported that strains within serotypes of DENV vary significantly in their binding to DENV-specific antisera (Halstead and Simasthien, 1970), and to DENV-crossreactive MAbs (Morens *et al.*, 1987). Variation among DENV strains and serotypes in antibody binding may have important implications for passive immunotherapy coverage of DENV infections.

One critical aspect in the discussion of immunotherapy and/or immunoprophylaxis in DENV infection is the epitope specificities of the most potent neutralizing MAbs. A significant number of studies suggest that antibodies that neutralize infection most efficiently bind to epitopes highly variable at the sequence level, but structurally conserved among flaviviruses on the lateral surface of E DIII (Beasley and Barrett, 2002; Gromowski and Barrett, 2007; Holzmann *et al.*, 1997; Lin and Wu, 2003; Mandl *et al.*, 1989; Oliphant *et al.*, 2005; Ryman *et al.*, 1998; Sanchez *et al.*, 2005; Sukupolvi-Petty *et al.*, 2007; Volk *et al.*, 2004; Wu *et al.*, 2003b). MAbs that bind outside the E DIII lateral surface are generally less neutralizing or non-neutralizing against flaviviruses (Crill and Chang, 2004; Goncalvez *et al.*, 2004b; Oliphant *et al.*, 2006; Stiasny *et al.*, 2006). On the other hand, all four serotypes of DENV have circulated regularly in endemic areas, and several countries have reported the simultaneous presence of more than one DENV serotype (Gubler and Meltzer, 1999; Kyle and Harris, 2008). Thus, whether a monovalent antibody therapy would protect in a specific period is uncertain. Therefore, designing a multivalent passive antibody strategy for prevention and/or treatment of DENV infection might involve the combination of potent type-specific MAbs with the ability to protect against two or more DENV serotypes. Alternatively, broadly neutralizing subcomplex- or complex-specific MAbs may be attractive candidates for further development of immunoprophylaxis against all serotypes of DENV (Gromowski *et al.*, 2008; Lisova *et al.*, 2007; Sukupolvi-Petty *et al.*, 2007; Thullier *et al.*, 2001). However, the low neutralizing activity that is characteristic of these antibodies could be a problem in the development of effective therapeutic antibodies. Approaches to improve the biological activity of these antibodies, such as affinity maturation followed by functional screening, may be a way to overcome this challenge (Wu *et al.*, 2005).

Additionally, even if an epitope is conserved among strains of DENV, another concern is the selection of antibody escape variants. Recently, passive immunization of rhesus monkeys with the neutralizing anti-DENV MAb 5H2 provided formal evidence of the *in vivo* selection of DENV escape variants (Lai *et al.*, 2007). Monkeys that received a higher dose of MAb 5H2 (2 mg/kg) and were challenged with 10 FFU of DENV-4, were protected. In contrast, monkeys that received 0.9 mg/kg and were challenged with 10^5 FFU were infected. The viruses present in several viremic samples from monkeys in the latter group were recovered; sequence analysis showed a single amino acid substitution, $Glu_{174}Lys$, on the E protein. DENV-4 selection experiments *in vitro* corroborated the *in vivo* finding that virus strains rapidly escape MAb 5H2 pressure (Lai *et al.*, 2007). Thus, the utility of antibody-mediated prevention of DENV infections may be complicated by the rapid selection of neutralization escape mutants *in vivo*. High concentrations of the antibody that provide complete neutralization may be the solution to this problem. Another possible strategy is the combination of MAbs against spatially separate neutralizing epitopes on E. This strategy could also be effective in covering a wider range of circulating DENV strains within a serotype. However, even cocktails of antibodies can encounter some problems. For instance, a clinical trial consisting of a passive-immunization study with a cocktail of three neutralizing MAbs (2G12, 2F5 and 4E10) against HIV was capable of suppressing or delaying viraemia in several individuals. However, activity was lost as soon as a 2G12-resistant virus emerged (Trkola *et al.*, 2005). The fact that no resistance to MAbs 2F5 and 4E10 was found may be explained by insufficient concentration of these MAbs when the 2G12-resistant variant emerged. Overall, these results suggest a necessity for detailed evaluations of the positive and negative aspects of specific strategies for passive antibody therapy.

Manufacturing cost

Over the last decade, MAbs have become a flourishing class of drugs that has played an important role in the advances in pharmacotherapy. However, MAbs are amongst the most expensive drugs. For instance, Synagis® (palivizumab) is a humanized MAb approved by the FDA for the prevention of serious lower respiratory tract disease caused by RSV in pediatric patients at high risk for disease. The acquisition cost of Synagis® is approximately ~$900.00 per dose or ~$4000.00 per child per season, which has caused much debate about whether this MAb is cost-effective (Nuijten *et al.*, 2007; Reeve *et al.*, 2006; Roeckl-Wiedmann *et al.*, 2003). These high prices reveal the high dose required for effectiveness, and consequently, the need for large-scale production capacity.

Most of the approved MAbs are manufactured using culture of mammalian cells and pu-

rification procedures based on chromatography with intermediate filtration and viral clearance processes (Farid, 2006); but this platform has limitations, such us high cost, low capacity, and the potential presence of cellular and adventitious contaminants. Consequently, large-scale production is not feasible for developing countries. Pressure to decrease costs of MAbs production has also encouraged the search for alternative manufacturing technologies, including the improvement in the productivity of mammalian cells cultured in bioreactors, and the use of transgenic expression systems or *E. coli* and yeast for production of antibody fragments (Farid, 2006). The alternative plant pharmaceutical platforms for manufacturing MAbs look promising for lowering the mass production cost (Ko and Koprowski, 2005; Ramessar *et al.*, 2008). To improve the commercial potential of antibodies, all efforts should focus on bringing down production costs.

Future market uncertainty

An important consideration in the development of therapeutic MAbs is the future market for this approach. It is difficult to predict at this stage what the commercial opportunity will be. However, a hypothetical exercise to this end can be carried out.

Infection with DENV causes a wide spectrum of diseases ranging from inapparent infection to severe and fatal DHF/DSS. The incubation period of dengue is typically 4 to 7 days, and viraemia can persist 4 to 5 days. DF is usually a self-limiting condition and death as a result is uncommon. The main concern is the development of DHF/DSS, which tends to occur after the fever subsides. The primary goal of passive immunotherapy would be to provide prompt protection in DENV-infected individuals so that development of DHF/DSS is prevented. In an infectious outbreak, identification of groups at high-risk of developing serious illness would be the ideal pathway for effective passive immunotherapy. However, the risk factors for the development of DHF/DSS are still uncertain. Currently, the most well-documented and widely accepted risk factor for DHF/DSS is a previous dengue infection with a heterologous strain (Halstead, 2003). However, serotyping of virus strains takes time and is not widely available, especially in developing countries.

Age seems to represent another risk factor for DHF/DSS. Classical DF is primarily a disease of older children and adults, while DHF is mostly limited to young children in Asia (WHO, 1997). Notwithstanding, older age groups are at risk for DHF in the Americas (Guzman and Kouri, 2003). Recent data indicate that elderly patients in some settings are apparently more likely than children and younger adults to develop severe illness when infected with DENV (Garcia-Rivera and Rigau-Perez, 2003). Comorbid conditions, including diabetes mellitus, serious bacterial infection and immunosuppression, have also been associated with a higher risk of death in dengue infections (Guzman *et al.*, 1999; Rigau-Perez and Laufer, 2006). Therefore, individuals with these conditions may be good candidates for passive immunotherapy.

It is difficult to predict progression from classic DF to severe DHF/DSS. Thus, rapid diagnosis and serotype identification would assume a paramount importance for preventing severe disease with antibodies. Future challenges in the diagnosis of DF and DHF include the application of techniques to avoid cross-reactivity among different serotypes of DENV and other flaviviruses (Chiou *et al.*, 2008). Moreover, speed and accuracy of diagnosis should be taken into consideration with cost and availability in endemic areas. In spite of our view of the potential for antibody therapy, we must take into account the reality of dengue infection in endemic areas. DENV infections represent a major health and economic concern to many developing countries in Asia and in the Americas (Gubler and Meltzer, 1999). In these countries, the public health system in charge of managing epidemic vector-borne infections has decayed over the past 40 years. The situation is further aggravated by ineffective and outdated health legislation. In addition, economic constraints present in most of these countries have dictated prioritized activities to limit epidemics, rather than to apply preventive measures. The potential for use of MAbs in endemic areas would depend on strategies to target improvements in surveillance, diagnosis, and development of cost-effective antibody manufacturing.

The current situation with DF and DHF in tropical developing countries makes the short-term possibility of resolving most of the underlying factors uncertain. Thus, prospects for control of dengue with passive immunoprophylaxis appear hopeless in the near future. Notwithstanding, in view of the persistent risk for individuals traveling to DENV endemic areas (Jelinek, 2000; O'Brien *et al.*, 2001), one could speculate that immunoprophylaxis in its first approach would be oriented to this group. Data from GeoSentinel, a global network of travel medicine providers (Freedman *et al.*, 1999), demonstrate that dengue is now more often detected than malaria in travellers returning from South America or Asia (Freedman *et al.*, 2006). Travellers from nonendemic countries were previously thought to be only at risk of DF primary infection. However, with the onset of globalization and marked increase in air travel, the risk of secondary infection is increasing (Gushulak *et al.*, 2007; Lawn *et al.*, 2003; Wichmann *et al.*, 2007). Although DHF/DSS is rare in travellers, fatal cases have been reported (CDC, 2006; Delgado *et al.*, 2008; Laferl *et al.*, 2006). Since dengue has become a leading emerging disease among travellers to the tropics and subtropical areas during the past 20 years, the development of a dengue vaccine and specific antiviral therapy should be expedited. Vaccination schedules with long intervals would be a major concern for travellers. Possible reactogenicity and interference with immunization in travellers needing vaccines against YFV and JEV, would be other obstacles. Thus, licensing dengue vaccine for travellers encounters particular challenges beyond the vaccine development for individuals living in endemic countries. The major advantage of passive immunoprophylaxis for travellers is that it would offer protection more rapidly during a period of high-risk exposure, thereby avoiding the potential side effects of vaccination that could cause morbidity and even mortality, as has been reported for the yellow fever vaccine (Martin *et al.*, 2001a; McMahon *et al.*, 2007).

Conclusions and future considerations for development of anti-DENV MAbs

Although dengue virus infections affect 50–100 million people yearly, with frequent and recurrent epidemics, progress in vaccine development has been slow due in part to the fact that there are four viral serotypes, all of which are equally infectious (see Chapter 12). Even with the development of live attenuated virus vaccines, it remains difficult to select vaccine candidates that are both highly immunogenic and sufficiently attenuated. The possible risk of ADE introduces another level to the complexity of designing vaccines. Therefore, alternatives to vaccines are needed to help control dengue virus-associated diseases.

It is clear that the attractive features of MAbs, such as the exquisite specificity of antigen recognition and effective recruitment of the immune system, make them potential candidates as anti-infective agents. A major advantage of MAbs is their long serum half-life that confers prolonged *in vivo* protection. Furthermore, it may be possible to refine particular properties of MAbs to improve their clinical potential. These properties include antigen-binding specificity and affinity, effector functions, pharmacokinetics and molecular architecture (Carter, 2006). These attributes motivate the attempts to develop anti-DENV MAbs as an option for the prevention of dengue infections.

Over the last several years, advances in antibody screening and isolation technologies, as well as in the construction and use of non-immune and immune antibody repertoires, have contributed significant insights to our understanding of the immune response to flaviviruses. Significantly, co-crystallographic studies of antibodies in complex with flavivirus E DIII protein have provided extensive information about the neutralizing mechanisms of flaviviruses. However, many questions still remain. For instance, the role of antibodies that recognize epitopes on DI and DII of E protein in protection and their mechanisms of neutralization, the function of antibodies against prM, M and non-structural proteins, and the importance of antibody-effector functions to the potency of anti-DENV MAbs *in vivo* all remain to be determined. Advances in these topics should open new possibilities for the design of antibody-based therapies and, at the same time, could contribute to the development and improvement of flavivirus vaccines.

Critical information gaps that require consideration in the development of anti-DENV MAbs for passive immunization include the following:

- There are still questions to be addressed with respect to the critical epitopes on DENV that are associated with pathogenesis, and which may be targets of neutralizing MAbs. There is a need to continue to delineate the epitopes recognized by type-specific neutralizing antibodies and those by cross-reactive non-neutralizing antibodies generated during natural infection.
- Much of the information required for epitope definition will need to be generated via induction of neutralizing MAbs against all four serotypes of DENV in humans, together with passive protection studies in a more suitable animal model. Especial emphasis should be given to the characterization of antibody responses, including their specificities and kinetics.
- In some cases, antibodies that do not neutralize *in vitro* could exhibit potent neutralizing activity *in vivo*. Thus, new approaches should be developed to test effector functions mediated by Fc, including ADCC and complement-mediated immune responses. Further exploration of these antibodies and their extensive evaluation in animal models would allow identification of the best candidates for potential therapeutics.
- Development of high-throughput assays for the screening of relevant and useful MAbs are needed to accelerate basic and clinical research.
- While non-human primates are the preferred model to test potential vaccines and antivirals prior to human clinical trials, this model is limited in terms of cost and accessibility for preliminary assessment of antibody neutralization potency. Establishment of small animal models of DENV disease is essential to facilitate the evaluation of MAb candidates for passive protection. Moreover, development of transgenic mice that express human Fc receptors would be worthwhile for testing antibodies that incorporate Fc variants.

Although MAbs are a long way from a practical stage, there is an ongoing need to assess their potential as therapeutic and/or prophylactic agents. A dialogue with public health authorities in dengue-endemic areas should help to select desirable target candidates. Certainly, communication between different entities, such as the FDA, industry and the scientific community would prove essential for the timely development of such antiviral agents.

Acknowledgements

This work has been supported by the Intramural Research Program of the National Institute of Allergy and Infectious Diseases, National Institutes of Health.

References

Agrawal, A.G., and Petersen, L.R. (2003). Human immunoglobulin as a treatment for West Nile virus infection. J. Infect. Dis. *188*, 1–4.

Alcon, S., Talarmin, A., Debruyne, M., Falconar, A., Deubel, V., and Flamand, M. (2002). Enzyme-linked immunosorbent assay specific to Dengue virus type 1 nonstructural protein NS1 reveals circulation of the antigen in the blood during the acute phase of disease in patients experiencing primary or secondary infections. J. Clin. Microbiol. *40*, 376–381.

Allison, S.L., Schalich, J., Stiasny, K., Mandl, C.W., and Heinz, F.X. (2001). Mutational evidence for an internal fusion peptide in flavivirus envelope protein Eur. J. Virol. *75*, 4268–4275.

Arakawa, M., Yamashiro, T., Uechi, G., Tadano, M., and Nishizono, A. (2007). Construction of human Fab (gamma1/kappa) library and identification of human monoclonal Fab possessing neutralizing potency against Japanese encephalitis virus. Microbiol Immunol *51*, 617–625.

Armour, K.L., Clark, M.R., Hadley, A.G., and Williamson, L.M. (1999). Recombinant human IgG molecules lacking Fcgamma receptor I binding and monocyte triggering activities. Eur. J. Immunol. *29*, 2613–2624.

Beasley, D.W., and Aaskov, J.G. (2001). Epitopes on the dengue 1 virus envelope protein recognized by neutralizing IgM monoclonal antibodies. Virology *279*, 447–458.

Beasley, D.W., and Barrett, A.D. (2002). Identification of neutralizing epitopes within structural domain III of the West Nile virus envelope protein. J. Virol. *76*, 13097–13100.

Bernasconi, N.L., Traggiai, E., and Lanzavecchia, A. (2002). Maintenance of serological memory by polyclonal activation of human memory B-cells. Science *298*, 2199–2202.

Bhardwaj, S., Holbrook, M., Shope, R.E., Barrett, A.D., and Watowich, S.J. (2001). Biophysical characterization and vector-specific antagonist activity of domain III of the tick-borne flavivirus envelope protein. J. Virol. *75*, 4002–4007.

Blitvich, B.J., Scanlon, D., Shiell, B.J., Mackenzie, J.S., Pham, K., and Hall, R.A. (2001). Determination of the intramolecular disulfide bond arrangement and biochemical identification of the glycosylation sites

of the nonstructural protein NS1 of Murray Valley encephalitis virus. J. Gen. Virol. *82*, 2251–2256.

Boonpucknavig, S., Vuttiviroj, O., and Boonpucknavig, V. (1981). Infection of young adult mice with dengue virus type 2. Trans. R. Soc. Trop. Med. Hyg. *75*, 647–653.

Bowley, D.R., Labrijn, A.F., Zwick, M.B., and Burton, D.R. (2007). Antigen selection from an HIV-1 immune antibody library displayed on yeast yields many novel antibodies compared to selection from the same library displayed on phage. Protein Eng. Des. Sel. *20*, 81–90.

Bressanelli, S., Stiasny, K., Allison, S.L., Stura, E.A., Duquerroy, S., Lescar, J., Heinz, F.X., and Rey, F.A. (2004). Structure of a flavivirus envelope glycoprotein in its low-pH-induced membrane fusion conformation. EMBO J. *23*, 728–738.

Broker, M., and Kollaritsch, H. (2008). After a tick bite in a tick-borne encephalitis virus endemic area: current positions about post-exposure treatment. Vaccine *26*, 863–868.

Burton, D.R., and Barbas, C.F., 3rd (1994). Human antibodies from combinatorial libraries. Adv. Immunol. *57*, 191–280.

Butrapet, S., Kimura-Kuroda, J., Zhou, D.S., and Yasui, K. (1998). Neutralizing mechanism of a monoclonal antibody against Japanese encephalitis virus glycoprotein E. Am. J. Trop. Med. Hyg. *58*, 389–398.

Cabezas, S., Rojas, G., Pavon, A., Alvarez, M., Pupo, M., Guillen, G., and Guzman, M.G. (2008). Selection of phage-displayed human antibody fragments on Dengue virus particles captured by a monoclonal antibody: application to the four serotypes. J. Virol. Methods *147*, 235–243.

Calvert, A.E., Huang, C.Y., Kinney, R.M., and Roehrig, J.T. (2006). Non-structural proteins of dengue 2 virus offer limited protection to interferon-deficient mice after dengue 2 virus challenge. J. Gen. Virol. *87*, 339–346.

Carter, P.J. (2006). Potent antibody therapeutics by design. Nat. Rev. Immunol. *6*, 343–357.

Casadevall, A., Dadachova, E., and Pirofski, L.A. (2004). Passive antibody therapy for infectious diseases. Nat. Rev. Microbiol. *2*, 695–703.

C.D.C., C.f.D.C.a.P. (2006). Travel-associated dengue – United States, 2005. MMWR Morb. Mortal Wkly Rep. *55*, 700–702.

Cecilia, D., and Gould, E.A. (1991). Nucleotide changes responsible for loss of neuroinvasiveness in Japanese encephalitis virus neutralization-resistant mutants. Virology *181*, 70–77.

Chanock, R.M., Crowe, J.E., Jr., Murphy, B.R., and Burton, D.R. (1993). Human monoclonal antibody Fab fragments cloned from combinatorial libraries: potential usefulness in prevention and/or treatment of major human viral diseases. Infect. Agents Dis. *2*, 118–131.

Chen, Y., Maguire, T., Hileman, R.E., Fromm, J.R., Esko, J.D., Linhardt, R.J., and Marks, R.M. (1997). Dengue virus infectivity depends on envelope protein binding to target cell heparan sulfate. Nat. Med. *3*, 866–871.

Chen, Y.C., Huang, H.N., Lin, C.T., Chen, Y.F., King, C.C., and Wu, H.C. (2007a). Generation and characterization of monoclonal antibodies against dengue virus type 1 for epitope mapping and serological detection by epitope-based peptide antigens. Clin Vaccine Immunol. *14*, 404–411.

Chen, Z., Earl, P., Americo, J., Damon, I., Smith, S.K., Yu, F., Sebrell, A., Emerson, S., Cohen, G., Eisenberg, R.J., Gorshkova, I., Schuck, P., Satterfield, W., Moss, B., and Purcell, R. (2007b). Characterization of chimpanzee/human monoclonal antibodies to vaccinia virus A33 glycoprotein and its variola virus homolog *in vitro* and in a vaccinia virus mouse protection model. J. Virol. *81*, 8989–8995.

Chen, Z., Earl, P., Americo, J., Damon, I., Smith, S.K., Zhou, Y.H., Yu, F., Sebrell, A., Emerson, S., Cohen, G., Eisenberg, R.J., Svitel, J., Schuck, P., Satterfield, W., Moss, B., and Purcell, R. (2006a). Chimpanzee/human mAbs to vaccinia virus B5 protein neutralize vaccinia and smallpox viruses and protect mice against vaccinia virus. Proc. Natl. Acad. Sci. U.S.A. *103*, 1882–1887.

Chen, Z., Moayeri, M., Zhou, Y.H., Leppla, S., Emerson, S., Sebrell, A., Yu, F., Svitel, J., Schuck, P., St Claire, M., and Purcell, R. (2006b). Efficient neutralization of anthrax toxin by chimpanzee monoclonal antibodies against protective antigen. J. Infect. Dis. *193*, 625–633.

Chiba, N., Osada, M., Komoro, K., Mizutani, T., Kariwa, H., and Takashima, I. (1999). Protection against tick-borne encephalitis virus isolated in Japan by active and passive immunization. Vaccine *17*, 1532–1539.

Chiou, S.S., Crill, W.D., Chen, L.K., and Chang, G.J. (2008). Enzyme-linked immunosorbent assays using novel Japanese encephalitis virus antigen improve the accuracy of clinical diagnosis of flavivirus infections. Clin. Vaccine Immunol. *15*, 825–835.

Choi, K.S., Nah, J.J., Ko, Y.J., Kim, Y.J., and Joo, Y.S. (2007). The DE loop of the domain III of the envelope protein appears to be associated with West Nile virus neutralization. Virus Res. *123*, 216–218.

Chu, J.J., Rajamanonmani, R., Li, J., Bhuvanakantham, R., Lescar, J., and Ng, M.L. (2005). Inhibition of West Nile virus entry by using a recombinant domain III from the envelope glycoprotein. J. Gen. Virol. *86*, 405–412.

Chung, K.M., Nybakken, G.E., Thompson, B.S., Engle, M.J., Marri, A., Fremont, D.H., and Diamond, M.S. (2006). Antibodies against West Nile Virus nonstructural protein NS1 prevent lethal infection through Fc gamma receptor-dependent and -independent mechanisms. J. Virol. *80*, 1340–1351.

Chung, K.M., Thompson, B.S., Fremont, D.H., and Diamond, M.S. (2007). Antibody recognition of cell surface-associated NS1 triggers Fc-gamma receptor-mediated phagocytosis and clearance of West Nile Virus-infected cells. J. Virol. *81*, 9551–9555.

Corver, J., Ortiz, A., Allison, S.L., Schalich, J., Heinz, F.X., and Wilschut, J. (2000). Membrane fusion activity of tick-borne encephalitis virus and recombinant subviral particles in a liposomal model system. Virology *269*, 37–46.

Crill, W.D., and Chang, G.J. (2004). Localization and characterization of flavivirus envelope glycoprotein cross-reactive epitopes. J. Virol. *78*, 13975–13986.

Crill, W.D., and Roehrig, J.T. (2001). Monoclonal antibodies that bind to domain III of dengue virus E glycoprotein are the most efficient blockers of virus adsorption to Vero cells. J. Virol. *75*, 7769–7773.

Crowe, J.E., Jr., Cheung, P.Y., Wallace, E.F., Chanock, R.M., Larrick, J.W., Murphy, B.R., and Fry, K. (1994). Isolation and characterization of a chimpanzee monoclonal antibody to the G glycoprotein of human respiratory syncytial virus. Clin. Diagn. Lab. Immuno.l *1*, 701–706.

Daffis, S., Kontermann, R.E., Korimbocus, J., Zeller, H., Klenk, H.D., and Ter Meulen, J. (2005). Antibody responses against wild-type yellow fever virus and the 17D vaccine strain: characterization with human monoclonal antibody fragments and neutralization escape variants. Virology *337*, 262–272.

Delgado, M.J., Gutierrez, J.M., Radic, L.B., Maretic, T., Zekan, S., Avsic-Zupanc, T., Aymar, E.S., Trilla, A., and Brustenga, J.G. (2008). Imported dengue hemorrhagic fever, Europe. Emerg. Infect. Dis .*14*, 1329–1330.

Despres, P., Dietrich, J., Girard, M., and Bouloy, M. (1991). Recombinant baculoviruses expressing yellow fever virus E and NS1 proteins elicit protective immunity in mice. J. Gen. Virol. *72 (Pt 11)*, 2811–2816.

Ehrlich, P.H., Moustafa, Z.A., Harfeldt, K.E., Is aacson, C., and Ostberg, L. (1990). Potential of primate monoclonal antibodies to substitute for human antibodies: nucleotide sequence of chimpanzee Fab fragments. Hum. Antibodies Hybridomas *1*, 23–26.

Endy, T.P., Nisalak, A., Chunsuttitwat, S., Vaughn, D.W., Green, S., Ennis, F.A., Rothman, A.L., and Libraty, D.H. (2004). Relationship of preexisting dengue virus (DV) neutralizing antibody levels to viremia and severity of disease in a prospective cohort study of DV infection in Thailand. J. Infect. Dis. *189*, 990–1000.

Falconar, A.K. (1999). Identification of an epitope on the dengue virus membrane (M) protein defined by cross-protective monoclonal antibodies: design of an improved epitope sequence based on common determinants present in both envelope (E and M) proteins. Arch. Virol. *144*, 2313–2330.

Falconar, A.K. (2008). Use of synthetic peptides to represent surface-exposed epitopes defined by neutralizing dengue complex- and flavivirus group-reactive monoclonal antibodies on the native dengue type-2 virus envelope glycoprotein. J. Gen. Virol. *89*, 1616–1621.

Falconar, A.K., and Young, P.R. (1991). Production of dimer-specific and dengue virus group cross-reactive mouse monoclonal antibodies to the dengue 2 virus non-structural glycoprotein NS1. J. Gen. Virol. *72 (Pt 4)*, 961–965.

Falconar, A.K., Young, P.R., and Miles, M.A. (1994). Precise location of sequential dengue virus subcomplex and complex B-cell epitopes on the nonstructural-1 glycoprotein. Arch. Virol. *137*, 315–326.

Falconar, A.K.I., and Young, P.R. (1989). Functional analysis of a panel of monoclonal antibodies generated against the nonstructural glycoprotein, NS1 of dengue virus 2 (PR1590). 2nd International Symposium on Positive Strand RNA Viruses Vienna, Austria.

Falgout, B., Bray, M., Schlesinger, J.J., and Lai, C.J. (1990). Immunization of mice with recombinant vaccinia virus expressing authentic dengue virus nonstructural protein NS1 protects against lethal dengue virus encephalitis. J. Virol. *64*, 4356–4363.

Farid, S.S. (2006). Established bioprocesses for producing antibodies as a basis for future planning. Adv. Biochem. Eng. Biotechnol. *101*, 1–42.

Freedman, D.O., Kozarsky, P.E., Weld, L.H., and Cetron, M.S. (1999). GeoSentinel: the global emerging infections sentinel network of the International Society of Travel Medicine. J. Travel Med. *6*, 94–98.

Freedman, D.O., Weld, L.H., Kozarsky, P.E., Fisk, T., Robins, R., von Sonnenburg, F., Keystone, J.S., Pandey, P., and Cetron, M.S. (2006). Spectrum of disease and relation to place of exposure among ill returned travelers. N. Engl. J. Med. *354*, 119–130.

Frodin, J.E., Lefvert, A.K., and Mellstedt, H. (1990). Pharmacokinetics of the mouse monoclonal antibody 17–1A in cancer patients receiving various treatment schedules. Cancer Res. *50*, 4866–4871.

Gao, G.F., Hussain, M.H., Reid, H.W., and Gould, E.A. (1994). Identification of naturally occurring monoclonal antibody escape variants of louping ill virus. J. Gen. Virol. *75 (Pt 3)*, 609–614.

Garcia-Rivera, E.J., and Rigau-Perez, J.G. (2003). Dengue severity in the elderly in Puerto Rico. Rev Panam Salud Publica *13*, 362–368.

Garcia, G., Vaughn, D.W., and Del Angel, R.M. (1997). Recognition of synthetic oligopeptides from non-structural proteins NS1 and NS3 of dengue-4 virus by sera from dengue virus-infected children. Am. J. Trop. Med. Hyg. *56*, 466–470.

Ghetie, V., and Ward, E.S. (2002). Transcytosis and catabolism of antibody. Immunol. Res *25*, 97–113.

Gollins, S.W., and Porterfield, J.S. (1986). A new mechanism for the neutralization of enveloped viruses by antiviral antibody. Nature *321*, 244–246.

Goncalvez, A.P., Chien, C.H., Tubthong, K., Gorshkova, I., Roll, C., Donau, O., Schuck, P., Yoksan, S., Wang, S.D., Purcell, R.H., and Lai, C.J. (2008). Humanized monoclonal antibodies derived from chimpanzee Fabs protect against Japanese encephalitis virus *in vitro* and *in vivo*. J. Virol. *82*, 7009–7021.

Goncalvez, A.P., Engle, R.E., St Claire, M., Purcell, R.H., and Lai, C.J. (2007). Monoclonal antibody-mediated enhancement of dengue virus infection *in vitro* and *in vivo* and strategies for prevention. Proc. Natl. Acad. Sci. U.S.A. *104*, 9422–9427.

Goncalvez, A.P., Men, R., Wernly, C., Purcell, R.H., and Lai, C.J. (2004a). Chimpanzee Fab fragments and a derived humanized immunoglobulin G1 antibody that efficiently cross-neutralize dengue type 1 and type 2 viruses. J. Virol. *78*, 12910–12918.

Goncalvez, A.P., Purcell, R.H., and Lai, C.J. (2004b). Epitope determinants of a chimpanzee Fab antibody that efficiently cross-neutralizes dengue type 1 and type 2 viruses map to inside and in close proximity to fusion loop of the dengue type 2 virus envelope glycoprotein. J. Virol. *78*, 12919–12928.

Gonzales, N.R., De Pascalis, R., Schlom, J., and Kashmiri, S.V. (2005). Minimizing the immunogenicity of antibodies for clinical application. Tumour Biol. *26*, 31–43.

Gould, E.A., and Buckley, A. (1989). Antibody-dependent enhancement of yellow fever and Japanese encephalitis virus neurovirulence. J. Gen. Virol. *70 (Pt 6)*, 1605–1608.

Gould, E.A., Buckley, A., Barrett, A.D., and Cammack, N. (1986). Neutralizing (54K) and non-neutralizing (54K and 48K) monoclonal antibodies against structural and non-structural yellow fever virus proteins confer immunity in mice. J. Gen. Virol. *67 (Pt 3)*, 591–595.

Gould, L.H., Sui, J., Foellmer, H., Oliphant, T., Wang, T., Ledizet, M., Murakami, A., Noonan, K., Lambeth, C., Kar, K., Anderson, J.F., de Silva, A.M., Diamond, M.S., Koski, R.A., Marasco, W.A., and Fikrig, E. (2005). Protective and therapeutic capacity of human single-chain Fv-Fc fusion proteins against West Nile virus. J. Virol. *79*, 14606–14613.

Green, L.L., Hardy, M.C., Maynard-Currie, C.E., Tsuda, H., Louie, D.M., Mendez, M.J., Abderrahim, H., Noguchi, M., Smith, D.H., Zeng, Y., David, N.E., Sasai, H., Garza, D., Brenner, D.G., Hales, J.F., McGuinness, R.P., Capon, D.J., Klapholz, S. and Jakobovits, A. (1994). Antigen-specific human monoclonal antibodies from mice engineered with human Ig heavy and light chain YACs. Nat. Genet. *7*, 13–21.

Gromowski, G.D., and Barrett, A.D. (2007). Characterization of an antigenic site that contains a dominant, type-specific neutralization determinant on the envelope protein domain III (ED3) of dengue 2 virus. Virology *366*, 349–360.

Gromowski, G.D., Barrett, N.D., and Barrett, A.D. (2008). Characterization of dengue virus complex-specific neutralizing epitopes on envelope protein domain III of dengue 2 virus. J. Virol. *82*, 8828–8837.

Gubler, D.J. (1998). Dengue and dengue hemorrhagic fever. Clin. Microbiol. Rev. *11*, 480–496.

Gubler, D.J., and Meltzer, M. (1999). Impact of dengue/dengue hemorrhagic fever on the developing world. In Adv Virus Res, K. Margniorosch, F.A. Murphy, and A.J. Shatkin, eds. (California, USA: Elsevier Academic Press), pp. 35–70.

Guirakhoo, F., Heinz, F.X., and Kunz, C. (1989). Epitope model of tick-borne encephalitis virus envelope glycoprotein E: analysis of structural properties, role of carbohydrate side chain, and conformational changes occurring at acidic pH. Virology *169*, 90–99.

Gupta, A.K., Lad, V.J., and Koshy, A.A. (2003). Protection of mice against experimental Japanese encephalitis virus infections by neutralizing anti-glycoprotein E monoclonal antibodies. Acta Virol. *47*, 141–145.

Gushulak, B., Funk, M., and Steffen, R. (2007). Global changes related to travelers' health. J. Travel Med. *14*, 205–208.

Guzman, M.G., Alvarez, M., Rodriguez, R., Rosario, D., Vazquez, S., Vald s, L., Cabrera, M.V., and Kouri, G. (1999). Fatal dengue hemorrhagic fever in Cuba, 1997. Int J. Infect. Dis. *3*, 130–135.

Guzman, M.G., and Kouri, G. (2003). Dengue and dengue hemorrhagic fever in the Americas: lessons and challenges. J. Clin. Virol. *27*, 1–13.

Haley, M., Retter, A.S., Fowler, D., Gea-Banacloche, J., and O'Grady, N.P. (2003). The role for intravenous immunoglobulin in the treatment of West Nile virus encephalitis. Clin. Infect. Dis. *37*, e88–90.

Hall, R.A., Brand, T.N., Lobigs, M., Sangster, M.Y., Howard, M.J., and Mackenzie, J.S. (1996). Protective immune responses to the E and NS1 proteins of Murray Valley encephalitis virus in hybrids of flavivirus-resistant mice. J. Gen. Virol. *77 (Pt 6)*, 1287–1294.

Hall, R.A., Kay, B.H., Burgess, G.W., Clancy, P., and Fanning, I.D. (1990). Epitope analysis of the envelope and non-structural glycoproteins of Murray Valley encephalitis virus. J. Gen. Virol. *71 (Pt 12)*, 2923–2930.

Halstead, S.B. (1970). Observations related to pathogenesis of dengue hemorrhagic fever. V.I. Hypotheses and discussion. Yale J. Biol. Med. *42*, 350–362.

Halstead, S.B. (1979). *In vivo* enhancement of dengue virus infection in rhesus monkeys by passively transferred antibody. J. Infect. Dis. *140*, 527–533.

Halstead, S.B. (2003). Neutralization and antibody-dependent enhancement of dengue viruses. In Adv Virus Res, T.J. Chambers, and T.P. Monath, eds. (California, USA: Elsevier Academic Press), pp. 421–467.

Halstead, S.B., Shotwell, H., and Casals, J. (1973). Studies on the pathogenesis of dengue infection in monkeys. I.I. Clinical laboratory responses to heterologous infection. J. Infect. Dis. *128*, 15–22.

Halstead, S.B., and Simasthien, P. (1970). Observations related to the pathogenesis of dengue hemorrhagic fever. I.I. Antigenic and biologic properties of dengue viruses and their association with disease response in the host. Yale J. Biol. Med. *42*, 276–292.

Hamdan, A., Green, P., Mendelson, E., Kramer, M.R., Pitlik, S., and Weinberger, M. (2002). Possible benefit of intravenous immunoglobulin therapy in a lung transplant recipient with West Nile virus encephalitis. Transpl. Infect. Dis. *4*, 160–162.

Harrison, S.C. (2008). Viral membrane fusion. Nat. Struct. Mo. Biol. *15*, 690–698.

Hasegawa, H., Yoshida, M., Shiosaka, T., Fujita, S., and Kobayashi, Y. (1992). Mutations in the envelope protein of Japanese encephalitis virus affect entry into cultured cells and virulence in mice. Virology *191*, 158–165.

Henchal, E.A., Henchal, L.S., and Schlesinger, J.J. (1988). Synergistic interactions of anti-NS1 monoclonal antibodies protect passively immunized mice from lethal challenge with dengue 2 virus. J. Gen. Virol. *69*, 2101–2107.

Henchal, E.A., Henchal, L.S., and Thaisomboonsuk, B.K. (1987). Topological mapping of unique epitopes on the dengue-2 virus NS1 protein using monoclonal antibodies. J. Gen. Virol. *68 (Pt 3)*, 845–851.

Hessell, A.J., Hangartner, L., Hunter, M., Havenith, C.E., Beurskens, F.J., Bakker, J.M., Lanigan, C.M., Landucci, G., Forthal, D.N., Parren, P.W., Marx, P.A., and Burton, D.R. (2007). Fc receptor but not complement binding is important in antibody protection against HIV Nature *449*, 101–104.

Hiramatsu, K., Tadano, M., Men, R., and Lai, C.J. (1996). Mutational analysis of a neutralization epitope on the dengue type 2 virus (DEN2) envelope protein: monoclonal antibody resistant DEN2/DEN4 chimeras exhibit reduced mouse neurovirulence. Virology *224*, 437–445.

Ho, M., Nagata, S., and Pastan, I. (2006). Isolation of anti-CD22 Fv with high affinity by Fv display on human cells. Proc. Natl. Acad. Sci. U.S.A. *103*, 9637–9642.

Holbrook, M.R., Wang, H., and Barrett, A.D. (2001). Langat virus M protein is structurally homologous to prM. J. Virol. *75*, 3999–4001.

Holmes, E.C., and Burch, S.S. (2000). The causes and consequences of genetic variation in dengue virus. Trends Microbiol. *8*, 74–77.

Holzmann, H., Mandl, C.W., Guirakhoo, F., Heinz, F.X., and Kunz, C. (1989). Characterization of antigenic variants of tick-borne encephalitis virus selected with neutralizing monoclonal antibodies. J. Gen. Virol. *70*, 219–222.

Holzmann, H., Stiasny, K., Ecker, M., Kunz, C., and Heinz, F.X. (1997). Characterization of monoclonal antibody-escape mutants of tick-borne encephalitis virus with reduced neuroinvasiveness in mice. J. Gen. Virol. *78*, 31–37.

Hoogenboom, H.R. (2005). Selecting and screening recombinant antibody libraries. Nat. Biotechnol. *23*, 1105–1116.

Hung, J.J., Hsieh, M.T., Young, M.J., Kao, C.L., King, C.C., and Chang, W. (2004). An external loop region of domain III of dengue virus type 2 envelope protein is involved in serotype-specific binding to mosquito but not mammalian cells. J. Virol. *78*, 378–388.

Huse, W.D., Sastry, L., Iverson, S.A., Kang, A.S., Alting-Mees, M., Burton, D.R., Benkovic, S.J., and Lerner, R.A. (1989). Generation of a large combinatorial library of the immunoglobulin repertoire in phage lambda. Science *246*, 1275–1281.

Iacono-Connors, L.C., Smith, J.F., Ksiazek, T.G., Kelley, C.L., and Schmaljohn, C.S. (1996). Characterization of Langat virus antigenic determinants defined by monoclonal antibodies to E., NS1 and preM and identification of a protective, non-neutralizing preM-specific monoclonal antibody. Virus Res. *43*, 125–136.

Ishida, I., Tomizuka, K., Yoshida, H., Tahara, T., Takahashi, N., Ohguma, A., Tanaka, S., Umehashi, M., Maeda, H., Nozaki, C., Halk, E., and Lonberg, N. (2002). Production of human monoclonal and polyclonal antibodies in TransChromo animals. Cloning Stem Cells *4*, 91–102.

Jacobs, S.C., Stephenson, J.R., and Wilkinson, G.W. (1992). High-level expression of the tick-borne encephalitis virus NS1 protein by using an adenovirus-based vector: protection elicited in a murine model. J. Virol. *66*, 2086–2095.

Jacobs, S.C., Stephenson, J.R., and Wilkinson, G.W. (1994). Protection elicited by a replication-defective adenovirus vector expressing the tick-borne encephalitis virus non-structural glycoprotein NS1. J. Gen. Virol. *75 (Pt 9)*, 2399–2402.

Jelinek, T. (2000). Dengue fever in international travelers. Clin. Infect. Dis. *31*, 144–147.

Jiang, W.R., Lowe, A., Higgs, S., Reid, H., and Gould, E.A. (1993). Single amino acid codon changes detected in louping ill virus antibody-resistant mutants with reduced neurovirulence. J. Gen. Virol. *74*, 931–935.

Johnson, A.J., and Roehrig, J.T. (1999). New mouse model for dengue virus vaccine testing. J. Virol. *73*, 783–786.

Kanai, R., Kar, K., Anthony, K., Gould, L.H., Ledizet, M., Fikrig, E., Marasco, W.A., Koski, R.A., and Modis, Y. (2006). Crystal structure of west nile virus envelope glycoprotein reveals viral surface epitopes. J. Virol. *80*, 11000–11008.

Kaufman, B.M., Summers, P.L., Dubois, D.R., Cohen, W.H., Gentry, M.K., Timchak, R.L., Burke, D.S., and Eckels, K.H. (1989). Monoclonal antibodies for dengue virus prM glycoprotein protect mice against lethal dengue infection. Am. J. Trop. Med. Hyg. *41*, 576–580.

Kaufman, B.M., Summers, P.L., Dubois, D.R., and Eckels, K.H. (1987). Monoclonal antibodies against dengue 2 virus E-glycoprotein protect mice against lethal dengue infection. Am. J. Trop. Med. Hyg. *36*, 427–434.

Kimura-Kuroda, J., and Yasui, K. (1988). Protection of mice against Japanese encephalitis virus by passive administration with monoclonal antibodies. J. Immunol. *141*, 3606–3610.

Klasse, P.J., and Sattentau, Q.J. (2002). Occupancy and mechanism in antibody-mediated neutralization of animal viruses. J. Gen. Virol. *83*, 2091–2108.

Kliks, S.C., Nimmanitya, S., Nisalak, A., and Burke, D.S. (1988). Evidence that maternal dengue antibodies are important in the development of dengue hemorrhagic fever in infants. Am. J. Trop. Med. Hyg. *38*, 411–419.

Ko, K., and Koprowski, H. (2005). Plant biopharming of monoclonal antibodies. Virus Res. *111*, 93–100.

Kohler, G., and Milstein, C. (1975). Continuous cultures of fused cells secreting antibody of predefined specificity. Nature *256*, 495–497.

Krieg, A.M. (2002). CpG motifs in bacterial DNA and their immune effects. Annu. Rev. Immunol. *20*, 709–760.

Kuhn, R.J., Zhang, W., Rossmann, M.G., Pletnev, S.V., Corver, J., Lenches, E., Jones, C.T., Mukhopadhyay, S., Chipman, P.R., Strauss, E.G., Baker, T.S., and Strauss, J.H. (2002). Structure of dengue virus: implications for flavivirus organization, maturation, and fusion. Cell *108*, 717–725.

Kyle, J.L., and Harris, E. (2008). Global spread and persistence of dengue. Annu. Rev. Microbiol. *62*, 71–92.

Laferl, H., Szell, M., Bischof, E., and Wenisch, C. (2006). Imported dengue fever in Austria 1990–2005. Travel Med. Infect. Dis. *4*, 319–323.

Lai, C.J., Goncalvez, A.P., Men, R., Wernly, C., Donau, O., Engle, R.E., and Purcell, R.H. (2007). Epitope determinants of a chimpanzee dengue virus type 4 (DENV-4)-neutralizing antibody and protection against DENV-4 challenge in mice and rhesus monkeys by passively transferred humanized antibody. J. Virol. *81*, 12766–12774.

Lai, C.Y., Tsai, W.Y., Lin, S.R., Kao, C.L., Hu, H.P., King, C.C., Wu, H.C., Chang, G.J., and Wang, W.K. (2008). Antibodies to envelope glycoprotein of dengue virus during the natural course of infection are predominantly cross-reactive and recognize epitopes containing highly conserved residues at the fusion loop of domain I.I. J. Virol. *82*, 6631–6643.

Lang, A.B., Cryz, S.J., Jr., Schurch, U., Ganss, M.T., and Bruderer, U. (1993). Immunotherapy with human monoclonal antibodies. Fragment A specificity of polyclonal and monoclonal antibodies is crucial for full protection against tetanus toxin. J. Immunol. *151*, 466–472.

Lanzavecchia, A., Corti, D., and Sallusto, F. (2007). Human monoclonal antibodies by immortalization of memory B-cells. Curr. Opin. Biotechnol. *18*, 523–528.

Laoprasopwattana, K., Libraty, D.H., Endy, T.P., Nisalak, A., Chunsuttiwat, S., Vaughn, D.W., Reed, G., Ennis, F.A., Rothman, A.L., and Green, S. (2005). Dengue Virus (DV) enhancing antibody activity in preillness plasma does not predict subsequent disease severity or viremia in secondary DV infection. J. Infect. Dis. *192*, 510–519.

Lawn, S.D., Tilley, R., Lloyd, G., Finlayson, C., Tolley, H., Newman, P., Rice, P., and Harrison, T.S. (2003). Dengue hemorrhagic fever with fulminant hepatic failure in an immigrant returning to Bangladesh. Clin Infect Dis *37*, e1–4.

Leyendeckers, H., Odendahl, M., Lohndorf, A., Irsch, J., Spangfort, M., Miltenyi, S., Hunzelmann, N., Assenmacher, M., Radbruch, A., and Schmitz, J. (1999). Correlation analysis between frequencies of circulating antigen-specific IgG-bearing memory B-cells and serum titers of antigen-specific IgG. Eur J. Immunol. *29*, 1406–1417.

Libraty, D.H., Young, P.R., Pickering, D., Endy, T.P., Kalayanarooj, S., Green, S., Vaughn, D.W., Nisalak, A., Ennis, F.A., and Rothman, A.L. (2002). High circulating levels of the dengue virus nonstructural protein NS1 early in dengue illness correlate with the development of dengue hemorrhagic fever. J. Infect. Dis. *186*, 1165–1168.

Lin, B., Parrish, C.R., Murray, J.M., and Wright, P.J. (1994). Localization of a neutralizing epitope on the envelope protein of dengue virus type 2. Virology *202*, 885–890.

Lin, C.W., and Wu, S.C. (2003). A functional epitope determinant on domain III of the Japanese encephalitis virus envelope protein interacted with neutralizing-antibody combining sites. J. Virol. *77*, 2600–2606.

Lin, C.W., and Wu, S.C. (2004). Identification of mimotopes of the Japanese encephalitis virus envelope protein using phage-displayed combinatorial peptide library. J Mol Microbiol Biotechnol *8*, 34–42.

Lin, Y.L., Chen, L.K., Liao, C.L., Yeh, C.T., Ma, S.H., Chen, J.L., Huang, Y.L., Chen, S.S., and Chiang, H.Y. (1998a). DNA immunization with Japanese encephalitis virus nonstructural protein NS1 elicits protective immunity in mice. J. Virol. *72*, 191–200.

Lin, Y.L., Liao, C.L., Chen, L.K., Yeh, C.T., Liu, C.I., Ma, S.H., Huang, Y.Y., Huang, Y.L., Kao, C.L., and King, C.C. (1998b). Study of Dengue virus infection in SCID mice engrafted with human K562 cells. J. Virol. *72*, 9729–9737.

Lin, Y.W., Wang, K.J., Lei, H.Y., Lin, Y.S., Yeh, T.M., Liu, H.S., Liu, C.C., and Chen, S.H. (2002). Virus replication and cytokine production in dengue virus-infected human B lymphocytes. J. Virol. *76*, 12242–12249.

Lindenbach, B.D., and Rice, C.M. (2003). Molecular biology of flaviviruses. In Adv Virus Res, T.J. Chambers, and T.P. Monath, eds. (California, USA: Elsevier Academic Press), pp. 23–61.

Lisova, O., Hardy, F., Petit, V., and Bedouelle, H. (2007). Mapping to completeness and transplantation of a group-specific, discontinuous, neutralizing epitope in the envelope protein of dengue virus. J. Gen. Virol. *88*, 2387–2397.

Littaua, R., Kurane, I., and Ennis, F.A. (1990). Human IgG Fc receptor II mediates antibody-dependent enhancement of dengue virus infection. J. Immunol. *144*, 3183–3186.

Lobigs, M., Dalgarno, L., Schlesinger, J.J., and Weir, R.C. (1987). Location of a neutralization determinant in the E protein of yellow fever virus (17D vaccine strain). Virology *161*, 474–478.

Lobo, E.D., Hansen, R.J., and Balthasar, J.P. (2004). Antibody pharmacokinetics and pharmacodynamics. J. Pharm. Sci. *93*, 2645–2668.

Lok, S.M., Kostyuchenko, V., Nybakken, G.E., Holdaway, H.A., Battisti, A.J., Sukupolvi-Petty, S., Sedlak, D., Fremont, D.H., Chipman, P.R., Roehrig, J.T., Diamond, M.S., Kuhn, R.J., and Rossman, M.G. (2008). Binding of a neutralizing antibody to dengue virus alters the arrangement of surface glycoproteins. Nat. Struct. Mol. Biol. *15*, 312–317.

Lok, S.M., Ng, M.L., and Aaskov, J. (2001). Amino acid and phenotypic changes in dengue 2 virus associated with escape from neutralisation by IgM antibody. J. Med. Virol. *65*, 315–323.

Lonberg, N. (2008). Human monoclonal antibodies from transgenic mice. In Handb Exp Pharmacol, Y. Chernajovsky, and A. Nissim, eds. (Heidelberg, Germany: Springer), pp. 69–97.

Lonberg, N., Taylor, L.D., Harding, F.A., Trounstine, M., Higgins, K.M., Schramm, S.R., Kuo, C.C., Mashayekh, R., Wymore, K., McCabe, J.G., Donna Munoz-O'Regan, D., O'Donnell, S.L., Lapachet, E.S.G., Bengoechea, T., Fishwild, D.M., Carmack, C.E., Kay, R.M. and Huszar, D. (1994). Antigen-specific human antibodies from mice comprising four distinct genetic modifications. Nature *368*, 856–859.

Long, L., Shee-Mei Lok, I-Mei Yu, Ying Zhang, Richard, J. Kuhn, Jue Chen, Michael, G. Rossmann (2008). The Flavivirus Precursor Membrane-Envelope Protein Complex: Structure and Maturation. Cell *319* 1830 – 1834.

Lutterbuese, P., Brischwein, K., Hofmeister, R., Crommer, S., Lorenczewski, G., Petersen, L., Lippold, S., da Silva, A., Locher, M., Baeuerle, P.A., and Schlereth, B. (2007). Exchanging human Fcgamma1 with murine Fcgamma2a highly potentiates anti-tumor activity of anti-EpCAM antibody adecatumumab in a syngeneic mouse lung metastasis model. Cancer Immunol. Immunother. *56*, 459–468.

Mandl, C.W., Guirakhoo, F., Holzmann, H., Heinz, F.X., and Kunz, C. (1989). Antigenic structure of the flavivirus envelope protein E at the molecular level, using tick-borne encephalitis virus as a model. J. Virol. *63*, 564–571.

Marasco, W.A., and Sui, J. (2007). The growth and potential of human antiviral monoclonal antibody therapeutics. Nat Biotechnol *25*, 1421–1434.

Martin, M., Tsai, T.F., Cropp, B., Chang, G.J., Holmes, D.A., Tseng, J., Shieh, W., Zaki, S.R., Al-Sanouri, I., Cutrona, A.F., Ray, G., Weld, L.H., and Cetron, M.S. (2001a). Fever and multisystem organ failure associated with 17D-204 yellow fever vaccination: a report of four cases. Lancet *358*, 98–104.

Martin, W.L., West, A.P., Jr., Gan, L., and Bjorkman, P.J. (2001b). Crystal structure at 2.8 A of an FcRn/heterodimeric Fc complex: mechanism of pH-dependent binding. Mol. Cell *7*, 867–877.

Mason, P.W., Zugel, M.U., Semproni, A.R., Fournier, M.J., and Mason, T.L. (1990). The antigenic structure of dengue type 1 virus envelope and NS1 proteins expressed in Escherichia coli. J. Gen. Virol. *71*, 2107–2114.

Mathews, J.H., and Roehrig, J.T. (1984). Elucidation of the topography and determination of the protective epitopes on the E glycoprotein of Saint Louis encephalitis virus by passive transfer with monoclonal antibodies. J. Immunol. *132*, 1533–1537.

McMahon, A.W., Eidex, R.B., Marfin, A.A., Russell, M., Sejvar, J.J., Markoff, L., Hayes, E.B., Chen, R.T., Ball, R., Braun, M.M., and Cetron, M. (2007). Neurologic disease associated with 17D-204 yellow fever vaccination: a report of 15 cases. Vaccine *25*, 1727–1734.

McMinn, P.C., Lee, E., Hartley, S., Roehrig, J.T., Dalgarno, L., and Weir, R.C. (1995). Murray valley encephalitis virus envelope protein antigenic variants with altered hemagglutination properties and reduced neuroinvasiveness in mice. Virology *211*, 10–20.

Mehlhop, E., and Diamond, M.S. (2006). Protective immune responses against West Nile virus are primed by distinct complement activation pathways. J .Exp. Med. *203*, 1371–1381.

Men, R., Yamashiro, T., Goncalvez, A.P., Wernly, C., Schofield, D.J., Emerson, S.U., Purcell, R.H., and Lai, C.J. (2004). Identification of chimpanzee Fab fragments by repertoire cloning and production of a full-length humanized immunoglobulin G1 antibody that is highly efficient for neutralization of dengue type 4 virus. J. Virol. *78*, 4665–4674.

Messer, W.B., Gubler, D.J., Harris, E., Sivananthan, K., and de Silva, A.M. (2003). Emergence and global spread of a dengue serotype 3, subtype III virus. Emerg. Infect. Dis. *9*, 800–809.

Mirick, G.R., Bradt, B.M., Denardo, S.J., and Denardo, G.L. (2004). A review of human anti-globulin antibody (HAGA., HAMA., HACA., HAHA) responses to monoclonal antibodies. Not four letter words. Q J Nucl Med Mol Imaging *48*, 251–257.

Modis, Y., Ogata, S., Clements, D., and Harrison, S.C. (2003). A ligand-binding pocket in the dengue virus envelope glycoprotein. Proc. Natl. Acad. Sci. U.S.A. *100*, 6986–6991.

Modis, Y., Ogata, S., Clements, D., and Harrison, S.C. (2004). Structure of the dengue virus envelope protein after membrane fusion. Nature *427*, 313–319.

Modis, Y., Ogata, S., Clements, D., and Harrison, S.C. (2005). Variable surface epitopes in the crystal structure of dengue virus type 3 envelope glycoprotein. J. Virol. *79*, 1223–1231.

Monath, T.P. (1994). Dengue: the risk to developed and developing countries. Proc. Natl. Acad. Sci. U.S.A. *91*, 2395–2400.

Morens, D.M., Venkateshan, C.N., and Halstead, S.B. (1987). Dengue 4 virus monoclonal antibodies identify epitopes that mediate immune infection enhancement of dengue 2 viruses. J. Gen. Virol. *68 (Pt 1)*, 91–98.

Morita, K., Tadano, M., Nakaji, S., Kosai, K., Mathenge, E.G., Pandey, B.D., Hasebe, F., Inoue, S., and Igarashi, A. (2001). Locus of a virus neutralization epitope on the Japanese encephalitis virus envelope protein determined by use of long PCR-based region-specific random mutagenesis. Virology *287*, 417–426.

Morrison, S.L., Johnson, M.J., Herzenberg, L.A., and Oi, V.T. (1984). Chimeric human antibody molecules: mouse antigen-binding domains with human constant region domains. Proc. Natl. Acad. Sci. U.S.A. *81*, 6851–6855.

Mukhopadhyay, S., Kim, B.S., Chipman, P.R., Rossmann, M.G., and Kuhn, R.J. (2003). Structure of West Nile virus. Science *302*, 248.

Mukhopadhyay, S., Kuhn, R.J., and Rossmann, M.G. (2005). A structural perspective of the flavivirus life cycle. Nat Rev Microbiol *3*, 13–22.

Nelson, S., Jost, C.A., Xu, Q., Ess, J., Martin, J.E., Oliphant, T., Whitehead, S.S., Durbin, A.P., Graham, B.S., Diamond, M.S., and Pierson, T.C. (2008). Maturation of West Nile virus modulates sensitivity to antibody-mediated neutralization. PLoS Pathog *4*, e1000060.

Niedrig, M., Klockmann, U., Lang, W., Roeder, J., Burk, S., Modrow, S., and Pauli, G. (1994). Monoclonal antibodies directed against tick-borne encephalitis virus with neutralizing activity *in vivo*. Acta Virol. *38*, 141–149.

Nimmerjahn, F., and Ravetch, J.V. (2008). Fcgamma receptors as regulators of immune responses. Nat. Rev. Immunol. *8*, 34–47.

Nuijten, M.J., Wittenberg, W., and Lebmeier, M. (2007). Cost effectiveness of palivizumab for respiratory syncytial virus prophylaxis in high-risk children: a UK analysis. Pharmacoeconomics *25*, 55–71.

Nybakken, G.E., Nelson, C.A., Chen, B.R., Diamond, M.S., and Fremont, D.H. (2006). Crystal structure of the West Nile virus envelope glycoprotein. J. Virol. *80*, 11467–11474.

Nybakken, G.E., Oliphant, T., Johnson, S., Burke, S., Diamond, M.S., and Fremont, D.H. (2005). Structural basis of West Nile virus neutralization by a therapeutic antibody. Nature *437*, 764–769.

O'Brien, D., Tobin, S., Brown, G.V., and Torresi, J. (2001). Fever in returned travelers: review of hospital admissions for a 3-year period. Clin. Infect. Dis. *33*, 603–609.

Ober, R.J., Radu, C.G., Ghetie, V., and Ward, E.S. (2001). Differences in promiscuity for antibody–FcRn interactions across species: implications for therapeutic antibodies. Int. Immunol. *13*, 1551–1559.

Oliphant, T., Engle, M., Nybakken, G.E., Doane, C., Johnson, S., Huang, L., Gorlatov, S., Mehlhop, E., Marri, A., Chung, K.M., Ebel, G.D., Kramer, L.D., Fremont, D.H., and Diamond, M.S. (2005). Development of a humanized monoclonal antibody with therapeutic potential against West Nile virus. Nat. Med. *11*, 522–530.

Oliphant, T., Nybakken, G.E., Austin, S.K., Xu, Q., Bramson, J., Loeb, M., Throsby, M., Fremont, D.H., Pierson, T.C., and Diamond, M.S. (2007). Induction of epitope-specific neutralizing antibodies against West Nile virus. J. Virol. *81*, 11828–11839.

Oliphant, T., Nybakken, G.E., Engle, M., Xu, Q., Nelson, C.A., Sukupolvi-Petty, S., Marri, A., Lachmi, B.E., Olshevsky, U., Fremont, D.H., Pierson, T.C., and Diamond, M.S. (2006). Antibody recognition and neutralization determinants on domains I and II of West Nile Virus envelope protein. J. Virol. *80*, 12149–12159.

Pasatiempo, A.M., Kroser, J.A., Rudnick, M., and Hoffman, B.I. (1994). Acute renal failure after intravenous immunoglobulin therapy. J. Rheumatol. *21*, 347–349.

Pennington, J.E. (1990). Newer uses of intravenous immunoglobulins as anti-infective agents. Antimicrob. Agents Chemother. *34*, 1463–1466.

Phillpotts, R.J., Stephenson, J.R., and Porterfield, J.S. (1987). Passive immunization of mice with monoclonal antibodies raised against tick-borne encephalitis virus. Brief report. Arch. Virol. *93*, 295–301.

Pierson, T.C., Fremont, D.H., Kuhn, R.J., and Diamond, M.S. (2008). Structural insights into the mechanisms of antibody-mediated neutralization of flavivirus infection: implications for vaccine development. Cell Host Microbe *4*, 229–238.

Pierson, T.C., Xu, Q., Nelson, S., Oliphant, T., Nybakken, G.E., Fremont, D.H., and Diamond, M.S. (2007). The stoichiometry of antibody-mediated neutralization and enhancement of West Nile virus infection. Cell Host Microbe *1*, 135–145.

Putnak, J.R., Charles, P.C., Padmanabhan, R., Irie, K., Hoke, C.H., and Burke, D.S. (1988). Functional and antigenic domains of the dengue-2 virus nonstructural glycoprotein NS-1. Virology *163*, 93–103.

Qu, X., Chen, W., Maguire, T., and Austin, F. (1993). Immunoreactivity and protective effects in mice of a recombinant dengue 2 Tonga virus NS1 protein produced in a baculovirus expression system. J. Gen. Virol. *74*, 89–97.

Ramessar, K., Rademacher, T., Sack, M., Stadlmann, J., Platis, D., Stiegler, G., Labrou, N., Altmann, F., Ma, J., Stoger, E., Capell, T., and Christou, P. (2008). Cost-effective production of a vaginal protein microbicide to prevent HIV transmission. Proc. Natl. Acad. Sci. U.S.A. *105*, 3727–3732.

Razumov, I.A., Kazachinskaia, E.I., Ternovoi, V.A., Protopopova, E.V., Galkina, I.V., Gromashevskii, V.L., Prilipov, A.G., Kachko, A.V., Ivanova, A.V., L'Vov, D.K., and Loktev, V.B. (2005). Neutralizing monoclonal antibodies against Russian strain of the West Nile virus. Viral Immunol *18*, 558–568.

Reeve, C.A., Whitehall, J.S., Buettner, P.G., Norton, R., Reeve, D.M., and Francis, F. (2006). Cost-effectiveness of respiratory syncytial virus prophylaxis with palivizumab. J Paediatr Child Health *42*, 253–258.

Reichert, J.M. (2007). Trends in the development and approval of monoclonal antibodies for viral infections. BioDrugs *21*, 1–7.

Rey, F.A., Heinz, F.X., Mandl, C., Kunz, C., and Harrison, S.C. (1995). The envelope glycoprotein from tick-borne encephalitis virus at 2 A resolution. Nature *375*, 291–298.

Rico-Hesse, R. (2003). Microevolution and virulence of dengue viruses. In Adv Virus Res, T.J. Chambers, and T.P. Monath, eds. (California, USA: Elsevier Academic Press), pp. 315–341.

Riechmann, L., Clark, M., Waldmann, H., and Winter, G. (1988). Reshaping human antibodies for therapy. Nature *332*, 323–327.

Rigau-Perez, J.G., and Laufer, M.K. (2006). Dengue-related deaths in Puerto Rico, 1992–1996: diagnosis and clinical alarm signals. Clin. Infect. Dis. *42*, 1241–1246.

Roeckl-Wiedmann, I., Liese, J.G., Grill, E., Fischer, B., Carr, D., and Belohradsky, B.H. (2003). Economic evaluation of possible prevention of RSV-related hospitalizations in premature infants in Germany. Eur J Pediatr *162*, 237–244.

Roehrig, J.T. (2003). Antigenic structure of flavivirus proteins. In Adv Virus Res, T.J. Chambers, and T.P. Monath, eds. (California, USA: Elsevier Academic Press), pp. 141–175.

Roehrig, J.T., Bolin, R.A., and Kelly, R.G. (1998). Monoclonal antibody mapping of the envelope glycoprotein of the dengue 2 virus, Jamaica. Virology *246*, 317–328.

Roehrig, J.T., Hunt, A.R., Johnson, A.J., and Hawkes, R.A. (1989). Synthetic peptides derived from the deduced amino acid sequence of the E-glycoprotein of Murray Valley encephalitis virus elicit antiviral antibody. Virology *171*, 49–60.

Roehrig, J.T., Johnson, A.J., Hunt, A.R., Bolin, R.A., and Chu, M.C. (1990). Antibodies to dengue 2 virus E-glycoprotein synthetic peptides identify antigenic conformation. Virology *177*, 668–675.

Roehrig, J.T., Staudinger, L.A., Hunt, A.R., Mathews, J.H., and Blair, C.D. (2001). Antibody prophylaxis and therapy for flavivirus encephalitis infections. Ann. NY Acad. Sci. *951*, 286–297.

Ryman, K.D., Ledger, T.N., Campbell, G.A., Watowich, S.J., and Barrett, A.D. (1998). Mutation in a 17D-204 vaccine substrain-specific envelope protein epitope alters the pathogenesis of yellow fever virus in mice. Virology *244*, 59–65.

Ryman, K.D., Ledger, T.N., Weir, R.C., Schlesinger, J.J., and Barrett, A.D. (1997a). Yellow fever virus envelope protein has two discrete type-specific neutralizing epitopes. J. Gen. Virol. *78 (Pt 6)*, 1353–1356.

Ryman, K.D., Xie, H., Ledger, T.N., Campbell, G.A., and Barrett, A.D. (1997b). Antigenic variants of yellow fever virus with an altered neurovirulence phenotype in mice. Virology *230*, 376–380.

Sanchez, M.D., Pierson, T.C., Degrace, M.M., Mattei, L.M., Hanna, S.L., Del Piero, F., and Doms, R.W. (2007). The neutralizing antibody response against West Nile virus in naturally infected horses. Virology *359*, 336–348.

Sanchez, M.D., Pierson, T.C., McAllister, D., Hanna, S.L., Puffer, B.A., Valentine, L.E., Murtadha, M.M., Hoxie, J.A., and Doms, R.W. (2005). Characterization of neutralizing antibodies to West Nile virus. Virology *336*, 70–82.

Sanna, P.P., Samson, M.E., Moon, J.S., Rozenshteyn, R., De Logu, A., Williamson, R.A., and Burton, D.R. (1999). pFab-C.M.V., a single vector system for the rapid conversion of recombinant Fabs into whole IgG1 antibodies. Immunotechnology *4*, 185–188.

Sarantopoulos, S., Kao, C.Y., Den, W., and Sharon, J. (1994). A method for linking VL and VH region genes that allows bulk transfer between vectors for use in generating polyclonal IgG libraries. J. Immunol. *152*, 5344–5351.

Schlesinger, J.J., Brandriss, M.W., Cropp, C.B., and Monath, T.P. (1986). Protection against yellow fever in monkeys by immunization with yellow fever virus nonstructural protein NS1. J. Virol. *60*, 1153–1155.

Schlesinger, J.J., Brandriss, M.W., and Walsh, E.E. (1985). Protection against 17D yellow fever encephalitis in mice by passive transfer of monoclonal antibodies to the nonstructural glycoprotein gp48 and by active immunization with gp48. J. Immunol. *135*, 2805–2809.

Schlesinger, J.J., Brandriss, M.W., and Walsh, E.E. (1987). Protection of mice against dengue 2 virus encephalitis by immunization with the dengue 2 virus non-structural glycoprotein NS1. J. Gen. Virol. *68*, 853–857.

Schlesinger, J.J., and Chapman, S. (1995). Neutralizing F(ab')2 fragments of protective monoclonal antibodies to yellow fever virus (YF) envelope protein fail to protect mice against lethal YF encephalitis. J. Gen. Virol. *76*, 217–220.

Schlesinger, J.J., Foltzer, M., and Chapman, S. (1993). The Fc portion of antibody to yellow fever virus NS1 is a determinant of protection against YF encephalitis in mice. Virology *192*, 132–141.

Schlesinger, R.W. (1977). Dengue viruses, 1977/01/01 edn (New York: Springer-Verlag).

Schofield, D.J., Glamann, J., Emerson, S.U., and Purcell, R.H. (2000). Identification by phage display and characterization of two neutralizing chimpanzee monoclonal antibodies to the hepatitis E virus capsid protein. J. Virol. *74*, 5548–5555.

Schofield, D.J., Satterfield, W., Emerson, S.U., and Purcell, R.H. (2002). Four chimpanzee monoclonal antibodies isolated by phage display neutralize hepatitis a virus. Virology *292*, 127–136.

Seif, S.A., Morita, K., Matsuo, S., Hasebe, F., and Igarashi, A. (1995). Finer mapping of neutralizing epitope(s) on the C-terminal of Japanese encephalitis virus E-protein expressed in recombinant *Escherichia coli* system. Vaccine *13*, 1515–1521.

Sekul, E.A., Cupler, E.J., and Dalakas, M.C. (1994). Aseptic meningitis associated with high-dose intravenous immunoglobulin therapy: frequency and risk factors. Ann Intern Med *121*, 259–262.

Serafin, I.L., and Aaskov, J.G. (2001). Identification of epitopes on the envelope (E) protein of dengue 2 and dengue 3 viruses using monoclonal antibodies. Arch Virol *146*, 2469–2479.

Shimoni, Z., Niven, M.J., Pitlick, S., and Bulvik, S. (2001). Treatment of West Nile virus encephalitis with intravenous immunoglobulin. Emerg. Infect. Dis. *7*, 759.

Simmons, C.P., Bernasconi, N.L., Suguitan, A.L., Mills, K., Ward, J.M., Chau, N.V., Hien, T.T., Sallusto, F., Ha do, Q., Farrar, J., de Jong, M.D., Lanzavecchia, A., and Subbarao, K. (2007). Prophylactic and therapeutic efficacy of human monoclonal antibodies against H5N1 influenza. PLoS Med 4, e178.

Stadler, K., Allison, S.L., Schalich, J., and Heinz, F.X. (1997). Proteolytic activation of tick-borne encephalitis virus by furin. J. Virol. 71, 8475–8481.

Stiasny, K., Brandler, S., Kossl, C., and Heinz, F.X. (2007a). Probing the flavivirus membrane fusion mechanism by using monoclonal antibodies. J. Virol. 81, 11526–11531.

Stiasny, K., Kiermayr, S., Holzmann, H., and Heinz, F.X. (2006). Cryptic properties of a cluster of dominant flavivirus cross-reactive antigenic sites. J. Virol. 80, 9557–9568.

Stiasny, K., Kossl, C., Lepault, J., Rey, F.A., and Heinz, F.X. (2007b). Characterization of a structural intermediate of flavivirus membrane fusion. PLoS Pathog 3, e20.

Sui, J., Li, W., Murakami, A., Tamin, A., Matthews, L.J., Wong, S.K., Moore, M.J., Tallarico, A.S., Olurinde, M., Choe, H., Anderson, L.J., Bellini, W.J., Farzan, M., and Marasco, W.A. (2004). Potent neutralization of severe acute respiratory syndrome (SARS) coronavirus by a human mAb to S1 protein that blocks receptor association. Proc. Natl. Acad. Sci. U.S.A. 101, 2536–2541.

Sukupolvi-Petty, S., Austin, S.K., Purtha, W.E., Oliphant, T., Nybakken, G.E., Schlesinger, J.J., Roehrig, J.T., Gromowski, G.D., Barrett, A.D., Fremont, D.H., and Diamond, M.S. (2007). Type- and subcomplex-specific neutralizing antibodies against domain III of dengue virus type 2 envelope protein recognize adjacent epitopes. J. Virol. 81, 12816–12826.

Takegami, T., Miyamoto, H., Nakamura, H., and Yasui, K. (1982). Biological activities of the structural proteins of Japanese encephalitis virus. Acta Virol 26, 312–320.

Tan, C.H., Yap, E.H., Singh, M., Deubel, V., and Chan, Y.C. (1990). Passive protection studies in mice with monoclonal antibodies directed against the non-structural protein NS3 of dengue 1 virus. J. Gen. Virol. 71 (Pt 3), 745–749.

Throsby, M., Geuijen, C., Goudsmit, J., Bakker, A.Q., Korimbocus, J., Kramer, R.A., Clijsters-van der Horst, M., de Jong, M., Jongeneelen, M., Thijsse, S., Smit, R., Visser, T.J., Bijl, N., Marissen, W.E., Loeb, M., Kelvin, D.J., Preiser, W., ter Meulen, J., and de Kruif, J. (2006). Isolation and characterization of human monoclonal antibodies from individuals infected with West Nile Virus. J. Virol. 80, 6982–6992.

Thullier, P., Demangel, C., Bedouelle, H., Megret, F., Jouan, A., Deubel, V., Mazie, J.C., and Lafaye, P. (2001). Mapping of a dengue virus neutralizing epitope critical for the infectivity of all serotypes: insight into the neutralization mechanism. J. Gen. Virol. 82, 1885–1892.

Timofeev, A.V., Butenko, V.M., and Stephenson, J.R. (2004). Genetic vaccination of mice with plasmids encoding the NS1 non-structural protein from tick-borne encephalitis virus and dengue 2 virus. Virus Genes 28, 85–97.

Traggiai, E., Becker, S., Subbarao, K., Kolesnikova, L., Uematsu, Y., Gismondo, M.R., Murphy, B.R., Rappuoli, R., and Lanzavecchia, A. (2004). An efficient method to make human monoclonal antibodies from memory B-cells: potent neutralization of SARS coronavirus. Nat. Med. 10, 871–875.

Trkola, A., Kuster, H., Rusert, P., Joos, B., Fischer, M., Leemann, C., Manrique, A., Huber, M., Rehr, M., Oxenius, A., Weber, R., Stiegler, G., Vcelar, B., Aceto, L., and Gunthard, H.F. (2005). Delay of HIV-1 rebound after cessation of antiretroviral therapy through passive transfer of human neutralizing antibodies. Nat. Med. 11, 615–622.

Ueda, S., Matsuda, F., and Honjo, T. (1988). Multiple recombinational events in primate immunoglobulin epsilon and alpha genes suggest closer relationship of humans to chimpanzees than to gorillas. J. Mol. Evol. 27, 77–83.

Vaughn, D.W., Green, S., Kalayanarooj, S., Innis, B.L., Nimmannitya, S., Suntayakorn, S., Endy, T.P., Raengsakulrach, B., Rothman, A.L., Ennis, F.A., and Nisalak, A. (2000). Dengue viremia titer, antibody response pattern, and virus serotype correlate with disease severity. J. Infect. Dis. 181, 2–9.

Vijh-Warrier, S., Murphy, E., Yokoyama, I., and Tilley, S.A. (1995). Characterization of the variable regions of a chimpanzee monoclonal antibody with potent neutralizing activity against HIV-1. Mol. Immunol. 32, 1081–1092.

Volk, D.E., Beasley, D.W., Kallick, D.A., Holbrook, M.R., Barrett, A.D., and Gorenstein, D.G. (2004). Solution structure and antibody binding studies of the envelope protein domain III from the New York strain of West Nile virus. J. Biol. Chem. 279, 38755–38761.

Wallis, T.P., Huang, C.Y., Nimkar, S.B., Young, P.R., and Gorman, J.J. (2004). Determination of the disulfide bond arrangement of dengue virus NS1 protein. J. Biol. Chem. 279, 20729–20741.

Warrier, S.V., Pinter, A., Honnen, W.J., Girard, M., Muchmore, E., and Tilley, S.A. (1994). A novel, glycan-dependent epitope in the V2 domain of human immunodeficiency virus type 1 gp120 is recognized by a highly potent, neutralizing chimpanzee monoclonal antibody. J. Virol. 68, 4636–4642.

WHO (1997). Dengue haemorrhagic fever: diagnosis, treatment, prevention and control, 2 edn (Geneva: World Health Organization).

Wichmann, O., Gascon, J., Schunk, M., Puente, S., Siikamaki, H., Gjorup, I., Lopez-Velez, R., Clerinx, J., Peyerl-Hoffmann, G., Sundoy, A., Genton, B., Kern, P., Callieri, G., de Gorgolas, M., Muhlberger, N., and Jelinek, T. (2007). Severe dengue virus infection in travelers: risk factors and laboratory indicators. J. Infect. Dis. 195, 1089–1096.

Winkler, G., Maxwell, S.E., Ruemmler, C., and Stollar, V. (1989). Newly synthesized dengue-2 virus nonstructural protein NS1 is a soluble protein but becomes partially hydrophobic and membrane-associated after dimerization. Virology 171, 302–305.

Wong, S.J., Boyle, R.H., Demarest, V.L., Woodmansee, A.N., Kramer, L.D., Li, H., Drebot, M., Koski, R.A., Fikrig, E., Martin, D.A., and Shi, P.Y. (2003). Immunoassay targeting nonstructural protein 5 to differentiate West Nile virus infection from dengue and St. Louis encephalitis virus infections and from flavivirus vaccination. J. Clin. Microbiol. 41, 4217–4223.

Wu, H., Pfarr, D.S., Tang, Y., An, L.L., Patel, N.K., Watkins, J.D., Huse, W.D., Kiener, P.A., and Young, J.F. (2005). Ultra-potent antibodies against respiratory syncytial virus: effects of binding kinetics and binding valence on viral neutralization. J. Mol. Biol. *350*, 126–144.

Wu, H.C., Jung, M.Y., Chiu, C.Y., Chao, T.T., Lai, S.C., Jan, J.T., and Shaio, M.F. (2003a). Identification of a dengue virus type 2 (DEN-2) serotype-specific B-cell epitope and detection of DEN-2-immunized animal serum samples using an epitope-based peptide antigen. J. Gen. Virol. *84*, 2771–2779.

Wu, K.P., Wu, C.W., Tsao, Y.P., Kuo, T.W., Lou, Y.C., Lin, C.W., Wu, S.C., and Cheng, J.W. (2003b). Structural basis of a flavivirus recognized by its neutralizing antibody: solution structure of the domain III of the Japanese encephalitis virus envelope protein. J. Biol. Chem. *278*, 46007–46013.

Wu, S.F., Liao, C.L., Lin, Y.L., Yeh, C.T., Chen, L.K., Huang, Y.F., Chou, H.Y., Huang, J.L., Shaio, M.F., and Sytwu, H.K. (2003c). Evaluation of protective efficacy and immune mechanisms of using a non-structural protein NS1 in DNA vaccine against dengue 2 virus in mice. Vaccine *21*, 3919–3929.

Xiao, X., and Dimitrov, D.S. (2007). Monoclonal antibodies against viruses and bacteria: a survey of patents. Recent Patents Anti-Infect. Drug Disc. *2*, 171–177.

Yamauchi, T. (1989). [Immunological analysis of Japanese encephalitis virus using anti-Kamiyama monoclonal antibodies]. Kansenshogaku Zasshi *63*, 387–399.

Yao, Z.J., Kao, M.C., Loh, K.C., and Chung, M.C. (1995). A serotype-specific epitope of dengue virus 1 identified by phage displayed random peptide library. FEMS Microbiol.l Lett. *127*, 93–98.

Young, P.R. (1990). Antigenic analysis of dengue virus using monoclonal antibodies. South-East Asian J Trop Med Public Health *21*, 646–651.

Young, P.R., Hilditch, P.A., Bletchly, C., and Halloran, W. (2000). An antigen capture enzyme-linked immunosorbent assay reveals high levels of the dengue virus protein NS1 in the sera of infected patients. J Clin Microbiol *38*, 1053–1057.

Zhang, M.J., Wang, M.J., Jiang, S.Z., and Ma, W.Y. (1989a). Passive protection of mice, goats, and monkeys against Japanese encephalitis with monoclonal antibodies. J. Med. Virol. *29*, 133–138.

Zhang, W., Chipman, P.R., Corver, J., Johnson, P.R., Zhang, Y., Mukhopadhyay, S., Baker, T.S., Strauss, J.H., Rossmann, M.G., and Kuhn, R.J. (2003a). Visualization of membrane protein domains by cryo-electron microscopy of dengue virus. Nat. Struct. Biol. *10*, 907–912.

Zhang, Y., Corver, J., Chipman, P.R., Zhang, W., Pletnev, S.V., Sedlak, D., Baker, T.S., Strauss, J.H., Kuhn, R.J., and Rossmann, M.G. (2003b). Structures of immature flavivirus particles. EMBO J. *22*, 2604–2613.

Zhang, Y., Kaufmann, B., Chipman, P.R., Kuhn, R.J., and Rossmann, M.G. (2007). Structure of immature West Nile virus. J. Virol. *81*, 6141–6145.

Zhang, Y.M., Falgout, B., Bray, M., Dubois, D.R., Eckels, K.H., Chanok, R.M., and Lai, C.J. (1989b). Use of subunit proteins and recombinant vaccinia viruses expressing nonstructural protein NS1 for immunisation against dengue virus infection Paper presented at: Development of Flavivirus Vaccines for Dengue and Japanese Encephalitis (Vienna, Austria, WHO).

Index

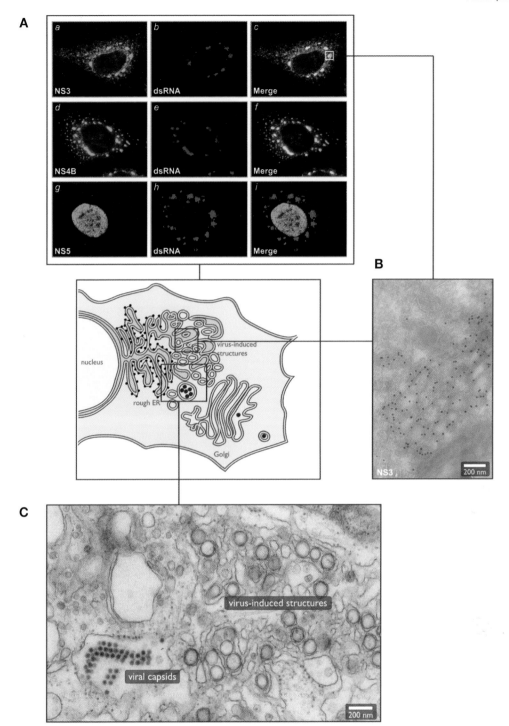

Plate 3.2 Microscopic analyses of Dengue virus-infected cells. **A** Colocalization studies of DENV NS proteins with dsRNA, an intermediate of viral RNA replication. At 24 h p.i. infected cells were fixed and processed for immunolabelling. Used antibodies are given in the lower left of each picture. Merged pictures are shown on the right. The dot-like structures in the cytoplasm of infected cells represent the sites of viral replication (adapted from Miller *et al.*, 2006). **B** Accumulations of virus-induced membrane alterations as shown by the ultrathin cryo-sections of cells 24 h p.i. NS3 was detected by immunogold labelling. Strong labelling of vesicle-accumulations is visible. **C** Overview of DENV-induced membrane alterations in the cytoplasm of infected cells. Resin-embedded sections of cells 24 h p.i. reveal double membrane vesicles and viral particles inside membranous structures (adapted from Bartenschlager and Miller, 2008).

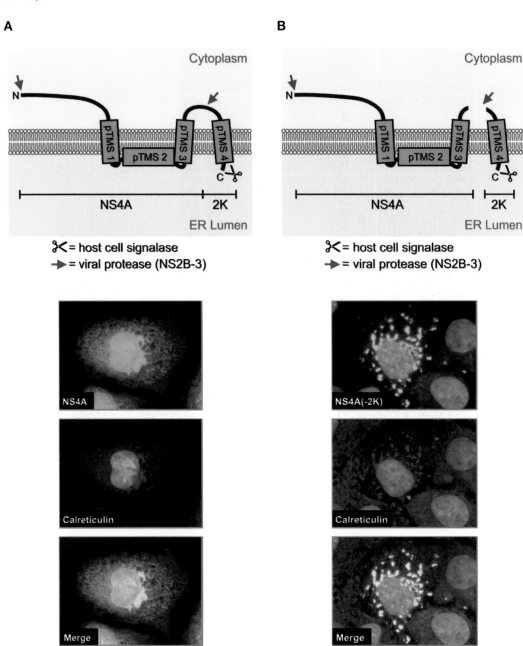

Plate 3.3 Dengue virus NS4A: a putative inducer of membrane alterations. **A** DENV NS4A-2K is a polytopic trans-membrane protein. Its N-terminal one-third is localized in the cytoplasm. The predicted trans-membrane segments (pTMS) 1 and 4 span the membrane from the cytoplasmic to the luminal site, whereas pTMS3 seems to span the lipid bilayer from the luminal to the cytoplasmic site. pTMS2 most probably does not span the membrane but is closely associated with the luminal site of the lipid bilayer. NS4A is cleaved off from the preceding NS3 protein during polyprotein processing by the viral NS2B-3 protease. It also cleaves at the NS4A-2K site within NS4A and removes pTMS4 (the '2K-fragment'). This cleavage is a prerequisite for processing by the host cell signalase at the 2K-4B site. Immunofluorescence studies have shown that expression of NS4A-2K in eukaryotic cells leads to a reticular staining pattern at ER membranes (lower panel). **B** Expression of the mature form of NS4A (lacking the 2K fragment) induces ER derived dot-like structures in the cytoplasm (lower panel). Since these structures resemble the membrane-alterations observed in DENV-infected cells, NS4A might be responsible for the induction of membrane structures that harbour the DENV RC (adapted from Miller *et al.*, 2007).

Noninfectious immature particle   Infectious mature particle

A   **B**   **C**

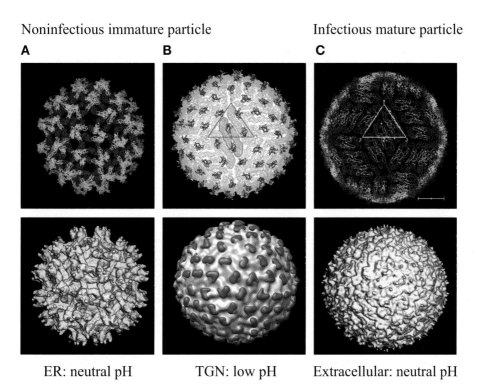

ER: neutral pH   TGN: low pH   Extracellular: neutral pH

Plate 13.3 Structures of immature and mature DENV. **A** Pseudo-atomic structure of the immature virus particle, at neutral pH (ER) conformation, with prM highlighted in light blue (upper panel) (Zhang *et al.* 2004). The three domains of E, I, II, and III are highlighted in red, yellow and blue respectively. Surface-shaded representation of immature particles at neutral pH (lower panel). **B** Pseudo-atomic structure of the immature virus particle in the low pH conformation. One icosahedral asymmetric unit and a representative E protein raft is highlighted (upper panel). Surface-shaded representation of immature particles at the trans-Golgi network – TGN low pH conformation (lower panel). prM is shaded in blue (Yu *et al.* 2008). **C** Pseudo-atomic structure of mature DENV particle at extracellular neutral pH, with each monomer shown in red, yellow and blue for domains I, II and III, respectively. One icosahedral asymmetric unit is outlined (upper panel) (Kuhn *et al.* 2002). Surface shaded electron micrograph of the mature infectious virus particle following disassociation of pr from the particle (lower panel).

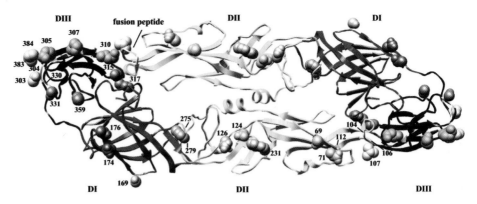

Plate 14.1 Mapping of representative MAb epitopes onto the DENV-2 E protein dimer. The three domains of DENV-2 E glycoprotein are shown: Domain I is red, domain II is yellow, and domain III is blue. The amino acid residues that significantly affect binding of MAbs, and/or were responsible for escape mutants, are in space-filling representation. Domain III: type-specific MAb 1F1 (residues 303, 304, 330, 383 and 384), subcomplex-specific MAb 1A1D-2 (residues 304, 305, 307 and 310) and cross-reactive MAb 13D4–1 (residues 315, 317, 331 and 359) (Sukupolvi-Petty *et al.*, 2007). Domain II: flavivirus cross-reactive MAb 1A5 (residues 106 and 107) (Goncalvez *et al.*, 2004b); flavivirus cross-reactive MAb 6B6C-1 (residues 104, 106, 126 and 231) and flavivirus group-reactive MAb 1B7–5 (residues 104, 126 and 231) (Crill and Chang, 2004); and MAb 10F2 (residues 69, 71, 112 and 124) (Lok *et al.*, 2001). Domain I: DENV-4-specific MAb 5H2 (residues 174 and 176) (Lai *et al.*, 2007); MAb 4G2 (residues 169 and 275) (Serafin and Aaskov, 2001); and MAb M10 (residue 279) (Beasley and Aaskov, 2001).

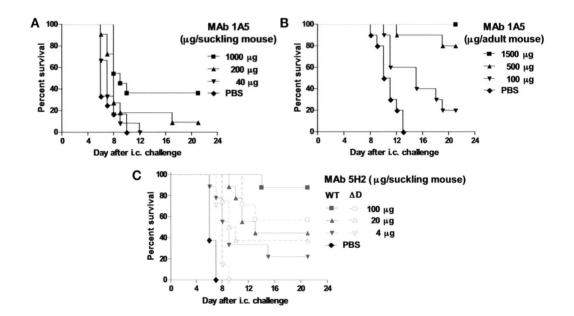

Plate 14.4 Protective activity of humanized DENV IgG1 antibodies (MAbs) using mouse DENV challenge models. Inbred BALBc mice ($n = 10$) were injected i.p. with MAb 1A5 (**A** and **B**) or MAb 5H2 (**C**) at various doses indicated. Unprotected control mice were administered PBS diluent. Twenty-four hours later mice were infected i.c. with DENV-2 strain NGB (**A** and **B**), or DENV-4 strain H241 (**C**). The animals were monitored daily and euthanized when clinical signs of infection appeared. Kaplan–Meier survival curves are shown.

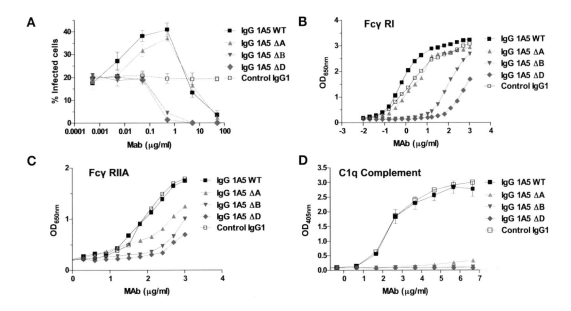

Plate 14.5 ADE of DENV-4 and binding to human Fcγ receptors and C1q by IgG 1A5 WT and its variants. **A** ADE of DENV-4 infection at 1 MOI in U937 cells mediated by IgG 1A5 Fc variants. IgG 1A5 variants ΔA and ΔB, which contain sequence substitutions of subclass IgG2 or IgG4 in the Fc, were described previously (Goncalvez *et al.*, 2007). These IgG subclasses have reduced Fcγ receptor-binding affinity (Armour *et al.*, 1999). IgG 1A5 ΔD contains a 9-aa deletion in the N-terminal Fc region. Dengue-negative human serum was used as control. **B**, **C** and **D** Binding of IgG 1A5 and variants to human Fcγ RI, Fcγ RIIA and C1q proteins. IgG 1A5 WT and the control IgG1 bind Fcγ RI, Fcγ RIIA, and C1q. ΔA ΔB and ΔD variants show no binding to C1q protein. The ΔA variant shows similar Fcγ RI binding to the IgG 1A5 WT, whereas the ΔB and ΔD variants show weak or no binding to human Fcγ Rs.

Other Books of Interest

Caister Academic Press   www.caister.com